Lecture Notes in Electrical Engineering

Volume 834

The book series *Lecture Notes in Electrical Engineering* (LNEE) publishes the latest developments in Electrical Engineering - quickly, informally and in high quality. While original research reported in proceedings and monographs has traditionally formed the core of LNEE, we also encourage authors to submit books devoted to supporting student education and professional training in the various fields and applications areas of electrical engineering. The series cover classical and emerging topics concerning:

- Communication Engineering, Information Theory and Networks
- Electronics Engineering and Microelectronics
- Signal, Image and Speech Processing
- Wireless and Mobile Communication
- Circuits and Systems
- Energy Systems, Power Electronics and Electrical Machines
- Electro-optical Engineering
- Instrumentation Engineering
- Avionics Engineering
- Control Systems
- Internet-of-Things and Cybersecurity
- Biomedical Devices, MEMS and NEMS

For general information about this book series, comments or suggestions, please contact leontina.dicecco@springer.com.

To submit a proposal or request further information, please contact the Publishing Editor in your country:

China

Jasmine Dou, Editor (jasmine.dou@springer.com)

India, Japan, Rest of Asia

Swati Meherishi, Editorial Director (Swati.Meherishi@springer.com)

Southeast Asia, Australia, New Zealand

Ramesh Nath Premnath, Editor (ramesh.premnath@springernature.com)

USA, Canada:

Michael Luby, Senior Editor (michael.luby@springer.com)

All other Countries:

Leontina Di Cecco, Senior Editor (leontina.dicecco@springer.com)

**** This series is indexed by EI Compendex and Scopus databases. ****

More information about this series at https://link.springer.com/bookseries/7818

Amit Kumar · Jacek M. Zurada ·
Vinit Kumar Gunjan · Raman Balasubramanian
Editors

Computational Intelligence in Machine Learning

Select Proceedings of ICCIML 2021

Springer

Editors
Amit Kumar
BioAxis DNA Research Centre
Hyderabad, India

Vinit Kumar Gunjan
Department of Computer Science
and Engineering
CMR Institute of Technology
Kandlakoya, India

Jacek M. Zurada
Department of Electrical and Computer
Engineering
University of Louisville
Louisville, KY, USA

Raman Balasubramanian
Department of Computer Science
and Engineering
Indian Institute of Technology Roorkee
Roorkee, Uttarakhand, India

ISSN 1876-1100 ISSN 1876-1119 (electronic)
Lecture Notes in Electrical Engineering
ISBN 978-981-16-8483-8 ISBN 978-981-16-8484-5 (eBook)
https://doi.org/10.1007/978-981-16-8484-5

This Springer imprint is published by the registered company Springer Nature Singapore Pte Ltd.
The registered company address is: 152 Beach Road, #21-01/04 Gateway East, Singapore 189721, Singapore

Preface

Information and data technologies are quickly advancing nowadays, and intelligent computers that can make decisions, analyse data, and perform other jobs that were previously done by humans are now much more efficient and competent than they used to be. The most important element of this next wave of intelligent computing is machine learning and cognitive science. Their work is underpinned by new advancements in computing power and is founded on mathematical, statistical, and data curation. And last but not least, these speedy advances are supported by the proliferation of free software, inexpensive data storage, and the massive social platform needs that have gone global. Every corporation today needs to integrate these strategies into their fabric to succeed. However, those ideas have just recently become accessible to organizations.

The aim of this book is to provide fresh input into the field of computational intelligence and machine learning, which is rapidly expanding. The editors have sought to produce a unified interpretation of the framework of this emerging research field in the novel literature by exploring current ideas in machine learning and computational intelligence and their applications. In addition to its focus on visualization, the book serves as a tool for promoting studies in machine learning and computational intelligence. Researchers and application scientists in machine learning, computational intelligence, and data technologies will find it very useful. Scholars who want to study these subjects should also use this book as a reference.

Because the material of this book is so interdisciplinary, it is light on claims regarding the reader's context. And it pulls ideas from other disciplines, when necessary, only taking up the most relevant ideas in statistics, artificial intelligence, information theory, computational intelligence, and the like. The book offers a detailed perspective on the many study directions by discussing case studies in chapters. The intent is to provide helpful information to readers seeking a comprehensive look at the application of machine learning, cognitive, and similar techniques.

This book gathers selected research papers presented at the International Conference on Computational Intelligence in Machine (ICCIML 2021), held virtually on 1–2 June 2021. The conference was jointly organized by CAIRO UTM, Malaysia; USTC, Bangladesh; and CMR College of Engineering and Technology, Hyderabad,

India. This publication is organized in a total of 51 chapters, which have been structured around their contributions to the overall area of the proceedings. The book's chapters describe the central algorithms and theories that are used in technologies and applications related to facial recognition, artificial neural networks, automotive applications, automation devices, business management systems, and modern speech processing systems. Additionally, the book explores contemporary breakthroughs in medical diagnostic systems, sensor networks, and VLSI domain systems. Wherever applicable, discussions of the learning process and software modules are included in deep learning algorithms.

Hyderabad, India
Louisville, USA
Kandlakoya, India
Roorkee, India

Amit Kumar
Jacek M. Zurada
Vinit Kumar Gunjan
Raman Balasubramanian

Contents

About the Editors

Amit Kumar is DNA Forensics Professional, Entrepreneur, Engineer, Bioinformatician, and an IEEE Volunteer. In 2005, he founded the first Private DNA Testing Company Bio Axis DNA Research Centre (P.) Ltd in Hyderabad, India, with a US Collaborator. He has vast experience of training 1000+ crime investigating officers and helped 750+ criminal and non-criminal cases to reach justice by offering analytical services in his laboratory. His group also works extensively on genetic predisposition risk studies of cancers and has been helping many cancer patients from 2012 to fight and win the battle against cancer. He was a member of the IEEE Strategy Development and Environmental Assessment Committee (SDEA) of IEEE MGA. He is a senior member of IEEE and has been a very active IEEE volunteer at Section, Council, Region, Technical Societies of Computational Intelligence and Engineering in Medicine and Biology and IEEE MGA levels in several capacities. He has driven several IEEE conferences, conference leadership programs, entrepreneurship development workshops, innovation, and internship-related events. Currently, he is Managing Director of BioAxis DNA Research Centre (P) Ltd and IEEE MGA Nominations and Appointments committee member.

Jacek M. Zurada is Professor of Electrical and Computer Engineering and Director of the Computational Intelligence Laboratory at the University of Louisville, Kentucky, USA, where he served as Department Chair and Distinguished University Scholar. He received his M.S. and Ph.D. degrees (with distinction) in electrical engineering from the Technical University of Gdansk, Poland. He has published over 420 journal and conference papers in neural networks, deep learning, computational intelligence, data mining, image processing, and VLSI circuits. He has authored or co-authored three books, including the pioneering text Introduction to Artificial Neural Systems, co-edited the volumes Computational Intelligence: Imitating Life, Knowledge-Based Neurocomputing, and co-edited twenty volumes in Springer Lecture Notes on Computer Science. In addition to his pioneering neural networks textbook, his most recognized achievements include an extension of complex-valued neurons to associative memories and perception networks; sensitivity concepts applied to multilayer neural networks; application of networks to clustering, biomedical image classification, and drug dosing; blind sources separation; and rule extraction as a tool for prediction of protein secondary structure.

Vinit Kumar Gunjan is an Associate Professor in the Department of Computer Science and Engineering at CMR Institute of Technology Hyderabad (affiliated to Jawaharlal Nehru Technological University, Hyderabad). An active researcher; published research papers in IEEE, Elsevier and Springer Conferences, authored several books and edited volumes of Springer series, most of which are indexed in SCOPUS database. Awarded with the prestigious Early Career Research Award in the year 2016 by Science Engineering Research Board, Department of Science and Technology Government of India. Senior Member of IEEE; an active Volunteer of IEEE Hyderabad section; volunteered in the capacity of Treasurer, Secretary and Chairman of IEEE Young Professionals Affinity Group and IEEE Computer Society. Was involved as organizer in many technical and non-technical workshops, seminars and conferences of IEEE and Springer. During his tenure, worked with top leaders of IEEE and was awarded with best IEEE Young Professional award in 2017 by IEEE Hyderabad Section.

Raman Balasubramanian is Professor in the Department of Computer Science and Engineering at the Indian Institute of Technology Roorkee. He received his Ph.D. in Mathematics (2001) from the Indian Institute of Technology Madras, India. He obtained his B.Sc. and M.Sc. in Mathematics from the University of Madras in 1994 and 1996. respectively. His research areas include computer vision, fractional transform theory, wavelet analysis, multimedia security, video skimming and summarization, medical imaging, machine learning, multilingual text recognition, EEG based pattern analysis, visualization, and volume graphics.

Machine Learning-Based Project Resource Allocation Fitment Analysis System (ML-PRAFS)

M. Rudra Kumar, Rashmi Pathak, and Vinit Kumar Gunjan

Abstract Many of the existing project automation tools can develop a project schedule with estimates of complete project summary time, budget, tracking possibilities, etc.; still one of the significant gaps in the system is about estimating how a particular resource has the necessary competence, experience, and the track record of completing the tasks more effectively. Taking cue of the gap in the system, using the fundamental approach of game theory model, a contemporary resource fitment prediction using the machine learning model is proposed. The datasets that are trained using the SVM classifiers have provided an accuracy level of around 97% which signifies the efficacy of the model. For the future research scope, if the auto-fitment of scores for the metrics are developed, it might lead to more sustainable solutions with a minimal set of human intervention to the system.

Keywords Project management · Resource allocation · Machine learning model for resource allocation · ML-PRAFS

1 Introduction

1.1 Outlook

Project management as a system is very important in ensuring timely completion of the projects, effective resource utilization, minimizing the cost overruns and toward attaining operational excellence. In the current competitive environment, it is highly important that the organizations ensure quick turnaround time in the completion of

M. Rudra Kumar (✉)
Department of CSE, GPCET, Kurnool,
Andhra Pradesh, India

R. Pathak
Department of CSE, National Institute of Technology, Silchar, India
e-mail: rashmi21_rs@cse.nits.ac.in

V. K. Gunjan
CMR Institute of Technology, Hyderabad, Telangana, India

A. Kumar et al. (eds.), *Computational Intelligence in Machine Learning*, Lecture Notes in Electrical Engineering 834, https://doi.org/10.1007/978-981-16-8484-5_1

projects, ensuring there is no lag in the project environment that might affect the desired outcome [1].

With the emergence of ICT-based project management solutions, there is a paradigm shift in the way organizations can focus on effective project management practices and leverage on the resource utilization to improve the project outcomes. However, one of the common issues that are imperative in the project management conditions could be attributed to high-cost overruns, a significant delay in the project trends, etc. But in addition to this, the other significant challenge in the case of project management is the effective utilization of resources [2].

There are many automated project management tools and solutions that are available online, which will help the organizations and the project management teams, in developing a robust project schedule. Right from auto-scheduling the tasks, to reflect on the baseline, milestones, efforts, and budget there are multiple factors that are handled by the automated tools. But one of the key areas wherein the task-related conditions are assessed for a resource utilization is not integral to many of the contemporary systems.

There is a need for a more comprehensive system that can support the project teams in understanding the composition of the resource utilization and how such task performance-based decision making can be empowered to allocate the work to right kind of resources executing the project.

In the other dimension, there are many contemporary solutions for project management that are being reviewed, wherein the AI-based solutions, predictive modeling-based WBS analysis, and other such significant models. While few such models are being more effective and is providing the necessary insights, still there is scope for improvements in terms of resource allocation for the project work break down structures [3].

1.2 Purpose of the Research

Project management as a process has three critical factors that have a significant impact on project planning.

- Task
- Time duration
- Resources required.

While these three are imperative for the process of developing a WBS, the other secondary factors that impact the WBS schedule are

- Predecessors
- Resource utilization ratio
- Cost of the resource
- Milestones
- Summary tasks.

While many of the existing project automation tools can develop a project schedule with estimates of complete project summary time, budget, tracking possibilities. However, one of the significant gaps in the system is about estimating how a particular resource has the necessary competence, experience, and the track record of completing the tasks more effectively [4].

In an illustrative scenario, it can be stated that in the case of a project "P," if for job A is one of the WBS tasks, and the team has four different resources who can be allocated to the job, the existing systems has no significance to the resources. It accepts any resource allocated to the respective task. But there are hardly any systems that can highlight as to how if resource x or y or z or m can be more effective for the specific kind of tasks and the related conditions.

If such resource advocation model can be integrated into the project management system, wherein the project team-related capability and success factor are analyzed by the system to predict the accuracy with which the schedule could be completed by the resources, it shall help in developing more comprehensive conditions [5].

Though some of the AI models are currently used in the project management toward developing a comprehensive system, still the factor of predicting the accuracy of timelines for completion of a project is relatively niche in the system planning.

Considering the abilities of the machine learning algorithms that can predict the accuracy of a system, if the similar kind of implementation plans is used for the WBS-related resource allocation classification, it can help in creating a more comprehensive and effective system, which can lead to sustainability in the project planning practices [1].

1.3 Outline of the Proposed Model

If a comprehensive system can be developed wherein the machine learning model is able to gauge the complexity score of the project task and accordingly take certain features as the competency score of the resource, impact factor score of the task and other certain parameters into account, toward estimating whether the specific resource could be able to complete the task within the timelines, it will help in choosing or allocation of right kind of resource for a project scenario.

If the machine learning model is developed for the above-mentioned condition, which can be used by the system for getting trained to identify whether an individual member is able to complete the assigned task, such model can lead to more accurate prediction of the schedule. However, the model that is proposed in this manuscript is independent of the levels of estimating whether the resource shall be able to complete the proposed task.

In furtherance, the system must be integrated into any of the existing project management tools which can support in achieving the desired input for more effective project management planning.

1.4 *Organizing of the Paper*

In Sect. 2 of this paper, the related work in terms of the distinct set of project management elements wherein the AI solutions are essential is discussed. Followed by in Sect. 3, the framework of the proposed model is discussed. Also, the test results of the model are discussed in Sect. 4. Followed by, in Sect. 5, the conclusion of the model is presented.

2 AI-Based PMP

Analysis, predictive modeling is paving its way into project management practices. While the role of automation in the process might increase, still the probabilities and accuracy of the project schedules can see significant improvement. Some of the key aspects that the usage of machine learning models can support in the aforementioned scenario of project management practices are explored in subsequent sections.

2.1 *Effective Utilization of Resources*

One of the common phenomena in the case of project success rates across the industrial verticals is about choosing the right kind of resources for a project scenario. Though many of the team members could be competent toward handling a specific task, one or few of the members among the group might be more competent for the specific contextual circumstances [6]. If the best of the resource for the specific job is chosen, the accuracy with which the job could be completed intime goes up.

2.1.1 The Analogy Drew to the Above Scenario

In the case of a sports team, all the members are equipped to handle all the tasks in the game. But one of few in the team can handle that specific process in the game more effectively than the others. The probability of estimating who among the members could handle a specific kind of circumstance or process effectively than the peer group is integral to game theory.

If the similar practice of game theory logic is extended into the machine learning analysis, it can help the project management team in choosing right kind of resources for the tasks and toward improvising the optimal performance conditions in a project environment [7].

2.2 More Effective Planning

Various elements of project planning are integral and interdependent over each of the other factors. For instance, in the project task, the time duration is one of the important factors for completion of the project. In the instance of a project requiring one week of one resource to be completed, the time frame for completion could vary based on the number of resources scheduled for the task at a time and also the budget of the task varies based on who is allocated for the project [8].

Taking such pragmatic factors into account, if the most appropriate resource for a task is selected, it is imperative that the probability of the tasks being completed within the scheduled time is more pragmatic.

2.3 Effective Cost Management

Duration of the task, human resource costs, infra requirement, and other factors have a direct impact on the cost overrun for a project. There are multiple models like activity-based costing, bottom-up costing, milestone-based costing models, etc., which have a direct impact on the project conditions. One of the critical parameters toward handling the project more effectively is about completion of the project within the estimated timelines and budget [1].

If there is significant overrun in the cost of the project, the cost–benefit viability of the model might get impacted and it might lead to more implications for the stakeholders associated with the process.

2.4 Project Dynamics

The project dynamics could be attributed to the agile environment of how the projects schedules and tasks are varying over the go. With more organizations using the path of the agile structure of project development, the onus in terms of using more competent team members toward a specific task has a high significance. But the challenge is about ensuring that the team composition is maintained in a balanced approach, where the resource cost and the resource utilization are handled below the threshold levels [2]. The ability, capacity, and the threshold values in the case of human competence toward measuring the process completion scope are essential. If such estimates are accurate, it will help in creating a comprehensive system [9].

Based on the inputs discussed above, it is evident that the model if implemented effectively, it can lead to a more integrated model which can develop sustainable features and the ones that can create significant impact. However, gaps in the system are to be addressed, in order to develop more effective project management systems that are empowered by the intelligence to assimilate the resource capabilities and deliver a comprehensive solution [4].

3 Framework for the Proposed Model

3.1 Fundamental Elements of the Model

The proposed solution is about developing an independent assessment system which can evaluate the resource mapping for a task based on the complexity ranking of the system and the competency scoring of the resource-based on various metrics that are integral to the model.

However, the fundamental principle behind the development of the system is the "game theory" model, which in parlance to the project environment stands more appropriate to the context discussed in this paper.

3.1.1 Game Theory

Game theory is a theoretical framework for conceiving social situations among competing players. In some respects, game theory is the science of strategy or at least the optimal decision making of independent and competing actors in a strategic setting [7].

In the parlance of the model to the project environment, every member of the group might be keen on working over a specific task or could have the necessary competence to execute a broad range of work. However, the optimization in terms of choosing the most appropriate to a task, among the available options, can lead to increased scope and probability of adherence to the schedule [8].

3.2 Metrics for the Models

Considering such a paradigm, some of the key aspects that are chosen in the proposed model of machine learning potential for task analysis system are.

The metrics chosen above are the key factors in the case of using the system for understanding whether the proposed model can have a significant solution. The metrics discussed in Table 1 are considered as the features for the proposed machine learning model.

Table 1 Features description

Metric	Dimension	Purpose
Task complexity score (TC)	Work break schedule task (WBS)	To understand the significance of the work complexity and the intricacies involved in the completion of the task
Task impact score (TI)	WBS	The impact score is about the rate of the effect that the specific task has on the schedule (if there are too many successors that rely on the completion of a specific task, in such condition, the score is high)
Project impact (PI)	WBS	A holistic impact of the task over the other aspects like leading to cost overruns, or risk register triggers, etc.
Resource capability score (RCS)	Resource end	The experience and efficiency of the resource toward executing the project are the score
Resource option scoring (ROS)	Resource end	The number of resources available as an option for executing the project is the weighted average score
Resource trust score (RTS)	Resource end	The trust by the team member in terms of belief about a member be able to complete the task as expected (probability score)

3.3 *Data Classifiers and Training Model*

In the machine learning models, there are numerous set of classifiers that are integral to the training models. It is imperative that the classifiers that are used for training the model have a significant impact on the outcome that can be envisaged in the system.

For the proposed solution, the emphasis is on multi-class classifiers that are used for training the model. The purpose of using the multi-class classifier is about understanding the probability of a resource competency defined for accuracy to a specific task.

3.3.1 Illustrated scenario

For a Task-A to be executed, the perceived impact of the task is High, and the resource to be allocated for the task too should be High or as the best alternative "Good." Any resource with lower than the "Good" ranking can be considered as In-effective allocation for the project task.

SVM is based on the fundamental that decision planes reflect the decision boundaries. The decision plane can be defined as a plane that classifies or segregate set of objects with different class membership (in the proposed project, the classes are Highly Suitable, Suitable, Moderately Suitable, In-effective) [10].

If the classification is presumed as green (Highly Suitable, Suitable) or red (Moderately Suitable, In-effective), as depicted in Fig. 1, the line defines the boundary between the green and red categories. However, in the case of machine training, the categories shall have four distinct sets.

The process followed in the above linear classifier model could be attributed as hyperplane classifiers, and the inherent reason behind the selection of the support vector machine (SVM) model is about its efficacy in handling such data classification across the training sets [11].

SVM primarily classifies the tasks based on the constructing hyperplanes over multidimensional space that categorizes the distinct class labels. As the SVM can support in handling both regression and classification tasks and can also handle the categorical and multiple continuous variables, the model is perceived as more effective for the proposed system.

For categorical variables, the dummy variables are created with case values as 0 or 1, and the categorical dependent variable consisting of four levels as A, B, C, D is depicted for three set of variables.

$$A : \{1000\},\ B : \{0100\},\ C : \{0010\},\ D : \{0001\}$$

In terms of developing an optimal hyperplane, SVM adapts the iterative training algorithm, which is used in the process of minimizing the error function. For the proposed training model, the C-SVM classifier is used.

The training is idealized for minimization of the error function.

$$\frac{1}{2}w^T w + C\sum_{i=1}^{N} \xi_i$$

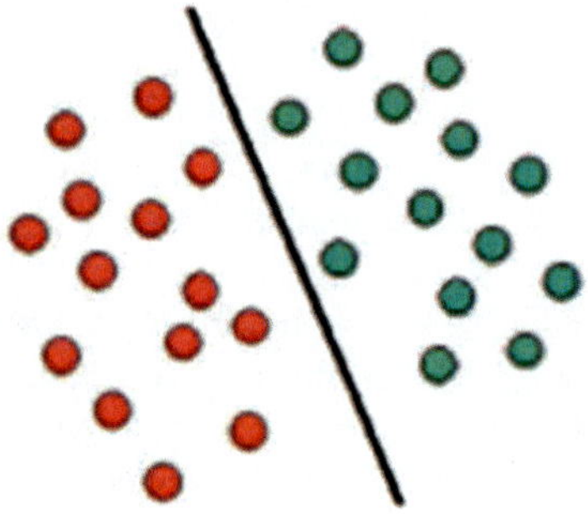

Fig. 1 Decision plane classification

subject to the constraints:

$$y_i\left(w^T\phi(x_i)+b\right)\geq 1-\xi_i \quad \text{and} \quad \xi_i\geq 0, i=1,\ldots,N$$

In the equation above, C is the capacity constant,
Vector of coefficients is denoted by w.
b reflects the constant.

ξ_i Signifies constraints for handling non-separable data (inputs).
ξ_i The index i tags the N training cases.

It is imperative that $y \in \pm 1$ denotes the class labels and x_i indicates the independent variables. The kernel ϕ is embraced for the transformation of the data from the input state which is independent of the feature space.

Selection of C holds critical importance in the training process toward minimizing the error conditions.

3.4 *Algorithm for the Proposed Model*

The algorithm used for the proposed model is about.

Let T be the task for which the resource R has to be reviewed for rating.

{

T is *task selected.*

R is *resource for rating.*

{

T_p is *task criticality score*

$$T_p = \sum(T_c + T_i + P_i)$$

whereas, individually all the sub-score (T_c, T_i, P_i) are significant for the system, and the value range is

$$Tc = \geq 1 \leq 10$$
$$Ti = \geq 1 \leq 10$$
$$Pi = \geq 1 \leq 10$$

If the $T_p \geq 27$, the task is rated as "Highly Complex" denoted as HC.
If the $T_p \geq 20 \leq 26$, the task is rated as "Moderate Complex" denoted as MC.
If the $T_p \geq 15 \leq 19$, the task is rated as "Low Complex" denoted as LC.
If the $T_p \geq 1 \leq 18$, the task is rated as "Normal Task" denoted as NT.

}

{

The resource score category is considered as

RCS (resource capability score)
ROS (resource option score)
RTS (resource trust score)

For all the RCS, ROS, and RTS, the scoring pattern is at the stream range of 1–10

$$\text{RCS} = \geq 1 \leq 10$$
$$\text{ROS} = \geq 1 \leq 10$$
$$\text{RTS} = \geq 1 \leq 10$$

$$R = \sum(\text{RCS} + \text{ROS} + \text{RTS})$$

If the $R \geq 27$, *the task is rated as* W.
If the $R \geq 20 \leq 26$, *the task is rated as* X.
If the $R \geq 15 \leq 19$, *the task is rated as* Y.
If the $R \geq 1 \leq 18$, *the task is rated as* Z.
$R =$ *either W or X or Y or Z.*
}
When the input "*I*" is tested for *T*-value allocation, the process followed is about
{
Eligibility score
If the TIp has the label of HC.... then only Rwmust be used for analysis.
If the TIp has the label of MC ... then R_x *or* R_w *or* R_y *can be considered for allocation accuracy prediction.*
If the TIp has a score of LC, only R_x *or* R_y *must be considered.*
If the TIp has a score of NT, R_x *or* R_y *can be considered.*
}
{
If the T_p score validation is effective, then the scoring output for task allocation accuracy is estimated as

$$T = 2 * ((T_p * R)/T_p + R)$$

}
If the $T \geq 27$, *the task is rated as highly suitable.*
If the $T \geq 20 \leq 26$, *the task is rated as suitable.*
If the $T \geq 15 \leq 19$, *the task is rated as moderately suitable.*
If the $T \geq 1 \leq 18$, *the task is rated as ineffective.*
}

The inputs are evaluated in the scoring pattern, it can be comprehended that the *T* score indicates the potential accuracy of the system.

Table 2 Datasets

Particulars	Dataset records	Remarks
Total data used for analysis	1700	
Three sets used for training	1200	Every set constituting 400 records for training purpose
Positive records in the training set	950	The records that can be classified for highly suitable or moderate suitable or suitable conditions
Records of ineffective, in the training set	250	Records that constitute the ineffective training records
Total records used for testing the system	500	Test data
Positive records in the test data	376	Records that are to be classified for positive category
Negative records in the test data	124	Records to be traced by the system as negative or ineffective for task

3.5 Datasets

To understand the efficacy of the proposed system, a synthesized dataset is generated with random score value for all the six-feature metrics, three of task complexity, and three of the resource abilities. Table 2 indicates the cumulative set of records that are integral to the testing process. Also, in terms of classification, the test is evaluated based on the metrics of efficacy which is defined for a model as precision, recall, $F1$ score, specificity, and the sensitivity for the project conditions.

For the purpose of accuracy estimations from the datasets, the ones that are categorized to the high scores constituting as Highly Suitable, Suitable, or Moderate Suitable as positive records, and the ones that are to be classified as to be Ineffective are chosen as negative records for the inference purpose.

Table 2 indicates the dataset used for the analysis.

4 Test Result Analysis

The tests are carried out for the data over the system application executed for the algorithm discussed in the earlier section. The test data results from the process generated are classified for the evaluation purpose (Table 3).

Based on the inputs that are garnered for the respective record conditions, the estimation of the accuracy levels for the system is evaluated. From the contextual conditions that are evaluated for the process, it can be stated that the scoring pattern of the model are (Table 4).

Table 3 Result classification

Metrics	Detection from test records
True positive	364
False positive	12
False negative	14
True negative	110

Table 4 Performance metrics

Metrics	Performance
Precision	96.81
Recall	96.30
Accuracy	94.80
$F1$ score	96.55
Specificity	90.16

When the metrics are reviewed, the precision rate of 96.81% is imperative in the case of the data tested. Though the data garnered for the evaluation parameters are synthesized data, still the rate of accuracy could be more pragmatic, considering the correlation estimated for the records chosen for the analysis (Fig. 2).

In terms of the recall rate, it depicts the cost of a false negative high. In the instances of false negative being high, it reflects the risks in embracing the model. However, the recall rate indicates that for the tested model, the rate of false negatives is well within the controllable levels of around 3%, which indicates that the model can be used in integration to many of the existing project management models.

The $F1$ score of the proposed model reflects the rate of 96% which indicates that the model has the overall accuracy, taking into account the precision and recall values in totality. With a good $F1$ score attained, the model is imperative for a low rate of false positives and low set of false negatives. Also, it can be inferred that the model

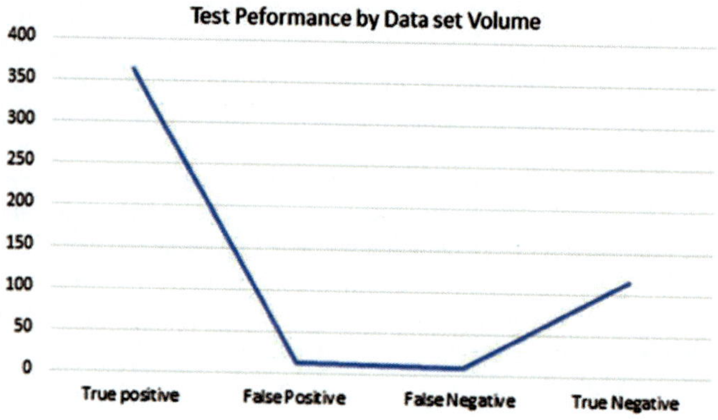

Fig. 2 Test performance by records

Table 5 Records classification for classes

Classification	Actual segregation	Positive tracing	Accuracy %
Highly suitable	176	169	96.0
Suitable	127	124	97.6
Moderately suitable	73	71	97.3
Ineffective	124	115	92.7

does not have much of real threats and do not be intrigued by certain levels of false alarms [12].

The other parameter in which the study is evaluated is to focus on the metrics of how accurately the system has identified the respective classes in the test records. The comprehensive Table 5 indicates the quantum of records that are considered for different classes in the testing sets and the accuracy of prediction for the respective class category (Fig. 3).

The model has performed ineffective ways of identifying most of the classes close to accurate parameters. This set of inputs can be inferred as the effectiveness of the SVM classifier that is used for the training of dataset patterns [12]. However, the model has certain level issues in the case of detecting the ineffective set, which can be focused upon by working on reinforced learning model of the training sets.

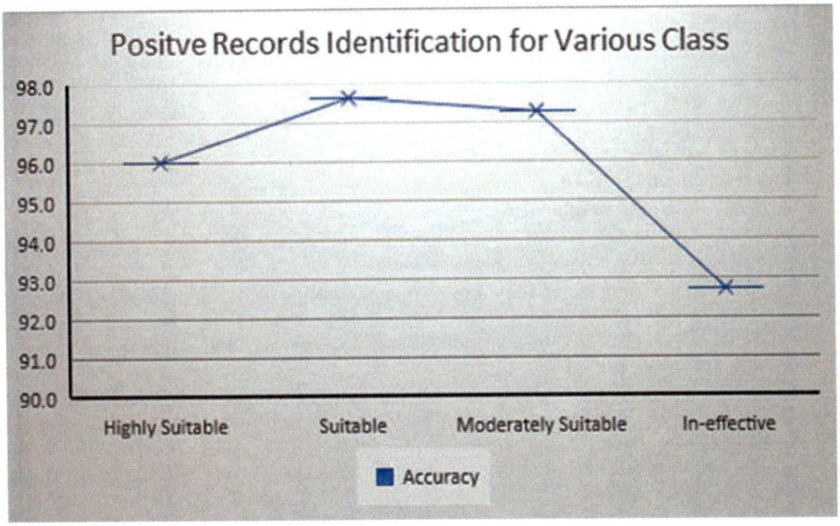

Fig. 3 Class performance

5 Conclusion

While many of the existing project automation tools can develop a project schedule with estimates of complete project summary time, budget, tracking possibilities. However, one of the significant gaps in the system is about choosing the resource allocation with a high probability of fitment for the task. Though some of the AI models are currently used in the project management toward developing a comprehensive system, still the factor of predicting the accuracy of timelines for completion of a project is relatively niche in the system planning. Comprehending on the game theory model wherein the scores of individual members capability play a vital role can lead to improving the accuracy of planning. The proposed machine learning model when trained with the datasets using SVM classifier provides considerable performance as discussed in the test results. For the future solutions, the model of focusing on autoscoring for the complexity of the task score and the resources could be used as the backward integrated solution, which might lead to more optimization of the services

References

1. J. Thomas, T. Mengel, Preparing project managers to deal with complexity—advanced project management education. Int. J. Proj. Manage. (2008)
2. C.P. Killen, K. Jugdev, N. Drouin, Y. Petit, Advancing project and portfolio management research: applying strategic management theories. Int. J. Proj. Manage. (2012)
3. J.K. Pinto, Project management, governance, and the normalization of deviance. Int. J. Proj. Manage. (2014)
4. S. Pemsel, A. Wiewiora, R. Müller, M. Aubry, K. Brown, A conceptualization of knowledge governance in project-based organizations. Int. J. Proj. Manage. (2014)
5. L. Project, *Rights-Based Approaches* (2009)
6. A. Perini, A. Susi, P. Avesani, A machine learning approach to software requirements prioritization. IEEE Trans. Softw. Eng. (2013)
7. R. Breen, Game theory, in *The Oxford Handbook of Analytical Sociology* (2017)
8. E. Kalai, Game theory: analysis of conflict. *Games Econ. Behav.* (1991)
9. K. Bredin, People capability of project-based organisations: a conceptual framework. Int. J. Proj. Manage. (2008)
10. K. Bennett, E. Bredensteiner, Duality and geometry in SVM classifiers. ICML (2000)
11. C.C. Chang, C.J. Lin, Training ν-support vector classifiers: theory and algorithms. Neural Comput. (2001)
12. D.N. Carvajal, P.C. Rowe, Sensitivity, specificity, predictive values, and likelihood ratios. Pediatr. Rev. **31**(12), 511–513 (2010)

Electric Theft Detection Using Unsupervised Machine Learning-Based Matrix Profile and *K*-Means Clustering Technique

Saddam Hussain, Mohd. Wazir Mustafa, Steve Ernest James, and Shadi Khan Baloch

Abstract Electric theft is the major issue faced by utility companies in different countries as it causes significant revenue losses and affects the power grid reliability. This paper presents a novel electric theft detection framework based on an unsupervised machine learning technique employing matrix profile and *K*-means clustering algorithm. The proposed framework is based on three stages to identify the fraudster consumers in a conventional electric consumption meter dataset acquired from Pakistan's power distribution company. Initially, the missing and inconsistent observations are filtered out from the acquired dataset. After that, the matrix profile from each consumer's consumption profile is computed to identify the irregular and sudden changes present in them. Later, the *K*-means clustering algorithm is used on the datasets divided based on their computed matrix profile values in order to label each consumer into "Healthy" and Theft." The developed framework is compared against the latest state of art machine learning algorithms and statistical-based outlier detection methods. The proposed technique achieved an accuracy of 93% and a detection rate of 91%, which is greater than all compared models.

Keywords Electric power distribution · Electric theft · Matrix profile · *K*-means clustering algorithm

S. Hussain (✉) · Mohd. W. Mustafa
School of Electrical Engineering, University Technology Malaysia, 81310 Skudai, Johor Bahru, Malaysia
e-mail: hussainsaddam@graduate.utm.my

S. E. James
Department of Actuarial Science and Risk Management, Institute of Business Management, Korangi Creek, Karachi, Sindh 75190, Pakistan

S. K. Baloch
Department of Mechatronics Engineering, Mehran University of Engineering and Technology, Jamshoro, Sindh 76062, Pakistan

A. Kumar et al. (eds.), *Computational Intelligence in Machine Learning*, Lecture Notes in Electrical Engineering 834, https://doi.org/10.1007/978-981-16-8484-5_2

1 Introduction

Restructuring of electrical distribution companies to meet standards of the modern grid has already begun in most of the advanced countries [1, 2], among numerous modernization in the old grid, especially smart meters has altered various outdated applications [3–6]. However, Pakistan still has an old conventional system of metering [7]. The biggest problem faced by power distribution companies is power theft [8], which worsens the economic state of the electric utilities [9, 10], restrains new funds for capacity expansion [11], which ultimately leads to electricity scarcity [12]. It happens due to bad governance [13], poor accountability, and an elevated level of corruption [14]; comprehensive studies on various electricity crisis in Pakistan and proposed solutions for those can be found here [15–17].

Pakistan electric distribution sector is based on public and private distribution companies [18], majority of electric power is distributed through the public sector [17], few statistics are shown in Fig. 1, and details can be seen in annual reports published by NEPRA Pakistan [19]. The existing process practiced by power distribution companies (PDC) for computing and registering electric consumption data through the electromechanical meter is illustrated in Fig. 2.

As from Fig. 2, it can be observed that the entire process is manual and subjected to numerous issues to falsify meter reading. For example, the bribe is given to meter reader employees by the consumer for registering wrong meter reading; the metering process is reversed or made a hole to slow down its counter by consumers, bogus meter installation, bypassing meter registering process, etc. The dedicated study and practical demonstration of various power theft approach in Pakistan can be found here [20]. Similarly, based on an on-site inspection conducted by the largest PDC,

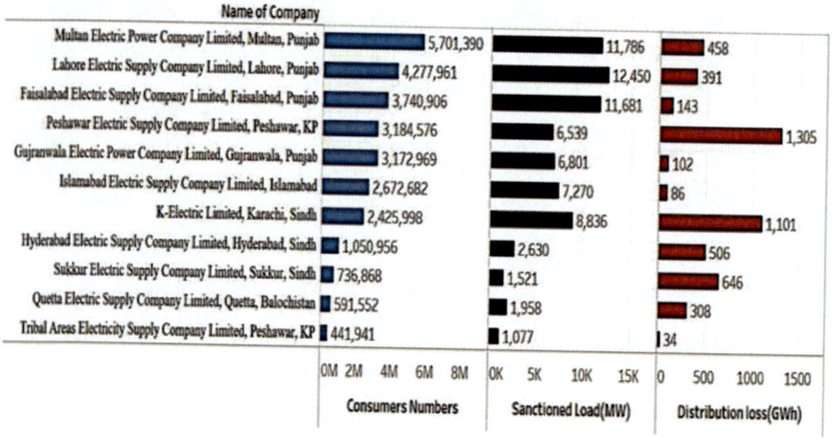

Fig. 1 Pakistan electricity public distribution companies, number of consumers, sanctioned load, and distribution loss

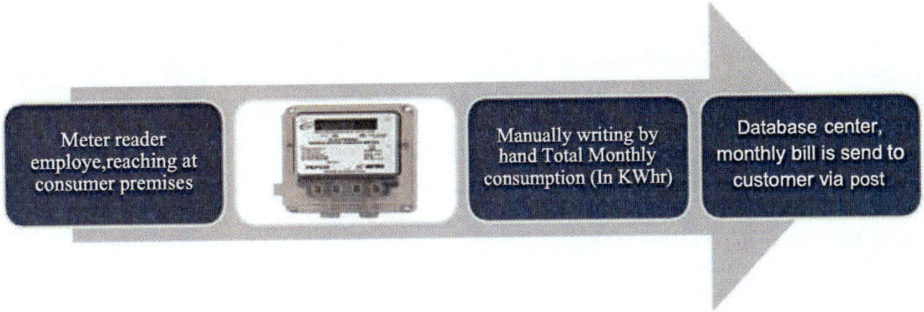

Fig. 2 Current process of data collection, processing, and registering in power distribution companies of Pakistan

Multan electric power supply company (MEPCO), in 2016, most occurring theft methods were computed and illustrated in Fig. 3.

Electricity theft is accomplished by manipulating consumption data in order to reduce electricity charges [21]. It is claimed that power theft conducted through electromechanical meters cannot be eliminated entirely. However, it can be controlled by increasing theft detection hit rate accuracy. In [22], author has asserted to increase accuracy from the labor-intensive procedure of theft detection (of 3% accuracy) to utilizing a combination of data mining approaches and SVM algorithm (achieving 60% theft detection hit rate accuracy) by choosing TNB Malaysia power utility company consumer's historical power consumption of 25 months as a dataset; the proposed method claimed to benefit also in reducing on-site inspection time by efficiently going only toward suspected labeled consumers rather than the manual approach of checking each and every consumer possible. In [23], author has taken Brazil Light SA company historical monthly consumption dataset in order to distinguish irregular consumption patterns from the pool of normal consumption patterns

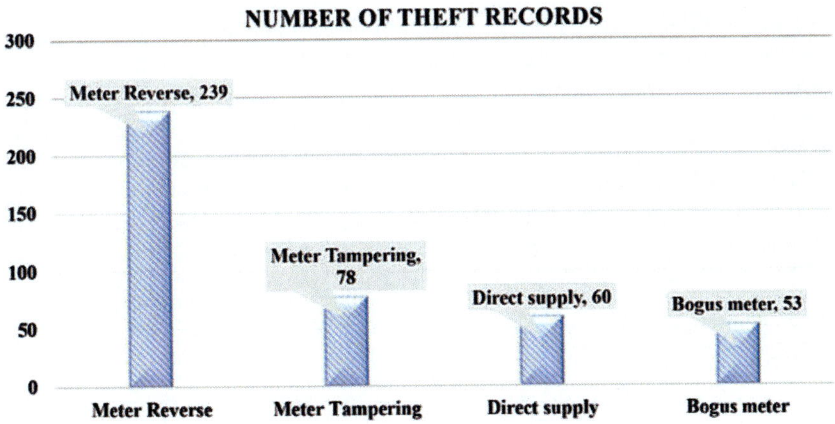

Fig. 3 Most occurring theft methods in MEPCO, PDC of Pakistan

Fig. 4 Existing method of on-site inspections practiced by PDCs in Pakistan

by using a combined approach of neural networks committee and a neuro-fuzzy hierarchical system claimed to achieve positive predictive value (PPV) of 51.2%, PPV is the percentage of verified abnormal consumers to all those who has been cataloged as accused of some irregularity and has been checked by the inspection team, their approach labels high irregular consumption as suspicious, and the investigation is initiated on those labeled consumers. Further reading on various recent NTL detection and solutions can be found here [24–26].

In order to prevent electric theft, the current process practiced by most PDCs in Pakistan is the executions of manual on-site inspections, as illustrated in Fig. 4. The existing process is conducted inefficiently, eventually identifying almost nothing or very few theft cases. As most of the theft conducting consumers get alert instantly and avoid conducting theft, particularly those who hoke direct lines or bogus meters. Since it is a tedious and expensive job and requires a considerable amount of time to seek approval from officials to reach consumer premises and finally label, i.e., "theft" or "healthy." Therefore, a highly selective approach to reach only suspected consumers is essentially required to save considerable time and limited resources.

2 Proposed Methodology

The unsupervised machine learning-based technique employing the K-means algorithm and matrix profile is used in this study to develop the electric theft detection framework. The historical consumption data of about four years is acquired from MEPCO, Pakistan. The acquired dataset is preprocessed to filter out any missing and inconsistent values initially. After that, each consumer's matrix profile is computed to identify the irregularity and sudden perturbations in consumption profiles. Later, the K-means clustering algorithm is employed separately on datasets divided based on computed matrix profile values in order to label each consumer healthy or theft.

2.1 Matrix Profile

The matrix profile (MP) is precise, threshold-free, and a scalable algorithm. It is used for annotating multidimensional and unidimensional time-series data. The time-series data is the order of values taken by a specific variable over a particular time. The underlying difference among other data types and TSD is its order of occurrence; by changing it, the entire significance of data will be altered [27]. The MP technique facilitates locating the nearest neighbor to every point from the collection of the data set given. A higher value of MP shows discord or outlier, and lower values show time-series motif (it is the most similar subsequence pair present in given TSD) [28].

The use of spotting outliers by computing MP value is demonstrated by generating a synthetic dataset and merging it with six anomalous datasets at various locations. The anomalous dataset is of a different length to the original dataset. As shown in Fig. 5, higher values of MP show that anomalies have occurred at this point. By checking the MP index value, we can obtain the anomaly's exact location. Simultaneously by computing discord discovery and plotting it as illustrated in Fig. 6, the exact shape of anomalous data can be examined. The computed MP value and discord discovery are made with the help of the time-series matrix profile (tsmp) package in R [29].

By computing the matrix profile of each consumer's consumption profile of the acquired dataset in this research work, it is observed that those consumers that attain MP values lower than 1.5 get a regular consumption pattern. In contrast to that, irregularity is observed in consumption profiles exceeding the MP value of 1.5, as illustrated in Fig. 7. For convenience purposes, only two consumers on a random basis are chosen to plot their actual consumption along with the computed matrix profile in Fig. 7. It can be observed that a normal consumption pattern (from 0 to 36, red data points) has lower matrix profile values, and irregular consumption patterns (from annotations 37 to 72) have deviating matrix profile values.

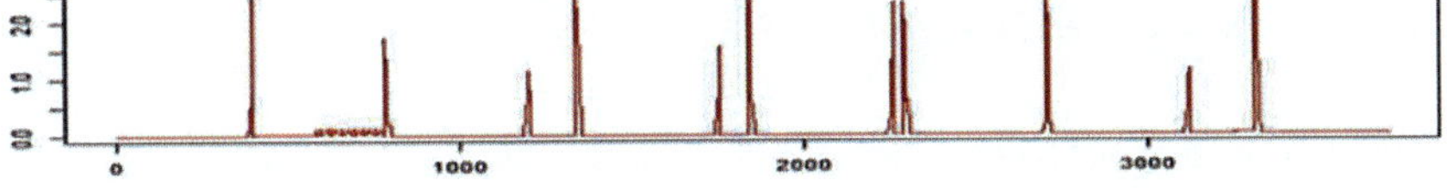

Fig. 5 Anomalous pattern identification in synthetic dataset

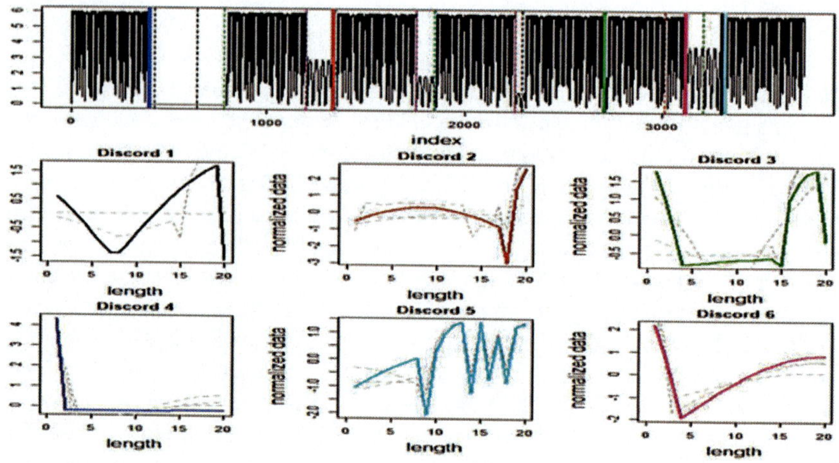

Fig. 6 Shape of anomalous data present in the synthetic dataset

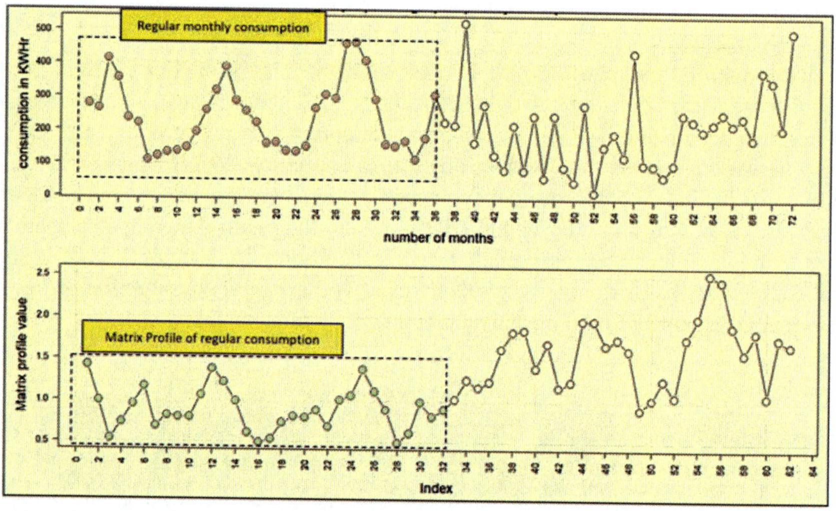

Fig. 7 Electric consumption data and matrix profile computed values

2.2 K-*Means Clustering Algorithm*

In this section, the final labeling process of consumer's consumption profiles is made. The computed matrix profile values of consumer's profiles assist in dividing the consumer's acquired dataset into two categories based on a cut-off MP value of 1.5, as stated (in Matrix profile section). By employing the *K*-means clustering algorithm on a divided dataset, those consumers forming a cluster of regular consumption profiles were labeled "healthy" and others as a “theft.” The consumers forming a regular consumption profiles cluster is illustrated in Fig. 8. Those consumers having irregular consumption patterns are grouped in similar types of clusters; few consumers’ consumption profiles are illustrated in Fig. 9.

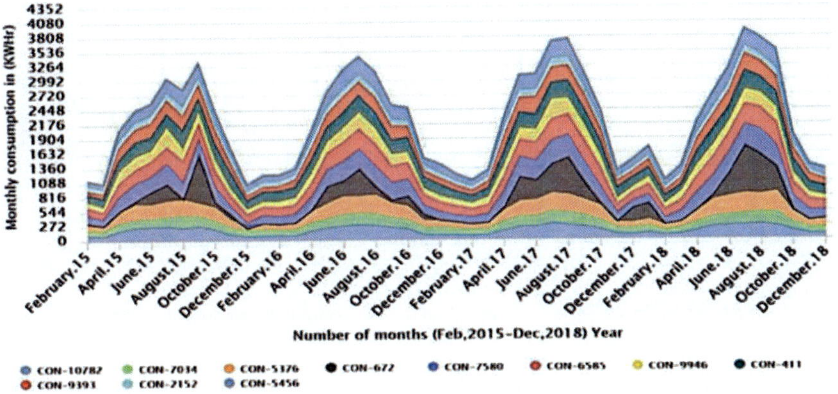

Fig. 8 Healthy or normal consumption profiles of consumers

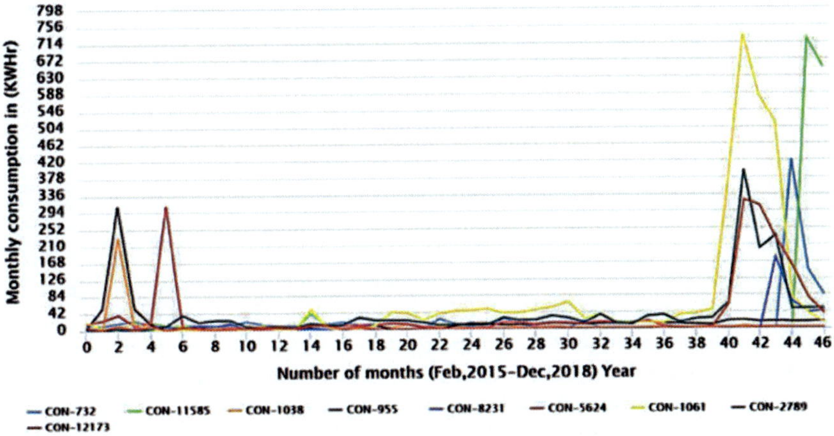

Fig. 9 Theft or irregular consumption profiles of consumers

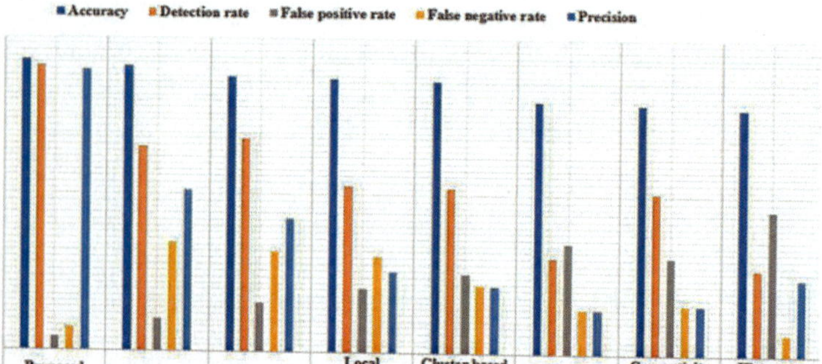

	Proposed technique	Isolation forrest	One-class SVM	Local correlation integral	Cluster based local outlier factor	Local outlier factor	Connectivity based outlier factor	Histogram based outlier score
Accuracy	93	91	87.65	87.25	86.5	80.15	79.35	78.25
Detection rate	91	65.31	68.03	53.06	52.36	30.34	51.29	27.21
False positive rate	4.36	10.34	15.63	20.55	25.4	34.92	30.85	45.9
False negative rate	7.48	34.69	31.97	30.37	21.7	14.25	15.86	6.65
Precision	90	51.61	42.37	25.94	21.47	14.25	15.85	24.5

Fig. 10 Performance metrics score attained by proposed technique against compared algorithms

2.3 *Results and Discussion*

The proposed matrix profile *K*-means, an unsupervised machine learning-based theft detection framework, is compared against the latest state-of-the-art unsupervised machine learning techniques. The outcomes of the technique against compared methods are depicted in Fig. 10. It can be observed from the results that the developed model attained high accuracy and detection rate of 92 and 91%.

3 Conclusion

A novel electric theft detection method using an unsupervised machine learning approach has been proposed in this research work. A unique combination of matrix profile and *K*-means algorithm has been made to identify electric frauds in the acquired real-world dataset of conventional energy consumption meters. The developed theft detection framework is compared against the latest machine learning techniques such as isolation forest, one-class support vector machine, cluster-based local outlier factor, and most frequent statistical-based techniques such as local outlier factor, histogram-based outlier detection, local correlation integral, and connectivity-based outlier factor. The outcomes attained by the proposed technique in terms of accuracy, detection rate, false-positive rate, and precision are far better than all compared models.

References

1. X. Fang, S. Misra, G. Xue, D. Yang, Smart grid—the new and improved power grid. IEEE Commun. Surv. Tutorials **14**(4), 944–980 (2012)
2. E. Hossain, I. Khan, F. Un-noor, S.S. Sikander, S.H. Sunny, Application of big data and machine learning in smart grid, and associated security concerns: a review. IEEE Access **7**, 13960–13988 (2019)
3. D. Alahakoon, X. Yu, Smart electricity meter data intelligence for future energy systems: a survey. IEEE Trans. Ind. Inform. **12**(1), 425–436 (2016)
4. W. Kong, Z.Y. Dong, D.J. Hill, F. Luo, Y. Xu, Short-term residential load forecasting based on resident behaviour learning. IEEE Trans. Power Syst. **33**(1), 1087–1088 (2018)
5. B. Stephen, A.J. Mutanen, S. Galloway, G. Burt, P. Jarventausta, Enhanced load profiling for residential network customers. IEEE Trans. Power Deliv. **29**(1), 88–96 (2014)
6. A. Maamar, K. Benahmed, Machine learning techniques for energy theft detection in AMI (2018), pp. 57–62
7. Z. Hussain, S. Memon, S. Zardari, Systematic analysis of smart meter technologies and their implantation challenges in Pakistan **28**(4), 113–120 (2016)
8. M. Imran, N. Amir, A short-run solution to the power crisis of Pakistan. Energy Policy **87**, 382–391 (2015)
9. M.M. Ra, S. Rehman, S. Asia, National energy scenario of Pakistan—current status, future alternatives, and institutional infrastructure : an overview **69**(Oct 2015), 156–167 (2017)
10. F. Jamil, E. Ahmad, The relationship between electricity consumption, electricity prices and GDP in Pakistan. Energy Policy **38**(10), 6016–6025 (2010)
11. G. Das, M. Aslam, H. Rahman, R. Samoo, N. Hussain, K. Harijan, Overcoming electricity crisis in Pakistan : a review of sustainable electricity options. Renew. Sustain. Energy Rev. **72**(Nov 2016), 734–745 (2017)
12. F. Shaikh, Q. Ji, Y. Fan, The diagnosis of an electricity crisis and alternative energy development in Pakistan **52**, 1172–1185 (2015)
13. K. Shahzad, S. Ullah, R.B. Ansted, D. Mamoon, Evaluating human resource management capacity for effective implementation of advanced metering infrastructure by electricity distribution companies in Pakistan. Util. Policy **41**, 107–117 (2016)
14. T.B. Smith, Electricity theft: a comparative analysis. Energy Policy **32**(18), 2067–2076 (2004)
15. I.N. Kessides, Chaos in power: Pakistan's electricity crisis. Energy Policy **55**, 271–285 (2013)
16. B.S. Aftab, Pakistan' s energy crisis: causes, consequences and possible remedies, no. Jan (2014), pp. 1–6
17. O. Rauf, S. Wang, P. Yuan, J. Tan, An overview of energy status and development in Pakistan. Renew. Sustain. Energy Rev. **48**, 892–931 (2015)
18. K. Zaman, M.M. Khan, M. Ahmad, R. Rustam, Determinants of electricity consumption function in Pakistan : old wine in a new bottle. Energy Policy **50**(March 2010), 623–634 (2012)
19. NEPRA, *National Electric Power Regulatory Authority Annual Report*, p. 108
20. Z. Hussain, S. Memon, R. Shah, Z.A. Bhutto, M. Aljawarneh, Methods and techniques of electricity thieving in Pakistan. J. Power Energy Eng. **04**(09), 1–10 (2016)
21. S. Mclaughlin, D. Podkuiko, P. Mcdaniel, Energy theft in the advanced metering infrastructure, vol. 6712, no. May (2011)
22. J. Nagi, K.S. Yap, S.K. Tiong, S.K. Ahmed, M. Mohamad, Nontechnical loss detection for metered customers in power utility using support vector machines. IEEE Trans. Power Deliv. **25**(2), 1162–1171 (2010)
23. C. Muniz, M. Vellasco, R. Tanscheit, K. Figueiredo, A neuro-fuzzy system for fraud detection in electricity distribution. Comput. Intell. 1096–1101 (2009)
24. G.M. Messinis, N.D. Hatziargyriou, Review of non-technical loss detection methods. Electr. Power Syst. Res. **158**, 250–266 (2018)

25. J.L. Viegas, P.R. Esteves, R. Melício, V.M.F. Mendes, S.M. Vieira, Solutions for detection of non-technical losses in the electricity grid: a review. Renew. Sustain. Energy Rev. **80**(May), 1256–1268 (2017)
26. Y. Wang, Q. Chen, T. Hong, C. Kang, Review of smart meter data analytics: applications, methodologies, and challenges. IEEE Trans. Smart Grid **10**(3), 3125–3148 (2019)
27. H. Lutkepohl, M. Kratzig, *Applied Time Series Econometric* (Cambridge University Press, Cambridge, 2004)
28. C.-C.M. Yeh, Towards a near universal time series data mining tool: introducing the matrix profile (2018)
29. Time series with matrix profile. Package, "Package 'tsmp'" (2019). GitHub link: https://github.com/matrix-profile-foundation/tsmp

Placement Analysis—A New Approach to Ease the Recruitment Process

Srishti Singh, Prashant Singh, Aniket Singh, Harsh Gupta, Nitin Rakesh, Neha Tyagi, Parma Nand, and Mandeep Kaur

Abstract The major sequential approach for getting placed in a good firm or organization is going through rigorous technical and interview rounds. This method is not lucrative but only time-consuming for both the students and the management with almost little output as compared to market standards. Currently, campus monitored and streamlined methods are used to segregate individual candidates possessing potential, but these methods are ineffective and not ecstatic for students. We developed a web application for both the students and admins to ease the process of placements. This analysis model is established with the aim in mind to keep the students acquainted with their weaknesses and areas with a scope of improvement. The users of either end keep track of their progress from time to time. Together, this website and the end-users and dedicated volunteer enthusiasm will coordinate several events over the year, raise awareness throughout the fraternity, and raise millions of talented enthusiasts to support countless companies and organizations impacted by lack of talented manpower to implement their efforts ideas.

Keywords Web application · Data · Placement · Azure · Database

1 Introduction

A typical placement process consists of a few rounds of technical assessment, which can vary from company to company, and a final interview round. The qualification rate of these rounds continuously decreases because it is very hard for the college's placement cell to keep track of each student's progress and provide necessary help at the right stage during the recruiting process [1]. According to the statistics, India reveals that students not getting the desired placement is very much linked to the opacity of the placement process and its rigorous and tiring continuities, and its standardized qualification rate in the college is a cause of worry [2]. The story is similar in almost all prestigious institutions. Besides, such a qualification rate continues to dip at a yearly rate of 1–2%. Approximately 400–500 students become ready

S. Singh (✉) · P. Singh · A. Singh · H. Gupta · N. Rakesh · N. Tyagi · P. Nand · M. Kaur
Sharda University, Greater Noida, India

A. Kumar et al. (eds.), *Computational Intelligence in Machine Learning*, Lecture Notes in Electrical Engineering 834, https://doi.org/10.1007/978-981-16-8484-5_3

for the recruitment process from each department each year. Among this population, nearly 4.77% of candidates could write the correct logic for a programmed minimum requirement for any programming job. And more than 60% could not even write code that compiles [3]. Only 1.4% could write functionally correct and efficient code. These erratic placement processes have brought severe irrational burdens to the students, college authorities, and recruiters. This situation emphasizes the urgency to bring a change in the placement procedures.

At present, the recruitment process which has been adopted by various institutions is very cumbersome. The early and enthusiastic use of this website immediately after the placement season starts could produce satisfactory outcomes. However, the former treatment is good and has provided the needed results in the past. It has helped the concerned authorities very much [4]. Yet it is having its flaws and loopholes. A major amount of time after the onset of the recruitment process is wasted in going through various courses, advising and doctored articles to and from the seniors, counselors, and in imaging tests; thus, only 1–8% of the entire student population is given the deserving jobs or internships. The management authorities also lose track of individual progress reports providing the necessary guidance at the right time. These management challenges urge the development of new and automated methods of the placement procedure that can keep the students, admins, and trusted counselors connected to each other and bridging the gap between the end-users [5].

Earlier in the background, we also discussed self-evaluation for distinguishing two potential candidates from each other among a whole lot of aspirants who have developed their skills over the years. Personalized reports can accurately identify the states of the aspirant's weak points (e.g., aptitude, technical or soft skills), which can be traced from his/her previous experiences at different mock or real-time interviews. Also getting timely relevant notifications and reminders to both the admin and students have been proposed as a possible feature to infer a more enhanced and interactive session on this placement analyzing model. After a company completes its placement drive, a report having details of students who got placed and those who were rejected after any round is generated. This report is accessible only to the Admins [6].

2 Literature Survey

2.1 PAT

Placement and Training (PAT) cell can be apprehended as a connecting tool between the companies and the students about to go for a recruitment drive. Their ambition is to reduce people physically interacting and irregularities faced during training processes. To achieve that automation, they developed an Android framework. The proposed system is an Android application to monitor mobile student information and keep them up to date on the latest activities at the college. The program will be used by the students, teachers, and parents [7].

2.2 *Workable*

The WORKABLE app is an applicant sourcing and tracking platform built for organizations of all shapes and sizes. But in this, there have been numerous instances where the students and the companies were stuck in the never-ending web-based systems with minimal productivity for either end-user. The servers were supposed to do the tedious task of dealing with just storing and retrieving data without caring about the ultimate goal of helping the students getting placed. They were more involved in placement 'Management' and paying the least attention to the placement 'Analysis' [8].

3 Proposed Methodology

The proposed project steps in to boost the student's confidence and cut off companies engaging time. Final placements and internships in colleges and universities is an important technical and skillful event that causes serious consequences when not taken seriously [9]. In this model, an automated application is proposed to keep track of the placement activities as well as the individual progress report by both the students and the admins [10]. Together this application along with the end-users and dedicated volunteers will look forward to coordinating several placement events over the year and raise millions of talented enthusiasts to provide support to countless companies and organizations impacted by lack of talented manpower to implement their ideas [11]. The change in trends of placements of two individuals of the same potential group is opposite and the variation in their performances in the mocks or other interviews are significantly larger than the normal student groups.

The proposed value-oriented website will provide the chance to both the admin and especially the students, through self-accessed and curated progress reports, to be well acquainted with their students' weaknesses and areas with the scope of improvement [12]. As a student, the user can view a history of the placement drives he has been a part of, and get an analysis of his/her performance in various rounds. As an admin, the user can view the placement history of an individual or the group and the individual progress analysis report.

The technology used for the application is given in Table 1.

The Login page allows both the admin as well as students to gain access to the portal by entering their Email ID and password [13]. Both the users who navigate to the website are presented with a login page as a way to gain access to the application. It not only ensures authentication but also credibility for our website. Authentication

Table 1 Technology used

Tech used	Specifications
Django	Django was used to create web applications and high-level websites. A single person could make it happen, thanks to a lot of features provided in its frameworks
Azure	Azure Database for PostgreSQL is a relational database service based on the open-source Postgres database engine
CSS3	CSS3 included support for additional selectors, drop shadows, rounded corners, multiple backgrounds, animation, transparency, and much more
JavaScript	JavaScript helped in the creation of HTML pages capable of responding to user events like mouse clicks, keypresses, and selection of elements in the form

is successful and the user is directed to the application home page [14]. If the authentication fails, the user is redirected to the login page. Our navigation is very intuitive and simple. Any other page on our site is just within two clicks of any other page on our site. A responsive site plays well on all screens and devices. It is mobile-friendly and caters to the device you are viewing it on.

While registering, the application displays a student to enter all his basic information from name, email id (which has to always be verified and authenticated to log in to keep the data secure and accessible at the time of need), followed by gender and academic information. Details included in this are the year of graduation, CGPA and later on, we also added the degree option so that we can open up our model to the aspirants from all spheres of academia.

We believe visuals affect everything from conversion rate to time on page, trustworthiness, and organic backlinks. With good readability, we look forward to making sure everything put up on the page is informative and relevant.

4 Results

The already registered students get filtered from each other. To give results about self-evaluation and integration with the placement process, the profile data during registration are equally separated into various successive parts. Then the significance of each part is determined. This processing method is essentially equivalent to obtaining one-unit data at every registration. Then the all group student data are averaged, and the improvement rooms are determined. For the result of each student in the eligible group, the personalized report's data generated during, say, mock or real-time interview is equally and honestly separated and notified to the candidate. This processing is essentially equivalent to giving the tests in the original placement drives along with more focus, attention to detail, smoother and more streamlined.

Then the entire group is constantly kept in touch with the previous year's students and their views. Figure 1 shows the landing page of our model which is tried to make very informative at the same time being concise, while these figures show the Student login and College Admin Login pages. The data on each page is redirected for further

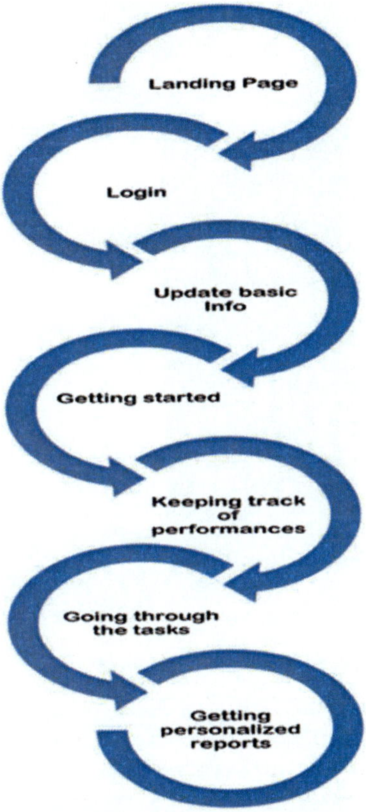

Fig. 1 A basic flow of application navigation

relevant basic information. Each student and admin data is kept obscure and safe. With a synchronized effort toward a single goal, our team was and continues to be absolutely responsive. Our aim was to seize every challenge as an opportunity and commit ourselves to excel in whatever we undertook. Each student and admin data is kept obscure and secure responsibly (Fig. 2).

Figures 3 and 4 show a general layout of our login screens. Their design and functionality are both common and similar. Just like our overall site, we have tried and made it very simple, attractive, and friendly. A user can input the Email ID and the password together to log in.

5 Conclusion

The analyses given in the previous sections indicate that the reduction of quick exits at early stages of placement drives is caused by the irregular effect of admins and students collectively, and the significant changes are mainly caused by the volume

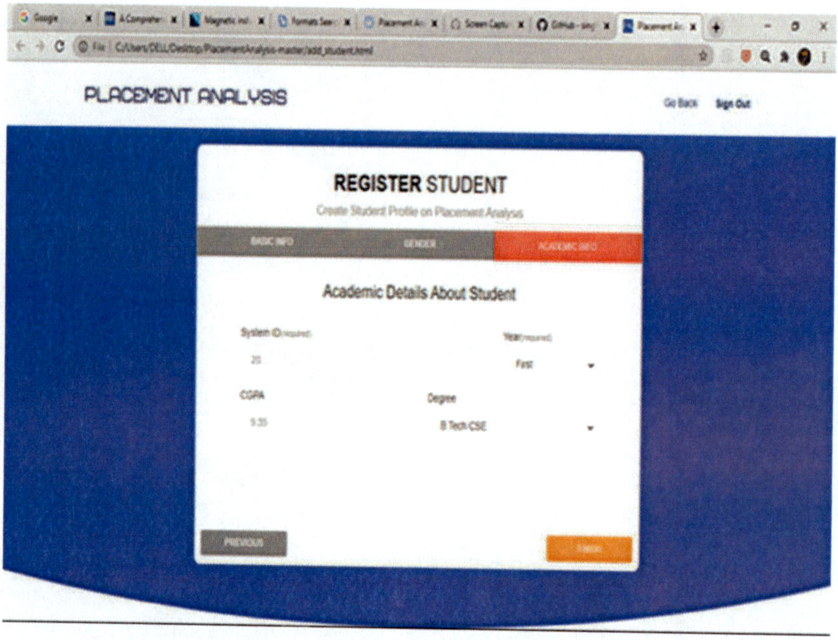

Fig. 2 Student registration page

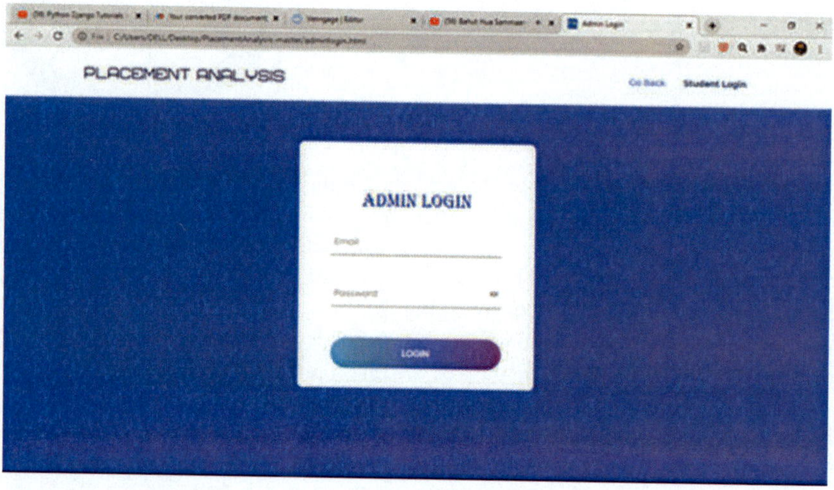

Fig. 3 Admin login page

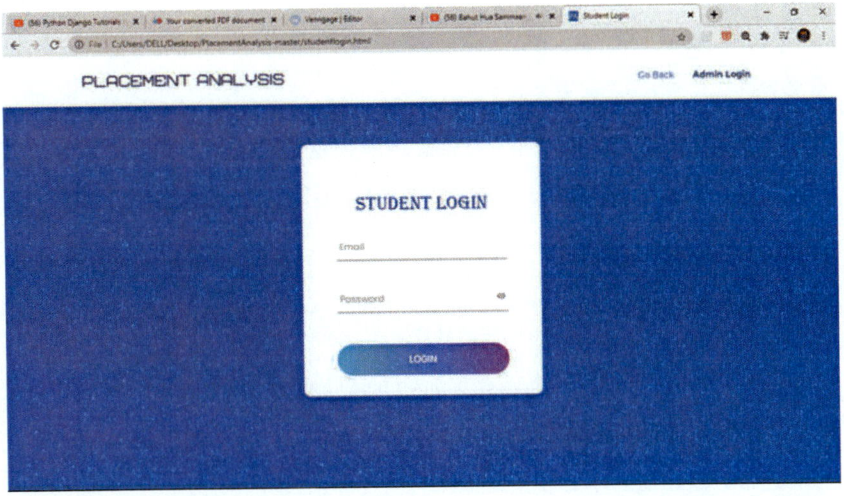

Fig. 4 Student login page

changes of very fundamental or root-based anomalies at early times of the process. This model has the capability to induce placement rise and is attributed to the changes of basic components [15]. At higher levels, the increase is mapped primarily by the change of the total passed out contemplated students in the college. With the model committing itself to quality of opportunities and time, this website is very much helpful for aspirants of all domains. The individual differences (e.g., aptitude) among the various personalized weak point diminishing reports used may be another reason for the high consistency of the model. In general, this is just a pilot study to validate the feasibility of this approach to discriminate the gap of contempt, and more work needs to be performed.

6 Future Scope

Though our project is itself matured enough but still betterment is always an open door. In this case, also we can add some features to this software to make it more reliable. We have made sure to keep the accessibility as quick as possible. The enrollment process and signing up to the portal is so convenient and user-friendly that they will love to be here every time. We are also planning to incorporate backup procedures to make sure of the database integrity. We are also trying to bring up a feature where we bridge the gap of virtual communication between student and company through virtual meetings [16–29].

References

1. B. Godiwala, Y. Doshi, B. Vora, A. Odhekar, Training and placement cell android application. Int. J. Comput. Sci. Mob. Comput. (2020)
2. R.A. Dhole, Web based automation of training & placement cell. Int. J. Sci. Eng. Res. **7**(2) (2016)
3. R. Ramya, G. Pallesami, S.K.V. Jayakumar, Analysis of campus recruitment parameters, but yet unpublished.
4. P.S. Sharma, R.R. Shetty, G.V. Yadkikar, D. Kanade, College automation system. IJIRST Int. J. Innov. Res. Sci. Technol. **2**(10) (2016). ISSN (online): 2349-60102
5. M. Kaur, A. Kaur, R.S. Sawhney, Integrated campus management system using cloud computing. Int. J. Comput. Appl. (0975-8887) (2012) (*International Conference on Computing, Communication and Sensor Network (CCSN)*)
6. S.B. Vanjale, R.K. Modi, S. Raj, A. Jain, Smart training & placement system. Int. J. Comput. Sci. Technol. **188**. www.ijcst.com. ISSN: 0976-8491 (Online)
7. N. Rathod, S. Shah, K. Shirsat, An interactive online training & placement system. Int. J. Adv. Res. Comput. Commun. Eng. **3**(12) (2013); Int. J. Res. Appl. Sci. Eng. Technol. (IJRASET) **6**(X) (2018). ISSN: 2321-9653; IC Value: 45.98; SJ Impact Factor: 6.887
8. In 2015, Lalit Mohan Joshi presented a paper on a web-based college management system for engineering colleges and schools based on Java Platform. The program being proposed is user friendly and meets the basic requirements, vol. 5(II), Mar 2017. IC Value: 45.98, ISSN: 2321-9653
9. A. Yadav, A. Tripathi, N. Rakesh, S. Pandey, Protecting composite IoT server by secure secret key exchange for Xen intra virtual machines. Int. J. Inf. Comput. Secur. (2019). https://www.inderscience.com/info/ingeneral/forthcoming.php?jcode=ijics
10. S.M. Singal, N. Rakesh, R. Matam, A.K. Adhikari, An optimal storage and repair mechanism for group repair code in a distributed storage environment. Int. J. Intell. Decis. Technol. **12**(4), 441–451 (2018). https://doi.org/10.3233/IDT-180347. https://content.iospress.com/articles/intelligent-decision-technologies/idt180347
11. D. Sarkar, N. Rakesh, K.K. Mishra, A. Gangwar, Population based clustering to enhance the utilization of surrogate in content delivery networks. Int. J. Intell. Decis. Technol. **12**(4), 453–460 (2018). https://content.iospress.com/articles/intelligent-decision-technologies/idt180348
12. D. Sarkar, N. Rakesh, K.K. Mishra, A. Mehra, Density-aware replica server placement for utilization enhancement. Int. J. Adv. Intell. Paradigms (2019). https://www.inderscience.com/info/ingeneral/forthcoming.php?jcode=ijaip
13. A. Yadav, A. Tripathi, N. Rakesh, S. Pandey, Optimizing the DTLS handshake design for trusted execution environment enabled sensor nodes. Int. J. Secur. Networks (2019)
14. G. Rathee, N. Rakesh, Resilient packet transmission (RPT) for the buffer based routing (BBR) protocol. J. Inf. Process Syst. **12**(1), 57–72 (2016). http://jips.jatsxml.org/upload/pdf/jips-12-1-57.pdf
15. A. Mundra, N. Rakesh, V. Tyagi, Query centric cyber physical system (QCPS) approach for multiple heterogeneous systems. Int. J. Comput. Sci. Technol. (IJCST) **4**(2), 23–27 (2013). http://arxiv-web3.library.cornell.edu/ftp/arxiv/papers/1306/1306.6397.pdf; http://ijcst.com/?page_id=4250
16. A. Mundra, B.K. Gupta, G. Rathee, M. Chawla, N. Rakesh, V. Tyagi, Validated real time middle ware for distributed cyber physical systems using HMM. Int. J. Distrib. Parallel Syst. (IJDPS) **4**(2), 23–33 (2013). https://doi.org/10.5121/ijdps.2013.4204. http://airccse.org/journal/ijdps/papers/4213ijdps03.pdf
17. A. Mundra, G. Rathee, M. Chawla, N. Rakesh, A. Soni, Transport information system using query centric cyber physical systems (QCPS). Int. J. Comput. Appl. **85**(3), 12–16 (2014). https://doi.org/10.5120/14820-3050. http://www.ijcaonline.org/archives/volume85/number3/14820-3050

18. N. Rakesh, V. Tyagi, Parallel architecture coding: link failure-recovery mechanism (PAC: LF-RM). Int. J. Syst. Assur. Eng. Manage. (2012). https://doi.org/10.1007/s13198-012-0119-4. http://link.springer.com/article/10.1007%2Fs13198-012-0119-4. © The Society for Reliability Engineering, Quality and Operations Management (SREQOM), The Division of Operation and Maintenance, Lulea University of Technology, Sweden, Springer, Aug 2012
19. B.K. Gupta, A. Mundra, N. Rakesh, Failure detection and recovery in hierarchical network using FTN approach. Int. J. Comput. Sci. Issues (IJCSI) **10**(2), 414–421 (2013). http://www.ijcsi.org/papers/IJCSI-10-2-2-414-421.pdf
20. N. Rakesh, V. Tyagi, Linear-code multicast on parallel architectures. Adv. Eng. Software **42**(12), 1074–1088 (2011). https://doi.org/10.1016/j.advengsoft.2011.07.012. http://www.sciencedirect.com/science/article/pii/S0965997811002079
21. N. Rakesh, V. Tyagi, Linear network coding on multi-mesh of trees using all to all broadcast. Int. J. Comput. Sci. Issues **8**(3), no. 1, 462–471 (2011). http://ijcsi.org/papers/IJCSI-8-3-1-462-471.pdf
22. A. Kapoor, H. Magotra, G. Bharti, S. Pandit, N. Rakesh, Implementing video data compression along with network coding in video multicast. Int. J. Adv. Comput. Networks Secur. Commun. Networks **1**(1), 153–156 (2011). http://seekdl.org/journal_page_papers.php?jourid=85&issueid=21&page=25&#seekdl
23. N. Tyagi, T. Gupta, R. Nagi, N. Rakesh, V. Tyagi, Naive parameter for the network erasure coding for video transmission over a wireless network. Int. J. Mob. Adhoc Network **1**(2), 199–206 (2011). http://www.ifrsa.org/index.php?option=com_content&view=article&id=41&Itemid=40
24. K. Shah, G. Dua, D. Sharma, P. Mishra, N. Rakesh, Transmission of successful route error message (RERR) in routing aware multiple description
25. S.P. Singh, S.S.P. Shukla, N. Rakesh, V. Tyagi, Problem reduction in online payment system using hybrid model. Int. J. Managing Inf. Technol. **3**(3), 62–71 (2011). http://arxiv.org/abs/1109.0689
26. N. Rakesh, N. Nitin, Analysis of multi-sort algorithm on multi-mesh of trees (MMT) architecture. J. Supercomput. **57**(3), 276–313 (2010). https://doi.org/10.1007/s11227-010-0404-4. http://link.springer.com/article/10.1007%2Fs11227-010-0404-4
27. N. Singh, P. Chauhan, N. Rakesh, N. Nitin, Team spirit model using standards for video delivery. Int. J. Commun. Network Syst. Sci. **3**(4), 364–368 (2010). https://doi.org/10.4236/ijcns.2010.34046. http://www.scirp.org/journal/PaperInformation.aspx?PaperID=1633
28. P.K. Verma, T. Gupta, N. Rakesh, N. Nitin, A mobile ad-hoc routing algorithm with comparative study of earlier proposed algorithms. Int. J. Commun. Network Syst. Sci. **3**(3), 289–293 (2010). https://doi.org/10.4236/ijcns.2010.33037. http://www.scirp.org/Journal/PaperInformation.aspx?paperID=1473
29. N. Rakesh, N. Nitin, Multi-sort algorithm on multi-mesh of trees MMT. J. Comput. **1**(1), 1–8 (Aug 09). http://www.masaumnet.com/archives/mjc/volume1/issue1/mjc010112.pdf

Continuous Assessment Analyzer Using Django

Shriyans, Swati Dhoundiyal, Nitin Rakesh, Neha Tyagi, Parma Nand, and Mandeep Kaur

Abstract This research paper addresses ongoing academic concerns about how assessment should be done in higher education. Globally, the outbreak of the COVID-19 has resulted into the shutdown of the educational institutions. These unprecedented challenges raise a need to adapt and find solutions to keep students motivated in their learning track. It has been observed that quizzes and assignments have become the focus of assessment that also supports student learning. To put an end to these challenges, we have tried to automate all the assessment processes possible through a web application created using Django, HTML, CSS, etc.

Keywords Assessment · Continuous assessment · Student-learning · Active engagement · Effective assessment · Active learning · CDN

1 Introduction

Assessment is the act of judging or deciding something quantitatively and qualitatively. Effective assessment promotes the active engagement of students in learning.

Assessments that are performed during a course module are referred to as "**continuous evaluation**". Continuous evaluation may be used in place of or in addition to the final assessment to determine a final score. That is why this type of assessment is also called as curriculum integrated, or embedded assessment. Continuous assessment, as part of a broader area of assessment, keeps up with the times, meeting the requirements of increasing learner-involvement in the educational process [1].

Shriyans (✉) · S. Dhoundiyal · N. Rakesh · N. Tyagi · P. Nand · M. Kaur
Department of Computer Science and Engineering, Sharda University, Greater Noida, Uttar Pradesh, India

A. Kumar et al. (eds.), *Computational Intelligence in Machine Learning*, Lecture Notes in Electrical Engineering 834, https://doi.org/10.1007/978-981-16-8484-5_4

Researchers claim that online learning increases knowledge retrieval and takes less time, signalling that the shifts brought on by COVID-19 are here to last. This method, as stated, helps to identify the desired results, providing resources for all students to attain those outcomes, and collecting, evaluating, interpreting, and reporting data in order to better explain and enhance student learning [2].

As a result, we created an automatic CA analyser with a user-friendly interactive GUI to make all of the work simple and straightforward. From making quizzes to marking all of the students' grades to calculating statistics, interacting with the application is a lot easier [3].

CA analyser provides various features including the following:

Functionality provided to a student:

- View his/her academic profile.
- Keeping a track of Subject-Wise attendance.
- Attempting objective quizzes.
- Analysis of the course outcome.

Functionality provided to a teacher:

- List of students inside a given course.
- Creating and assigning quizzes to students.
- Marking attendance & record of every student.
- Automatic CA generation for every student.

All manual labour, including all of the above-mentioned features, is done automatically by this application, saving time for both students and teachers [4]. It facilitates an effective learning process and encourages students to exercise their learning activities over the course, leaving more time to study rather than just memorizing the course content before the final examination [5].

2 Existing System

For assessment of students at the end of the course, the traditional examination system with a long, time-consuming conducting and evaluating process is followed till date [6]. The traditional methods require the physical campus, availability of the faculties physically and the systematic seating arrangement with some pens and papers. For calculating the results of students, counting the number of backlogs, etc. we are using the manual framework directly. Repeated updates are needed to be done as there are the revaluation of results due to the supplementary exams.

Keeping the current scenario in mind, it is not possible for educational institutions to conduct pen paper exams for students to attempt. The analysis of results is a tedious task. In order to overcome the constraints of the insecurity, loss of data and time consumption, the necessity of a better system emerges [7]. The existing system requires a lot of human effort which in fact is also time-consuming too. Paperwork, uncertainty and time-consuming forms are a major restriction of the existing structure [8]. Henceforth, there is a need to minimize the efforts of the faculties and other non-academic staff in analysing the result.

3 Proposed Methodology

Effects of COVID-19 pandemic led to the interference in the learning journey of the students. And in case of the final year students, it was the final straw for at-risk students who struggled to get education at the best of times [7].

How will the education sector and educators deal to overcome these challenges? How will students continue to learn, even as schools, by necessity, becomes a digital space?

Proposed system is a solution to the above raised queries, also a step towards e-learning and assessment system. Online assessment is the excellent way of narrowing down the undying efforts of the teachers as well as the other non-academic staff who are working towards simplification and modernization of the traditional examination and assessment methods. Cumulative assessment is becoming one of the most effective and easy solutions for conducting assessment tests for students [9]. Though e-learning is still in its developing stages, learners and faculties at institutions have the technology and the resources to use video conferencing, email, and other technologies to continue their learning experience. The convenience of learning at home and at his own pace encourages students to participate in online assessments. These assessments are highly interactive, customizable, trustworthy, secure, and also reflects the student efforts.

The proposed system maintains the continuity of information of all through the framework [10]. It includes generalized information that is simple to store and retrieve. This application produces quiz reports, topic smart reports, class-wise reports, subject-wise and overall attendance reports. In addition to this, graphs of all the above data with percentage are generated. It is secure and quick. With the help of the proposed framework, a part of the time is spared which is the major benefit. The login and incognito pages secure your structure appropriately. The fundamental calculation is done accurately. Thus, precision adds to major advantage. To automate the storing and analysis of subjects' marks, we have used the local database, here we use MySQL, and to interact with the database we have used Django (Python) [11, 12] and for the user-friendly interface we have used HTML5, CSS3, Bootstrap, JS, Ajax, Chart.js and other CDN service [13].

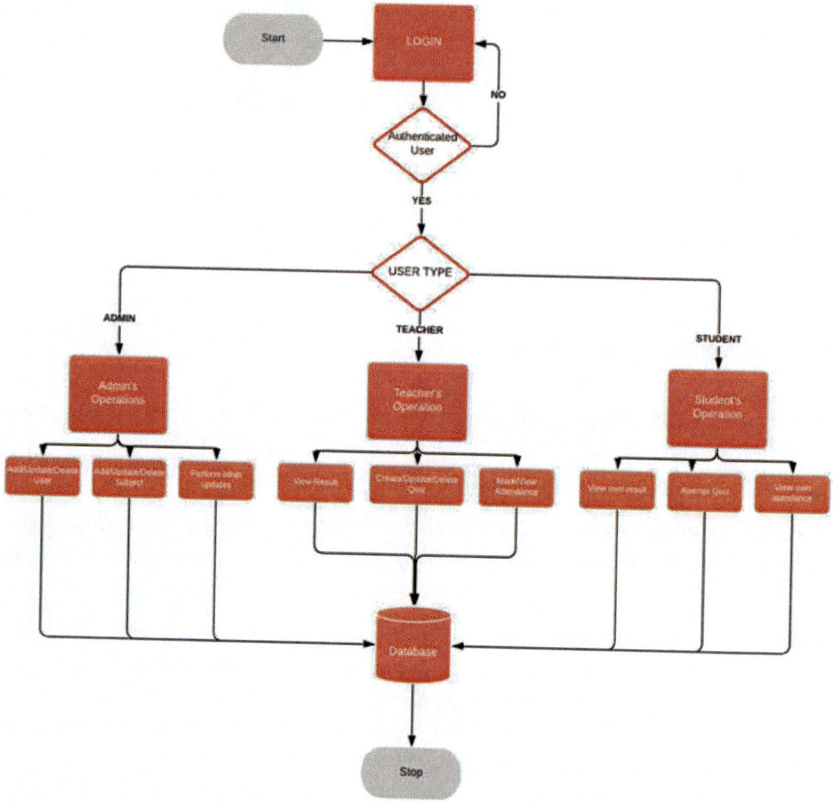

Fig. 1 Flow of the data in the application

As soon as student attempts any quiz his score is updated and the analysis part is also performed automatically. The database is updated and, in the result section necessary updates in the graphics and tables are done for the both teacher and student side automatically [14]. After attempting the quiz, the student's score is updated in the database, and it can be viewed both by the teacher and the student (Fig. 1).

4 Result of the Interface and Functionality

The detailed functionality of the project can be vividly explained through a series of results attached (Figs. 2, 3, 4, 5, 6, 7 and 8).

Fig. 2 The homepage of the project which is also the login page where user can login using proper username and password

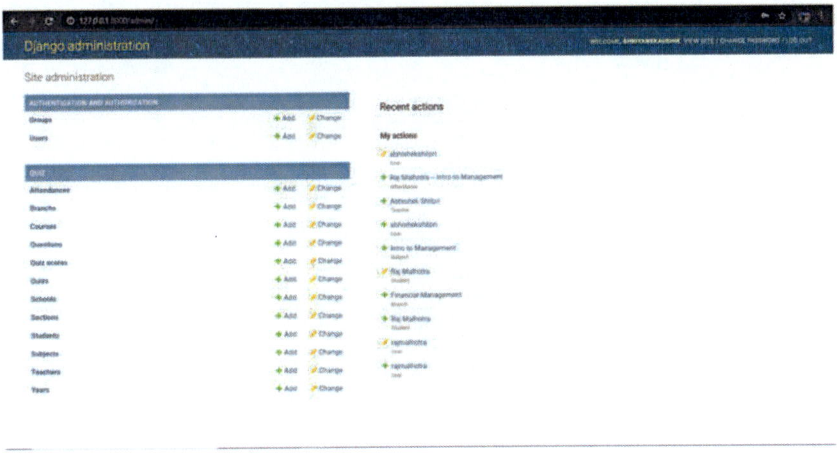

Fig. 3 Admin dashboard—admin page where only the admin can have access and can add/remove the data in the project

5 Conclusion

The main motive of the project is to analyse the students' result and show it in the most possible user-friendly way. By user-friendly, we mean showing subject-wise results in a graphical way, as it is easy to read and interpret, compared to simply showing results in a tabular form. In addition to this, we also have the platform for assigning and attempting the quiz. Teachers can assign the quiz to the students and the students can attempt the quiz using their login credentials. However, these do not

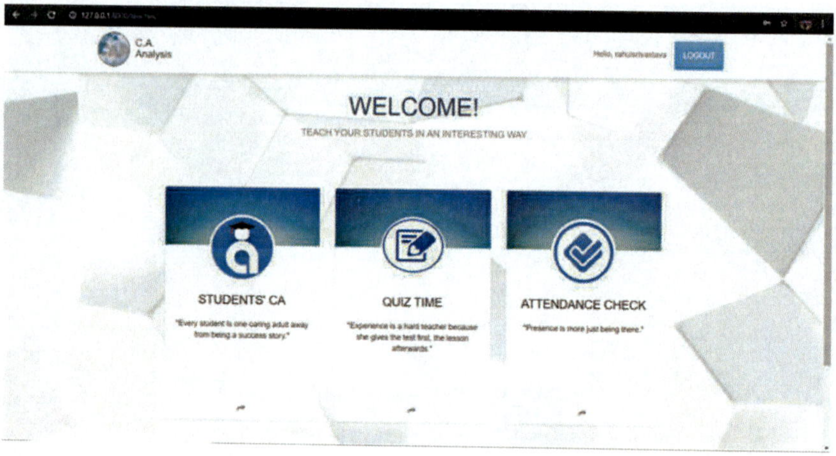

Fig. 4 Teacher dashboard—teacher's landing page where only the teacher can have access and can do tasks like creating and assigning a quiz, mark/view the attendance and analyse student's result

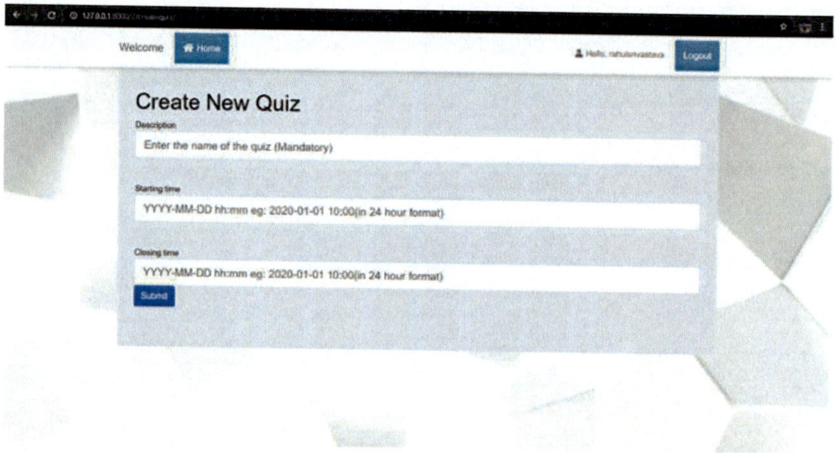

Fig. 5 Demonstration of creating new quiz which can be later attempted by the students

directly measure student learning. It is best to have a variety of measures that lead to the same conclusions [15].

6 Future Scope

As everyone, once again, falls in the jaws of COVID-19 outbreak, but this time it is more serious and infectious than the previous year-long pandemic. We believe that

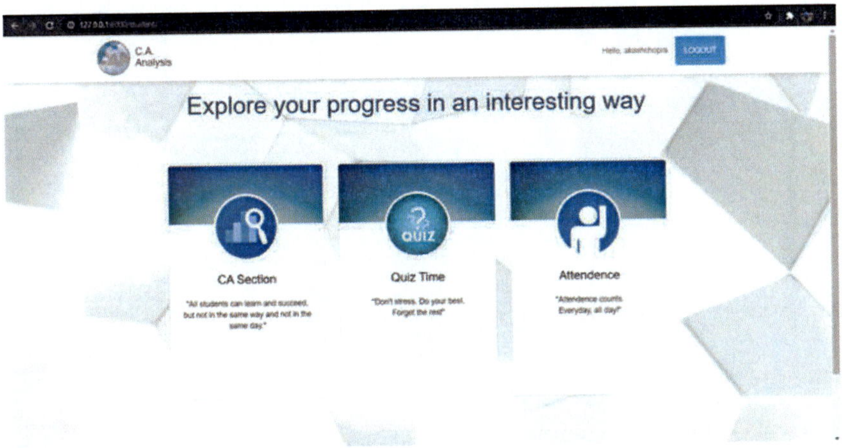

Fig. 6 Student dashboard—student can attempt quiz, view the score of the attempted quiz, view his attendance

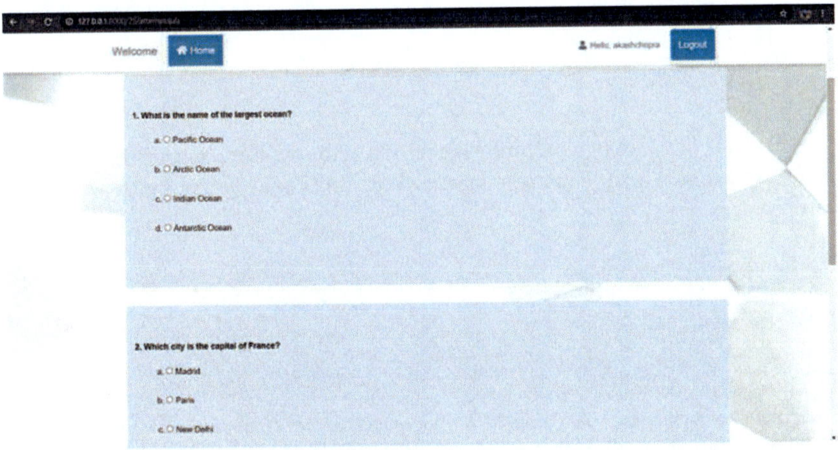

Fig. 7 Demonstration of student attempting a quiz

traditional learning and online learning will become one and the same, and online education will finally become an integral part of education.

Further improvisation in the project can be done by adding a new user type in the project called Management. This user type will include people who are at the HOD, Dean, Registrar, Chancellor and Vice-Chancellor position. The management group could look at the progress of not only students but also teachers, that too on single screen [16]. Through this, the management can have a look at the progress report and pace of the course.

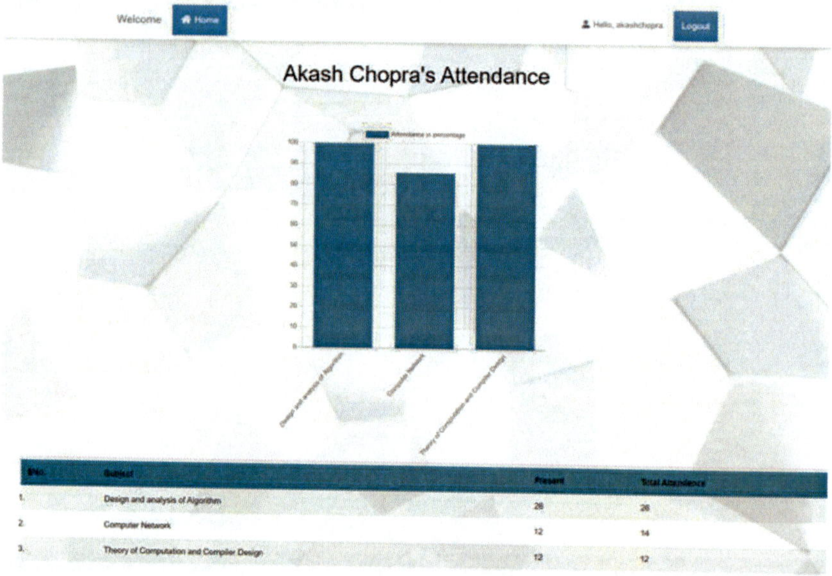

Fig. 8 Student's attendance shown in graphical and tabular form

At present, we are only monitoring the results of the current semester; however, we would like to add a feature that allows us to keep track of a student's past semesters' performance.

References

1. A. Manurung, A. Fildzah, J. Rajagukguk, Computer based test (CBT) system for student academic examination, in *AISTSSE 2018: Proceedings of the 5th Annual International Seminar on Trends in Science and Science Education, AISTSSE 2018*, 18–19 Oct 2018, Medan, Indonesia (European Alliance for Innovation, 2019), p. 31
2. K. Timmerman, T. Doom, Infrastructure for continuous assessment of retained relevant knowledge. ACM Inroads **8**(2), 73–77 (2017)
3. A.T. Korucu, H. Cakir, The effect of dynamic web technologies on student academic achievement in problem-based collaborative learning environment. Malays. Online J. Educ. Technol. **6**(1), 92–108 (2018)
4. M. Neary, Supporting students' learning and professional development through the process of continuous assessment and mentorship. Nurse Educ. Today **20**(6), 463–474 (2000)
5. R. Hernández, Does continuous assessment in higher education support student learning? High. Educ. **64**(4), 489–502 (2012)
6. B. Christudas, MySQL, in *Practical Microservices Architectural Patterns* (Apress, Berkeley, CA, 2019), pp. 877–884
7. C. Bell, *Introducing the MySQL 8 Document Store* (Apress, New York, 2018)
8. J.M. Santos, E. Ortiz, S. Marín, Variation indexes of marks due to continuous assessment. Empirical approach at university/Índices de variación de la nota debidos a la evaluación

continua. Contrastación empírica en la enseñanza universitaria. Cultura Educación **30**(3), 491–527 (2018)
9. H. Da Rocha, *Learn Chart js: Create Interactive Visualizations for the Web with Chart.js 2* (Packt Publishing Ltd., Birmingham, 2019)
10. S. Aryal, Bootstrap: a front-end framework for responsive web design (2019)
11. D. Ghimire, Comparative study on Python web frameworks: Flask and Django (2020)
12. J. Vainikka, Full-stack web development using Django REST framework and react (2018)
13. D. Dinh, Z. Wang, Modern front-end web development: how libraries and frameworks transform everything (2020)
14. M. Madsen, O. Lhoták, F. Tip, A model for reasoning about JavaScript promises. Proc. ACM Program. Lang. **1**(OOPSLA), 1–24 (2017)
15. I.N. Day, F.M. van Blankenstein, P.M. Westenberg, W.F. Admiraal, Explaining individual student success using continuous assessment types and student characteristics. High. Educ. Res. Dev. **37**(5), 937–951 (2018)
16. F. Kereki, *Modern JavaScript Web Development Cookbook: Easy Solutions to Common and Everyday JavaScript Development Problems* (Packt Publishing Ltd., 2018)

Fuzzy Logic in Battery Energy Storage System (BESS)

Mohd Azali Bin Zainal Abidin and Kumeresan A. Danapalasingam

Abstract The battery energy storage system (BESS) is a portable device that consists of batteries, controllers, sensors, relays, and other elements that are vital for battery charging and electricity supply operations. A BESS can be used in on-grid or off-grid applications to supply electricity as the main source or a backup source of energy and to eliminate power factor surcharge by utility companies. This research seeks to design, apply, and validate the use of fuzzy logic controller as a battery management system to regulate the simultaneous charging and discharging processes of batteries. In this project, BESS hardware is constructed to include two 24 V battery sets, an electrical load, proper electrical connections, switches, relays, current sensors, voltage measurements, LED indicators, and a microcontroller. Fuzzy logic controllers, one for each battery set, are designed, simulated and analyzed in MATLAB before implemented in the actual BESS hardware. The experiment involves repeatedly running the BESS for several hours, through a number of charging and discharging cycles, and analyzing current and voltage measurements. By comparing the performances of the fuzzy logic controllers and that of a conventional sequential algorithm, it can be concluded that the fuzzy logic controller can be applied as a safe and efficient battery management system and may improve the battery life in a BESS.

Keywords BESS · Fuzzy logic controller

1 Introduction

The battery energy storage system (BESS) is an electrical energy storage system produced using fuzzy logic controller. The basic idea is that it is possible to use those saved resources at a later time. Extensive research in this field has resulted in developments in batteries that have developed the idea of the battery energy storage system (BESS) into a commercial reality. The battery energy storage system (BESS) plays an important role in the incorporation and enhancement of the usage of renewables.

M. A. B. Z. Abidin (✉) · K. A. Danapalasingam
Electrical Engineering Department, Politeknik Ibrahim Sultan, Pasir Gudang, Malaysia
e-mail: mohdazali@pis.edu.my

A. Kumar et al. (eds.), *Computational Intelligence in Machine Learning*, Lecture Notes in Electrical Engineering 834, https://doi.org/10.1007/978-981-16-8484-5_5

Battery management devices enable the monitoring and control of battery output. Many battery control parameters can be used [1], such as:

i. The battery voltage
ii. Current
iii. State of charge (SOC)
iv. State of health
v. Depth of discharge (DOD).

For each type of battery, the voltage and current of the charging cycle are different. A battery charger or charger controller designed to handle one battery type cannot usually be used with another. To ensure the optimum recharge and discharge status of the battery, a fuzzy control technique is developed. Trial and error is one method to design the fuzzy logic control rule [2]. Fuzzy logic is based on the experimental results of experts in the relevant application area. Hence, the use of fuzzy rules and inference engines helps fuzzy controllers to make decisions similar to humans while working in the field.

1.1 Problem Statement

To ensure the capacity of nano-gel batteries using current and voltage, measurement of the accuracy of each battery is required. This is a challenge because different characteristics during charging and discharging are generated by current and voltage. Relays are needed to operate under the state of the batteries to control the system of charging or discharging based on the capacity of the batteries.

The temperature of a battery is very critical for preserving the output of the battery during charging and discharging. The temperature affects the lifetime of a battery. One solution for avoiding overcharge is to employ a battery management control device, which can be done through the knowledge of the available energy in the battery.

The current and voltage levels should be running at the battery energy storage system (BESS). Different characteristics during charge and discharge are generated by current and voltage. Overcharging can reduce battery life and cause substantial battery damage.

1.2 Research Goal

The goal of the research is to design the battery energy storage system (BESS) with a fuzzy logic controller using MATLAB/Simulink. Through current and voltage behavior on the system, the battery performance can be monitored and controlled. Battery performance will impact the fuzzy logic rules. Battery energy storage system (BESS) hardware is constructed to include two 24 V battery sets.

1.3 Objective

The objectives of the research are:

- To control the charging and discharging processes of a battery energy storage system (BESS) using battery voltage and current measurements.
- To build the battery energy storage system (BESS) with a fuzzy logic controller.
- To compare the performance of conventional sequential algorithm and fuzzy logic controller.

1.4 Scope of the Project

- Develop a battery energy storage system (BESS) that consists of two 24 V 4 Ah battery sets of nano-gel batteries.
- Simulate the battery energy storage system (BESS) using MATLAB/Simulink.
- To monitor and control the charging and discharging process of the batteries using a conventional sequential algorithm and a fuzzy logic controller (FLC).
- Compare the performance of conventional sequential algorithm and fuzzy logic controller (FLC).

2 Literature Review

2.1 Introduction

A battery energy storage system (BESS) is used to store large-scale electricity in electric and power grids. The lithium-ion battery is the most commonly used type of battery for this.

In the last few years, several battery measurement techniques have been introduced, varying between direct and model-based calculation of current and voltage. The open-circuit voltage and Coulomb counting phase are considered to be popular direct measurement techniques [3].

Today, many storage technologies are available other than those used to implement battery energy storage systems (BESS), based on various system requirements such as superconducting magnetic energy storage (SME), compressed air energy storage (CAES), pumped hydroenergy storage (PHES), the hydrogen-based energy storage system (HES), and flow battery energy storage (FBES). The battery energy storage system (BESS), however, is preferred for versatile localization, performance, scalability, and other features based on technology [4].

Besides using state of charge (SOC), many other different approaches can also be used for battery management and control. The Coulomb counting method is the most commonly used method of estimating SOC, but due to the initial SOC, the Coulomb counting algorithm at many times leads to an aggregated error [5].

A rechargeable battery is an important component because it is compact and flexible. The charging of the battery is kept within a reasonable (20–80%) cap to ensure that overcharge and discharge are prevented [6].

A good battery capacity estimate, longer battery life, and higher battery efficiency delivery provide a number of benefits [7]. When charging a battery, the charger should not exceed 3 A [8]. While a fully charged battery provides maximum life, overloading will greatly reduce their lifetime. The longer the battery is being charged, the more the electrolytes decrease in volume, and power increases. A battery should not be overloaded via charging, as electrolyte volume decreases with too much boost/charge.

2.2 Battery

The capacity of a battery is the key parameter for assessing its efficiency [9]. SOC is another way of measuring the strength of the battery, as shown in Eq. 1.

$$\mathrm{SOC}(t) = \mathrm{SOC}(t-1) + \int_0^t \frac{I}{C_{\mathrm{bat}}} \tag{1}$$

where SOC(t) = current state of charge, SOC($t - 1$) = initial state of charge, I = current during charge/discharge, and C_{bat} = capacity of the battery.

2.2.1 Battery Self-discharge

When a battery is not connected to a charging device, it undergoes self-discharge. The rate of self-discharge depends on the temperature and type of the battery. The rate of self-discharge also varies based on battery technology. In the first 24 h, a Li-ion battery releases 5% of its charge, followed by 2–3% every month [10]. Self-discharge is also faster at temperatures above 55 °C. These temperatures can be higher if the battery is stored in a garage or hot place [8].

Another common misconception when it comes to storing batteries is that storing them on a concrete floor will cause them to auto-discharge rapidly. Battery cases were made of hard rubber about 30 years ago, which allowed the batteries to discharge from the battery moisture directly into the concrete floor [11].

However, modern battery cases are made of plastic polypropylene, and as such, unnecessary self-discharge can be ignored even if stored on concrete surfaces. There are various methods used for measuring self-discharge of a battery. The indirect approach is based on cell voltage control during rest. Supercapacitors can work with this approach [12]; however, it is not suitable for lithium-ion cell measurements with high accuracy.

2.2.2 Performance and State of a Battery

The use of Coulomb counting charges can be inaccurate and may lead to errors due to the battery's nonlinear nature being triggered. Kalman filter systems are used to increase precision in battery management [13–15]. However, SOC measurements alone cannot be relied upon even with these advances. Implementation of fuzzy logic with current and voltage measurement can quantify a new parameter called battery working state (BWS). BWS is determined by a value within a certain range (0 is idle, − 1 is discharging, and 1 is charging) and is represented by few membership functions. This prevents overcharging and discharging, thus, prolonging battery life [16].

2.3 *Fuzzy Logic Controller*

Zadeh invented the fuzzy logic controller as an alternative approach to the classical control technique in the beginning of 1960, which has since been gradually integrated into the field of control theory [17]. The idea of fuzzy logic is to deal with the input as a partial set of membership functions instead of crisp values. Mamdani [18] converted the algorithm of fuzzy logic into a controller when he used it to control a dynamic system.

Fuzzy means fluffy, broken, or uncertain. Fuzziness is a measure of deterministic ambiguity and is related more to the extent to which events occur, rather than the chance of their likelihood. Fuzzy logic also refers to a type of several valued logics in which variables can vary in truth between 0 and 1, and the definition of partial facts is used, in which the value of the facts may be fully valid and completely incorrect [19]. The fuzzy logic control is a knowledge base which can determine its parameters without an accurate model device [20].

Fuzzy logic has been applied in various fields from control theory to artificial intelligence. It can handle some level of uncertainty and imprecision [21].

2.3.1 Battery State of Charge (SOC) Estimation Algorithm Using Fuzzy Adaptive Federated Filtering

In this setup, a battery pack is connected in series and consists of a series of connected cells as shown in Fig. 1. The solution to the problem is to correctly and securely use any cell state of charge (SOC) to get a full packaging state of charge (SOC). A fusion multi-source algorithm collects sensor information for efficient estimation of processing information [22].

State of charge (SOC) incoherence affects the precision of the battery state of charge (SOC). Based on battery cell anomalies and measurement errors, predictions of accuracy and reliability have a significant effect on state of charge (SOC). The fuzzy system is designed to increase cell incoherence and adjustment in charging

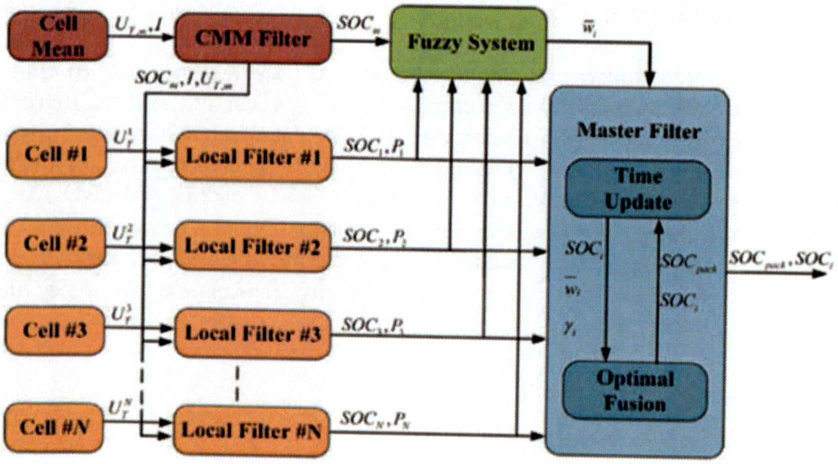

Fig. 1 Battery pack SOC structure

status estimates. The approximate status of the battery average and the state of charge (SOC) are combined with a fuzzy method to assess the fusion weight [23].

2.3.2 Battery Management Using Fuzzy Logic Controller

Control of batteries was developed and implemented for renewable energy sources (solar panel, generator) using battery management and fuzzy logic controllers (FLC). By using MATLAB/Simulink, the modeling, energy storage devices (ESD), and generator can be evaluated and handled. A fuzzy logic controller (FLC) has been designed to control the charging/discharging mode of the battery to increase its life.

The fuzzy controller is designed to be in a charge or discharge mode with respect to the system. The ΔSOC and Power (*P*) equations displayed below are configured as inputs to the fuzzy logic controller, and the output variable is the current input to the battery model, as shown in Fig. 2 [24].

2.3.3 Fuzzy Logic Controller Energy Management

An energy management system controls the transfer of energy between different parts to satisfy load demand. Effective control of power exchange between the components allows for a major increase in efficiency, and the usage of renewable energy sources leads to a reduction of harmful emissions.

In this system as shown in Fig. 3, the wind turbine and solar system design required maximum performance. The difference was charged and discharged between the real battery capacity. Battery charging and discharge time were based on the state of

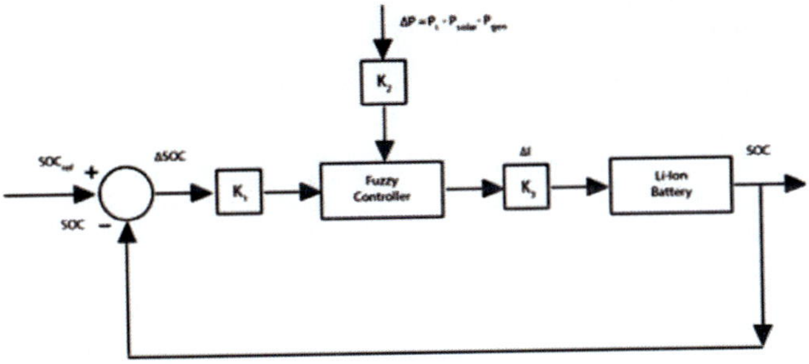

Fig. 2 Battery management using fuzzy logic controller

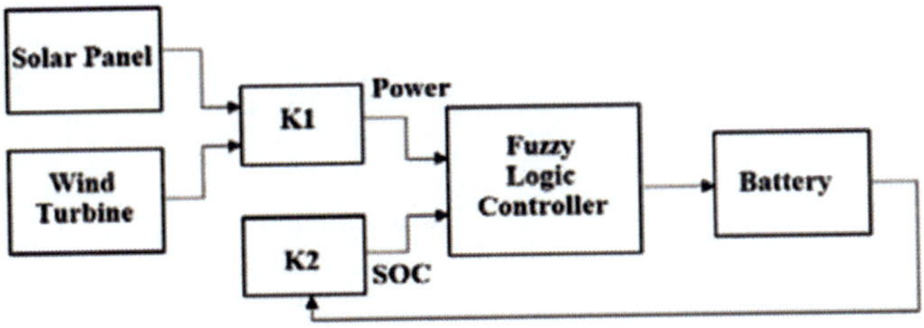

Fig. 3 Fuzzy logic in battery management

charge (SOC). A fuzzy controller was used in order to raise the battery to preserve the SOC level of the battery [25].

3 Methodology

3.1 Battery

Several methods were evaluated in order to achieve the objectives, and relevant methods were carefully selected that fulfilled our demands. The research was mainly aimed at developing a battery energy storage system (BESS) that protects the battery against deterioration and improves battery life as much as possible. The entire system was built with this goal in mind. Figure 4 shows a simplified view of the overall workflow of the project.

The battery energy storage system (BESS) specifications have been evaluated and simulated in several process platforms for achieving the objectives. Proteus and MATLAB/Simulink were used in this project to design and develop the battery energy storage system (BESS). The battery energy storage system (BESS) hardware was constructed to include two 24 V battery sets, an electrical load, proper electrical

Fig. 4 Project flowchart

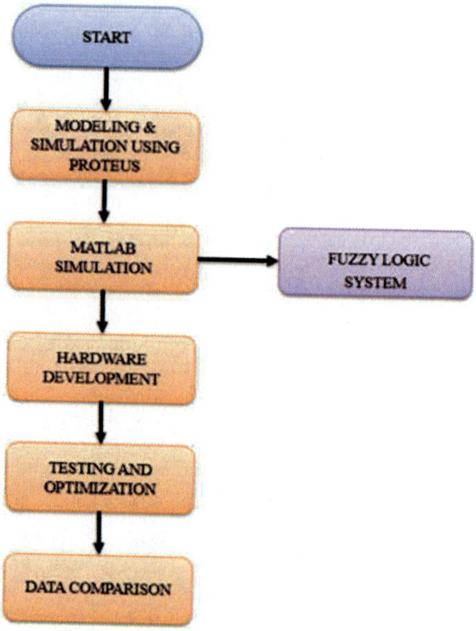

connections, switches, relays, current sensors, voltage measurements, LED indicators, and an Arduino Mega microcontroller. The algorithm was designed using the Arduino IDE software and was implemented into the system to control it. MATLAB/Simulink was used to apply fuzzy logic to the circuit. Two inputs (current and voltage) and an output battery work state (BWS) were selected in the implementation of the fuzzy method. The output for both approaches will be evaluated and compared by several experiments of a conventional sequential algorithm and a fuzzy logic controller.

3.2 *Proteus Simulation*

Two parallel battery sets were each composed of two relay-operated series of batteries. Proteus was used for this circuit design. The battery charger is shown as a battery, and the load/inverter is shown as a resistor. The circuit diagram is shown in Fig. 5.

The charger or the load can be attached to each battery set or the correct relays can be left floating by enabling/disabling them. The microcontroller used here is the Arduino Mega 2456. The conventional sequential algorithm tests the battery sets' current and terminal voltage continuously and decides the indications to be sent to the relays. Voltage dividers are used to lower the voltage so that the voltage remains below 5 V, which can be read by the Arduino analog pins. For the measurement of current, a 1-Ω shunt resistance is used.

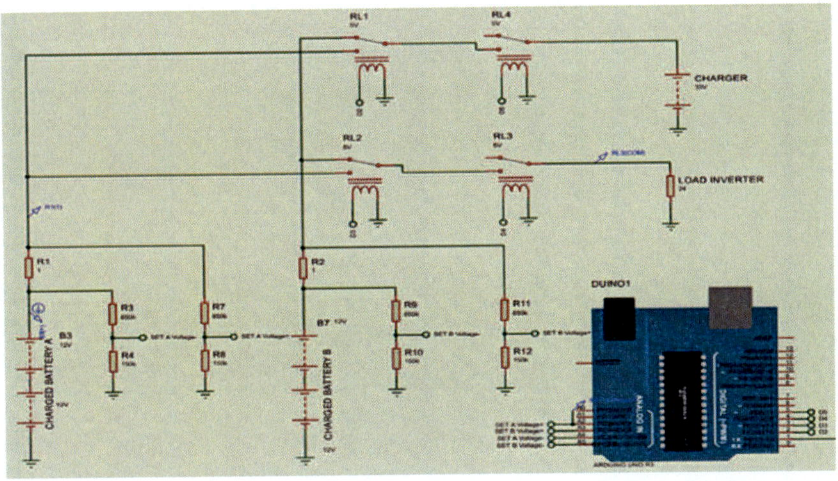

Fig. 5 BESS circuit diagram using Proteus

Proteus was used to develop the circuit battery charging and discharging mechanism. The voltage dividers were designed according to the following derivation.

$$I * R_1 = \frac{R_4}{R_4 + R_3} * V_{A_+} - \frac{R_8}{R_7 + R_8} * V_{A\_} \tag{2}$$

Here R_1 is worth 1 Ω, which can be found quickly. The voltage on the Arduino pins should also never exceed 5 V. When the charger is charged and the battery is charged fully, the maximum voltage is V_{A_+}. The maximum charge current for redundant protection is 5 A for measurement. The maximum current is based on experimental data. The voltage will be:

$$28.8 \text{ V} + 5 \text{ V} = 33.8 \text{ V} \tag{3}$$

Hence, the voltage divider will have a voltage of 5 V when the voltage is 33.8 V

$$\frac{R_4}{R_4 + R_3} * 33.8 = 5 \tag{4}$$

Therefore, the 150k and 850k values were selected. The model using Proteus is a very basic model that is essentially a voltage source with resistance of series. This is not a precise simulation. The behavior depicted here is very different from that of a practical battery. There is also no charging or discharge behavior shown in the Proteus battery model. The Arduino serial output and the operation of relays were shown via a virtual terminal.

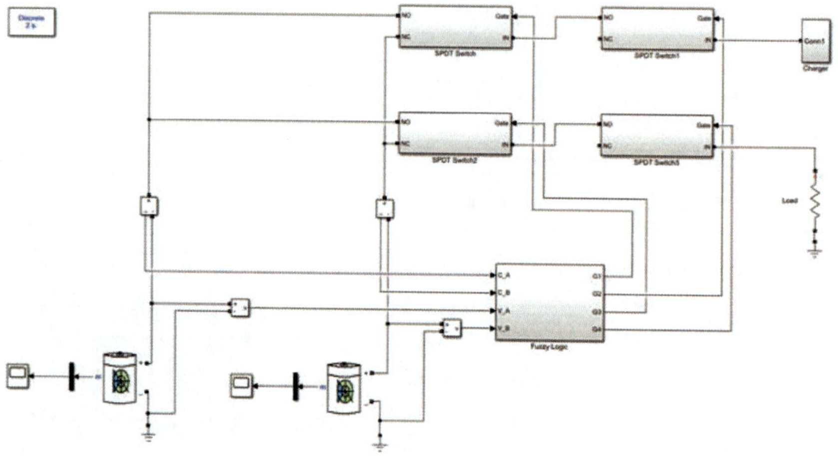

Fig. 6 BESS circuit diagram using MATLAB/Simulink

3.3 *MATLAB/Simulink Simulation*

The diagram below was generated by the MATLAB library. This library consists of more precise battery displays. The overall block diagram of the proposed structure is presented in this diagram, as shown in Fig. 6.

Batteries in MATLAB/Simulink have better indicators like voltage, current, and state of charge (SOC). MATLAB/Simulink allows changing of the capacity, nominal voltage, and also the type of the batteries.

Battery energy storage system (BESS) circuit was designed using MATLAB/Simulink. The actual performance of batteries can be found in MATLAB/Simulink. The battery at MATLAB/Simulink has three parameters (SOC, current, and voltage) and can be used for simulation purposes. The battery characteristic can be adjusted easily in MATLAB/Simulink. The fuzzy logic controller was designed using MATLAB/Simulink as shown in Fig. 7.

3.4 *Fuzzy Logic Designer*

In this project, the fuzzy logic system uses the Mamdani method. The inputs of the fuzzy logic system are current and voltage, and the output is battery work state (BWS). The voltage ranges from 23 V (low) to 29 V (high), and the current ranges from − 1 (discharging) to 1 (charging).

The general principles of the design are that, when the current is low and the voltage is also low, this indicates that the battery work state is also very low. Different

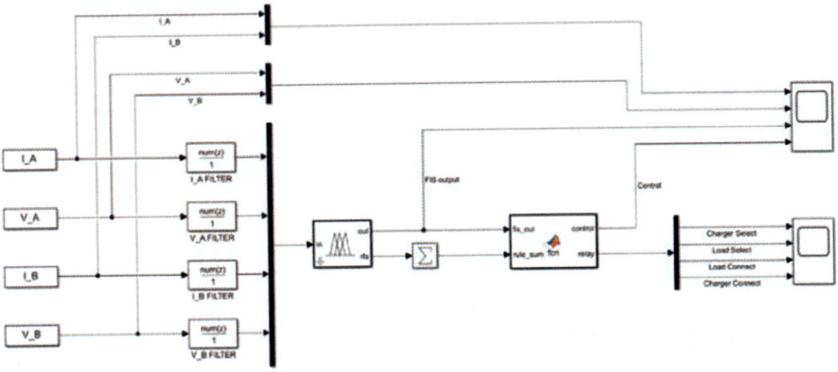

Fig. 7 Fuzzy logic controller using MATLAB/Simulink

rule base designs can be added until the requirements of the system are met. The rules can be added.

3.5 *Hardware Development*

In this part of the study, the real hardware was chosen. All the hardware was first constructed, and then the battery energy storage system (BESS) was run to get the data and analysis of the performance of the system, as shown in Fig. 8. Four batteries, two current sensors, a relay module, an inverter, a microcontroller, and a charger form the main components of a battery energy storage system (BESS). A proper electrical connection is needed in this part to ensure that all measurements are accurate and efficient.

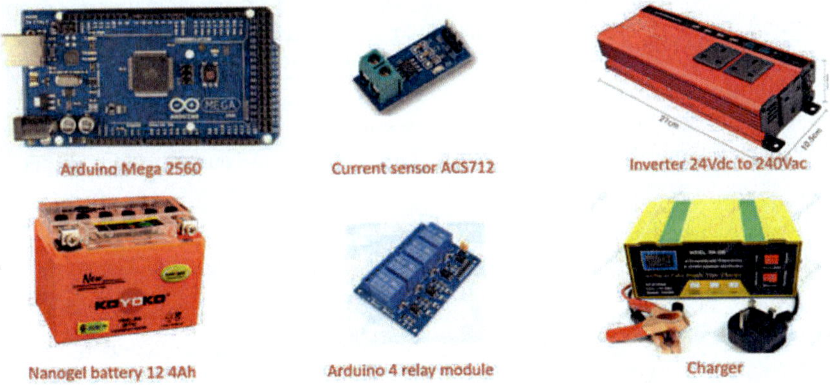

Fig. 8 Hardware component in BESS

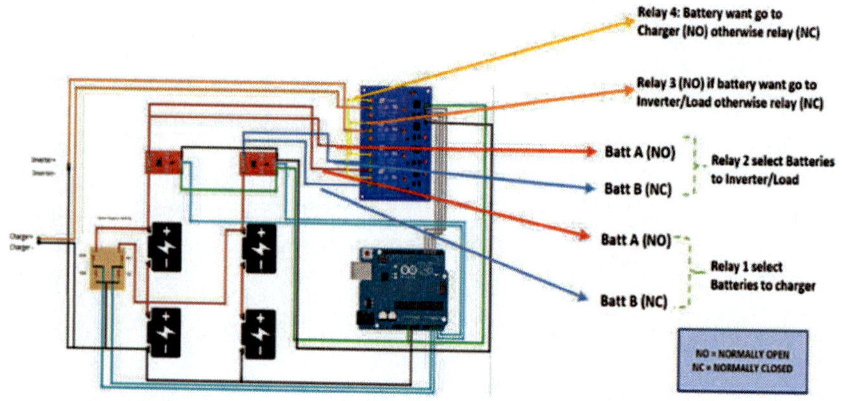

Fig. 9 BESS connection diagram

The proposed battery energy storage system (BESS) is shown in Fig. 9. The figure shows two sets of batteries, A and B, which are connected along with a voltage divider to pins A2 and A3 of the Arduino board. The voltage divider makes sure that the Arduino board is safe and not overloaded. The current sensor connects to relay 1 and relay 2, where relay 1 helps select the batteries for the charger, and relay 2 helps select batteries to connect to the inverter or load. Relay 3 and relay 4 are connected to the load/inverter and the charger itself, respectively.

This system includes an Arduino Mega, 4 relay modules, a current sensor, 4 batteries, resistors (820k and 150k), a charger, and an inverter. The battery energy storage system (BESS) consists of 4 relays to control the charging and discharging of the batteries. The function of relay 1 is to select the batteries for the charger, where this relay then connects to relay 4. For example, if battery A needs to charge, the relay will be in a normally open (NO) state, while if battery B needs to charge, the relay will be in a normally closed (NC) state.

Relay 2 works in a manner similar to relay 1, where it selects batteries for the inverter/load. If battery A is full, relay 2 will be in a normally open (NO) state, and battery A will connect to inverter/load, and if battery B is full, the relay will be normally closed (NC). At relay 3, the relay will be in a normally open (NO) state if the batteries want to connect to the inverter/load. Relay 4 will be in a normally open (NO) state if the batteries are in low condition and need to connect to the charger.

3.6 *Box Storage for the Battery Energy Storage System*

All the components of the battery energy storage system (BESS) are fitted into a box for better safety, as shown in Fig. 10.

Fig. 10 BESS box

3.7 Battery Energy Storage System Flowchart

Figure 11 shows the battery energy storage system (BESS) flowchart. The battery energy storage system (BESS) will start by checking the battery work state (BWS).

Fig. 11 BESS flowchart

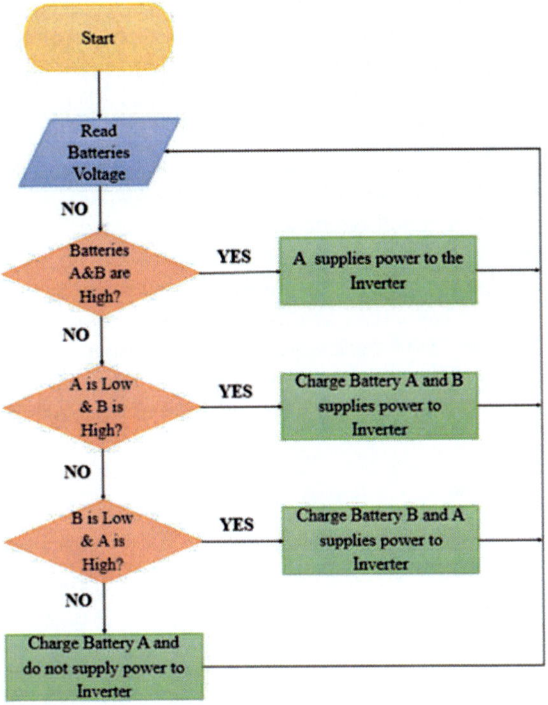

If the states of both the batteries are high, the charger will connect battery set A, while battery set B will be in idle mode.

When the state of battery set A is in a lower charge condition, it will connect to the charger, while battery set B will connect to the load. In a situation where both the batteries are low on charge, Battery A will charge, and the system will not connect to the load. The system is designed to continuously test the battery states without stopping.

4 Results and Discussion

4.1 *Battery Energy Storage System (BESS) with Conventional Sequential Algorithm*

Battery energy storage system (BESS) was run with the conventional sequential algorithm using Arduino Mega. There are 4 states which are used to determine the charging and discharging sequential of the batteries, as shown in Table 1. A battery is considered low when the battery voltage ($V_B < 21$ V). Normally, the system will start in state 3, where battery A has connected to the load and battery B is in idle condition, and both are high. After battery A becomes low, battery B is connected to the load/inverter and the system now reaches state 2. The process continues without there being a stop in the system, except when both batteries are at high or low status.

In a battery energy storage system (BESS), there are two 24 V battery sets which are connected to the charger or load/inverter depending on the battery state. Four relay modules control the battery by either sending it to the charger or load/inverter, which has 4 pins on Arduino mega to control the system, as shown in Table 2. If battery A is in state 3, relay 2 and relay 3 would be in a normally open (NO) state and battery A would be connected to load/inverter. Battery B in this case will be in idle condition, where the rest of the relays would be in normally closed (NC) condition.

There are a few conditions that contribute to the success of the battery energy storage system (BESS) charging and discharging process. Two conditions are considered here, where one is voltage measurement during discharging or when battery is connected to a load, and another one is current measurement once the battery is connected to a charger. This is further illustrated in Fig. 12.

Table 1 Status of batteries

Battery status		
State	Battery A	Battery B
State 0	Low	Low
State 1	High	Low
State 2	Low	High
State 3	High	High

Table 2 Relay configurations

Pin	Relay no	Relay status	Details
Pin 2	1	NO to battery A and NC to battery B	Select batteries to charge
Pin 3	2	NO to battery A and NC to battery B	Select batteries to load
Pin 4	3	NO to load/inverter and NC is not connected	Connect to load/inverter
Pin 5	4	NO to charger and NC is not connected	Connect to charger

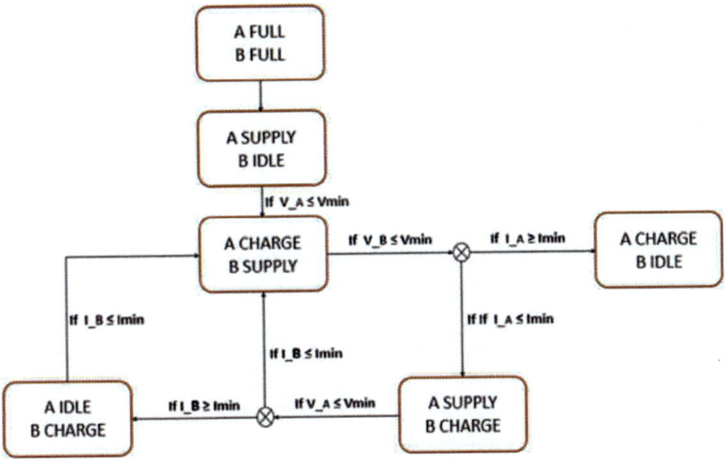

Fig. 12 Charging and discharging condition of the batteries

During the voltage measurement activity, if it is found that battery A has less than or equal to the minimum voltage (V_{min}), then battery A is connected to a charger and battery B is connected to the load/inverter. If on the other hand, the voltage in battery B becomes less than or equal to the minimum voltage, then battery B changes its state based on battery A's charging condition. If the current in battery A is more than the minimum current value (I_{min}), then battery A continues to be charged, while battery B enters an idle state. If the current in battery A is less than the I_{min} value, then battery A is connected to load/inverter and battery B is put to charge.

4.2 Results Using Conventional Sequential Algorithm

Referring to Fig. 13, it can be said that the conventional sequential algorithm can control the battery energy storage system (BESS). Battery voltage is one of the

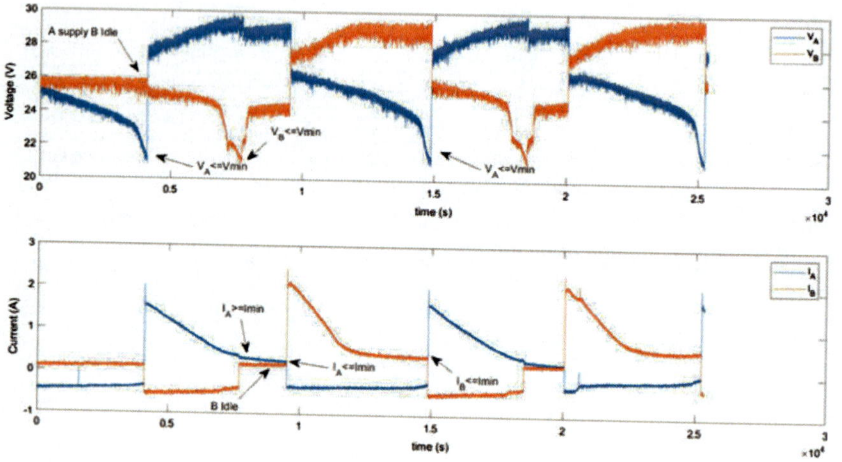

Fig. 13 Current and voltage measurement

methods used to control the batteries during charging and discharging. Both batteries start with a high state, where the voltage is 25 V.

Battery A starts discharging and battery B remains in an idle condition. Once battery A reaches the minimum voltage at 21 V, battery B starts discharging and battery A starts charging. During charging, the charging will stop when current measurement reaches the minimum value of 0.2 A. This experiment is essential and helps in designing the rules and optimizing the fuzzy logic controller.

4.3 Fuzzy Logic in Battery Energy Storage System (BESS)

Fuzzy logic is a very important part of this project. The data must be enough to design the rules base and to define the range for each state of the batteries.

4.3.1 State Identification

The system has two batteries with the same characteristics. Each battery has three main states (idle, charging, and discharging) and two events (ChargeEnd and DischargeEnd). The states using the current and voltage were identified as shown in Table 3.

The low limit of voltage at the discharging state is 21 V, and the low limit of current at the charging state is 0.2 A. From this data, the input membership functions were designed.

One experiment was conducted to collect the data. The measurements of current and voltage were analyzed. A histogram of the current and voltage was plotted to get

Table 3 The value range of current and voltage

State	Current (A)	Voltage (V)
Discharging	−0.56–0.00	21.0–26.8
Idle	−0.01–0.25	
Charging	0.20–2.20	27.0–29.9

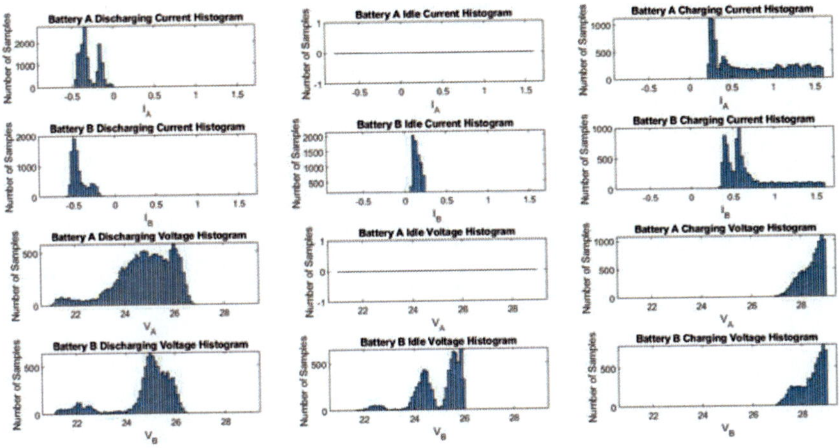

Fig. 14 Analysis of the first data

the value range of the current and voltage at every state, as shown in Fig. 14. In the graph, *Y*-axis refers to the number of samples and *X*-axis contains the current and voltage measurements. Every state of a battery such as idle condition, charging, and discharging, was measured.

4.3.2 Fuzzy Input Membership Function Design

The fuzzy logic system has 4 inputs (currents and voltages of batteries A and B) and 2 outputs (A and B controls) as shown in Fig. 15.

The membership function for current is shown in Fig. 16. In these functions, the battery has four states, which are discharging, charging, idle, and ChargeEnd. ChargeEnd gives information on whether a battery is fully charged or not.

The membership function for voltage is shown in Fig. 17. The parameters used here are DischargeEnd, idle, and charging. DischargeEnd is used to notify if a battery is already in low condition.

Table 4 shows the design parameters of the input membership functions and their value for each of the current and voltage parameters.

The fuzzy logic functions in battery energy storage system (BESS) can detect the end of the charging state by using the current and not the voltage; and detect the end

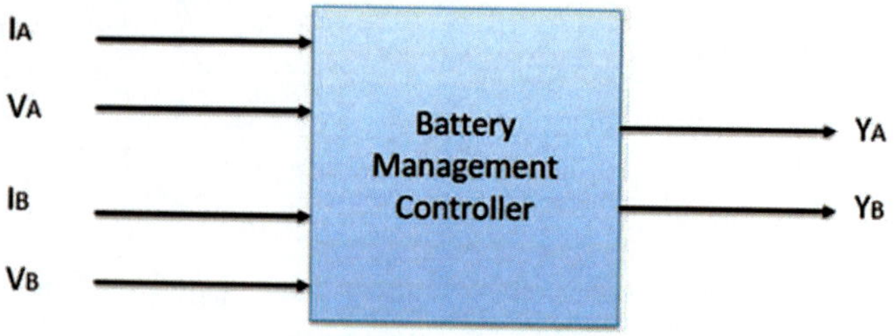

Fig. 15 Structure of the fuzzy logic system

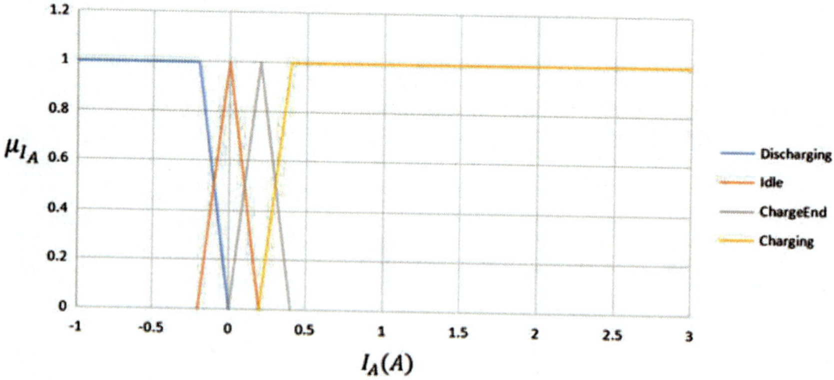

Fig. 16 Current input membership function

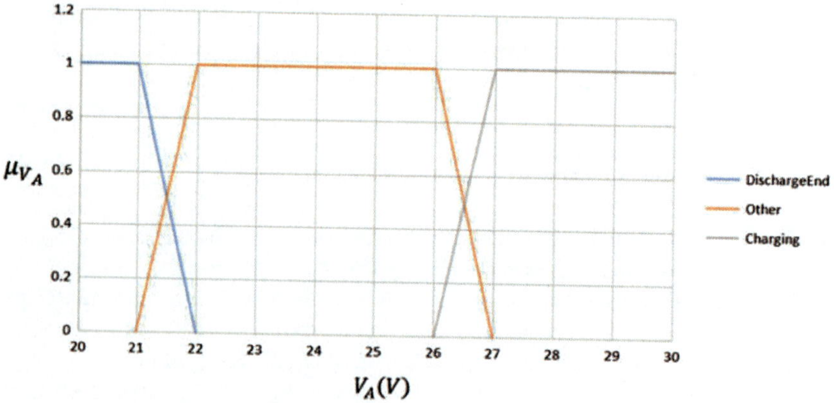

Fig. 17 Voltage input membership function

Table 4 Parameters of the input membership functions

State	Current (A)	State	Voltage (V)
Discharging	−1.00–0.00	DischargeEnd	20–22
		Other (discharging/idle)	21–27
Idle	−0.20–0.20		
ChargeEnd	0.00–0.40	Charging	26–30
Charging	0.20–3.00		

of the discharging state by using the voltage and not the current. Idle states can be identified by using the current but cannot be identified by using the voltage.

4.3.3 Fuzzy Output Membership Function Design

Table 5 shows the design parameters of the output membership functions. The fuzzy logic controllers are used to read the currents and voltages of batteries A and B, identify the states of the batteries, and make controls for them.

The output membership function for battery A and battery B is shown in Fig. 18. The X-axis is a value that represents the state of the batteries.

This part has two outputs to control batteries A and B. The output from each battery cannot be combined with another, so as to ensure that there is no conflict between the batteries, resulting in the maximizing of control.

Table 5 Parameters of the output membership functions

Control	Value
Discharge	−1
Idle	0
Charge	1

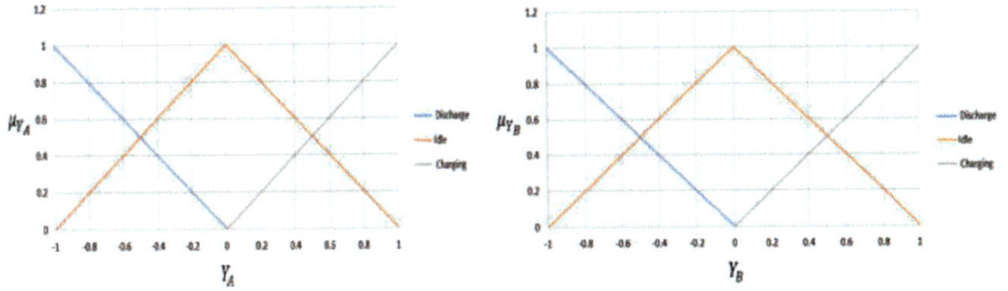

Fig. 18 Output membership function for battery A and battery B

Table 6 Exact identification of state

State	Symbol	Meaning	Current	Voltage
DischargeEnd	NB	Negative big	Discharging	DischargeEnd
Discharging	NS	Negative small	Discharging	Other
Idle	ZO	Zero	Idle	Other
Charging	PS	Positive small	Charging	Charging
ChargeEnd	PB	Positive big	ChargeEnd	Charging

Table 7 The state of ouput

State	Symbol	Meaning
Discharge	N	Negative
Idle	Z	Zero
Charge	P	Positive

4.3.4 The Design of the Rule Base of the Fuzzy Logic System

State Notification

Combining the current and voltage can result in the identification of 3 states and 2 events. Table 6 shows this in detail.

The details of the state of output can be seen in Table 7. Using the notification of the states, the rules base is designed.

Design the Rules Base

The principles of the design are detailed below.

(a) When the discharging has ended:

If the other battery is idle, then charge.
If the other battery is charging, then keep idle while waiting for charging.
If the other battery is at charging end, then charge.

(b) If discharging is ongoing, then continue discharging.
(c) If battery is idle:

If the other battery is at discharging end, then discharge.
If the other battery is discharging, then continue being idle.
If the other battery is charging, then continue being idle.
If the other battery is at charging end, then charge.

(d) If charging is ongoing, then continue charging.
(e) If charging has ended:

Table 8 Designed rule

A		B					B		B				
		NB	NS	ZO	PS	PB			NB	NS	ZO	PS	PB
A	NB			P	Z	P	A	NB			N	P	N
	NS			N	N	N		NS			Z	P	Z
	ZO	N	Z	P	Z	P		ZO	P	N	N	P	N
	PS	P	P	P				PS	Z	N	Z		
	PB	N	Z	N				PB	P	N	P		

The table items can be read as follows:
"If A is ZO and B is NS, then A is Z and B is N"

If the other battery is at discharging end, then discharge.
If the other battery is discharging, then continue being idle while waiting for the discharging.
If the other battery is idle, then discharge.

(f) If both are idle, then A will undergo charging and B will undergo discharging.

The application of fuzzy logic in battery energy storage system (BESS) leads to the designing of rules as shown in Table 8. The row is the state of battery A, and the column is that of B. The left is the control rule for A, and the right is the same for B.

In other words, if A is idle and B is discharging, then A continues to be idle and B continues discharging. In the rules table, the empty positions are the unfeasible states. The system is designed such that more rules can be added to it without any issues. Rules can be added or removed based on the project. The important point to note here is that the system can run efficiently based on the rules created. The complete rule base is:

a. If (IA is Discharging) and (VA is DischargeEnd) and (IB is Idle) and (VB is Other), then (YA is Charge) (YB is Discharge).
b. If (IA is Discharging) and (VA is DischargeEnd) and (IB is Charging) and (VB is Charging), then (YA is Idle) (YB is Charge).
c. If (IA is Discharging) and (VA is DischargeEnd) and (IB is ChargeEnd) and (VB is Charging), then (YA is Charge) (YB is Discharge).
d. If (IA is Discharging) and (VA is Other) and (IB is Idle) and (VB is Other), then (YA is Discharge) (YB is Idle).
e. If (IA is Discharging) and (VA is Other) and (IB is Charging) and (VB is Charging), then (YA is Discharge) (YB is Charge).
f. If (IA is Discharging) and (VA is Other) and (IB is ChargeEnd) and (VB is Charging), then (YA is Discharge) (YB is Idle).
g. If (IA is Idle) and (VA is Other) and (IB is Discharging) and (VB is DischargeEnd), then (YA is Discharge) (YB is Charge).
h. If (IA is Idle) and (VA is Other) and (IB is Discharging) and (VB is Other), then (YA is Idle) (YB is Discharge).

i. If (IA is Idle) and (VA is Other) and (IB is Charging) and (VB is Charging), then (YA is Idle) (YB is Charge).
j. If (IA is Idle) and (VA is Other) and (IB is ChargeEnd) and (VB is Charging), then (YA is Charge) (YB is Discharge).
k. If (IA is Charging) and (VA is Charging) and (IB is Discharging) and (VB is DischargeEnd), then (YA is Charge) (YB is Idle).
l. If (IA is Charging) and (VA is Charging) and (IB is Discharging) and (VB is Other), then (YA is Charge) (YB is Discharge).
m. If (IA is Charging) and (VA is Charging) and (IB is Idle) and (VB is Other), then (YA is Charge) (YB is Idle).
n. If (IA is ChargeEnd) and (VA is Charging) and (IB is Discharging) and (VB is DischargeEnd), then (YA is Discharge) (YB is Charge).
o. If (IA is ChargeEnd) and (VA is Charging) and (IB is Discharging) and (VB is Other), then (YA is Idle) (YB is Discharge).
p. If (IA is ChargeEnd) and (VA is Charging) and (IB is Idle) and (VB is Other), then (YA is Discharge) (YB is Charge).
q. If (IA is Idle) and (VA is Other) and (IB is Idle) and (VB is Other), then (YA is Charge) (YB is Discharge).

4.3.5 Construction of the Control System

The relevant measurements were collected from the battery energy storage system (BESS). The analog pin on the Arduino board measured the current and voltage for each battery and sent it to the battery energy storage controller in MATLAB/Simulink as shown in Fig. 19. The digital pin at Arduino sends the signal to the relay for performing the charging and discharging processes.

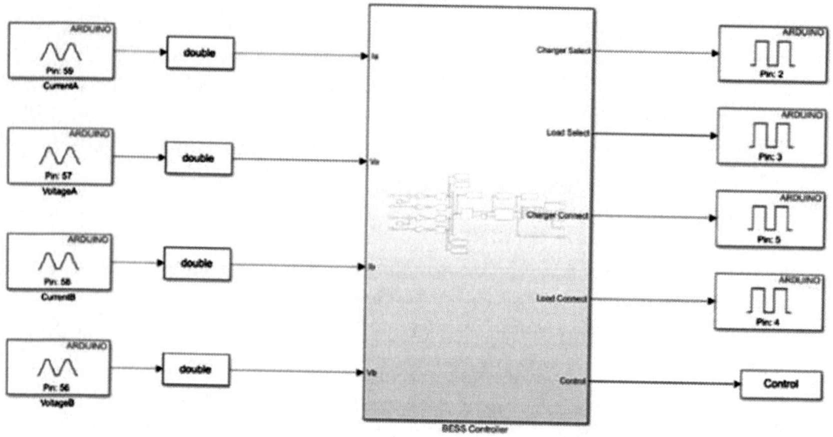

Fig. 19 BESS controller

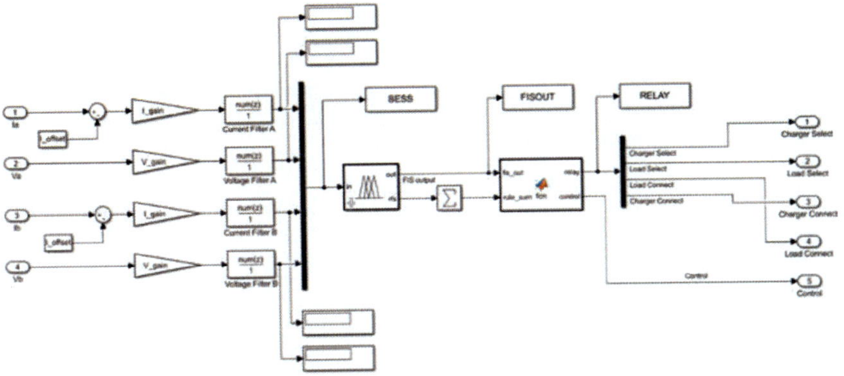

Fig. 20 Configuration of the control system

All the data was processed using fuzzy block and MATLAB functions, and it was saved for further analysis.

The overall configuration of the control system using fuzzy logic in the battery energy storage system (BESS) is shown in Fig. 20.

4.3.6 Construction of the Control System

The output of the fuzzy block is an analog value with a range between − 1 and 1. The analog value was converted to the ON/OFF signal. The MATLAB function block is used for this purpose. A lookup table was used to find the suitable control signal for the 4 relays, as shown in Table 9.

From the lookup table above, the output from the fuzzy logic system (A out and B out) will determine the row for controlling the relay, using Eq. 5.

Table 9 Lookup table for controlling the relay

Y_A	Y_B	Row ID	Charger select	Load select	Load connect	Charger connect
−1	−1	1	1	0	0	0
−1	0	2	0	1	1	0
−1	1	3	0	1	1	1
0	−1	4	1	0	1	0
0	0	5	1	0	0	0
0	1	6	0	1	0	1
1	−1	7	1	0	1	1
1	0	8	1	0	0	1
1	1	9	0	1	0	0

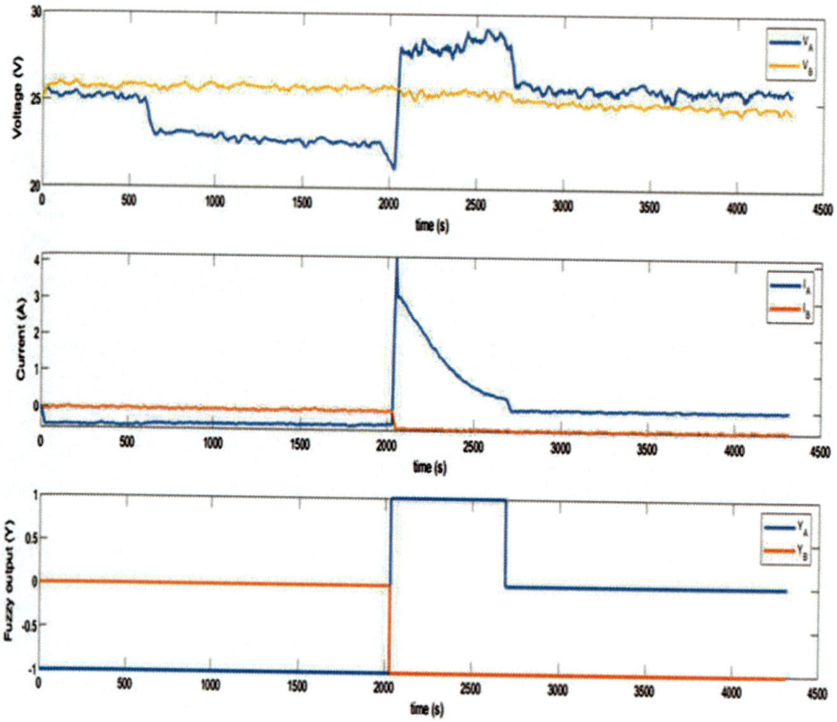

Fig. 21 Measurement data using fuzzy logic

$$\text{Row ID} = 3 * (Y_A + 1) + (Y_B + 1) + 1 \tag{5}$$

4.3.7 Result Using Fuzzy in Battery Energy Storage System (BESS)

Figure 21 shows the measurement of current and voltage for both batteries along with the fuzzy output, where 1 is charging, 0 is idle and −1 is discharging.

Figure 22 shows the relay control of the BESS for better performance.

4.3.8 Conventional Sequential Algorithm Versus Fuzzy Logic Battery Charging Controller

It can be seen that battery charging is considered to be one of the most sensitive processes for different devices. For any battery charging mechanism, there is a need to use a reliable and efficient battery charging controller that can charge the battery

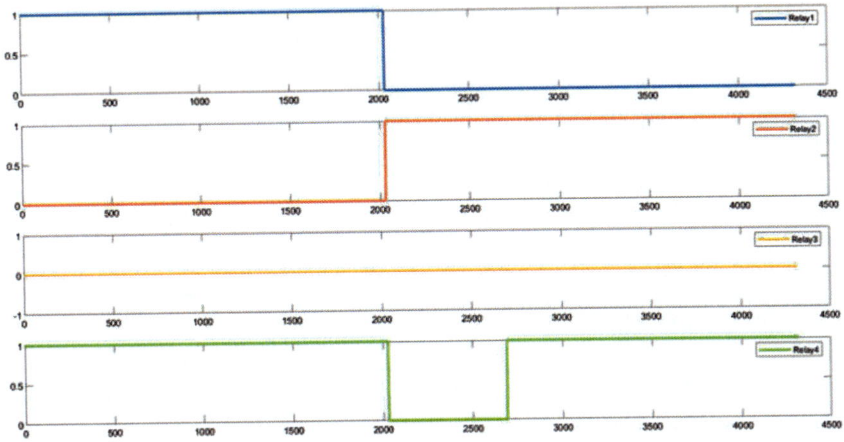

Fig. 22 Relay control the system

without any errors. For this purpose, mainly two kinds of controllers are used for charging and discharging the battery.

One is a conventional sequential algorithm and the other one is related to a fuzzy logic battery charging controller. The main fact to note here is that the fuzzy logic battery controller is far better than the conventional sequential algorithm. For this purpose, a deep analysis of the results is conducted, with comprehensive information about the comparison between these two controllers. Figures 23 and 24 display relevant results from the conventional sequential algorithm and fuzzy logic controller methods.

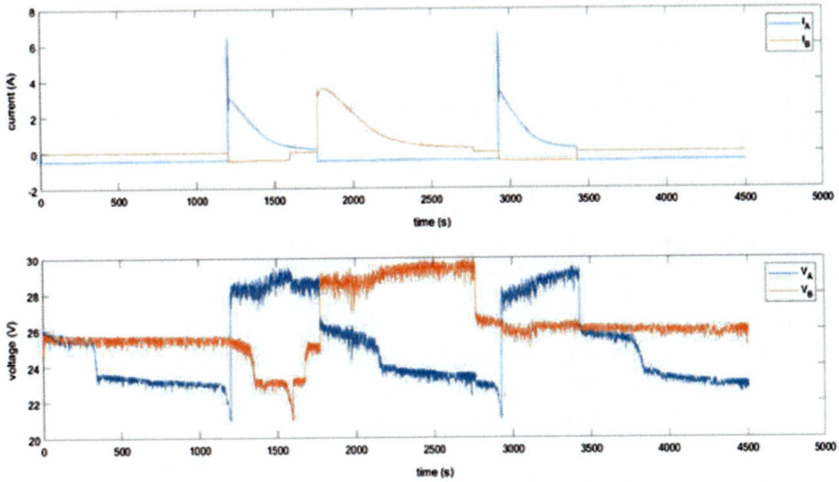

Fig. 23 Conventional sequential algorithm

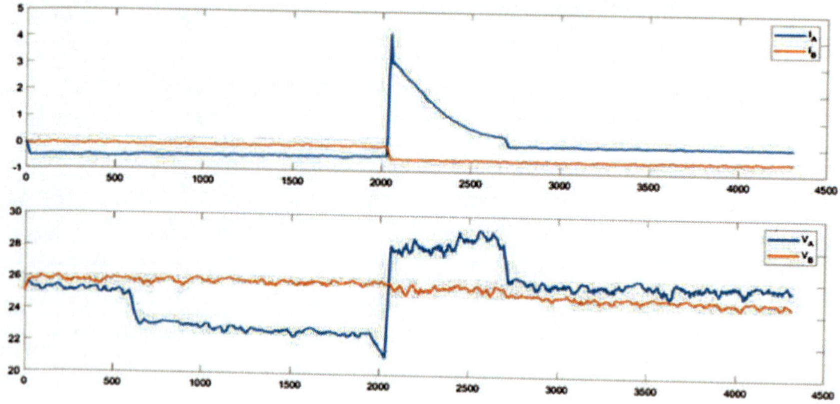

Fig. 24 Fuzzy logic controller

Table 10 Batteries efficiency

Energy efficiency	Conventional sequential algorithm	Fuzzy logic controller
Battery A	0.28	0.43
Battery B	0.05	0.23

In terms of the battery voltage, current, and power, energy efficiency can be represented as Eq. 6.

$$\eta = \frac{\int_{t_0}^{t_{\text{dis}}} V_{\text{dis}}(t).I_{\text{dis}}(t)\text{d}t}{\int_{t_0}^{t_{\text{chg}}} V_{\text{chg}}(t).I_{\text{chg}}(t)\text{d}t} \tag{6}$$

where V_{dis}, I_{dis}, and t_{dis} are the battery discharge voltage, current, and time in seconds, respectively. V_{chg}, I_{chg}, and t_{chg} are the battery charge voltage, current, and time in seconds, respectively. A comparison of the efficiency of these batteries was done, with results as shown in Table 10.

5 Conclusion

Battery energy storage system (BESS) was used to carry out a simultaneous battery charging and electricity supply with the fuzzy logic controller, and this was achieved by fuzzy logic. The current and voltage of the battery can control and maintain the process of battery charging and discharging. Discharge time will see an increase due to our fuzzy logic controller, as compared to a conventional sequential algorithm. Although it is the same battery and the same load, the system is optimized by the fuzzy logic controller. The system can control and manage the charging and discharging of different sets of batteries with a fuzzy logic controller.

References

1. R. Xiong, W. Shen, *Advanced Battery Management Technologies for Electric Vehicles* (2019)
2. Q. Sun et al., A new design of fuzzy logic control for SMES and battery hybrid storage system. Energy Procedia **105**, 4575–4580 (2017)
3. M. Khalid et al., Performance validation of electric vehicle's battery management system under state of charge estimation for lithium-ion battery, in *2018 International Conference on Computing, Electronic and Electrical Engineering (ICE Cube)* (2018)
4. E. Banguero et al., A review on battery charging and discharging control strategies: application to renewable energy systems. Energies (2018)
5. D. Saji, P.S. Babu, K. Ilango, SoC estimation of lithium ion battery using combined coulomb counting and fuzzy logic method, in *2019 4th International Conference on Recent Trends on Electronics, Information, Communication & Technology (RTEICT)* (2019)
6. B.M. Prabhakar, J. Ramprabhakar, V. Sailaja, Estimation and controlling the state of charge in battery augmented photovoltaic system, in *2016 Biennial International Conference on Power and Energy Systems: Towards Sustainable Energy (PESTSE)* (2016)
7. T.O. Ting et al., State-of-charge for battery management system via Kalman filter. Eng. Lett. **22**, 75–82 (2014)
8. G.Y.B.E. Ltd., Battery discharging & charging characteristics for motorcycle and powersport batteries, in *Battery Discharging & Charging Characteristics* (2014). Available from: https://www.yuasa.it/en/information/motorcycle-and-powersport/battery-discharging-charging-characteristics/
9. A.D. Jadhav, S. Nair, Battery management using fuzzy logic controller. J. Phys. Conf. Ser. **1172**, 012093 (2019)
10. K. Sibi Krishnan, P. Pathiyil, R. Sunitha, Generic battery model covering self-discharge and internal resistance variation, in *2016 IEEE 6th International Conference on Power Systems (ICPS)* (2016)
11. T. Martin, *Motorcycle Electrical Systems, Troubleshooting and Repair* (2007)
12. Y. Diab et al., Self-discharge characterization and modeling of electrochemical capacitor used for power electronics applications. IEEE Trans. Power Electron. **24**(2), 510–517 (2009)
13. G. Plett, Extended Kalman filtering for battery management systems of LiPB-based HEV battery packs: Part 1. Background. J. Power Sour. **134**, 252–261 (2004)
14. G. Plett, Extended Kalman filtering for battery management systems of LiPB-based HEV battery packs, Part 2: Modeling and identification, Part 3: State and parameter estimation. J. Power Sour. **134**, 252–292 (2004)
15. G. Plett, Extended Kalman filtering for battery management systems of LiPB-based HEV battery packs. Part 3. State and parameter estimation. J. Power Sour. **134**, 277–292 (2004)
16. S.G. Li et al., Energy and battery management of a plug-in series hybrid electric vehicle using fuzzy logic. IEEE Trans. Veh. Technol. **60**(8), 3571–3585 (2011)
17. M.V. Carlos Dualibe, P. Jespers, *Design of Analog Fuzzy Logic Controllers in CMOS Technologies, Implementation, Test and Application* (2003)
18. E.H. Mamdani, Application of fuzzy algorithms for control of simple dynamic plant. Proc. Inst. Electr. Eng. **121**(12), 1585–1588 (1974)
19. I. Perfilieva, V. Novák, J. Močkoř, *Mathematical Principles of Fuzzy Logic* (Kluwer Academic, Dordrecht, 1999)
20. C.-L. Liu, J.-H. Chen, Y.-H. Liu, Z.-Z. Yang, An asymmetrical fuzzy-logic-control-based MPPT algorithm for photovoltaic systems. Energies **7**(4), 2177–2193 (2014)

21. N.A. Carlson, Federated filter for fault-tolerant integrated navigation systems, in *IEEE PLANS'88, Position Location and Navigation Symposium, Record—'Navigation into the 21st Century'* (1988)
22. L. Hu, X. Hu, Y. Che, F. Feng, X. Lin, Z. Zhang, Reliable state of charge estimation of battery packs using fuzzy adaptive federated filtering (2020)
23. T.R.B. Kushal, M.F. Karin, Fuzzy logic controller for lithium-ion battery in standalone DC microgrid, in *2015 IEEE International WIE Conference on Electrical and Computer Engineering (WIECON-ECE)* (2015)
24. W.V.H. Hasaranga et al., A fuzzy logic based battery SOC level control strategy for smart micro grid (2017), pp. 215–222
25. W.V.H. Hasaranga, R.D.T.M. Hemarathne, M.D.C.P.K. Mahawithana, M.G.A.B.N. Sandanuwan, D.Hettiarachchi, K.T.M.U. Hemapala, A fuzzy logic based battery SOC level control strategy for smart Micro grid (2017)

Fault Classification of Cooling Fans Using a CNN-Based Approach

Labib Sharrar and Kumeresan A. Danapalasingam

Abstract In industries, cooling fans are vital in a wide range of machines to ensure a tolerable temperature for their intricate electronic components. Therefore, to avoid machine failure, a fault condition monitoring (FCM) system for cooling fans can be highly valuable. One way to monitor defects in rotational equipment is to analyze the machine vibration, which varies as the components wear off. Hence, this paper presents a technique to diagnose faults in cooling fans by analyzing the vibration data. In this conference paper, convolutional neural networks (CNNs) are used to classify the faults based on the vibration. The vibration data are collected from an experimental test bench that consists of a fan, an accelerometer, and a microcontroller, among others. The data were used to train the VGG16 and ResNet50 CNN architectures. The accuracy and effectiveness of these two architectures for vibration analysis are compared in this paper.

Keywords Artificial neural network · Convolutional neural network · Fault condition monitoring · Predictive maintenance

1 Introduction

In industries, it is important to keep machines under stable conditions, they are required to operate continuously to produce a high yield. The cooling fans are responsible for preventing the delicate electronic or mechanical components from overheating, as it may result in machine instability and damage [1]. Since they play such an important role in industries, having a fault condition monitoring (FCM) system can be truly valuable for predictive maintenance (PdM) of cooling fans [2]. The concept of PdM has become increasingly popular. In accordance with Kumar

L. Sharrar (✉) · K. A. Danapalasingam
Centre of Artificial Intelligence and Robotics, Universiti Teknologi Malaysia, Johor Bahru, Malaysia
e-mail: sharrar@graduate.utm.my

K. A. Danapalasingam
e-mail: kumeresan@utm.my

A. Kumar et al. (eds.), *Computational Intelligence in Machine Learning*, Lecture Notes in Electrical Engineering 834, https://doi.org/10.1007/978-981-16-8484-5_6

et al. [3], vibration analysis is an effective PdM method, since the vibration varies along with the state of the mechanical components. As such, vibration analysis can be paramount for fault detection in rotational machinery like industrial fans.

2 Research Background

FCM systems usually include three main phases: collection of data, extracting features, and data classification. Feature extraction is a crucial step that transforms input data into a set of features which can then be classified by an algorithm [4]. Wavelet transform (WT) is a prominent feature extraction method that has been deployed successfully in conjunction with several conventional classification algorithms, such as fuzzy logic, *k*-nearest neighbor (KNN), or support vector machine [5–7]. For example, Konar and Chattopadhyay [8] proposed a hybrid WT and SVM method for detecting possible defects in ball bearings of motors. Nowadays, the application of deep neural networks to PdM systems has become increasingly popular. These algorithms can autonomously learn the features and be trained to classify or predict data [9]. Most recently, convolutional neural networks (CNNs) have gained prominence in finding faults through vibration analysis. Vibration data to two-dimensional (2D) images, and then the features can be extracted from these images to gain insight into the state of the mechanical parts. A work done by Sanchez et al. [10] used CNNs to analyze vibration signals to detect bearing faults. Since CNN and image encoding algorithms have proven to be effective for vibration analysis, they are used in this project for fault diagnosis of cooling fans.

3 Methodology

3.1 System Overview

As the main objective of this paper is to present a CNN model that detects faults through vibration analysis, an experimental setup with an accelerometer was developed. The steps of the experiments are (1) to data collect vibration data from the experimental setup, (2) categorize the data into different classes, (3) convert the data into colored images through an encoding technique, and (4) use the encoded data to train and validate the CNN model before deployment.

Once the model's validation is sufficient, it will be deployed on the real-time system. Overall, the vibration data collected by the circuit are serially fed to the trained machine learning model. Based on the model's output, the system will decide whether there is a fault in the cooling fan. The system overview is shown in Fig. 1.

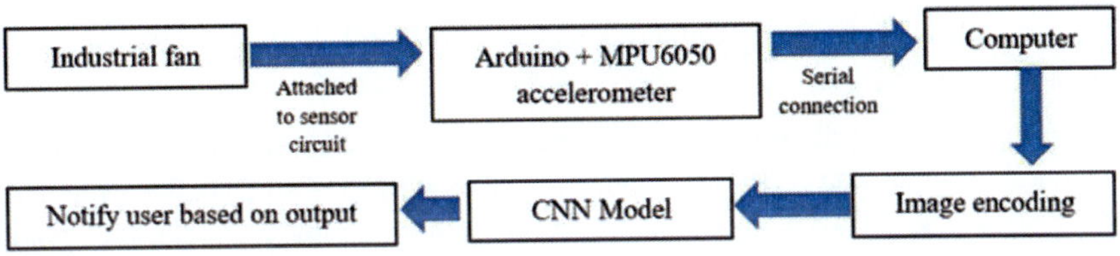

Fig. 1 The overview of the entire system

3.2 *Experimental Setup*

While there are a plethora of microcontroller boards in the market, an Arduino is chosen due to its user-friendly nature and ease of programming. It is connected to an MPU6050 accelerometer and programmed to sample data at 50 Hz. Data acquisition requires a storage device. For simplicity, the microcontroller was interfaced with a single computer to store the data. As the circuit samples the vibration data, the sampled data would be serially sent to a single board computer (Raspberry Pi) to be saved.

It is important to remember that when a cooling fan suffers from a defect, its vibration will change [1]. With the intent to experiment, we manually changed the fan speed to simulate the failure conditions. The table fan has three different modes of speed: 0, 1, and 2. Just for experimenting, speed modes 0 and 1 were considered normal speed, while speed mode of 2 was selected as the faulty mode. To begin data acquisition, the circuit board, as shown in Fig. 2, is attached to the outer metal casing of the fan.

The vibration data measured by the accelerometer consisted of acceleration along *x*, *y*, and *z* axes in the unit mm/s^2. Once all the data collection had been complete, the vibration data from the fan were used to train machine learning models on the two vibration analysis techniques mentioned earlier.

Fig. 2 The experimental setup was carried out on a regular fan as shown in this image

3.3 Fault Classification Using Image Encoding and CNN

CNNs can classify time series data by working together with image encoding techniques [11], which can convert sequential data to colored images. Using convolution, the notable internal features from those images can be extracted and fed into neural network layers for classification [12]. In this paper, an image encoding technique proposed by Hur et al. [13] has been applied to obtain the colored images. First and foremost, to go through the image encoding phase, the array of vibration data from each of the three accelerometer axes is normalized. The normalization equation is shown in Eq. (1). The steps for image encoding are illustrated in this subsection. It should be noted that all the formulas mentioned in this section were referenced by Hur et al. [13].

Step 1—Normalize the data from each accelerometer axes based on Eq. (1). The x represent the x-axis vibration data at a particular instant while the x − array is the array of x-axis vibration data over a certain interval:

$$\overline{x} = \frac{x - \min(x - \text{array})}{\max(x - \text{array}) - \min(x - \text{array})} \tag{1}$$

In the equation above, $\overline{x}$ represents the normalized data points along the x-axis and the same equation will be used to calculate the normalized data points for y and z.

Step 2—Convert each normalized data point into three integers that correspond to pixel values in red, green, and blue colored channels. Using Eqs. (2–4):

$$R\overline{x} = |\overline{x}| \tag{2}$$

$$G\overline{x} = \left|(\overline{x} - |\overline{x}|) \times 10^2\right| \tag{3}$$

$$B\overline{x} = \left|\left(\overline{x} \times 10^2 - \left|\overline{x} \times 10^2\right|\right) \times 10^2\right| \tag{4}$$

The variables $R\overline{x}$, $G\overline{x}$, and $B\overline{x}$ in the above equations represent the colors red, blue, and green for the x-axis values. The same operation would be carried out for the corresponding values of the y and z axes.

Step 3—The colored pixels generated over a certain interval would be stored in matrices for all three colors: red (R), green (G), and blue (B), like the example shown in equation Eq. (5).

$$R = \begin{bmatrix} Rx_1 & Ry_1 & Rz_1 \\ \vdots & \vdots & \vdots \\ Rx_N & Ry_N & Rz_N \end{bmatrix} \tag{5}$$

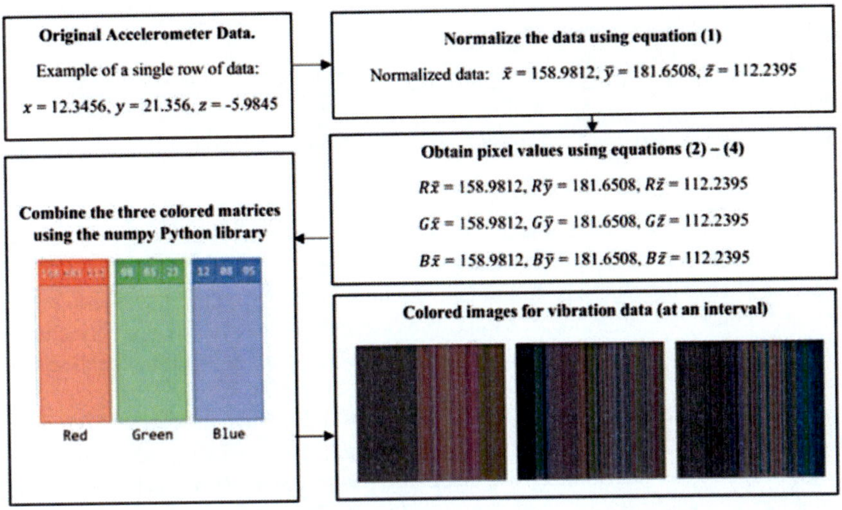

Fig. 3 Overview of the image encoding method

Step 4—Finally, the three colored matrices are concatenated into a single image. The overall process of image encoding has been explained in Fig. 3.

During the real-time implementation, the accelerometer will be attached to the cooling fan and the vibration data will be serially compiled on a computer. After sufficient data are collected, they will be categorized to go through the image encoding phase. The images obtained will be classified by a trained CNN model. The convolutional neural network in this project was created using the PyTorch deep learning framework.

Just as there are many categories of machine learning techniques, there are several CNN architectures. Among them, the ResNet and the VGG series are the most powerful algorithms for image classification. Therefore, the compiled RGB images were trained on the ResNet50 and the VGG16 architectures. Under the *results and analysis* section, their accuracy will be compared.

4 Results and Analysis

4.1 Application of Convolutional Neural Network

After data collection from the accelerometer had been complete, the data were categorized into three separate files to better analyze the vibrations at three different fan speeds. Firstly, as mentioned in the methodology section, the vibration readings were converted into colored images using formulas from the Iss2Image image encoding method presented by Hur et al. [13].

Table 1 Results of the two CNN architectures

CNN model	Epochs	Training loss	Testing loss	Testing accuracy
Resnet50	100	0.0817	0.0954	0.8117
VGG16	100	0.0085	0.0063	0.9865

After all the raw accelerometer data had been converted to colored images, the output results were prepared for training. The image data were split into three directories as always necessary for training a machine learning model: *train*, *validation*, and *test*. Once that was done, the data were trained on the two CNN architectures: ResNet50 and VGG16. The overall performance of each architecture has been shown in Table 1.

Based on Table 1, the VGG16 model performed better than the ResNet50 model, as it has a higher testing accuracy along with much lower training and testing loss. If we compare the graphs of testing accuracy for both models, we can see that the testing accuracy mostly remained steady for the VGG16 model, unlike for ResNet50, where the testing accuracy fluctuated tremendously over the 100 epochs. However, training VGG16 took longer since it is much deeper compared to the ResNet model. Nevertheless, the models may have overfitted. Improvements to training should be made in future. Overall, when using the trained model to make predictions on testing data, VGG16 displayed much higher accuracy. The testing was done in batches of 4 images at a time. The output displays for each model are shown in Table 2.

As seen from the display, for this batch, the VGG16 predicted all the samples correctly while the ResNet50 made a few mistakes. Although VGG16 did display high accuracy, it should be noted that neural networks work best when trained with large datasets [14]. The dataset for this project was relatively small. Therefore, results for both CNN architectures may have been better when trained on more data.

5 Conclusion

The primary purpose of this study is to present a vibration analysis technique for fault diagnosis of industrial cooling fans. In this paper, a CNN-based approach along with an image encoding algorithm is used to get insight into the vibration data and classify the state of the fan. The data are collected by an electronic circuit that consisted of an MPU6050 accelerometer and an Arduino Nano board. Due to the inaccessibility of an industrial type of fan or blower, an experimental fan was used for this study. The three different speed modes (0, 1, and 2) of the fan were used to represent the healthy and faulty conditions (at speed 2, the fan's condition is assumed to be faulty). The data collected are categorized into separate classes before being fed to the Iss2Image image encoding algorithm to obtain the colored images. Once the colored image dataset is complete, it is used to train a CNN model. In this paper, two different CNN models are trained: the VGG16 and ResNet50. Based on the results, the VGG16

Table 2 Comparison of output results

ResNet50 output
Actual image: ["two," "one," "zero," "two"]

Prediction: ["one," "one," "zero," "two"]

VGG16 output
Actual image: ["zero," "one," "two," "one"]

Prediction: ["zero," "one," "two," "one"]

displayed a higher accuracy than the ResNet50. Nevertheless, the accuracies of both architectures were significantly high. It is, however, possible that the dataset from the experimental setup is small compared to other databases. In future, these two CNN architectures should be trained on a larger dataset for better performance. Only then can we determine, which is the better architecture for vibration analysis and fault detection. Furthermore, we also plan to compare the best CNN model with conventional fault classification algorithms such as fuzzy logic and moving average, and ultimately prove which technique is most suitable for fault diagnosis of cooling fans.

Acknowledgements A significant portion of the credit for this project should go to the Center of Artificial Intelligence and Robotics (CAIRO) lab at Universiti Teknologi Malaysia (UTM), which permitted us to carry out the necessary experiments.

References

1. H. Taplak, E. Kurt, M. Parlak, Fault diagnosis for exhaust fan using experimental predictive maintenance method. Int. J. Acoust. Vib **21** (2016)
2. P. Zhang, T. Yang, J. Li, S. Huang, Flexible and smart online monitoring and fault diagnosis system for rotating machinery, in *2012 International Conference on Computer Distributed Control and Intelligent Environmental Monitoring* (2012), pp. 343–347. https://doi.org/10.1109/CDCIEM.2012.88
3. S. Kumar, M. Lokesha, K. Kumar, K.R. Srinivas, Vibration based fault diagnosis techniques for rotating mechanical components: review paper. IOP Conf. Ser. Mater. Sci. Eng. **376** (2019). https://doi.org/10.1088/1757-899X/376/1/012109
4. Z. Huo, Y. Zhang, P. Francq, L. Shu, J. Huang, Incipient fault diagnosis of roller bearing using optimized wavelet transform based multi-speed vibration signatures, in *Special Selection on Complex System Health Management Based on Condition Monitoring and Test Data* (2017). https://doi.org/10.1109/ACCESS.2017.2661967
5. H.F. Azgomi, J. Poshtan, Induction motor stator fault detection via fuzzy logic, in *2013 21st Iranian Conference on Electrical Engineering (ICEE)* (2013). https://doi.org/10.1109/IranianCEE.2013.6599711
6. S. Samanta, J.N. Berta, G. Sarkar, KNN based fault diagnosis system for induction motor, in *2016 2nd International Conference on Control, Instrumentation, Energy & Communication (CIEC)* (2016), pp. 304–308. https://doi.org/10.1109/CIEC.2016.7513791
7. C. Li, S. Liu, H. Zhang, Y. Hu, Machinery condition prediction based on wavelet and support vector machine, in *2013 International Conference on Quality, Reliability, Risk, Maintenance, and Safety Engineering (QR2MSE)* (2013)
8. P. Konar, P. Chattopadhyay, Bearing fault detection of induction motor using wavelet and support vector machines (SVMs). Appl. Soft Comput. **11**, 4203–4211 (2011)
9. D. Lee, V. Siu, R. Cruz, C. Yetman, Convolutional neural net and bearing fault analysis, in *International Conference on Data Mining (DMIN)* (2016)
10. L.A.P. Sanchez, D.A.M. Ravell, C.A.C. Monsivais, Vibration analysis in bearings for failure prevention using CNN. J. Braz. Soc. Mech. Sci. Eng. **42** (2020)
11. B. Zhao, H. Lu, S. Chen, J. Liu, D. Wu, Convolutional neural networks for time series classification. J. Syst. Eng. Electron. **28**, 162–169 (2017). https://doi.org/10.21629/JSEE.2017.01.18
12. P. Swietojanski, A. Ghoshal, S. Renals, Convolutional neural networks for distant speech recognition. IEEE Signal Process. Lett. **21**, 1120–1124 (2014). https://doi.org/10.1109/LSP.2014.2325781
13. T. Hur, J. Bang, T. Huynh, J. Lee, J.I. Kim, S. Lee, Iss2Image: a novel signal-encoding technique for CNN-based human activity recognition. Sensors (Basel) **18** (2018). https://doi.org/10.3390/s18113910
14. B. Fasel, Robust face analysis using convolutional neural networks. Object recognition supported by user interaction for service robots **2**, 40–43 (2002). https://doi.org/10.1109/ICPR.2002.1048231

Violence Recognition Using Convolutional Neural Networks

Shuzlina Abdul-Rahman, Yuzi Mahmud, and Muhammad Nasrullah

Abstract Livestream videos are becoming the norm in this digital era. With the advancement of streaming technology, real-time sharing has been effortless for many of the global population. There are many platforms that allow anyone to create or view livestream videos. This situation, however, creates a high possibility of showing inappropriate videos as well. Violent actions explicit languages and even crimes can be streamed, intentionally or otherwise. The impact on viewers may vary, though mostly negative. Thus, this research is motivated to help censor these unwanted scenes in real-time so that the negative impact can be reduced or avoided. By implementing the convolutional neural networks (CNNs) and training on three different off-the-shelf models in a restricted environment, it was found that the VGG19 model is the best for this research to be deployed in the censoring application compared to the MobileNetV2 and InceptionResNetV2 models.

Keywords Convolutional neural networks (CNNs) · VGG19 model · MobileNetV2 model · InceptionResNetV2 model · Livestream videos · Violence recognition

1 Introduction

CISCO forecasted that live video traffic is expected to grow 15-fold by 2022 than it was in 2017, which accounts for 82% of all the traffic [1]. A survey conducted by Interactive Advertising Bureau (IAB) in 2018 shows that 70% of digital device

S. Abdul-Rahman · Y. Mahmud (✉)
Faculty of Computer and Mathematical Sciences, Universiti Teknologi MARA, 40450 Shah Alam, Selangor, Malaysia
e-mail: yuzi@uitm.edu.my

S. Abdul-Rahman
e-mail: shuzlina@uitm.edu.my

M. Nasrullah
Riajati Sdn. Bhd., Suite 5-1G, Level 1, Jalan AU 1A/4E, Taman Keramat Permai Business Center, 54200 Kuala Lumpur, Malaysia

A. Kumar et al. (eds.), *Computational Intelligence in Machine Learning*, Lecture Notes in Electrical Engineering 834, https://doi.org/10.1007/978-981-16-8484-5_7

owners around the globe watched digital video content at least once a day for more than 30 min compared to 67% of digital device owners who prefer to watch livestream videos [2]. These statistics indicate that watching livestream videos are increasingly becoming a norm in today's world. Livestream video industry itself is estimated to be worth USD 70.5 billion by 2021 [3]. Thus, it is logical to deduce that livestream video is increasingly becoming a popular choice of content to be viewed in the near future.

Nowadays, there are many livestream platforms such as Facebook, Instagram, and Twitch.tv. Due to easy access to the Internet, practically, anyone can start to do live content or view it easily with unobtrusive access to real-life situations, processes, and events. This ease of access to anyone's livestream videos could pose great risks of viewing violent acts such as murders, shootings, and suicides. From February 2016 until June 2017, Facebook confirmed that 45 incidents of violence livestreamed videos were shown through Facebook Live [4]. For example, in April 2017, Steve Stephens conducted a livestreaming video to kill a 74-year-old man. Sadly, this video was watched by 1.6 million viewers worldwide for three hours before Facebook decided to remove it [5].

The World Health Organization (WHO) defines violence as the use of physical force with the intention toward oneself, another person, or in a group of people that it is most likely to cause death, injury, psychological damage, maldevelopment, or deprivation [6]. A meta-analysis conducted in 1994 concluded that there was a strong positive correlation between exposure to media violence and aggressive behavior [6]. The dangers of viewing violent livestream videos could cause negative impacts to the human mind. This situation would lead to negative impacts especially among children and adolescents such as following violent acts of bullying, murdering, and to suicide attempts. Thus, the aim of this study is to detect violent actions using the Convolutional Neural Networks (CNNs) in which three prominent models, namely the VGG19, MobileNetV2, and InceptionResNetV2, were experimented.

The structure of this paper is organized as follows: Section 2 discusses the related work from the previous literatures, Sect. 3 presents the methodology of this research, Sect. 4 presents the development of the model design, Sect. 5 discusses the analysis and findings of VGG19, MobileNetV2, and InceptionResNetV2 models, and lastly, Sect. 6 conclude the research.

2 Related Works

This section presents the related works regarding machine learning techniques proposed by the previous researchers.

2.1 Deep Learning

Deep learning is a sub-field of machine learning that is very popular in real-time applications. From the previous studies, some of the experimental results on image processing, computer vision, and pattern recognition have proved that deep learning systems performed well compared to conventional machine learning systems [7, 8]. The development of several deep learning techniques with improved performance for addressing problems such as object recognition, detection, and testing has been developed as a result. However, very few deep learning techniques have been proposed to tackle the problem of violence detection from videos [9, 10]. Most of the current techniques generate visual representations of videos by relying on the handcraft features [10]. The problem of using handcraft features is that it could not handle unseen data in a more effective way. Hence, this situation presents a major advantage of using deep learning techniques compared to traditional handcrafted-based techniques, which are the ability to achieve a high degree of generalization. Besides, in deep neural network, no prior information about the data is required and they can be inputted with raw pixel values without much complex preprocessing. Another advantage of deep learning technique is that it could be easily applied for different tasks without any significant changes to the architecture. The previous studies [10, 11] showed that deep learning could lead to better accuracy and speeds in recognizing violent videos. Owing to these reasons, we choose to develop a deep neural network for performing violent video recognition.

2.2 Convolutional Neural Networks (CNNs)

The architecture of the classic ANN limits the use of a deeper hidden layer [12] and not suitable for image classification as this network easily overfits due to the enormous amount of data for each image [13]. On the hand, the positive aspect of CNN is that it could reduce the number of parameters in ANN. This could help researchers to approach larger models with complex tasks [14], which was not possible in classic ANN. In addition, CNN could generate excellent performance in machine learning problems especially applications that dealt with image data [12, 15–17]. Currently, CNN is a popular type of DNN [18] since this technique could directly extract the features from a set of raw image data. Related features are not pre-trained; however, they will learn during training a group of images on the network. The automated way of feature extraction is the most accurate learning model for computer vision tasks including object detection, classification, and recognition. The problem with traditional machine learning approaches is that the feature extractions need to be done manually, and the objects need to classify separately during the classification of algorithms. Due to highlighted advantages of CNN, this technique is chosen to classify and recognize violent actions in videos.

2.3 The Previous Studies on Action Classification Frameworks

The progress of action detection matches object detection and has several significant improvements. Action detection is challenging as there are additional information, we need to consider before the learning process such as temporal variable. Actions are also not well defined; an action class might overlap with other action classes. Action classification also requires significantly more processing power than image classification [19]. Current existing action detection models are mostly categorized into four types of frameworks. The first framework is where a model performs classification at frame or segment-level, and then merge the predictions into detection results, such as the model proposed by Yuan et al. [20], and Montes et al. [21]. The second framework is based on region-based convolutional neural networks (R-CNNs) and its variants. The "proposal then classification" framework is implemented by generating or proposing a set of boxes on frames that are likely to contain an action; then classify the proposed action. Shou et al. [22] proposed a model under this framework.

The third framework is the "single-stream" framework, which is based on single-shot object detectors such as the YOLO model by Redmon and Farhadi [23], and the single-shot multibox detector (SSD) model by Liu et al. [24]. This framework could view the whole image during the process of training and testing time. After the process, it encodes the information associated to classes objects based on their appearance [25]. The fourth framework is based on the deconvolution layers [26] in the full-convolutional network-based object and object detection. It undergoes deconvolution for temporal up sampling to guarantee that the time or temporal length of output to be the same as the input video duration and perform predictions at the frame level. As for this research, the first framework of action classification is applied at a frame level, after which it will merge the predictions to get the results.

3 Research Methodology

This research used kinetics 700 datasets of human action images, a collection of 10 s videos of human actions [27, 28]. The dataset covers 700 classes of human actions taken from YouTube videos. The YouTube links and durations of human actions are stored in.csv format. The classes that focus on "punching bag," "punching person (boxing)," "wrestling," "side kick," "high kick," and "drop kicking" are extracted. The next step is to download the videos. Each class consists of 100 videos, and the total videos that have been downloaded are around 600. The downloading process is aided by Python script written by Ivan Pachenko on GitHub [29]. The script helps to load.csv file, read the link, download the respective video, and trim the video based on the timestamp given which is 10 s.

In this research, videos with "punching bag," "punching person (boxing)," and "wrestling" action are grouped as a single "punching" class. While "side kick," "high kick," and "drop kicking" videos are grouped as "kicking" class. Next, the frames of the videos need to be extracted and for this activity, a reference of Python script is coded from Adrian Rosebrock's script in GitHub repository [30]. This script helps to select the number of frames that could be extracted from each second of the video. Ten frames per second have been chosen for this activity, and each image has been renamed according to their respective class for convenience purposes. Then, these frames need to be filtered manually after the extraction process. The relevance of this process is to avoid data redundancy and to exclude any images that have no human action in them which will have adverse impact when training the deep learning models later. Images that are blank, too wordy, and too blurry, or the action is not visible also needed to be removed. After this cleaning process, a final amount of 1281 "kicking" and 1309 "punching" images is collected from kinetic 700 datasets.

A third class named "others" is created so that the system hypothetically would not blur videos without any violent action in them. For this class, human action image datasets called Stanford 40 Actions [31] are downloaded. The datasets cover 40 human action classes and consist of a total of 9532 images with up to 300 images per action class. A total of 1221 images covering all classes that are not associated with punching and kicking are chosen. The final amount of 3811 images (1281 kicking images, 1309 punching images, and 1221 non-related human actions images) was ready to be used for training and validation of the CNN models.

4 Model Design and Development

The process for this research starts with input of image batches of a human action preprocessed using Python code. The input is then split into training and validation with the ratios of 75% and 25%, respectively. Then, the CNN models were trained over 34 batches of 32 images and update the weights of the classification layers for every epoch of 150 epochs. The trained models were used to predict videos, classify each frame, and blur the image if the act of violence (punching and kicking classes) is detected in the frame. The study utilizes Keras model API [31] to use feature extraction from VGG19, MobileNetV2, and InceptionResNetV2 models. Same custom classification layers are used as the head of each model. The architecture design is shown in Fig. 1.

The class labels named "punching," "kicking," and "others" were given. Every model will undergo training and validation using 75% and 25% of supplied input for 150 epochs, respectively. At the end of the training, a confusion matrix, training loss, training accuracy, validation loss, and validation accuracy are plotted to be analyzed and compared. Two activities were conducted during the model development phase. The first activity was to train the designed model in the architecture. The sources for the scripts were from GitHub [29, 30] and Keras documentation [32]. To promote learning, Adam optimizer was used in all experimental models with the following variables: learning rate = 0.001, beta_1 = 0.9, beta_2 = 0.999, and epsilon = 1e−08, which are the recommended settings [33]. Adam optimizer was chosen as it is the

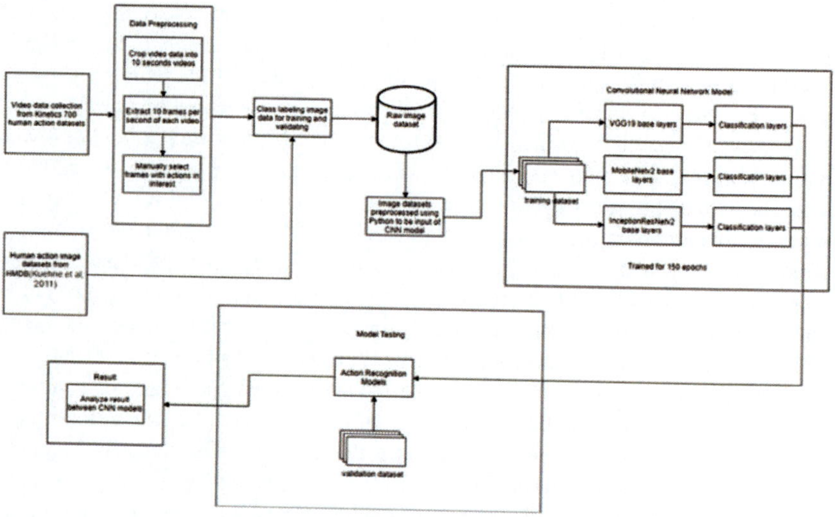

Fig. 1 Architecture design of training and validating images

most used optimizers in academic papers [34]. The batch size was also kept constant at size 32. The final weights and models were then saved in *.h5 format for convenience purposes. A learning rate scheduler was also implemented to reduce the learning rate by 0.1 factor when no best new validation loss is found after 20 epochs.

Two activities were conducted during the model development phase. The first activity was to train the designed model in the architecture. The sources for the scripts were from GitHub [29, 30] and Keras documentation [32]. To promote learning, Adam optimizer was used in all experimental models with the following variables: learning rate = 0.001, beta_1 = 0.9, beta_2 = 0.999, and epsilon = 1e−08, which are the recommended settings [33]. Adam optimizer was chosen as it is the most used optimizers in academic papers [34]. The batch size was also kept constant at size 32. The final weights and models were then saved in *.h5 format for convenience purposes. A learning rate scheduler was also implemented to reduce the learning rate by 0.1 factor when no best new validation loss is found after 20 epochs.

Training and testing were done on three CNN off-the-shelf models with transfer learning strategy to keep the feature extractor of the models and introduce classification layers that are suitable for this research requirements. The parameters for the deep learning have been the same for all experiments on the different models for fair results and comparison. The value for each parameter is shown in Table 1.

Table 1 The parameters throughout the training of all models

Parameter	Value
Initial learning rate	0.001
Beta_1 (Adam)	0.9
Beta_2 (Adam)	0.999
Epsilon (Adam)	1e-08
Batch size	32
Epochs	150

5 Discussions and Result

This section analyzes the training loss and accuracy of VGG19, MobilenetV2, and InceptionResNetV2 models.

5.1 Analysis on VGG19 Model

Figure 2 shows the training loss and accuracy on datasets plotted using Keras API. The accuracy for both training and validation starts at 0.779 and 0.810, respectively, and slowly increasing, converge, and halt at around epoch 70. The best validation accuracy VGG19 model could achieve was 0.929. For the loss of evaluation metric, the training loss began at a high value of 3.821, while evaluation loss began at 0.765. Both figures show that the training loss was able to converge around epoch 120, with the best value of 0.331. However, the validation loss was fluctuating throughout the epochs, with the best value of 0.072 around epoch 118, and the worst value of 1.446 around epoch 65. The learning rate scheduler reduced the learning rate from 0.001

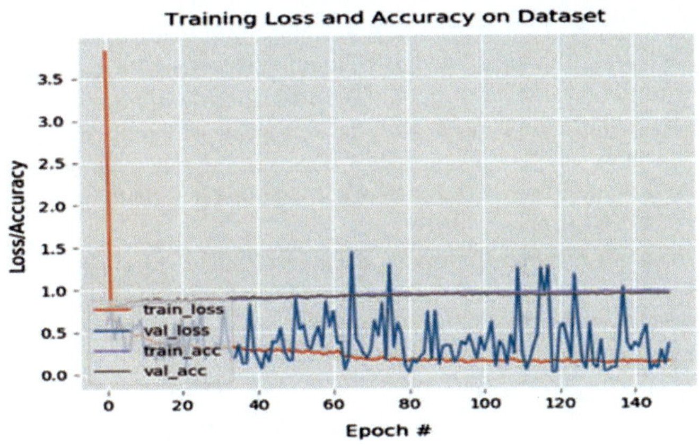

Fig. 2 The loss and accuracy of VGG19 model during training

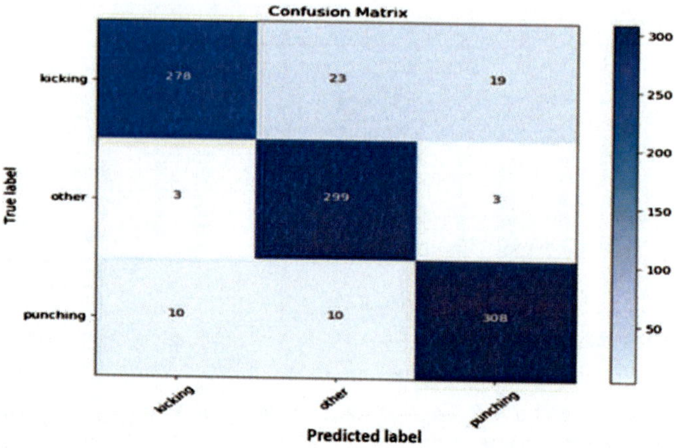

Fig. 3 The confusion matrix on the trained VGG19 model

to 0.0001 at epoch 62, 0.0001 to 0.00001 at epoch 115, and 0.00001 to 0.000001 at epoch 139. Next, Fig. 3 shows that the trained model was able to correctly classify most of the validation datasets. The model was most confused at the classifying "kicking" images compared to "other" and "punching" images.

5.2 *Analysis on MobileNetV2 Model*

Based on Figs. 4 and 5, the accuracy for both training and validation starts at 0.613 and 0.642, respectively. Training accuracy stopped increasing at epoch 45 with a value 0.922. The validation accuracy, however, has been observed as not improving throughout the epochs, with best accuracy value of 0.711 at early epochs. For the loss evaluation metric, the training loss began at value of 0.901 while evaluation loss began high at 1.20. From Figs. 4 and 5, the training loss was able to converge around epoch 40 with the best value of 0.349. The validation loss, however, was heavily fluctuating throughout the epochs, with the best value of 0.488 around epoch 125 and the worst value of 2.212 at epoch 119. The learning rate scheduler reduced the learning rate from 0.001 to 0.0001 at epoch 23, 0.0001 to 0.00001 at epoch 97, and 0.00001 to 0.000001 at epoch 146.

Figure 6 shows that the trained model had difficulty in correctly classifying actions. The model was most confused at classifying "other" images with "kicking" and "punching" images. It also had difficulty differentiating between "punching" and "kicking" images.

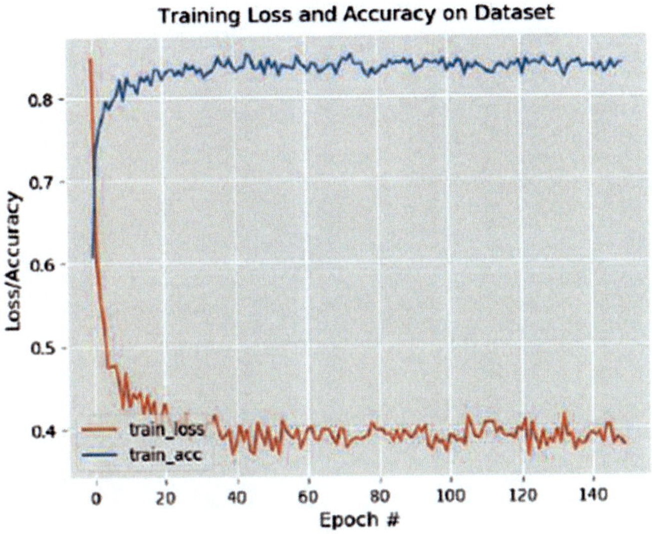

Fig. 4 The loss and accuracy of MobileNetV2 model during training

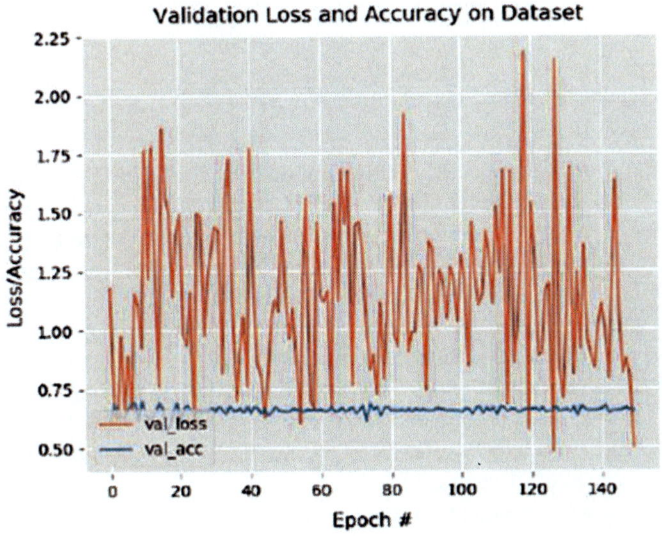

Fig. 5 The loss and accuracy of MobileNetV2 model during validation

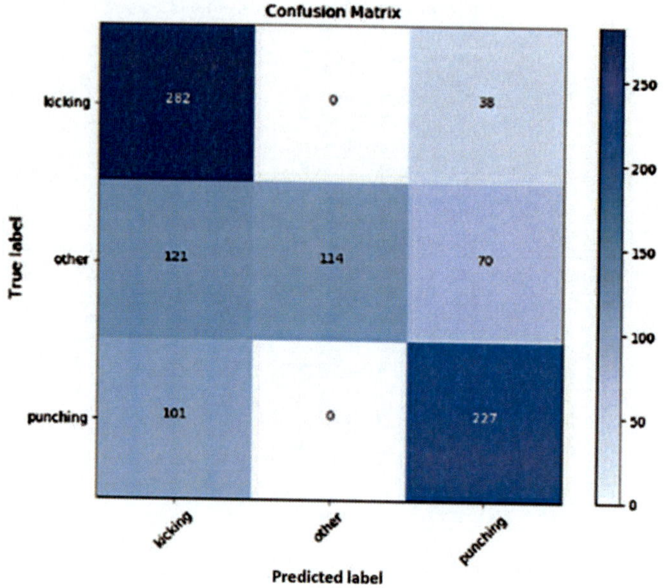

Fig. 6 The confusion matrix on the trained MobileNetV2 model

5.3 *Analysis on InceptionResNetV2 Model*

According on Fig. 7, the accuracy for both training and validation starts similarly at 0.609 and 0.591, respectively. The training accuracy converged and halts at around epoch 40 with mean value of 0.8550, while stopped earlier at around epoch 30. The best validation accuracy InceptionResNetV2 model that was able to achieve was 0.829. For the loss metric, the training loss began at high value of 1.177, while evaluation loss began at 1.098. As can be seen in Fig. 7, the training loss was able to converge around epoch 80 with the best value of 0.331. However, the validation loss fluctuated throughout the training, with the best value of 0.096 around epoch 61 and the worst value of 1.871 around epoch 65. The learning rate scheduler had reduced the learning rate from 0.001 to 0.0001 at epoch 33, 0.0001 to 0.00001 at epoch 53, and 0.00001 to 0.000001 at epoch 82.

Figure 8 shows the trained model had some trouble in classifying the actions correctly. It shows that the model was most confused at differentiating between “punching” and “kicking” images.

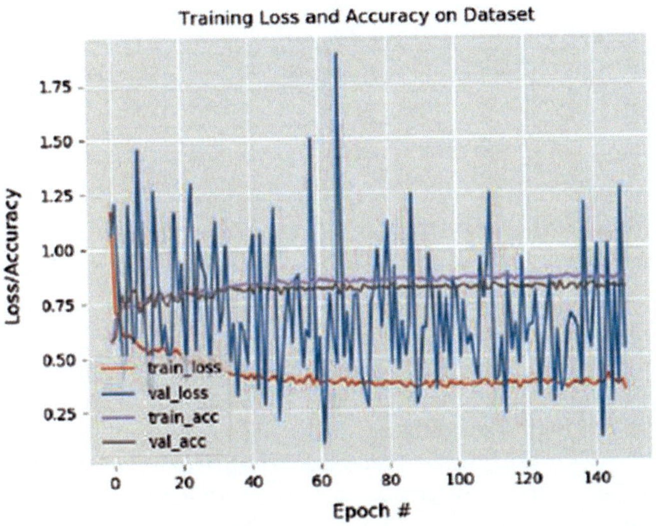

Fig. 7 The loss and accuracy of InceptionResNetV2 model during training

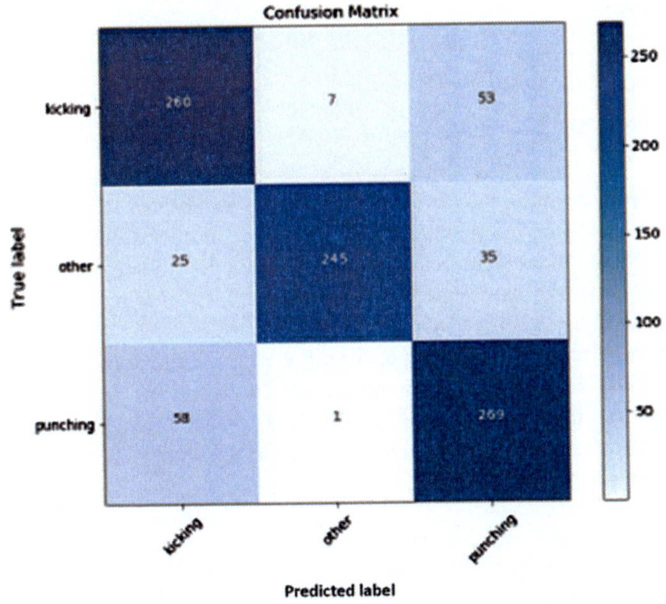

Fig. 8 The confusion matrix on the trained InceptionResNetV2 model

Table 2 Comparison of accuracy and loss among VGG19, MobileNetV2, and InceptionResNetV2

Metrics/Model	VGG19	MobileNetV2	InceptionResNetV2
Accuracy	0.929	0.654	0.812
Loss	0.344	1.157	0.698

5.4 Comparison of Analysis Between VGG19, MobileNetV2, and InceptionResNetV2

After all models have been duly trained, they were tested with test datasets that were different from what have been used during the training process. The final accuracy and loss of each model are shown in Table 2. Based on the table, VGG19 model comes as the best for this research task, followed by InceptionResNetV2 and lastly, MobileNetV2. There are several theories that explain these results: (1) Every neural network architecture has different optimal hyper parameters to be initialized for good results. (2) Neural networks with high number of layers are good for high number of class problems. Hence, for this research, VGG19 model is chosen as the backbone of the proposed system and integrate it with dashboard to develop the act of violence censoring system.

6 Conclusion

This paper demonstrated the development of CNN models by experimenting three models, namely VGG19, MobileNetV2, and InceptionResNetV2. It was found that the VGG19 model gave better results in terms of accuracy and loss. Several recommendations on this research are highlighted. First, it is recommended to use computer vision approach of image classification with localization. This research detects violent actions and blurring of the whole frame instead of the violent action doer. Second, it is recommended to train models with various parameters tuning to find optimal training environment. Lastly, it is also recommended to explore three-dimensional convolutional neural networks (3D CNNs) which would give the benefit of temporal aspects of video datasets.

Acknowledgements The authors would like to thank the conference reviewers and Faculty of Computer and Mathematical Sciences, Universiti Teknologi MARA, Shah Alam, Selangor, for sponsoring this research.

References

1. Cisco, & Cisco Systems, I., Cisco visual networking index: forecast and trends, 2017–2022 white paper. *Cisco Forecast and Methodology* (2019), pp. 2017–2022. http://www.cisco.com/en/US/solutions/collateral/ns341/ns525/ns537/ns705/ns827/white_paper_c11-481360_ns827_Networking_Solutions_White_Paper.html
2. K. Sruoginis, Live Video Streaming—A Global Perspective (2018). https://www.iab.com/wp-content/uploads/2018/06/IAB-Live-Video-Streaming-Trends.pdf
3. N. Patel, Why You Should Care About Live Streaming in 2018 (2019). https://neilpatel.com/blog/live-streaming-importance-2018/
4. A. Kantrowitz, Here's a Near Comprehensive List of All the Violence Aired on Facebook Live (2017). https://www.buzzfeednews.com
5. J. Morice, Man Accused of Broadcasting Easter Sunday Killing on Facebook Live (2017). https://www.cleveland.com/metro/2017/04/man_accused_of_broadcasting_ea.html#incart_2box
6. L.R. Huesmann, The impact of electronic media violence: scientific theory and research. J. Adolesc. Health **6**, 6–13 (2007). https://doi.org/10.1016/j.jadohealth.2007.09.005
7. S.T. Krishna, H.K. Kalluri, Deep learning and transfer learning approaches for image classification. Int. J. Recent Technol. Eng. **7**, 427–432 (2019)
8. I.A.M. Zin, Z. Ibrahim, D. Isa, S. Aliman, N. Sabri, N.N.A. Mangshor, Herbal plant recognition using deep convolutional neural network. Bull. Electr. Eng. Inf. **9**(5), 2198–2205 (2020)
9. E.B. Nievas, O.D. Suarez, G.B. Garcia, R. Sukthankar, Violence detection in video using computer vision techniques, in *International Conference on Computer Analysis of Images and Patterns* (2011), pp. 332–339
10. S. Sudhakaran, O. Lanz, Learning to detect violent videos using convolutional long short-term memory, in *14th International Conference on Advanced Video and Signal Based Surveillance (AVSS)* (2017), pp. 1–6
11. D. Tran, L. Bourdev, R. Fergus, L. Torresani, M. Paluri, Learning spatiotemporal features with 3D convolutional networks, in *Proceedings ICCV* (2015)
12. S. Albawi, T.A. Mohammed, S. Al-Zawi, Understanding of a convolutional neural network, in *2017 International Conference on Engineering and Technology* (IEEE Press, Turkey, 2017). https://doi.org/10.1109/ICEngTechnol.2017.8308186
13. S. Kumar, I. Gogul, Flower species recognition system using convolution neural networks and transfer learning, in *2017 Fourth International Conference on Signal Processing, Communication and Networking* (2017). https://doi.org/10.1109/ICSCN.2017.8085675
14. B.A. Bakri, Z. Ahmad, S.M. Hatim, Butterfly family detection and identification using convolutional neural network for lepidopterology. Int. J. Recent Technol. Eng. **8**(2), 635–640 (2019)
15. M.M. Saufi, M.A. Zamanhuri, N. Mohammad, Z. Ibrahim, Deep learning for roman handwritten character recognition. Indonesian J. Electr. Eng. Comput. Sci. **12**(2), 455–460 (2018)
16. R. Roslan, I.N.M. Razly, N. Sabri, Z. Ibrahim, Evaluation of psoriasis skin disease classification using convolutional neural network. IAES Int. J. Artif. Intell. **9**(2), 349–355 (2020)
17. I. Ibrahim, S. Abdul-Rahman, N.H.A. Hamid, M.R. Shamsuddin, Performance analysis of pre-trained residual neural network on micro-expressions recognition. Int. J. Adv. Trends Comput. Sci. Eng. **9**(1.4), 8–17
18. N. Sharma, V. Jain, A. Mishra, An analysis of convolutional neural networks for image classification. Procedia Comput. Sci. **132**, 377–384 (2018)
19. K. Duarte, Y.S. Rawat, M. Shah, VideocapsuleNet: a simplified network for action detection. *Advances in Neural Information Processing Systems* (2018), pp. 7610–7619
20. J. Yuan, B. Ni, X. Yang, A. Kassim, Temporal action localization with pyramid of score distribution features, in *Proceedings of the IEEE Conference on Computer Vision and Pattern Recognition* (2016), pp. 3093–3102

21. A. Montes, A. Salvador, S. Pascual, X. Giro, Temporal activity detection in untrimmed videos with recurrent neural networks, in *1st NIPS Workshop on Large Scale Computer Vision Systems* (2017)
22. Z. Shou, D. Wang, S.-F. Chang, Temporal action localization in untrimmed videos via multi-stage CNNS, in *Proceedings of the IEEE Conference on Computer Vision and Pattern Recognition* (2016), pp. 1049–1058
23. J. Redmon, A. Farhadi, YOLO9000: better, faster, stronger, in *Proceedings—30th IEEE Conference on Computer Vision and Pattern Recognition* (2017), pp. 6517–6525. https://doi.org/10.1109/CVPR.2017.690
24. W. Liu, D. Anguelov, D. Erhan, C. Szegedy, S. Reed, C.-Y. Fu, A.C. Berg, SSD: single shot multibox detector, in *European Conference on Computer Vision*
25. M.A.M.M. Asri, Z. Ahmad, I.A. Mohtar, S. Ibrahim, A real time Malaysia sign language detection algorithm based on YOLOv3. Int. J. Recent Technol. Eng. **8**(2), 651–656 (2019)
26. Z. Shou, J. Chan, A. Zareian, K. Miyazawa, S.-F. Chang, CDC: convolutional-de-convolutional networks for precise temporal action localization in untrimmed videos, in *IEEE Conference on Computer Vision and Pattern Recognition (CVPR)*
27. Kinetic 700 Dataset. https://deepmind.com/research/open-source/kinetics
28. A. Zisserman, J. Carreira, K. Simonyan, W. Kay, B. Zhang, C. Hillier, S. Vijayanarasimhan, F. Viola, T. Green, T. Back, P. Natsev, M. Suleyman, The Kinetics Human Action Video Dataset (2017). arXiv:1705.06950v1
29. Panchenko, I GitHub. https://github.com/ivan-pan
30. Rosebrock A Github. https://github.com/jrosebr1
31. B. Yao, X. Jiang, A. Khosla, A.L. Lin, L.J. Guibas, L. Fei-Fei, Human action recognition by learning bases of action attributes and parts, in *International Conference on Computer Vision (ICCV)*, Spain, 6–13 Nov 2011
32. Python Deep Learning API, Keras. https://keras.io/
33. D. Kingma, B.J. Adam, A Method for Stochastic Optimization (2014). arXiv:1412.6980
34. A. Karpathy, A Peek at Trends in Machine Learning (2017). https://medium.com/@karpathy/a-peek-at-trends-in-machine-learning-ab8a1085a106

Automated Grading of *Citrus suhuiensis* Fruit Using Deep Learning Method

Faris Azizi Mohmad, Anis Salwa Mohd Khairuddin, and Noraisyah Mohamed Shah

Abstract An automated grading system is important in assisting the farmers to perform quality inspection in a more effective manner as compared to manual approach. Besides that systematic fruit grading is a requirement for effective fruit and vegetable marketing. This is because delivering immature, and bruised fruits will lead to lower market price. Hence, this work proposed an automated *Citrus suhuiensis* fruit grading system based on image processing that can detect multi-index simultaneously such as maturity, quality and size of a local fruit. The fruits are classified according to the grading specification provided by Federal Agricultural Marketing Authority (FAMA). A convolutional neural network method is adopted to perform the classification process. A total of 303 training images and 75 test images were used in maturity dataset, whilst total of 283 training images and 68 test images were used in quality dataset. Experimental results showed that the proposed classification model able to classify the fruits into 6 classes of maturity and 3 classes of quality.

Keywords Classification · Convolutional neural network · Fruit grading · Machine learning

1 Introduction

Malaysia is one of the countries with steady economic growth and a high-human population. In addition to the growth of the economy, Malaysia has produced numerous vegetable and fruit varieties. In Malaysia, the popular form of orange is *Citrus suhuiensis* or Limau Madu, well known to the local population [1]. Often, due to their sweet flavour, they are called Limau Manis. It is a loose-peel mandarin with spherical, glossy green or greenish yellow peel-shaped fruits. More than 2000 ha of *C. suhuiensis* are cultivated in Malaysia, mostly in Terengganu [2]. Limau Madu is grown commercially for local consumption in Malaysia. One of the problems that has thwarted the progress of the citrus industry in Malaysia is low output and quality of

F. A. Mohmad · A. S. Mohd Khairuddin · N. Mohamed Shah (✉)
Department of Electrical Engineering, Faculty of Engineering, University of Malaya, Kuala Lumpur, Malaysia
e-mail: noraisyah@um.edu.my

A. Kumar et al. (eds.), *Computational Intelligence in Machine Learning*, Lecture Notes in Electrical Engineering 834, https://doi.org/10.1007/978-981-16-8484-5_8

fruit. Therefore, having an efficient grading system during the post-harvest handling stage is really important to solve this problem. The main and the most vital stage in post-harvest handling is grading [3]. Subjectivity, tediousness, labour requirements, affordability, inconsistency, etc., are numerous disadvantages associated with such manual grading. Such issues can be alleviated by implementing automation in the process of grading.

Previous works have applied machine learning techniques to classify and recognize fruit ripening and grading such as bananas [4], citrus fruits [5], mulberry [6], date fruit [7] and apples [8]. Grading can be made using single as well as multi-index, whilst the grading index can be based on colour, size, shape or texture. However, research is lacking on how this technology can be applied efficiently onto our local Malaysian fruits. Deep learning (DL) approach is one of the machine visions that has been widely use because of its high degree of accuracy. With their multi-level structures, DL models are very helpful in extracting complicated data from the input images [9]. Therefore, in this project, we proposed the use of deep learning approach for the automated local fruit grading system. In addition, this project focus on fruit grading based on FAMA grading specification to ensure the agricultural products graded is following the Malaysian standard.

2 Proposed Methodology

2.1 Data Acquisition

In this work, the local fruit selected is *C. suhuiensis* or Limau Manis, well known to the local population [1]. A 12-megapixel camera on iPhone 7 is used to capture the fruit images. The dimension of the image taken from the camera module is 4032 × 3024 pixel. In order to perform fruit grading, two different datasets are created which are maturity dataset and quality dataset. The maturity dataset consists of 303 training images and 75 test images, whereas the quality datasets consist of 283 training images and 68 test images. Each citrus image is manually classified based on different quality level and maturity stages with the help of the orchard owner and [10] as reference. The sample image taken for each class for maturity datasets is shown in Table 1.

3 Classification Using Convolutional Neural Network

In this project, a convolutional neural network is selected as the network architecture to train a deep learning model. CNNs make use of convolutional layers, Rectified Linear Unit (ReLU) layers, pooling layers, loss layers, and fully connected layers in its architecture. Each convolutional layer will be, followed by ReLU layer, then a pooling layer. After the pooling layer another one or more convolutional layer will be

Table 1 Sample of fruit images for different maturity

Sample image	Description	Sample image	Description
	Immature The whole surface of the fruit skin covered with dark green colour		40% of the surface of the skin covered with light green colour and 60% covered with yellow colour
	Mature start The surface of the fruit skin covered with light green colour		80–100% of the surface of the skin covered with yellow colour
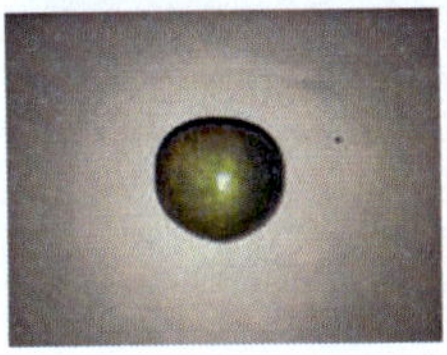	Mature 85% of the surface of the skin covered with light green colour and 15% covered with yellow colour	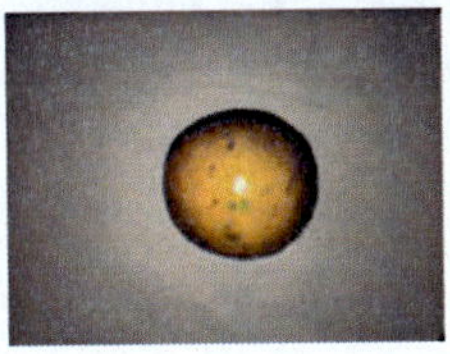	The of the surface of the skin covered with blackish yellow colour

used and finally followed by a fully connected layer. Convolutional layer is used to extract low-level features such as blobs, colours and edges by using a filter. Besides, pooling layer is used to reduce the size of the input image and to reduce the number of computations in the network. Feature that separates CNN from a typical neural network is that CNN will take into account the structure of the image during the image processing. RGB images of size $640 \times 480 \times 3$ is used as an input to this network. The structure of CNN used in this project is shown in Table 2. Meanwhile, Fig. 1 shows the effect of pooling layer towards feature map.

Table 2 Configuration of CNN used in this work

Sequence	Layer type	Strides	Filter shape	Number of filters	Dimension	Output
1	Convolutional	1	3×3	16	$3 \times 3 \times 4$	16
2	Max pooling	2	2×2	–	2×2	
3	Convolutional	1	3×3	32	$3 \times 3 \times 16$	32
4	Max pooling	2	2×2	–	2×2	
5	Convolutional	1	3×3	64	$3 \times 3 \times 32$	64
6	Max pooling	2	2×2	–	2×2	
7	Fully connected	–	–	–	$3 \times 3 \times 64$	1024
8	Fully connected	–	–	–	1024	256
9	Softmax	–	–	–	256	6

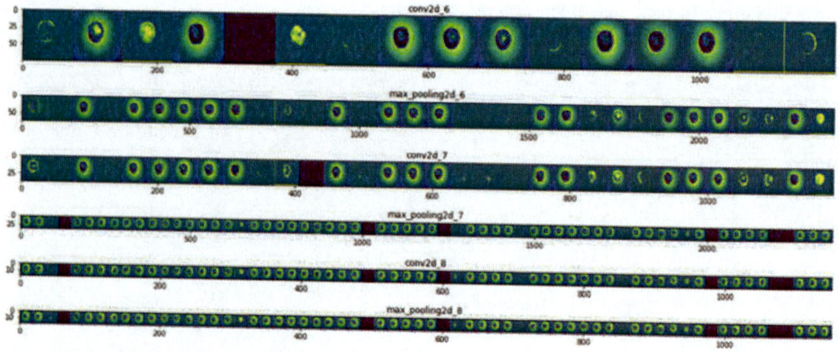

Fig. 1 Feature map for Limau Manis

A fully connected layer is a finite number of neurons which includes a vector as input and returns another one. In general, consider the jth node of the ith layer in the following equations:

$$z_j^{[i]} = \sum_{l=1}^{n_i-1} w_{j.l}^{[i]}, a_l^{[i-1]} + b_j^{[i]} \tag{1}$$

The input $a^{[i-1]}$ is the output of a convolution or pooling layer with the dimensions $(n_H^{[i-1]}, n_W^{[i-1]}, n_C^{[i-1]})$ where the learned parameters at the lth layer are weights: $w_{j,l}$ with $n_{l-1} \times n_l$ parameters and bias with n_l parameters.

The classification accuracy is computed based on Eq. (2):

$$\text{Model accuracy} = \frac{\text{Correctly classified test images}}{\text{Total test images}} \times 100\% \tag{2}$$

4 Results and Discussion

To determine the best number of convolutional layers that can be used for our model, a test is conducted by varying the number convolution layer. For this test, we are using 4 convolutional layers in the first test, 3 convolutional layers for second test and 2 convolutional layers for third test. The filter size, number of filters, number of epoch and batch size is kept constant. Figures 2 and 3 show the plot of model accuracy versus number of epochs, respectively.

The evolution of model accuracy and loss during training process is shown in Figs. 3 and 4. From the figure, we can see that the training accuracy rapidly improves over the first 10 epochs, whereas the training loss rapidly decrease over the first 3 epochs, and there is only small improvement in term of accuracy and loss for

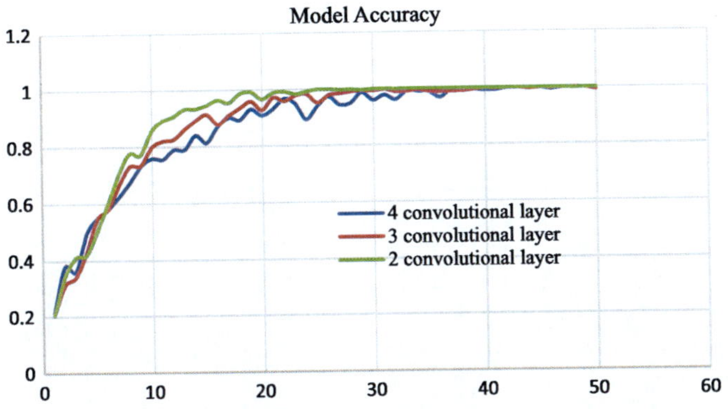

Fig. 2 The model accuracy versus number of epoch plot

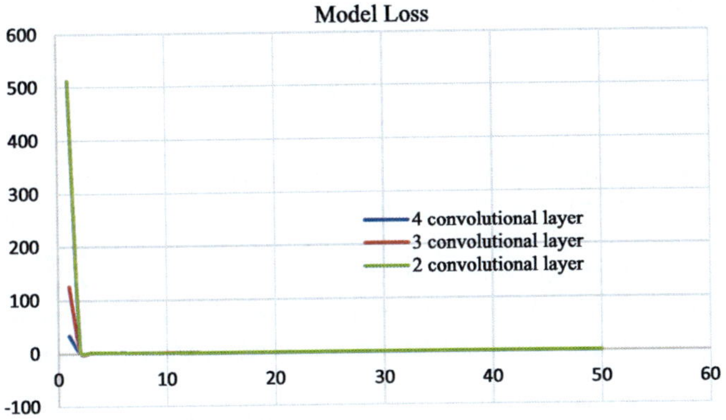

Fig. 3 The model loss versus number of epoch plot

the rest of epochs. Apart from that it can be seen that the accuracy for the model with 2 convolutional layer increase at the highest rate, followed by model with 3 convolutional layers and model with 4 convolutional layers. In contrast, the loss for the model with 2 convolutional layers decreases at the lowest rate, followed by model with 3 convolutional layers and model with 4 convolutional layers. This may be cause by a high number of parameters obtain when training the model smaller number of convolutional layers. Hence, we can conclude that the most suitable number of convolutional layers for our CNN model is 3 layers.

Several experiments were conducted by using different batch size and number of epochs to determine the best value for both parameters. The number of patterns displayed to the network before the weight matrix is modified is called batch size. If the batch size is smaller, patterns would be less frequent and the weights would then be all over the place and it would be difficult to converge. If batch size is high, learning will become sluggish, as the batch size would adjust only after several iterations. On

(a) (b)

Fig. 4 **a** Example prediction output for quality dataset where the output is 'poor quality' and **b** example prediction output for maturity dataset where the output is 'immature'

the other hand, the number of cycles for the entire dataset shown to a model is called epoch. This plays an important role in how well the model blends into the training data. High numbers of epochs may over-fit the data and may have problems with generalization on the test and validation dataset, as well as causing problems with vanishing and exploding gradients. A lower number of epochs can limit the model's potential. Different value of epoch needs to be tune based on the time and computing power available.

Six combinations of batch size and epoch were studied. The classification accuracy of the test images achieved is 100% with F_1 score of 1.0. This means that the proposed classification model managed to classify the test images accurately according their respective classes. From the result, combination of 10 batch size and 50 epochs yield the smallest value in term of test loss. Therefore, the model with combination of 10 batch size and 50 epochs is chosen as our final model. Meanwhile, the same parameters were used to classify quality dataset, and the classification model achieved 99.84% training accuracy, 100% test accuracy, 0.329% training loss, 0.0203% test loss and 100% F_1 score. Figure 4 presents example of output for the grading system: quality dataset and maturity dataset.

5 Conclusion

This work proposed an automated fruit grading system based on image processing that can classify the fruits based on the quality and maturity. In this work, the proposed CNN model adopted 3 convolutional layers to classify two datasets which are quality dataset and maturity dataset. In order to obtain best results, the model needs to be train with 10 batch size and total of 50 epochs. By using this CNN architecture, the maturity dataset achieves accuracy of 100%, 0.0132% test loss, 0.0479% training loss and 100% F_1 score. On the other hand, the quality dataset achieves 99.84% training accuracy, 100% test accuracy, 0.329% training loss, 0.0203% test loss and 100% F_1 score.

Acknowledgements The research funding was provided by RU Grant—Faculty Programme by Faculty of Engineering, University of Malaya with project number GPF042A-2019.

References

1. H. Ghazali, R. Rosman, Z. Sharif, Soluble solid content determination of Limau Madu using microwave sensing technique at 2.0–2.6 GHz. IOP Conf. Ser. Mater. Sci. Eng. **341**, 0–6 (2018). https://doi.org/10.1088/1757-899X/341/1/012006
2. N. Baiti, M. Kluang, P.O. Box, Preliminary growth performance of Limau Madu accessions 51–55 (2018)
3. P.R. Rokaya, D.R. Baral, D.M. Gautam, A.K. Shrestha, K.P. Paudyal, Effect of postharvest treatments on quality and shelf life of Mandarin (*Citrus reticulata* Blanco). Am. J. Plant Sci. **07**, 1098–1105 (2016). https://doi.org/10.4236/ajps.2016.77105
4. F.M.A. Mazen, A.A. Nashat, Ripeness classification of bananas using an artificial neural network. Arab. J. Sci. Eng. **44**, 6901–6910 (2019). https://doi.org/10.1007/s13369-018-03695-
5. M.R. Fiona, S. Thomas, I.J. Maria, B. Hannah, Identification of ripe and unripe citrus fruits using artificial neural network. J. Phys. Conf. Ser. **1362** (2019). https://doi.org/10.1088/1742-6596/1362/1/012033
6. H. Azarmdel, A. Jahanbakhshi, S.S. Mohtasebi, A.R. Muñoz, Evaluation of image processing technique as an expert system in mulberry fruit grading based on ripeness level using artificial neural networks (ANNs) and support vector machine (SVM). Postharvest Biol. Technol. **166**, 111201 (2020). https://doi.org/10.1016/j.postharvbio.2020.111201
7. A. Magsi, J. Ahmed Mahar, S.H. Danwar, Date fruit recognition using feature extraction techniques and deep convolutional neural network. Indian J. Sci. Technol. **12**, 1–12 (2019). https://doi.org/10.17485/ijst/2019/v12i32/146441
8. M.O. Al-Shawwa, Classification of apple fruits by deep learning. Int. J. Acad. Eng. Res. **3**, 1–6 (2019)
9. L. Deng, D. Yu, Deep learning: methods and applications. Found. Trends Signal Process. **7**, 197–387 (2013). https://doi.org/10.1561/2000000039
10. FAMA: Limaumanis.pdf (2001). http://www.fama.gov.my/documents/20143/0/Limaumanis.pdf/c33addc6-bd1c-12be-10f4-272fcbd6ecff

The Future of Car Automation Field with Smart Driverless Technologies

Shreyas Shivajirao Thombare

Abstract This study investigates the challenges and opportunities particularly in current transportation policies that may arise to change existing mobility systems with the help of intelligent autonomous vehicle (IAV) technologies. India is already one of the highest levels of population, so all this leading to a massive increase in road traffic which leads to serious issues such as permanent traffic jams increased air pollution at harmful levels, accidents, or even human losses. Hence, technology and transportation infrastructure providers are needed to provide safer, large scalable, better flexible, and low-cost effective solutions to all these problems. This paper provides a comprehensive review of the relevant literature and explores intelligent transportation systems (ITS) which is one of the best solutions, and this offers traffic monitoring, guidance, or alerting by communication capabilities in between vehicles to avoid any accidents in terms of safety elements (Litman in Autonomous vehicle implementation predictions implications for transport planning, 2018; Pino et al. in IEEE Access 6:17527–17532, 2018; Bagloee et al. in J Mod Transp 24(4):284–303, 2016; Aazam et al. in IEEE Commun Mag, 2018 (in press); Gong et al. in IEEE Trans Intell Transp Syst 19:390–401, 2017; Miettinen in Evolutionary algorithms in engineering and computer science: recent advances in genetic algorithms, evolution strategies, evolutionary programming, GE. Wiley, New York, 1999; Making Innov E-J, makingofinnovation.com, Oct 2015; Campioni et al. in IEEE Trans Veh Technol J Latex Class Files 14(8), 2015; El Zoghby et al. in IEEE 68(2) ITVTAB, 2019; IEEE International symposium on technology in society (ISTAS) proceedings, 2015; NCSL—National conference of state legislatures, database on autonomous vehicles legislation; Cai et al. in An empirical air-to-ground channel model based on passive measurements in LTE, 1140; Qin et al. in IEEE Trans Veh Technol; Meng et al. in IEEE Trans Control Syst Technol 25:1480–1487, 2017; Zeng et al. in Joint communication and control for wireless autonomous vehicular platoon systems. CoRR, 2018) [1–15]. So it is important to do the improvement in new vehicular automation methods. It must have its organizations, specialists, and foundations to the center for their safe and successive journey under enhancing street security and traveling comfort which is the prime need of nowadays (Lian et al. in

S. S. Thombare (✉)
Department of Electrical Engineering, Dr. D. Y. Patil Institute of Engineering and Technology, Pune, Maharashtra, India

A. Kumar et al. (eds.), *Computational Intelligence in Machine Learning*, Lecture Notes in Electrical Engineering 834, https://doi.org/10.1007/978-981-16-8484-5_9

Channel models a non-stationary 3-D wideband GBSM for HAP-MIMO communication systems, 1128; Du et al. in Resource allocation in vehicular networks based on dual-side cost minimization, 1079; Gonzalez-Martín et al. in Analytical models of the performance of C-V2X mode 4 vehicular communications, 1155; Wei et al. in IEEE Trans Veh Technol; Wang et al. in Optimizing content dissemination for real-time traffic management in large-scale internet of vehicle systems, 1093; Jo et al. in Image-to-image learning to predict traffic speeds by considering area-wide spatio-temporal dependencies; Vehicle automation IEEE 2019: collision avoidance intelligent vehicles system vehicle-to-everything; Du et al. in Design and assessment of an electric vehicle power train model based on real-world driving and charging cycles, 1178; Peng et al. in Connected vehicle series vehicular communications: a network layer perspective, 1064; Wei et al. in An integrated longitudinal and lateral vehicle following control system with radar and vehicle-to-vehicle communication, 1116) [16–25]. All over the world, currently automobile manufacturers are mainly focusing on developing, exhibiting, producing, and promoting new vehicle features with advance controlled strategies that could make possible the exchange of information with self-organized robotics and try to interfaces with new algorithms into the smart automobile world (Self-driving cars: the next revolution kpmg.com | cargroup.org CAR (car centre for automotive research); Hasan et al. in IEEE Access 6:20371–20389, 2018; Guerrero-ibanez et al. in IEEE Wirel Commun 22:122–128, 2015; Li et al. in IEEE Trans Veh Technol, 2017; Abdelhamid et al. in IEEE Trans Intell Transp Syst 99:1–14, 2017; Jo et al. in Image-to-image learning to predict traffic speeds by considering area-wide spatio-temporal dependencies, 1188; Tan and Hu in IEEE Trans Veh Technol; Garcia-Garcia et al. in Appl Soft Comput 70:41–65, 2018; Xiao et al. in Inf Sci 432:543–558, 2018; Favarò et al. in Presented at PSAM Topical 2017 on human reliability, quantitative human factors, and risk management, Munich, Germany, 7th–9th June 2017) [26–35]. The vision of this paper is to provide a multimodal transportation system that could overview all the recent research work in the field of interconnected transportation environment giving help to millions of vehicles that are facing traffic problems and safety issues. To this end, we propose smart automated connected vehicles with the help of intelligent transport systems (ITS) and can provide various application services to improve the safety, efficiency, reliability, and comfort of the driving system (Autonomous cars: past, present and future—a review of the developments in the last century, the present scenario and the expected future of autonomous vehicle technology, Jan 2015 with 15,903; Autonomous vehicles UF: automated guided vehicles autonomous cars autonomous driving autonomous trucks unmanned autonomous vehicle BT: autonomous systems intelligent vehicles RT: artificial intelligence mechatronics multi-agent systems vehicular automation NT: unmanned autonomous vehicles, by The Institute of Electrical and Electronics Engineers (IEEE), 2019; Int J Inf Commun Comput Technol, Jagan Institute of Management Studies, New Delhi Article, Jan 2016; Proceedings of 2014 RAECS UIET, Panjab University Chandigarh, 06–08 Mar 2014; Future Internet J, Published: 24 Jan 2019; Martin-Vega et al. in IEEE Trans Veh Technol 67:3069–3084, 2018; Bazzi et al. in IEEE Trans Wirel Commun 17:2402–2416, 2018; Noor-A-Rahim et al. in IEEE Access 6:23786–23799, 2018;Technical report, IEEE, Piscataway, NJ, USA, 2006; Bazzi et al. in IEEE Access 6:71685–71698, 2018) [36–45].

Keywords Ant colonies · Bee's colonies · Kilobit · Swarm intelligence · Volvo cars · Tesla autopilot · Google self-driving car · Artificial intelligence · IoT · Practical actual accidents · ITS · V2V · DSRC · VA NET · Wireless sensor networks · Accident prevention · Post-accident investigation

1 Introduction

Today peoples are living in the twenty-first century which is a highly advanced technological world, and humans have created cars that can reach 130 miles per hour, but we still drive them at the same slow speed as nineteenth-century horse carriages.

In the US-Alone, it spends 29.6 billion hours commuting in 2014 only in traffic problems and with that amount of time ancient Egyptians could have built 26 pyramids of Giza. So, we can compare how much amount of a monumental waste of time, energy, and human potential.

For decades' human's remedy for congestion was simple, building new roads or enlarge existing ones and it works admirably for some countries like Paris when the city tore down hundreds of historical buildings to create 85 miles of transportation friendly for human society, and this system still works today for fast-growing emerging cities, but in more established urban centers, significant network expansions are almost impossible. Because habitat is just too dense, real estate too expensive and public finances are too fragile. Our city's vascular system is getting clogged, and it is getting sick. Hence it is time to pay attention. Our current way of thinking is not working for getting our transportation to grow we need a new source of inspiration [1].

1.1 Need

Nature already is having the solution of our all problems and we could get our inspiration from the rules of nature which may be under biological or environmental creatures, insects, birds, and animals' behavior. Figure 1 shows how all this can be implemented into the vehicles also for making them automated. Take an ex-human's vascular system inside the body having specific properties all the veins and arteries making miracles of logistics every day. So this could conclude that biology has been billions of years. It has been testing countless solutions to move nutrients, gases, and proteins throughout the body. It is the world's greatest cultured conveyance laboratory. So, what if the solution to our traffic challenges was hidden inside. Every human is having nearly 60,000 miles' blood vessels in our bodies which is 2.5 times

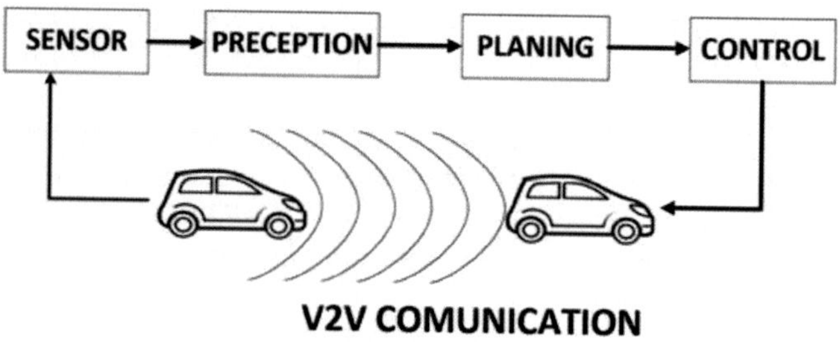

Fig. 1 Basic working principle block diagram

the earth's circumference inside our body. Hence if we compare this with our cities, we also have some subway systems and some tunnels and bridges, and also some planes and helicopters in the sky.

But most and the vast majority of our traffic is focused on the ground, on the surface. One of the few solutions that had been already suggested is as follows:

Humans are always trying to suspend transportation as it did with modern electrical grids. Research says "Tel Aviv and Abu Dhabi" are talking about testing these futuristic networks of suspended magnetic pod's, and we can keep diming and fly the fact that like a company "Airbus" are seriously working on flying urban taxis is telling us something. Flying cars are finally moving from science fiction to attractive business-case territory. So, building this 3-D transportation network is one of the ways that we can mitigate and solve traffic jams.

It is estimated that up to 30% of urban traffic is generated because drivers are looking for parking. But biology is never facing problems like this space inside our arteries is fully utilized. At every heartbeat, a higher blood pressure compacts millions of red drops of blood quickly flows throughout the body, and every tiny space inside the red blood cells is not wasted either. Healthy condition is fully utilized, and it is possible to do something same logic in transportation. Because blood is both a collective and individual form of transportation. But for cities, people are completely stuck in an endless debate between creating a car-centric society or an extensive mass transit system.

So, this paper can describe to us that there is a possibility that the system could combine the convenience of cars and the efficiencies of trains and buses. The current generation of driverless cars is just trying to break through the traffic network made by and for humans. This car is trying to learn traffic rules and cope with human unpredictability. Which is more challenging? [2].

1.2 *What Would Happen When the Whole City's Car and Transportation Become Driverless?*

As is known that red blood cells in our body are not flowing in lanes. They never stop with any indications as in comparison with cars stop at red light signals. Similarly, in a driverless city, it had no red lights and no lanes and when all the cars are driverless and connected everything is get predictable and reaction time, minimum. They can drive much quicker and can take any coherent initiative that can speed them up or the cars around them.

So, this paper wants to describe instead of rigid traffic rules, and the flow will be regulated by a mesh of dynamic and constantly self-improving algorithms shown in Fig. 2. The Result: A strange traffic that mixes the fast and smooth rigor of

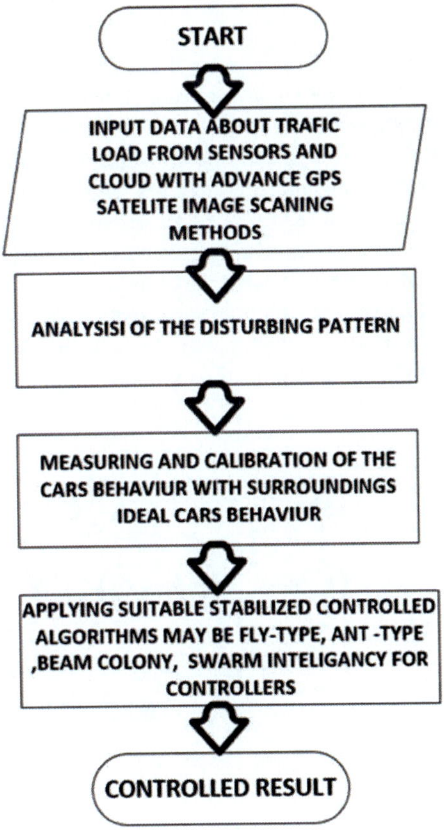

Fig. 2 Basic algorithm for traffic problems with solution

German Auto loans and creative vitality of the intersections of Mumbai traffic will be completely functionally exuberant. It will be liquid-like human blood [3].

Co-ordination of Automated—Hybrid Vehicles with Artificial Intelligence and Collective

When humans look around nature, it is having an amazing example all over the planet which is work under collective intelligence individuals that are working together [4].

Ex. (1) Ant Colonies—Many groups of ants are capable of searching through this huge territory for food making straight-line trails and their tracks just like the "One Big—Highways" to get to their food sources in common platform [5].

(2) Bee's colonies—The bees' colonies are also doing the same things in the air instead of ground and one of the other exes. Are groups of insects can have built a huge self-working network with precise accuracy just an ex with bird—flocks or group of fish school working and moving quickly in the sea or the sky [6].

(3) Even the smallest parameters like a cell's inside an organism are cells with identical DNA that can co-operative to create a complex organism like a starfish. All are these are part of the collective intelligence mechanism. So, under this paper, all these well-organized natural rules are going to use all these terms and research for creating advanced collectives automated robotic cars. Just like the low-cost robot "Kilobit" which of 1024-swarm robots [7].

Here in this technology, we are simply going to use an artificial collective behavior that has never existed in nature on its own, but it can work with advanced programming so that cars may self-organize for system synchronization movement pattern formation to catch up with the behavior of insects which work in nature. Just shown in Fig. 3 to improved patterns in Fig. 4 with the help as the algorithm's shown in Fig. 2 [8, 9].

Fig. 3 Cars traffic jam conditions before swarm intelligence

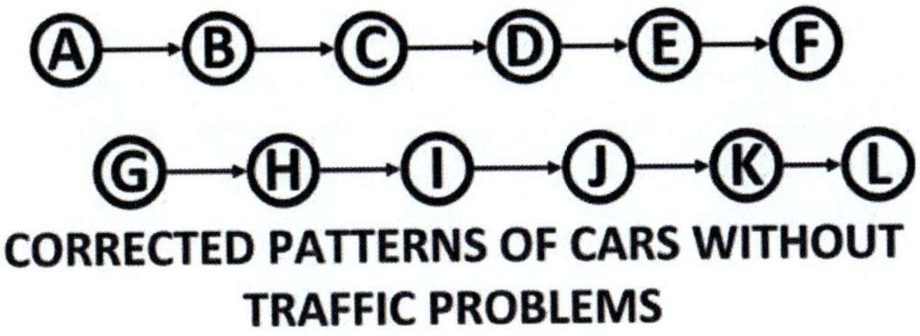

Fig. 4 Cars traffic-free conditions after swarm intelligence

2 The Five Principles of Swarm Intelligence

2.1 **Awareness**—Each member must be aware of their surroundings and abilities [10].
2.2 **Autonomy**—To self-co-ordinate, each member must operate as an autonomous master (not as slave or leader) [11].
2.3 **Solidarity**—When a task is completed, members should autonomously look for a new task [12].
2.4 **Expandability**—The system must permit dynamic expansion where members are seamlessly aggregated [13].
2.5 **Resiliency**—When members are removed the system must be self-healing [14].

3 Major Advance Work in Real-Time Top Autonomous Car's and Its Advance Features Areas

3.1 Mercedes-Benz F015 Luxury in Motion

A fully automated car just coming in the future time up to 2030. Advance features monitor all activities around itself and Handel every drastic condition by own self-intelligent programmed system. Self-parking and safe driving ability [15, 46].

3.2 Volvo Cars Unified Interface for Self-driving Cars

This car uses interfaced sensors which ready to take the center stage ground broking driverless job activity, till 100 Volvo cars had been already on roads. The car interface is designed to be simple intuitive and easy to use, which saves loss time throughout traffic by keeping the driver well-informed, relax with completely controlled features [47–49]. All sections of the roots are available under autonomous driving are displayed when the journey begins and well-organized programs and navigation system helps to monitor all 360° view of its surrounding and helps driver about what happening in the traffic environment, such as smooth overtaking activity on highways safely when autonomous driving is no longer available the driver gets requested to take control the journey [16, 50].

3.3 *Tesla Autopilot*

The car is combined a newly streamlined interface with enhanced control and versatility of your daily surrounded environment. Over the continuously updated information used to run a system with autopilot mode [17]. This model is designed especially for highway use which also helps in daily driving in urban areas. Traffic monitoring is done throughout GPS signals continuously. The car is designed in such a way that traffic cruse adjusting it for better response to other vehicles. Also, a car is having auto-park feature [18].

3.4 *Google Self-driving Car*

This vehicle uses two basic rules: First, it must know its location where it is in the world and then what it with surrounding it so that it makes its self-driving and automated, into the real world with all practice driving situations [19]. To locate his location instead of GPS, it can use lasers which identity more detailed like objects, their size, its speed, and its presides location. All this was done using sensors which can rotate almost 360° and all the information coming through different sensors was utilized by controlled. Then it cross-checks the information to make the decisions [20].

The Google car is having followed important components with its function's:

- **GPS Receiving antenna**: This is used to match the destination position customize Google maps and select short roads to save fuel.
- **Light over remote sensing system**: This is for controlling arrangement with sensor technology.
- **Video cameras**: This is used to read different signs of traffic roads and headlights.
- **Radar**: This is provided on both front and back sides by which system keeps the vehicle at correct track and without accidentally to other cars by recognizing their movements. Also, this sensor is rotated at 360° to select the distance between surrounding objects and the car.

4 Artificial Intelligence System in Car Automation Field

As in today's technological world, AI system is here and even from more than decades of years but just that it was seeing this terminology used lots nowadays. Already we have lots of common automation applications where this system is available for more than 10 years with smartphones and the Internet [21].

Ex.—As per the Honda car automation system, there is a new feature named "HANA"; the new artificial intelligence (AI) system for Honda cars, and it is a valuable assistant in drivers' daily life. This system is capable of a self-learning machine system where it can be physically observed the driver's mental state. It also studies the driver's emotions by his facial expressions, like confused, happy, angry, frustrated all these conditions. Sometimes driver choices define logic but as per the driver's mood. Just like this system also works with logic and helps to give suggestions to other people. It rewards drivers every day's activities and stores these memoirs in a cloud storage system, for future purpose analysis. So, the main three objects of this system are to [22].

- Understand driver conditions with accuracy and note down smaller changes in his attitude.
- Make suggestions to driver in all circumstances.
- Assisting driver properly to avoid any bad drastic driving conditions.

For better home and daily life experience of the driver, this assistant will work in the following ways of examples which is an explanation of Fig. 5 as below

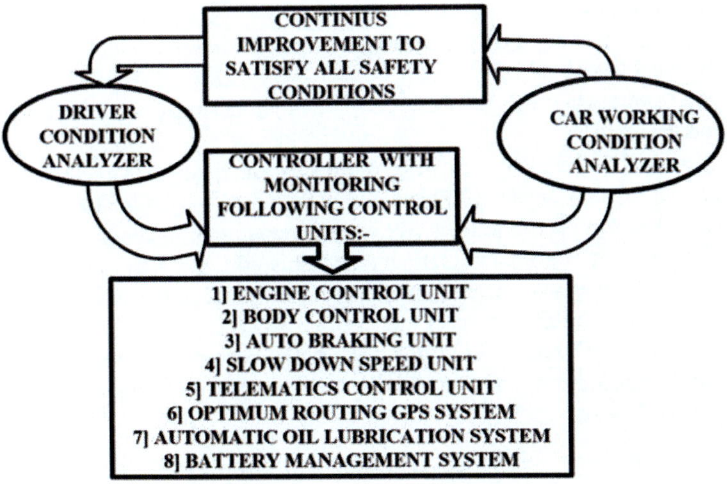

Fig. 5 Control automation cycle under artificial intelligence system

4.1 With starting when a car is on, the assistant AI-the system will start communicating with a driver as per schedule of time, it will start the conversation with "Good morning!"

4.2 As per the driver schedule of daily driving rewards AI-the system will recognize all next action like, if the driver is having really important work for that day in his office, the AI-the system will measure his face scan and record his stress level.

4.3 To reduce his stress, the car AI-System will have recommended him the low-traffic route for driver's destination location by GPS map's system and give an up-to-date status of all traffic conditions [23].

4.4 In the second stage, the system will try to use all the logics that the driver likes to do to reduce the stress. That is, it may play drivers favorite music, suggest drivers' favorite coffee shop on divers' route, and suggest proper enough time to have a cup of coffee.

4.5 Next important focus of the system is "Safety." So when the driver sets the destination location, the AI-the system will first check all previous records of ideal driving skill's abilities, and accordingly, this system will suggest chasing safe path for driving.

4.6 If the car is designed to assist two or more drivers in a particular family, then the system may connect to everyone's mobile phone. So, every time the AI-the system will calculate exactly how much energy every activity and rest of energy management will be utilized for secondary driver purpose.

4.7 Also, the system will calculate the proper time required to charge up the battery as per destination location and after full battery charring, it will message to driver, to ready for driving. This car will be able to save money and time [24].

4.8 The system at night will start charging when the traffic of charging electricity cost is low enough.

4.9 Also, when the car is fully charged, it will go into sleep mode and save's the power.

4.10 Now with reacting after the destination location, if the car is having additional power enough to take him back home, then the AI system also has bidirectional energy trading meters. Which will transfer all its remaining unwanted energy back to gold and make money. This will give new opportunities for making money when it has great prices for demand, and transportation system is high, so the modern cars are pacified with an electronic system which constantly collects all kind of information about the on the going journey. And recording and analyzing this data will create huge business opportunities with AI-systems [25].

Here is the way how will be this data will be the source of business?

- Artificial intelligence can collect first all the telemetry data from the user's car predict breakdown or schedule the next maintenance visit.
- AI- can analyze the user's lifestyle like places he goes to, the way it drives, the driving behavior, and so on.

- Based on this analysis, AI can make personalized offers withal possibilities of additional features or accessories like, e.g., wheel's chain, additional lighting order, traffic jams, and Wi-Fi-Hotspot
- AI- system also allows to target offers precisely to specific users, bring clients back to dealer's network and reduce the number of breakdowns due to time maintained [26].

5 IOI-Based Car Automation System

Imagine a future where vehicles will talk to each other, and with all these critical infrastructures, we would have connected vehicles technology; this can change the transportation system as it is known today, by enabling safe and intractable wireless communicating networking among vehicles and personal communication devices [27].

Safety:

Connected vehicles provide drivers 360° awareness with surrounding vehicles approximately 360 m. As shown in Fig. 6, drivers will receive warnings that inform them of a potential hazard with the vehicles around them through the visual display or seat vibration or tones, these are only warning signals to keep the driver's attention and avoiding him from slipping possibilities while driving. However, this also helps to avoid drastic and quick crashes or accidents [28].

For Ex. "Intersection movement Assist" application will warn drivers when it is unsafe to enter intersection areas with other cars' speed profile and movement sensations.

"DO NOT PASS WARNING ASSIST" will help to warn the driver when it is not safe to overtake or pass over front slower or faster moving vehicles.

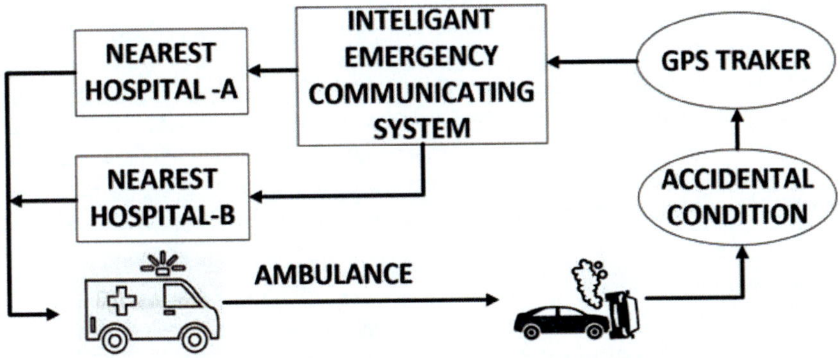

Fig. 6 Emergency ambulance management system by Internet of things system

"EMERGENCY ELECTRONIC BRAKE LIGHT WARNING" application will notify the driver when outside vehicles are applying brakes simultaneously and take quick actions themselves.

"BLINDSPOT WARNING" Application will allow a driver to virtually see all the things On-Screen what's happening in its surrounding blind spots.

"RAIL CROSSING VIOLATION WARNING".

The application will communicate with smart roadside infrastructures like railway crossings which alert drivers when a train is coming even the driver is unable to see hear or approaching trains.

"ROAD WEATHER MOTORIST WARNING".

The application will sense and monitor all weather conditions and collect data from weather base stations and helps for special case conditions like black eyes problems due to heavy rain sometimes roads become slippery and the weather is becoming appears to be dangerous. Hence, this entire information collected through multiple vehicles connections can help to determine the potential hazard exist possibility like

"Icy Roads AHEAD".

So warm's driver is to slow down speed before getting into any trouble. All this data also may send through vehicles back to traffic management centers (TMC) providing detailed real-time information that helps in monitoring and manage transportation system performance. Then the canters will take an action like industrial traffic signals and speed limits adjustment and alert maintenance system management for dispatching sand trucks and broadcasting all this from Readies to car's back all-weather conditions. It may also send alert signals to those cars that furthered are going to drive the car in the same weather conditions.

"ECO-LANE AVAILABLE" This application even helps us to reduce the car fuel and fascinates to green transportation systems. Echo-lens are like today's "DYNAMIC-V-LANES"; this helps in driving eco-friendly vehicles with less fuel consumption road spots monitoring which helps for "Eco-speed HARMONIZATION" with adjustment in engine power, fuel consumption and speed limits adjustments.

"ECO-APPROACH AND DEPARTURE AT SIGNALIZED INTERSECTIONS" Application helps for unnecessary stoppage and wasting of time by communicating with smart traffic signals. All these traffic signals will broadcast the data about the current signal facing, and timing system inside the vehicles uses this data to determine speed advice for drivers. Hence, the driver could adjust their vehicle's speed to pass the next traffic signals ON green or slow down when RED signal to stop in eco-friendly manner, fuel and emission rate of vehicles and driver efforts will get saved.

- **Emergency**

  This system with connected vehicles enhances unexpected roadside incidents like disabled and fast-moving vehicles, ambulances, police cars, and car crashes.

  "INCIDENT ZONE WARNING" Application will alert the driver about all the above-mentioned activities and allows the driver to change their path with first responders; all this will help in control traffic flow and reduce the conjunction while having accidents before gets out of car hand's [29].

- **Mobility**

  With millions of connecting vehicles system having an ability to share robust data within continuously to improve traffic flow and enhance the co-ordinations with transit and make out the community even more safe and capable "CONNECTION PROTECTION." Application will help in public transportation providers and travelers to the ability to communicate in two-way; the connected networks will give passengers real-time transient information where they can predict more accurately about next transport facility, their arriving times and reach into destination successively. This technology helps physically challenged peoples mostly for [30]. Ex. The mobile accessible pedestrian system application will send automated pedestrian calls with a smartphone to a traffic controller which creates changes in traffic signals until the peoples or streetwalkers can cross the roads safely. So, all these above things making our transportation safer and faster with an advanced result of accuracy [31].

6 Limitation's with Self-driving Cars and Real Practical Actual Accidents till Now

Not only the autonomous car industry but every other advanced technology-related thing having limitations would just impossible. So here are some crazy thought experiments where the autonomous car—the system will fail:

(I) Suppose passengers are seating in one of these advanced car systems and get a block with 4 surrounding other cars from front, back, right and left from every direction and supposed to move at the same rate of speed and acceleration at which yourself driving the car is moving. Now imagine a sudden heavy object falls out from the truck in front of yourself driving the car. Now your self-driving car cannot reignite all this drastic change in seconds even if it is able to trace the fault but it needs its own processing time to decide to avoid an accident [32]. But Now it has no option either there are four following possibilities:

- It will directly his to the front of the object
- It will take left and get crash with the left sidecar
- It will take right and get crash to right sidecar
- Or it will slow down its speed while the car from its back will hit at dangerous speed from the backside.

So, here is this common example where a car system will get confused between preprogramed reactions to the actual decision process [33]. So the one solution behind all this is to allow car with some programming logic to instruct the car to take the decisions by knowing the possibility of the future, and it looks like a premeditated futuristic machine, which could predict the accidence and their rate of percentage of damage to the human inside the car with the rise of other peoples surrounding at such instants. Now to be fare self-driving cars more advance required to be designed to reduce traffic accidents and fatalities by removing errors from normal driving equations plus self-decisions to achieve other benefits like easy road conjunctions, fuel-saving, minimize stressful driving, but accident can and will happen when they do the outcomes may be already determined may be before the actual time [34]. This principle can minimize harm, so ethical consideration will be so much difficult and more complicated both scenarios of the targeting algorithms are not that Mach advanced. In other words, it is systematically not able to differentiate between actual preprogramed algorithm to a real practical negative impact on the real world surrounded by us. So, our new technology must be able to analyze all the possibilities and capable of taking random [35].

A solution to avoid accidents and traffic load monitoring on highways using V2V communication:

This paper overviews the process of finding problems with traffic load detection on highways and developing a system to detect and avoid the possibility of accidents on a highway using vehicular-to-vehicular communication network [36, 37]. The most common method is to implement a V2V communication system along with a weighted cluster algorithm (WCA) and calibrating its performance on different aspects of a network [38].

The main objective of using this algorithm in a V2V system is to enhance the performance of the network which is complicated and unexpected accidental conditions. Here in the WCA method system need to recognize the quality of communicating nodes (V) through which automatically the stability of network parameters becomes also stable. For all this, WCA method needs to consider different factors that are related to the direction of speed and vehicles [40]. The main purpose of this review is to carry out examination and assess the results for the relevant operation of systems based on its applications, and its defined communication protocols which would distinguish the future road infrastructures which could be used by the vehicles this shown with example in Fig. 7. The most common standardization of the Third-Generation Partnership Project (3GPP) helps to implement effective message distribution to multiple users over a wide geographical area near to that vehicles so, the third generation of mobile cellular systems is well known as the Universal Mobile Telecommunication System (UMTS) which is worldwide used, were the fourth generation is denominated long-term evolution (LTE) is the recent trade-in network communications. This system proposes a hybrid its safety architecture that will help to combine vehicle-to-vehicle communication and vehicle-to-roadside sensor communication all this is used for effective installation and maintenance, roadside wireless sensor, and networking technology that represents a cost-effective solution [42].

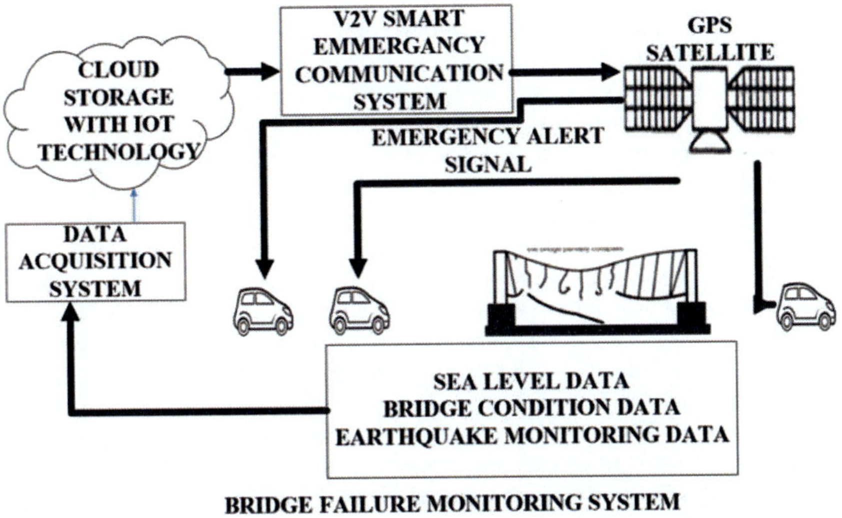

Fig. 7 Warning indication control management system by V2V communication system

7 Conclusion

The paper also focuses on generating new vehicular rules with advanced robotics in auto-car system inspired by insect's behavior through nature; hence, it will be getting closer to an idea that it can possible to create the kind of artificial systems that fascinate us in nature. This paper describes self-driving cars in the future will safely handle an aspect of the driving while the driver is focusing on some other activities. From an engineering perspective, designer, must use different logic in such a way that all automation must alone with "safety." This is having almost top priority so that vehicles can take many things like how close it is to other objects or matching speed with traffic or intersecting with other cars while overtaking. The paper talks about all limitations and advantages of nearly 17 crashes all over the world that took in past even with such advance automatic vehicle systems [36]. From this paper, it could get to the result that the more robotized our traffic grid will be the more organic and alive its movement will feel so biology has all the attributes for transportation genius today. So humanity has the dreams, concepts and technology to create 3-D transportation networks, and inventing new vehicles, and change the flow into the cities.

So, the car of the future is in integrate with IOI technology which is a rapidly growing field capable of transforming cars into artificial intelligence devices. Already 39% of drivers are in full consideration of automation technology, so it is very soon that cars will be able to drive themselves with no starring wheel, no brakes, no escalators paddle the software is in the trial now for self-driving mode. Not only driving safety also it will allow networking to network connectivity platforms which enables the driver to keep themselves in touch with the outside world enables menus of application and enjoy the entire at their car journey. We have also highlighted the huge potential of tomorrow's cars will be enabling the drivers to access higher demands services such as insurance extensions, with high-quality engine power requirement services so all this will be done with real and practice traffic problems on the roads. So, the future car will take you on your journey, so at the front of the office, then park the car and coming to pick-up you when you are ready to go. So finally, it is all about the lots of connecting programming and advance logic which will record integrate this with high processing automobiles system. So, the world is investing on large scale for this optimistic future in-car automation [39]. So finally, in this paper—for the first time we try to provide a methodological framework in the existing system to have a proper efficient vehicle-to-vehicle (V2V) communication protocol for futuristic automated cars to avoid the collisions in VANETs. This could help to reduce the number of road accidents by providing early warning indications by efficient sensing and proper communication system, and it must able to reduce the latency for delivering emergency warnings indications in various road conditions in VANETs [38, 39]. Vehicle-to-vehicle (V2V) [40], vehicle-to-infrastructure (V2I) [41], and vehicle-to-everything (V2X) [42] technologies are three advanced communication areas which are needed to be analyzed separately to highlight in more detail the distinctiveness of each of them. Hence, it is a need to design vehicle communication management protocols using vehicle-to-vehicle communication to focus on the core issues of safety [43]. It is observed through research results from an international study on automobile automation industries that installation of the wireless environment at every cross point would be costly [44]. Hence, V2V-based methodology is becoming more reasonable in economic status for implementation [45].

Acknowledgements It gives us great pleasure to present the paper on car autonomous systems. I would like to take this opportunity to acknowledge my guide Prof. Arun Sonar (Limgaokar), HOD Instrumentation for giving me all the help and guidance that needed.

References

1. T. Litman, Autonomous vehicle implementation predictions implications for transport planning (2018)
2. T. Pino, S. Choudhury, F. Al-Turjman, Dominating set algorithms for wireless sensor networks survivability. IEEE Access **6**, 17527–17532 (2018); R. Hussain, S. Zeadally, Autonomous cars: research results, issues and future challenges. IEEE Commun. Surv. Tutorials (2018)
3. S.A. Bagloee, M. Tavana, M. Asadi, T. Oliver, Autonomous vehicles: challenges, opportunities, and future implications for transportation policies. J. Mod. Transp. **24**(4), 284–303 (2016). https://doi.org/10.1007/s40534-016-0117-3
4. M. Aazam, S. Zeadally, K. Harras, Fog computing as¸ architecture, evaluation, and future research directions. IEEE Commun. Mag. (2018) (in press)
5. Y.-J. Gong, E. Chen, X. Zhang, L.M. Ni, J. Zhang, Ant mapper: an ant colony-based map matching approach for trajectory-based applications. IEEE Trans. Intell. Transp. Syst. **19**(2), 390–401 (2017)

6. K. Miettinen (ed.), *Evolutionary Algorithms in Engineering and Computer Science: Recent Advances in Genetic Algorithms, Evolution Strategies, Evolutionary Programming, GE* (Wiley, New York, 1999). ISBN 0471999024
7. A review and analysis of literature on autonomous driving e-journal: making-of innovation. Making Innov. E-J. makingofinnovation.com, Oct 2015
8. F. Campioni, S. Choudhury, K. Salomaa, S.G. Akl, Improved recruitment algorithms for vehicular crowd sensing networks. IEEE Trans. Veh. Technol. J. Latex Class Files **14**(8) (2015). https://doi.org/10.1109/TVT.2018.2885403
9. N. El Zoghby, V. Loscri, E. Natalizio, V. Cherfaoui, Robot cooperation and swarm intelligence. IEEE **68**(2) ITVTAB (2019). ISSN 0018-9545
10. Driving the self-driving vehicle expanding the technological design horizon, in *2015 IEEE International Symposium on Technology in Society (ISTAS) Proceedings* (2015). ISBN: 978-1-4799-8282-0
11. NCSL—National Conference of State Legislatures, database on Autonomous Vehicles Legislation. Available from, http://www.ncsl.org/research/transportation/autonomous-vehicles-legislative-database.aspx. Query executed on 03/27/2017
12. X. Cai, J. Rodríguez-Piñeiro, X. Yin, N. Wang, B. Ai, G.F. Pedersen, A.P. Yuste, An empirical air-to-ground channel model based on passive measurements in LTE, 1140
13. H. Qin, Z. Meng, W. Meng, X. Chen, H. Sun, F. Lin, M.H. Ang Jr., Autonomous exploration and mapping system using heterogeneous UAVs and UGVs in GPS-denied environments. IEEE Trans. Veh. Technol. https://doi.org/10.1109/TVT.2018.2890416
14. W. Meng, Z. He, R. Su, P. Yadav, R. Teo, L. Xie, Decentralized multi-UAV flight autonomy for moving convoys search and track. IEEE Trans. Control Syst. Technol. **25**(4), 1480–1487 (2017)
15. T. Zeng, O. Semiari, W. Saad, M. Bennis, Joint communication and control for wireless autonomous vehicular platoon systems. CoRR (2018). abs/1804.05290
16. Z. Lian, L. Jiang, C. He, D. He, Channel models a non-stationary 3-D wideband GBSM for HAP-MIMO communication systems, 1128
17. J. Du, F.R. Yu, X. Chu, J. Feng, G. Lu, Resource allocation in vehicular networks based on dual-side cost minimization, 1079
18. M. Gonzalez-Martín, M. Sepulcre, R. Molina-Masegosa, J. Gozalvez, Analytical models of the performance of C-V2X mode 4 vehicular communications, 1155
19. S. Wei, Y. Zou, X. Zhang, T. Zhang, X. Li, An integrated longitudinal and lateral vehicle following control system with radar and vehicle-to-vehicle communication. IEEE Trans. Veh. Technol. https://doi.org/10.1109/TVT.2018.2890418
20. X. Wang, Z. Ning, X. Hu, L. Wang, B. Hu, J. Cheng, V.C.M. Leung, Optimizing content dissemination for real-time traffic management in large-scale internet of vehicle systems, 1093 (Smith Predictor)
21. D. Jo, B. Yu, H. Jeon, K. Sohn, Image-to-image learning to predict traffic speeds by considering area-wide spatio-temporal dependencies. https://doi.org/10.1109/TVT.2018.2885366
22. Advanced driver assistance systems. Vehicle automation IEEE 2019: collision avoidance intelligent vehicles system vehicle-to-everything
23. G. Du, W. Cao, S. Hu, Z. Lin, T. Yuan, Design and assessment of an electric vehicle power train model based on real-world driving and charging cycles, 1178
24. H. Peng, L. Liang, X. Shen, G.Y. Li, Connected vehicle series vehicular communications: a network layer perspective, 1064 (Computation Offloading)
25. S. Wei, Y. Zou, X. Zhang, T. Zhang, X. Li, An integrated longitudinal and lateral vehicle following control system with radar and vehicle-to-vehicle communication, 1116
26. Self-driving cars: the next revolution kpmg.com I cargroup.org CAR (car centre for automotive research)
27. M.Z. Hasan, F. Al-Turjman, H. Al-Rizzo, Analysis of cross-layer design of quality-of-service forward geographic wireless sensor network routing strategies in green internet of things. IEEE Access **6**, 20371–20389 (2018)

28. J.A. Guerrero-ibanez, S. Zeadally, J. Contreras-Castillo, Integration challenges of intelligent transportation systems with connected vehicle, cloud computing, and internet of things technologies. IEEE Wirel. Commun. **22**, 122–128 (2015)
29. L. Li, K. Ota, M. Dong, Human-like driving: empirical decision-making system for autonomous vehicles. IEEE Trans. Veh. Technol. (2017). https://doi.org/10.1109/TVT.2018.2822762
30. S. Abdelhamid, H.S. Hassanein, G. Takahara, Reputation-aware, trajectory-based recruitment of smart vehicles for public sensing. IEEE Trans. Intell. Transp. Syst. **99**, 1–14 (2017)
31. D. Jo, B. Yu, H. Jeon, K. Sohn, Image-to-image learning to predict traffic speeds by considering area-wide spatio-temporal dependencies, 1188
32. L.T. Tan, R.Q. Hu, Mobility-aware edge caching and computing in vehicle networks: a deep reinforcement learning. IEEE Trans. Veh. Technol. https://doi.org/10.1109/TVT.2018.2867191
33. A. Garcia-Garcia, S. Orts-Escolano, S. Oprea, V. Villena-Martinez, P. Martinez-Gonzalez, J. Garcia-Rodriguez, A survey on deep learning techniques for image and video semantic segmentation. Appl. Soft Comput. **70**, 41–65 (2018)
34. L. Xiao, R. Wang, B. Dai, Y. Fang, D. Liu, T. Wu, Hybrid conditional random field based camera-LIDAR fusion for road detection. Inf. Sci. **432**, 543–558 (2018)
35. F.M. Favarò, N. Nader, S.O. Eurich, M. Tripp, N. Varadaraju, Analysis of autonomous vehicles accidents and considerations on human-machine interactions. Presented at PSAM Topical 2017 on Human Reliability, Quantitative Human Factors, and Risk Management, Munich, Germany, 7th–9th June 2017
36. Autonomous cars: past, present and future—a review of the developments in the last century, the present scenario and the expected future of autonomous vehicle technology, Jan 2015 with 15,903. https://doi.org/10.5220/0005540501910198
37. Autonomous vehicles UF: automated guided vehicles autonomous cars autonomous driving autonomous trucks unmanned autonomous vehicle BT: autonomous systems intelligent vehicles RT: artificial intelligence mechatronics multi-agent systems vehicular automation NT: unmanned autonomous vehicles, by The Institute of Electrical and Electronics Engineers (IEEE) (2019)
38. Int. J. Inf. Commun. Comput. Technol., Jagan Institute of Management Studies, New Delhi Article, Jan 2016
39. Efficient vehicle to vehicle communication protocol for VANETs, in *Proceedings of 2014 RAECS UIET*, Panjab University Chandigarh, 06–08 Mar 2014
40. Overview of vehicular communications. Future Internet J., Published: 24 Jan 2019
41. F.J. Martin-Vega, B. Soret, M.C. Aguayo-Torres, I.Z. Kovacs, G. Gomez, Geolocation-based access for vehicular communications: analysis and optimization via stochastic geometry. IEEE Trans. Veh. Technol. **67**, 3069–3084 (2018)
42. A. Bazzi, C. Campolo, B.M. Masini, A. Molinaro, A. Zanella, A.O. Berthet, Enhancing cooperative driving in IEEE 802.11 vehicular networks through full-duplex radios. IEEE Trans. Wirel. Commun. **17**, 2402–2416 (2018)
43. M. Noor-A-Rahim, G.G.M.N. Ali, H. Nguyen, Y.L. Guan, Performance analysis of IEEE 802.11p safety message broadcast with and without relaying at road intersection. IEEE Access **6**, 23786–23799 (2018)
44. IEEE, IEEE standard for information technology—telecommunications and information exchange between systems—local and metropolitan area networks—specific requirements. Part 15.4: Wireless medium access control (MAC) and physical layer (PHY) specifications for low-rate wireless personal area networks (WPANs). Technical Report, IEEE, Piscataway, NJ, USA (2006)
45. A. Bazzi, G. Cecchini, A. Zanella, B.M. Masini, Study of the impact of PHY and MAC parameters in 3GPP C-V2V mode 4. IEEE Access **6**, 71685–71698 (2018)
46. J. Zhao, B. Liang, Q. Chen, The key technology toward the self-driving car. Int. J. Intell. Unmanned Syst. https://doi.org/10.1108/IJIUS-08-2017-0008
47. C. Wang, Y. He, F.R. Yu, Q. Chen, L. Tang, Integration of networking, caching and computing in wireless systems: a survey, some research issues, and challenges. IEEE Commun. Surv. Tutorials **20**(1), 7–38 (2018)

48. H. Xing, J. Ploeg, H. Nijmeijer, Compensating for vehicle actuator delays in cooperative ACC systems, 1106
49. Y. Shin, K. Hwang, J. Park, D. Kim, S. Ahn, Communication circuits and systems precise vehicle location detection method using a wireless power transfer (WPT) system. Transp. Syst. 1167
50. Y. Guo, Q. Yang, F.R. Yu, V.C.M. Leung, Cache-enabled adaptive video streaming over vehicular networks: a dynamic approach. IEEE Trans. Veh. Tech., Mar 2018 (accepted)

Diagnosis and Medicine Prediction for COVID-19 Using Machine Learning Approach

M. Rudra Kumar, Rashmi Pathak, and Vinit Kumar Gunjan

Abstract The COVID-19 outbreak has thrown the entire world into an unanticipated unpleasant scenario, bringing the lives of people all over the world to a pandemic level and claiming thousands of lives. According to WHO, COVID-19 has spread to 220 nations and territories, with the number of infected cases and deaths reaching 167 million and 3 million (as of May 25, 2021) (WHO dashboard, https://covid19.who.int [1]) and has a serious impact on the public health system. The key hurdles in containing the present COVID-19 outbreak are early detection and diagnosis. As a result, it is critical to screen COVID-19-affected individuals as soon as possible. Otherwise, it will spread quickly. In this case, screening can determine whether or not a patient has COVID-19 pneumonia. One of the most effective methods for reaching this goal is through chest X-ray diagnosis. It is the one that is most easily detected. This study provided a deep learning model of a convolutional neural network solution for detecting COVID-19 pneumonia patients using chest X-ray pictures. We used a publicly accessible chest X-ray dataset from Kaggle (Dataset link, https://www.kaggle.com/tawsifurrahman/covid19-radiography-database [2]) to train the model, which included 10,006 photos with COVID-19 pneumonia and normal images separated into train, test, and validation sets. This proposed model has a classification precision of 96% on the test set and 97% on the validation set, which is rather good for classifying COVID-19 pneumonia and normal patients. Along with this, we add the functionality of suggesting medicine based on symptoms. We build this suggesting model using machine learning model with accuracy 79% of 120 rows of dataset developed by our own to show it is possible to suggest medicine based on symptoms. We have developed an application using the flask framework. This application may be used on any computer by any medical professional to detect COVID positive and negative patients automatically using chest X-ray images in a matter of seconds and

M. Rudra Kumar (✉)
Department of CSE, GPCET, Kurnool, Andhra Pradesh, India

R. Pathak
Department of CSE, National Institute of Technology, Silchar, Silchar, India

V. K. Gunjan
CMR Institute of Technology, Hyderabad, Telangana, India

A. Kumar et al. (eds.), *Computational Intelligence in Machine Learning*, Lecture Notes in Electrical Engineering 834, https://doi.org/10.1007/978-981-16-8484-5_10

recommends some medicine that currently threatens the COVID-19. This application can reduce the number of false positives and false negatives in the detection of COVID-19.

Keywords COVID-19 · Pneumonia · CNN model · Chest X-ray · Medicine suggestion

1 Introduction

Coronavirus disease-2019, shortly called COVID-19, is an infectious disease. It is caused by a severe acute respiratory coronavirus 2 (SARS-CoV-2). It is causing pandemics, whereas its first case had reported in Wuhan, China. This virus spreads through direct contact with people. It affects mainly the respiratory system, whereas it can affect other systems too. Its symptoms include fever, cough, cold, body pains, and breathing problems. It occurs in mild, moderate, and severe stages. Mild and moderate symptoms can recover from the disease. Severe conditions may lead to death. Thus, World Health Organization (WHO) released the precautions to maintain self-distance, wear masks, and many more.

COVID-19 affects the lungs and leads to pneumonia in a person. After affecting, it shows symptoms like dropping oxygen levels, breathing problems based on the severity of the disease. Pneumonia is an inflammatory illness of the human lung that affects the microscopic air sacs known as alveoli. Dry cough, fever, chest pain, and difficulty breathing are some of the symptoms. Pneumonia is caused by viruses or bacteria, as well as, less frequently, other microorganisms, certain drugs, or disorders such as autoimmune illnesses. Some risk factors for pneumonia-like chronic obstructive pulmonary disease include diabetes, asthma, a smoking history, and cough after a stroke. Symptoms and a physical examination are used to make a diagnosis. Blood tests, chest X-ray, and the sputum culture may help confirm the diagnosis of COVID-19 pneumonia. This disease may be classified by where it occurred, such as community or healthcare-associated or hospital-acquired pneumonia [3–6]. It also leads to cardiovascular disease [7].

The detection of opacities in chest X-ray images is challenging, even for experienced practitioners. In clinical labs, however, RT-PCR and antigen COVID-19 tests are utilized to diagnose the COVID-19. However, the disadvantage of these tests is that the turnaround times for COVID-19 findings range from 3 to more than 48 h and are costly. In general, not all rural parts of a country will have access to testing kits that provide results based on saliva or a samples obtained from the nose. We combine artificial intelligence, deep learning, and computer vision techniques to automate diagnosis, which have been widely used as adjunct tools in pneumonia diagnosis in recent years. X-rays, CT scans, and a variety of other medical procedures can be used to diagnose it. These images of afflicted and non-affected patients are fed into the models as input. To diagnose, the current effort employs artificial intelligence (AI) and deep learning technologies. Artificial intelligence (AI) is critical in establishing and improving health-care systems around the world [8]. It has been increasing research toward solving complex problems, including engineering [9–13], medicine [14–18], economy [19], and psychology [20].

Here, in the proposed system, diagnosis of COVID-19 can be made using convolution neural networks (CNNs), which have two parts, i.e., feature extraction and fully connected layers for classification, which are used to extract the features in multiple hidden layers. Based on the CNN model, the screening of COVID-19 is processed and makes a prediction. Also, we add the functionality of medicine prediction based on symptoms data is created using a machine learning model called decision tree. Decision tree is used chiefly machine learning algorithm for classification of structured data. The main objective of this project is to detect COVID-19 pneumonia using chest X-ray images. Here, the Web application takes the patient image and returns the output as if the patient is affected with COVID-19 pneumonia or not. Another main objective is to give medicine prediction based on the symptoms and seriousness of the condition.

The rest of this paper is organized as follows: Sect. 2 explains related work of AI techniques to solve the problem. Section 3 explains the material and proposed model with an algorithm. Section 4 explains the experiment results are discussed and the conclusion in Sect. 5.

2 Related Work

In this related work, we discuss the importance of AI techniques in medical imaging or diagnosis.

Computer vision has found widespread application in medical diagnosis. It is mostly employed in medical fields that demand visual requirements and image diagnosis through the use of various AI algorithms. A computer vision algorithm is used to determine whether a skin abnormality is an early warning sign of skin cancer. Computer vision employs a variety of medical imaging techniques, including chest X-ray (CXR), positron emission tomography scan (PET scan), magnetic resonance imaging (MRI), ultrasound, and CT scan images. In Python, computer vision is one of the AI techniques, and we use it with the OpenCV library.

According to several research, medical images of various ailments aid in the detection of viruses in the lungs. Deep learning-based models have been created in numerous works to identify pneumonia [21, 22], other forms of thoracic disorders [23–25], skin cancer [26], and other diseases from medical photos. Some of these experiments produced promising results using relatively simple CNN architecture [26].

Feature or knowledge extraction from medical imaging data can help detect diseases in the early stages [27]. Many researchers worked on medical images using the convolutional neural network (CNN) algorithm. A definite type of improved neural network for image classification is called convolutional neural network. Some researchers work related to COVID-19 patient detection using chest X-ray images [28–33]. Regarding these works, we are going to build a convolutional neural network model for classification. Here, we seek to prove that applying a deep learning algorithm in the early detection of COVID-19 infection is useful and determines whether the algorithm could develop an accurate deep learning model to help medical personnel to identify COVID-19 patients by using chest X-ray images.

In this paper, the CNN was constructed and trained from scratch, rather than using a pre-trained model for applying transfer learning. This study aims to give the maximum performance of the proposed convolutional neural network to classify chest X-ray images. A convolutional neural network (CNN) model is applied to the dataset of chest X-ray images collected from Kaggle [2] to build a model that detects COVID-19 pneumonia or not. With CNN classification, we also add medicine prediction functionality. This functionality is going to give medicine prediction based on the patient's symptoms.

3 Materials and Methodology

3.1 Dataset Used

We gathered the images for this study from a publicly available dataset on Kaggle. In Kaggle, this dataset is known as COVID-19 radiography database (winner of the COVID-19 dataset award by Kaggle community). In Kaggle, a team of researchers collaborated with medical specialists to construct a database of chest X-ray images for COVID-19 positive cases, as well as normal and viral pneumonia images. The dataset is skewed. As a result, we balanced the dataset and obtained 10,000 photos of the COVID-19 club with viral pneumonia and normal sets. Because our primary goal is to analyze the lungs, this data comprise of frontal chest X-ray images of various patients. For class 0, 5046 images are identical to the image shown in Fig. 1 (left), and for class 1, 4960 images are similar to the images shown in Fig. 1 (right). We balanced the dataset into 7898 train image sets, 1912 test image sets, and 196 validation image sets. For medicine suggestion, a dataset contains cough, cold, fever, pains, and breathing problem is prepared with around 120 rows.

3.2 Methodology

This section describes the CNN model, which includes the phases used for COVID-19 detection. The steps are shown in Fig. 2 and discussed hereafter.

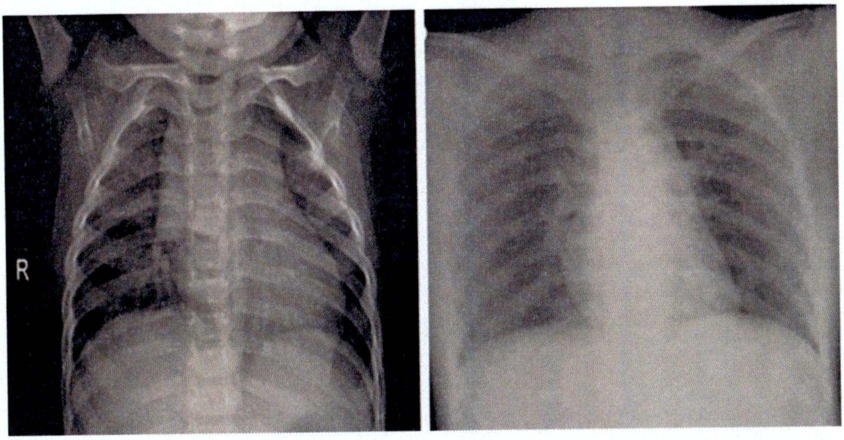

Fig. 1 Chest X-ray images of normal patient (left) and COVID-19 pneumonia patient (right)

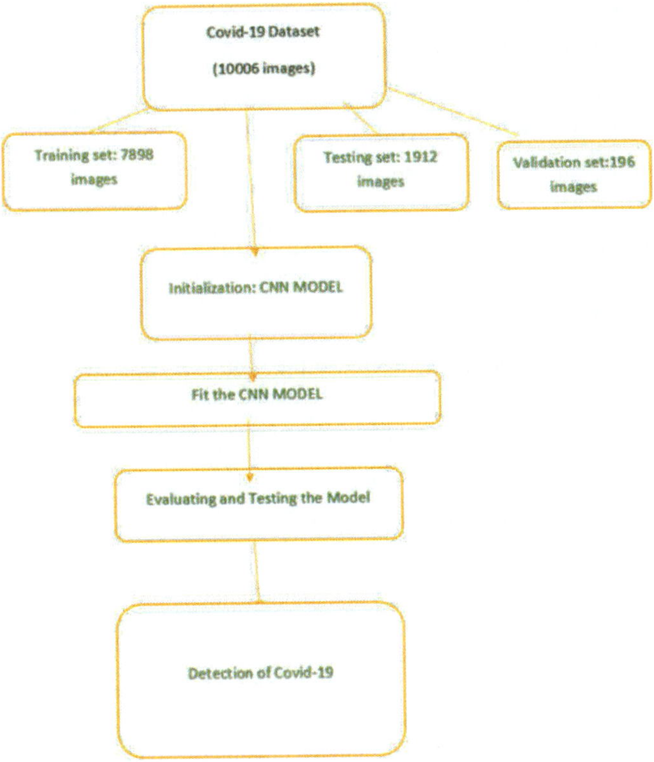

Fig. 2 The CNN prediction model for COVID-19 pneumonia

1. Dataset

   - The collected chest X-ray images dataset contains 10,006 X-ray images divided into 7898 images for training, 1912 for testing, and 196 for validation. The X-ray images were collected from the Kaggle Website [2]. Next, importing the necessary keras libraries in [34].

2. Initialization

   - The simple sequential model is used here, beginning with two convolutional layers of kernel size (7, 7) and Max pooling with pool size (3, 3), then two convolutional layers of kernel size (5, 5) and the same pool size, and finally two convolutional layers of kernel size (3, 3) with Max pooling and pool size (3, 3). (2, 2) grayscale image input size (224 * 224) is used, and the activation function ReLU is used.
   - Adding Flattening: converts the matrix into a single array.
   - Adding full connection (128 final layer of outputs, activation = ReLU and dense layer, activation = Softmax).

3. Fit the CNN Model

   - We fit the model using a fit method with batch size = 32 and epochs = 100 to train the model and get accuracy around 96%, which is pretty good.

4. Evaluating and Testing the Model

   - In evaluation, we check the loss and accuracy of epochs and generate the confusion matrix and classification report for test and validation sets.

Proposed algorithm

The algorithm that was used to implement the proposed model is detailed further below.

Step 1: Load the required libraries, such as keras and OpenCV.
Preprocess the image, for example, image = X.

(I) Resize the image (X) to (224, 224, 1)
(II) Range of random rotation = 10
(III) Range of random rotation = 10
(IV) Horizontal flip = false
(V) Zoom range = 0.1

Step 2: Apply the preprocessed image to a CNN model input.
Step 3: Retrieve the model's last convolution layer's output.
Step 4: Flatten the dimensions of the final convolution layer.
Step 5: Create a dense layer with unit = 128.
Step 6: Execute the ReLU activation function.
Step 7: For inference, apply a dense layer.
Step 8: Use Softmax to classify your data.

Step 9: Once the model is initialized, fit the training data into the model to build the proposed model.
Step 10: Evaluate the model with accuracy and loss of model based on epochs and generate confusion matrix and classification report on test and validation data.

We fit the symptoms dataset into a decision tree algorithm for medicine suggestion and achieved 79% accuracy.

The following section shows the evaluation part of the model.

4 Results and Discussion

An exact 10,006 chest X-ray image dataset was used to train and test the model in this experiment, with 7898 pictures for training, 1912 images for testing, and 196 images for validation. This model was trained using 3995 normal people's chest X-rays and 3903 infected people's chest X-rays, and it was tested on 1051 normal people's chest X-rays and 1057 infected people's chest X-rays. The dataset is eventually cycled through the CNN model several times in order to improve the learning process. Initially, 50 epochs were employed for training. The data showed that roughly 96% of the time was correct, and 11% of the time was incorrect. Figure 3 shows the trained model's accuracy and loss.

Now, we evaluate the test and validation data on the trained model using the confusion matrix and classification report. We get 96% of $F1$-score on test data and 97% on validation data in the classification report. Here, $F1$-score is nothing but calculating model accuracy on new data on the trained model. Figure 4 shows the confusion matrix of test and validation data. Table 1 shows the results of the classification report of the test, and Table 2 for validation data that consist of precision, recall, and $F1$-score.

A simple desktop application using flask to detect COVID-19 positive and negative cases has been developed based on the proposed model. The application can show in Fig. 5. Along with the proposed model, we add the functionality of suggesting medicine to COVID-19 patients based on symptoms shown in Fig. 6.

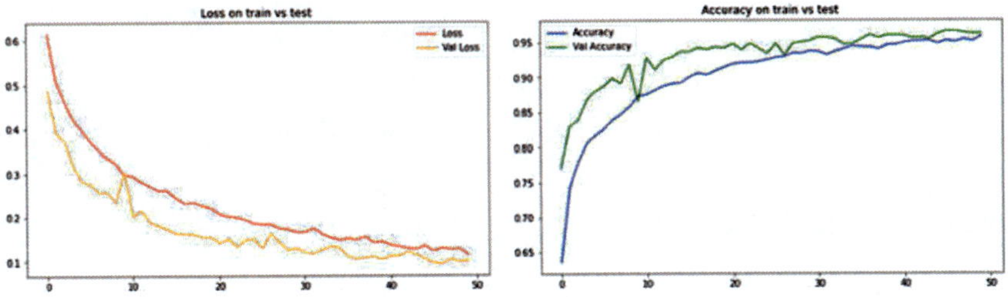

Fig. 3 Accuracy and loss of trained CNN model

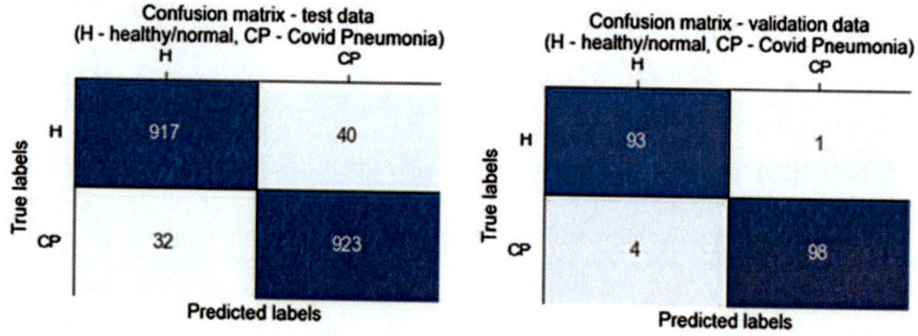

Fig. 4 Confusion matrix of test data (left) and validation data (right)

Table 1 Classification report on test data

	Precision	Recall	$F1$-score	Support
0	0.97	0.96	0.96	957
1	0.96	0.97	0.96	955
Accuracy				1912
Macro avg	0.96	0.96	0.96	1912
Weighted avg	0.96	0.96	0.96	1912

Table 2 Classification report on validation data

	Precision	Recall	$F1$-score	Support
0	0.96	0.99	0.97	94
1	0.99	0.96	0.98	102
Accuracy				196
Macro avg	0.97	0.98	0.97	196
Weighted avg	0.97	0.97	0.97	196

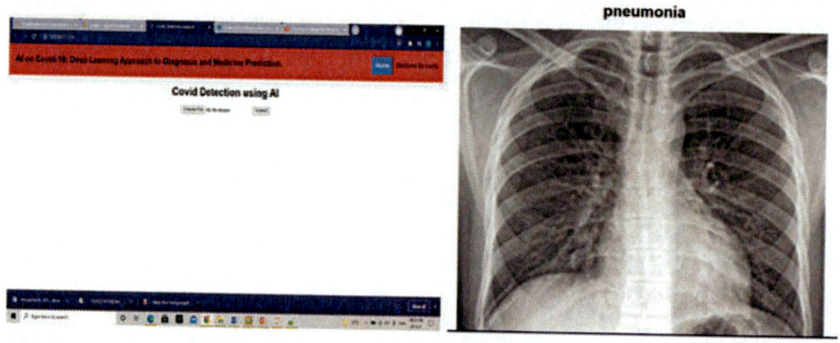

Fig. 5 Desktop application of proposed model

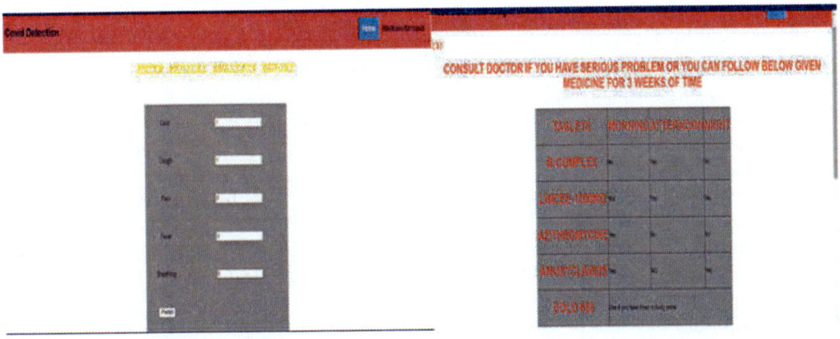

Fig. 6 Suggesting medicine based on symptoms

5 Conclusion and Future Work

In conclusion, this study demonstrates a very accurate artificial intelligence method for automatically identifying COVID-19 patients, which could be useful and successful in combating the current COVID-19 outbreak. In the absence of a radiologist or physical tests, we suggested the CNN model, which has an excellent score for the accuracy of detecting chest X-rays and can be utilized as an effective AI tool for the quick detection of infectious diseases such as COVID-19 pneumonia.

In future investigations, we hope to create CNN models with more dataset images to detect infections without producing false results and to recommend treatments for more diseases if the dataset is collected.

References

1. WHO dashboard, https://covid19.who.int
2. Dataset link, https://www.kaggle.com/tawsifurrahman/covid19-radiography-database
3. C. Huang et al., Clinical features of patients infected with 2019 novel coronavirus in Wuhan, China. Lancet **395**(10223), 497–506 (2020)
4. N. Chen, M. Zhou, X. Dong, J. Qu, F. Gong, Y. Han, Y. Qiu, J. Wang, Y. Liu, Y. Wei, J. Xia, T. Yu, X. Zhang, L. Zhang, Epidemiological and clinical characteristics of 99 cases of 2019 novel coronavirus pneumonia in Wuhan, China: a descriptive study. Lancet **395**(10223), 507–513 (2020)
5. D. Wang, B. Hu, C. Hu, F. Zhu, X. Liu, J. Zhang, B. Wang, H. Xiang, Z. Cheng, Y. Xiong, Y. Zhao, Clinical characteristics of 138 hospitalized patients with 2019 novel coronavirus–infected pneumonia in Wuhan, China. JAMA **323**(11), 1061–1069 (2020)
6. K. Liu, Y.-Y. Fang, Y. Deng, W. Liu, M.-F. Wang, J.-P. Ma, W. Xiao, Y.-N. Wang, M.-H. Zhong, C.-H. Li, G.-C. Li, H.-G. Liu, Clinical characteristics of novel coronavirus cases in tertiary hospitals in Hubei province. Chin. Med. J. **133**(9), 1025–1031 (2020)
7. T. Guo et al., Cardiovascular implications of fatal outcomes of patients with coronavirus disease 2019 (COVID-19). JAMA Cardiol., Early Access (2020). https://doi.org/10.1001/jamacardio.2020.1017
8. P. Hamet, J. Tremblay, Artificial intelligence in medicine. Metabolism **69**, S36–S40 (2017)

9. M. Jamshidi, A. Lalbakhsh, S. Lotfi, H. Siahkamari, B. Mohamadzade, J. Jalilian, A neuro-based approach to designing a Wilkinson power divider. Int. J. RF Microw. Comput.-Aided Eng. **30**(3), Art. no. e22091 (2020)
10. M. Jamshidi, A. Lalbakhsh, B. Mohamadzade, H. Siahkamari, S.M.H. Mousavi, A novel neural-based approach for the design of microstrip filters. AEU-Int. J. Electron. Commun. **110**, Art. no. 152847 (2019)
11. M.R. Kumar, V.K. Gunjan, Review of machine learning models for credit scoring analysis. Ingen. Solid. **16**(1) (2020)
12. D.J. Reddy, M.R. Kumar, Crop yield prediction using machine learning algorithm, in *2021 5th International Conference on Intelligent Computing and Control Systems (ICICCS)* (IEEE, 2021)
13. M.B. Jamshidi, N. Alibeigi, A. Lalbakhsh, S. Roshani, An ANFIS approach to modelling a small satellite power source of NASA, in *Proceedings of IEEE 16th International Conference on Networking, Sensing and Control (ICNSC)*, May 2019, pp. 459–464
14. Y. Mintz, R. Brodie, Introduction to artificial intelligence in medicine. Minim. Invasive Ther. Allied Technol. **28**(2), 73–81 (2019)
15. R.B. Parikh, Z. Obermeyer, A.S. Navathe, Regulation of predictive analytics in medicine. Science **363**(6429), 810–812 (2019)
16. V.K. Gunjan, M.R. Kumar, Predictive analytics for OSA detection using non-conventional metrics. Int. J. Knowl.-Based Org. (IJKBO) **10**(4), 13–23 (2020)
17. A. Becker, Artificial intelligence in medicine: what is it doing for us today? Health Policy Technol. **8**(2), 198–205 (2019)
18. N.J. Schork, Artificial intelligence and personalized medicine, in *Precision Medicine in Cancer Therapy* (Springer, Cham, 2019), pp. 265–283
19. M.B. Jamshidi, M. Gorjiankhanzad, A. Lalbakhsh, S. Roshani, A novel multiobjective approach for detecting money laundering with a neuro-fuzzy technique, in *Proceedings of IEEE 16th International Conference on Networking, Sensing and Control (ICNSC)*, May 2019, pp. 454–458
20. M.B. Jamshidi, N. Alibeigi, N. Rabbani, B. Oryani, A. Lalbakhsh, Artificial neural networks: a powerful tool for cognitive science, in *Proceedings of IEEE 9th Annual Information Technology, Electronics and Mobile Communication Conference (ICON)*, Nov 2018, pp. 674–679
21. V. Chouhan, S. Singh, A. Khamparia, D. Gupta, P. Tiwari, C. Moreira, R. Damaeviuus, V. Albuquerque, A novel transfer learning-based approach for pneumonia detection in chest X-ray images. Appl. Sci. **10**, 559 (2020)
22. A. Jaiswal, P. Tiwari, S. Kumar, D. Gupta, A. Khanna, J.J. Rodrigues, Identifying pneumonia in chest X-rays: a deep learning approach (2019)
23. P. Rajpurkar, J. Irvin, K. Zhu, B. Yang, H. Mehta, T. Duan, D. Ding, A. Bagul, C. Langlotz, K. Shpanskaya et al., Chexnet: radiologist-level pneumonia detection on chest X-rays with deep learning (2017)
24. X. Wang, Y. Peng, L. Lu, Z. Lu, M. Bagheri, R.M. Summers, Chestx-ray8: hospital-scale chest X-ray database and benchmarks on weakly supervised classification and localization of common thorax diseases, in *Proceedings of the IEEE Conference on Computer Vision and Pattern Recognition* (2017), pp. 3462–3471
25. E. Ranjan, S. Paul, S. Kapoor, A. Kar, R. Sethuraman, D. Sheet, Jointly learning convolutional representations to compress radiological images and classify thoracic diseases in the compressed domain, in *Proceedings of the 11th Indian Conference on Computer Vision, Graphics and Image Processing* (2018)
26. E. Andre, K. Brett, A. Roberto et al., Dermatologist-level classification of skin cancer with deep neural networks. Nature **542**(7639), 115–118 (2017)
27. A.H. Shurrab, A.Y.A. Maghari, Blood diseases detection using data mining techniques, in *2017 8th International Conference on Information Technology (ICIT)* (2017), pp. 625–631
28. A. Makris, I. Kontopoulos, K. Tserpes, COVID-19 detection from chest X-ray images using deep learning and convolutional neural networks. medRxiv (2020). Accessed 29 Mar 2020

29. L. Wang, Z.Q. Lin, A. Wong, COVID-Net: a tailored deep convolutional neural network design for detection of COVID-19 cases from chest X-ray images. Sci. Rep. **10** (2020)
30. R. Jain, M. Gupta, S. Taneja, D.J. Hemanth, Deep learning-based detection and analysis of COVID-19 on chest X-ray images. Appl. Intell. 1–11 (2020). Accessed 29 Mar 2020
31. S. Minaee, R. Kafieh, M. Sonka, S. Yazdani, G.J. Soufi, Deep-COVID: predicting COVID-19 from chest X-ray images using deep transfer learning. Med. Image Anal. **65**, 101794 (2020)
32. A.I. Khan, J. Shah, M. Bhat, CoroNet: a deep neural network for detection and diagnosis of COVID-19 from chest X-ray images. Comput. Methods Programs Biomed. **196**, 105581 (2020)
33. T. Ozturk, M. Talo, E.A. Yildirim, U.B. Baloglu, O. Yildirim, U.R. Acharya, Automated detection of COVID-19 cases using deep neural networks with X-ray images. Comput. Biol. Med. **121**, 103792 (2020)
34. Keras documentation, https://keras.io/

Automated Guided Vehicle Robot Localization with Sensor Fusion

Marvin Dares, Kai Woon Goh, Ye Sheng Koh, Che Fai Yeong, Eileen L. M. Su, and Ping Hua Tan

Abstract Robot localization is vital for the operation of an automated guided vehicle (AGV) but is susceptible to problems such as wheel slip. With more sensors fused together, the more environmental information can be collected by the AGV, which helps with the localization of AGV. Inertial measurement unit (IMU) and global positioning unit (GPS) are usually implemented to improve robot localization but are susceptible to noise and are effective outdoors. Indoors, however, are more suitable with light detection and ranging (lidar) device. This paper implements extended Kalman filter (EKF) and unscented Kalman filter (UKF) for robot localization on AGV. AGV localization was tested with EKF and UKF on three different test tracks with different turn conditions. The performance of the EKF and UKF was compared to each other. Different sensors were implemented along with sensor fusion. UKF generates better odometry estimation than EKF with 24.07% better accuracy. With the usage of lidar, wheel slip was compensated.

Keywords Automated guided vehicle · AGV · Robot localization · Sensor fusion

---

M. Dares (✉) · K. W. Goh · Y. S. Koh · C. F. Yeong · E. L. M. Su
School of Electrical Engineering, Universiti Teknologi Malaysia, Johor Bahru, Malaysia
e-mail: dmarvin2@live.utm.my

K. W. Goh
e-mail: kwgoh4@live.utm.my

Y. S. Koh
e-mail: yskoh@graduate.utm.my

C. F. Yeong
e-mail: cfyeong@utm.my

E. L. M. Su
e-mail: eileensu@utm.my

P. H. Tan
DF Automation and Robotics Sdn. Bhd., Johor Bahru, Malaysia
e-mail: phtan@dfautomation.com

A. Kumar et al. (eds.), *Computational Intelligence in Machine Learning*, Lecture Notes in Electrical Engineering 834, https://doi.org/10.1007/978-981-16-8484-5_11

1 Introduction

Automated guided vehicle (AGV) is an automated solution applied in a variety of industries. Navigation and path planning are challenging, and performance reliability is important [1]. They also stated the importance of sensor fusion which is reduction in uncertainty, increase in accuracy and reliability, extended spatial and temporal coverage, improved resolution, and reduce system complexity. Kalman filter is commonly used to combine data from multiple sensor in real time [2]. With multiple sensors, robot perception can be improved, hence better interaction with environment and localization [3].

Extended Kalman filter (EKF) is a step above classical Kalman filter. EKF handles nonlinearities by linearizing the system using Jacobian. EKF was utilized for mobile robot localization and can handle nonlinear model but prone to divergence under bad initial estimate and noise conditions [4]. EKF is also known to have false triggering and inaccuracies on highly unpredictable input [5] and the EKF can only work with differentiable system. Light detection and ranging (lidar), inertial measurement unit (IMU), and camera were fused with EKF in for enhanced pose estimation of unmanned aerial vehicle (UAV) [6]. Lidar was used as it is effective indoors for pose estimation while global positioning unit (GPS) is only effective outdoors.

Unscented Kalman filter (UKF) utilized unscented transformation where sigma points are passed through a nonlinear function, yielding transformed sets of points to compute the mean and covariance. UKF was applied in localization of mobile robot to filter wheel slip errors on encoder data by sensor fusion and obtained accurate positioning of robot when compared with an accurate laser sensor result in linear motion only [7] as the laser sensor does not perform well with rotational motion. For mobile robot localization, UKF could achieve better estimation accuracy than EKF.

2 Methodology

The AGV [8] used in this paper is a two wheeled differential drive mobile robot as shown in Fig. 1 with additional line follower module, IMU, and lidar sensors. Arduino Uno uses ATmega328 while Arduino Mega uses ATmega2560 which provides more available input output (I/O) pins and communication ports. Currently, for main controller, Arduino Mega as shown in Fig. 1, was used instead of the previous Arduino Uno to accommodate more I/O connections and the added LSA08 line follower module. The line follower module is a phototransistor-based sensor array meant to detect light reflection contrast between bright and dark surface, enabling the distinction of lines on surface so the AGV may track. There are sensors on the AGV to monitor the AGV's performances. Those sensors consist of rotary encoders, current sensors, lidar, and IMU. For this paper, sensors used are focused on robot localization which is lidar, IMU, and encoder. LSM9DS1, a 9 degree of freedom (9-DOF) IMU, as shown in Fig. 1, was added to the AGV to provide orientation and

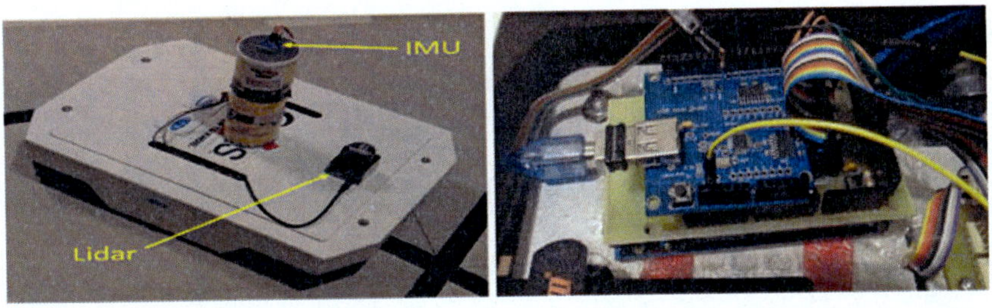

Fig. 1 On the left, the AGV used with added 9-DOF IMU and lidar and on the right, an Arduino Mega microcontroller with additional I/O connected to sensors

motion data. A 9-DOF IMU consists of accelerometer, gyroscope, and magnetometer which are used to generate orientation and motion data. The IMU was placed at an elevated position on the AGV to avoid electromagnetic interference from the motors which can affect magnetometer readings. The IMU communicates with the main controller using inter-integrated circuit (I2C) interface.

The lidar, also shown in Fig. 1, is a Hokuyo URG04LX-UG 01 and was added to provide measurement for position. Rotary encoder located at the back of the motor attached to the shaft of the motor to measure the angular velocity of the motor. The rotary encoders communicate with the main controller using quadrature encoder interface (QEI). Figure 2 shows the updated system architecture of the AGV. The data from sensors were passed to the microcontroller then to the Raspberry Pi via serial universal asynchronous receiver/transmitter (UART) for data processing and storage. Data processed through sensor fusion algorithm was done with Python programming.

Figure 2 shows the kinematic model used for the AGV. As described in [8], where it resembles that of Qbot from [9] and can be represented as equation:

$$\begin{bmatrix} \dot{x}(t) \\ \dot{y}(t) \\ \dot{\phi}(t) \end{bmatrix} = \begin{bmatrix} \cos\phi(t) & 0 \\ \sin\phi(t) & 0 \\ 0 & 1 \end{bmatrix} \begin{bmatrix} v(t) \\ \omega(t) \end{bmatrix} \quad (1)$$

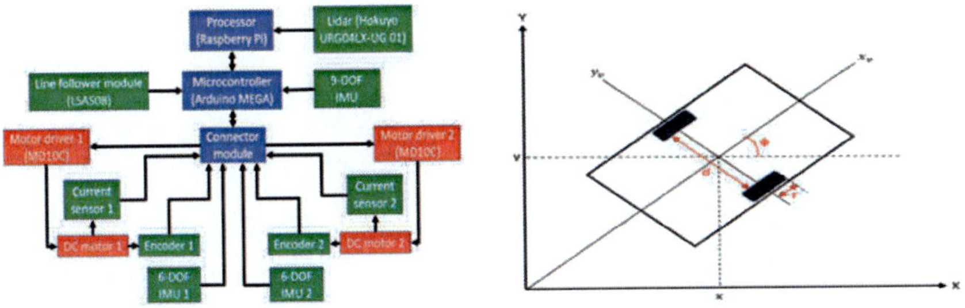

Fig. 2 On the left, the updated system architecture of AGV and on the right, the kinematic model of the AGV

where (x, y) is the center position of wheel axel and ϕ is the AGV's heading. $v(t)$ and $\omega(t)$ are represented by linear velocity and angular velocity, respectively, which are derivable using equation from [10]:

$$v(t) = \frac{r_R}{2}\omega_R(t) + \frac{r_L}{2}\omega_L(t) \tag{2}$$

$$\omega(t) = \frac{r_R}{d}\omega_R(t) + \frac{r_L}{d}\omega_L(t) \tag{3}$$

where radius of the right and left wheel is represented by r_R and r_L, respectively. Angular velocity of right and left wheel, ω_R and ω_L can be calculated using

$$\omega = \frac{2\pi n}{Nt} \tag{4}$$

where n, N, and t are the number of pulses, number of pulses per rotation, and sampling time, respectively. Thus, mathematical model for encoder is

$$\begin{bmatrix} x(k+1) \\ y(k+1) \\ \phi(k+1) \end{bmatrix} = \begin{bmatrix} x(k) \\ y(k) \\ \phi(k) \end{bmatrix} + \begin{bmatrix} \cos\phi(k) & 0 \\ \sin\phi(k) & 0 \\ 0 & 1 \end{bmatrix} \begin{bmatrix} v \\ \omega \end{bmatrix} \tag{5}$$

$$u = \begin{bmatrix} v \\ \omega \end{bmatrix} \tag{6}$$

where T is sampling time. The discrete kinematic equation is used to find the trajectory of the AGV. The state matrix x for the Kalman filters will be based on the kinematic equation. Control input is matrix u which consists of forward velocity and heading angle data from AGV. With IMU, AGV velocity, v can be calculated using Eq. (7) and heading angle ϕ using Eqs. (8) and (9).

$$v(k+1) = v(k) + Tax \tag{7}$$

$$\phi = \arctan 2(mx, my) \tag{8}$$

$$\text{Heading, } \phi \begin{cases} \phi - 2\pi, & \text{if } \phi > \pi \\ \phi + 2\pi, & \text{if } \phi < \pi \end{cases} \tag{9}$$

where ax is the accelerometer output in x-axis, magnetometer output in x-axis and y-axis is represented by mx and my, respectively. Heading angle is normalized using (9). Roll and pitch will not be used as the robot localization of this AGV does not involves roll and pitch. Hence, using (7)–(9), the mathematical model for IMU can be described as

$$\begin{bmatrix} x(k+1) \\ y(k+1) \\ \phi(k+1) \end{bmatrix} = \begin{bmatrix} x(k) \\ y(k) \\ 0 \end{bmatrix} + \begin{bmatrix} \cos\phi(k) & 0 \\ \sin\phi(k) & 0 \\ 0 & 1 \end{bmatrix} \begin{bmatrix} v \\ \phi \end{bmatrix} \tag{10}$$

$$u = \begin{bmatrix} v \\ \phi \end{bmatrix} \tag{11}$$

where the input u is velocity and heading angle from IMU. For lidar, the mathematical model will be

$$\begin{bmatrix} x(k+1) \\ y(k+1) \\ \phi(k+1) \end{bmatrix} = \begin{bmatrix} x(k) \\ y(k) \\ \phi(k) \end{bmatrix} + \begin{bmatrix} \cos\phi(k) \\ \sin\phi(k) \\ 1 \end{bmatrix} \begin{bmatrix} \Delta x & \Delta y & \Delta\phi \end{bmatrix} \tag{12}$$

$$u = \begin{bmatrix} \Delta x & \Delta y & \Delta\phi \end{bmatrix} \tag{13}$$

where the input for this model is Δx and Δy which are the displacement in x-axis and y-axis, respectively, and $\Delta\phi$ which is the heading angle difference. Residual values are generated using Eqs. (14) and (15), respectively, for EKF and UKF.

$$y = z - h(x) \tag{14}$$

$$y = z - \mu_z \tag{15}$$

where μ_z is the sum of weight and measurement of sigma points, $h(x)$ is the Jacobian of the measurement and z is last measurement. There are three tracks tested for AGV trajectory in this paper. These tracks are lines for AGV to follow during navigation using the line follower module. Figure 3 shows track A, track B, and track C, respectively. Track A is a square loop and it only involves 4 right turns with large turn radius. Track B is a modified track A which involves 1 left turn and 4 right turns for AGV. The turnings on track B include turns with smaller turn radius than the ones found in track A and turns with large turn radius. Track C is a figure 8 shaped track where there are 2 left turns and 2 right turns with small turn radius. These tracks are used instead of a straight line or a curve line track to introduce the various turn conditions that an AGV might encounter in a practical environment. Track A being

(a) (b) (c)

Fig. 3 **a** Track A, **b** track B, and **c** track C

the simplest, followed by track B, and finally track C as the most challenging track for the AGV. These test tracks are indoors with constant room lighting, since lidars are effective indoors but are sensitive to illumination and light changes [9].

3 Experiments and Result

Figure 4 shows AGV trajectory on track A and B, respectively, with encoder only considered in sensor fusion algorithm to test and compare EKF and UKF. Blue line represents odometry estimate generated by UKF, while the red line represents odometry estimate generated by EKF. Orange dotted line represents actual track. On track A, there are deviations toward right starting from the second turn. Same applies on track B except the deviations are more obvious as there are more turns involved in track B. These deviations during right turn are likely due to wheel slip. Observation on odometry from track B suggests that greater deviations are generated from turn with smaller turn radius. The deviations of these Kalman filter estimates can be measured by the residual generated by these algorithms. Figure 5 shows the residual generated by both EKF and UKF algorithm for track A and track B, respectively. For both track, residual values generated by UKF is less than EKF. However, on track B, residual values generated are higher than on track A. Due to track B having more turns and involves both left and right turn, more worked is required by the Kalman filters to adapt to the changes in AGV trajectory, causing more deviations

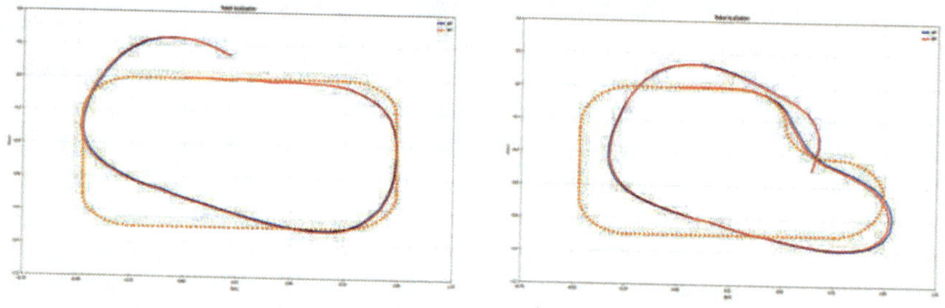

Fig. 4 From left to right, AGV estimated trajectory for track A and track B, respectively

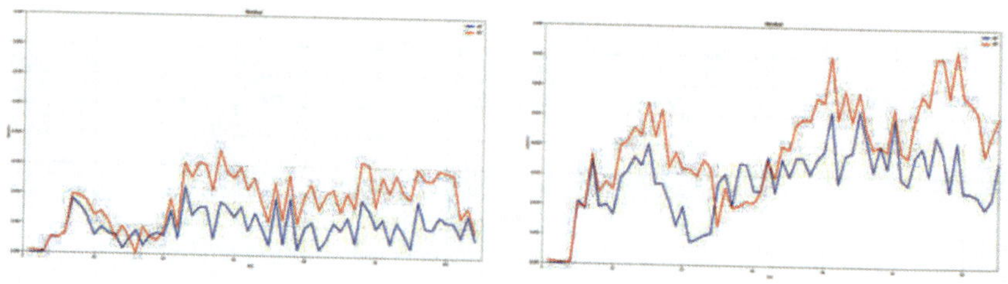

Fig. 5 From left to right, residual of Kalman filter comparison for track A and track B, respectively

in estimations. The average of residual for track A is 0.00469 and 0.00888 for UKF and EKF, respectively, whereas for track B, the average residual value generated is 0.011956 and 0.016949 for UKF and EKF, respectively. Residual values generated from UKF 30.87% less than EKF on track A and 17.27% less on track B. Based on the residual values from both tracks, UKF is 24.07% more accurate than EKF.

Figure 6 shows the trajectory of AGV on track A, track B, and C, respectively, with IMU and lidar output considered in the sensor fusion algorithm. Blue line represents odometry estimate generated by UKF while the red line represents odometry estimate generated by EKF. Black line, green line, and magenta line represent encoder, IMU output, and lidar output, respectively. Lidar was used as the measurement, z since it gives the most accurate indoor odometry data. Orange dotted line represents actual track. Based on observation on the trajectory of AGV, deviations of Kalman filter estimations are least on track A, followed by track B then C. On track B and C, compared to track A, there were turns with smaller turn radius which caused the turning to be harder and challenging for the AGV. Harder turn with smaller turn radius leads to greater deviations on odometry plotting, and this is most apparent on odometry plotting with encoder data, where deviations were caused by wheel slip at each turn made by AGV. For IMU, however, deviations were caused by inherent susceptibility of IMU to noise and drifts over time problem. The drift also worsens at every turn where the steering angle was greater than or equal 90°. However, IMU output has influence on the output of the Kalman filter estimation and can potentially influence to improve robot localization, provided the IMU output can be improved. IMU drifts are expected on cheap MEMS-based IMU. True for both with and without IMU, the residual of UKF is lesser than EKF. Figure 7 shows the generated residual values by sensor fusion on track A, B, and C, respectively. Based on the residual values generated, UKF generates lesser residual values than EKF, indicating that UKF generates more accurate estimations compared to EKF. The average residual values generated by both UKF and EKF are least on track A followed by track B then C, which corresponds to the deviations as observed on odometry plotting.

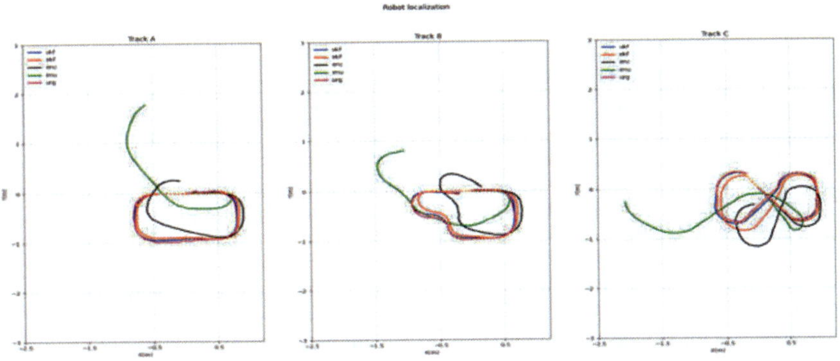

Fig. 6 AGV estimated trajectory with lidar and IMU for track A, track B, and track C

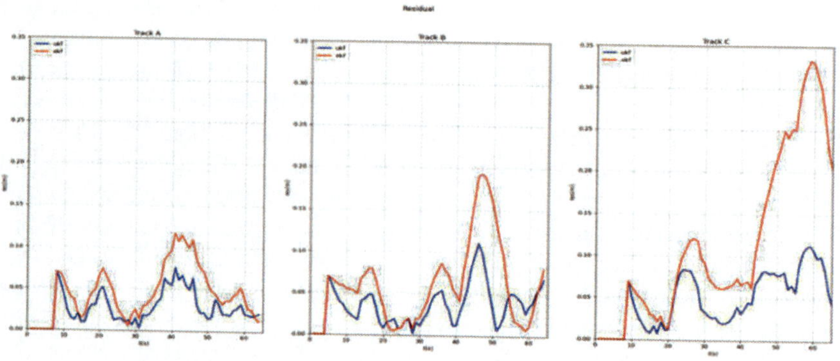

Fig. 7 Residual of Kalman filter with lidar and IMU comparison for track A, track B, and track C

4 Conclusion

All EKF and UKF have been tested on the AGV. In this paper, UKF performs better than EKF, generating better estimation for AGV trajectory. Deviations in odometry data on wheel slip in encoder readings can be compensated with the usage of lidar sensor as reference.

Phototransistor-based line follower module has been added to AGV for autonomous navigation, and more sensors have been added to improve odometry data. However, the IMU used in this paper is susceptible to inaccuracies. For future work, perhaps using better quality IMU sensor could reduce these inaccuracies. A well calibrated and industrial grade IMU can improve quality of generated data from IMU, hence better estimation accuracies.

Acknowledgements This work is supported by Universiti Teknologi Malaysia Research grant [01M85, 4B402] and Collaborative Research in Engineering, Science and Technology center (CREST) R&D grant [T20C2-18]. This work is also a collaboration with DF Automation Sdn. Bhd.

References

1. M.B. Alatise, G.P. Hancke, A review on challenges of autonomous mobile robot and sensor fusion methods. IEEE Access **8**, 39830–39846 (2020). https://doi.org/10.1109/access.2020.2975643
2. Y. Qingmei, S. Jianmin, A data fusion method applied in an autonomous robot (2008)
3. S. Jusoh, S. Almajali, A systematic review on fusion techniques and approaches used in applications. IEEE Access **8**, 14424–14439 (2020). https://doi.org/10.1109/ACCESS.2020.2966400
4. A. Eman, H. Ramdane, Mobile robot localization using extended Kalman filter, in *2020 3rd International Conference on Computer Applications & Information Security (ICCAIS)*, 19–21 Mar 2020, pp. 1–5

5. M. Thirumarimurugan, N. Bagyalakshmi, P. Paarkavi, Comparison of fault detection and isolation methods: a review, in *2016 10th International Conference on Intelligent Systems and Control (ISCO)*, 7–8 Jan 2016, pp. 1–6
6. H. Deilamsalehy, T.C. Havens, Sensor fused three-dimensional localization using IMU, camera and LiDAR, in *2016 IEEE SENSORS*, 30 Oct–3 Nov 2016, pp. 1–3
7. M.L. Anjum, J. Park, W. Hwang, H. Kwon, J. Kim, C. Lee, K. Kim, D.D. Cho, Sensor data fusion using unscented Kalman filter for accurate localization of mobile robots, in *ICCAS 2010*, 27–30 Oct 2010, pp. 947–952
8. M. Dares, K.W. Goh, Y.S. Koh, C.F. Yeong, E. Su, P.H. Tan, Development of AGV as test bed for fault detection, in *2020 6th International Conference on Control, Automation and Robotics (ICCAR)*, 20–23 Apr 2020, pp. 379–383
9. P. Yazdjerdi, N. Meskin, Actuator fault detection and isolation of differential drive mobile robots using multiple model algorithm, in *2017 4th International Conference on Control, Decision and Information Technologies (CoDIT)*, 5–7 Apr 2017, pp. 0439–0443
10. G. Ziwei, L. Rong, 2D range flow-based odometry fusing LiDAR and IMU, in *2019 IEEE International Conference on Robotics and Biomimetics (ROBIO)*, 6–8 Dec 2019, pp. 2761–2765

Implementation of Industrial Automation Water Distribution System Utilizing PLC: A Laboratory Setup

Aiman Najmi Anuar, Shafishuhaza Sahlan, Khairul Ijlal Rosli, Muhamad Fadzil Ab Rahim, Mohd Amri Md. Yunus, and Mohamad Hafis Izran Ishak

Abstract In this paper, an automated process control utilizing programmable logic controller (PLC) is implemented to a laboratory setup of a water distribution system. A common implementation of water transference in the local water utility company in a water distribution network is controlled manually, the process of distributing clean water to customer to meet the customers' demand is not efficient, thus difficult to achieve. Therefore, in this paper, an automated water distribution system is proposed by controlling the water level of the reservoir tank, representing demand from customers. The automation process is achieved through the utilization of OMRON CP1H PLC, the central processing unit of the system. The water level is measured by water-level sensors and monitored by flow rate meter for high accuracy measurement. The speed of the water pumps is regulated by the variable speed drive (VSD) based on several set-points of the water level of the reservoir tank for low, medium, and high level. The automated system is then monitored and controlled through a SCADA system using OMRON CX-Supervisor. From the results obtained, it can be observed that the proposed system able to control the water level automatically, continuously thus improving the efficiency of the water distribution system.

Keywords Programmable logic controller · Variable speed drives · Automation · Control algorithm · Supervisory control · Data acquisition system

---

A. N. Anuar · S. Sahlan (✉) · M. F. Ab Rahim · M. A. Md. Yunus · M. H. I. Ishak
School of Electrical Engineering, Universiti Teknologi Malaysia, UTM, Skudai, Johor, Malaysia
e-mail: shafis@utm.my

K. I. Rosli
ZEC Engineering Sdn Bhd, No 23B, Jalan Sagu 22, Taman Daya, Johor Bahru, Johor, Malaysia

M. F. Ab Rahim
Ranhill SAJ Sdn Bhd, Jalan Garuda, Larkin, Johor Bahru, Johor, Malaysia

A. Kumar et al. (eds.), *Computational Intelligence in Machine Learning*, Lecture Notes in Electrical Engineering 834, https://doi.org/10.1007/978-981-16-8484-5_12

1 Introduction

In Malaysia, the water treatment plants are commonly controlled manually and hence provide an inefficient water transference process. The manual control of water treatment process is not efficient since they are prone to human errors and wastage. Hence, if this continues without a properly installed control and monitoring system, water wastage will continue to occur and water deficiency is imminent; hence, insufficient water will be supplied to consumers, thus unable to meet the demand of clean water.

Automation or automatic control is the use of control systems to operate systems, processes, and equipment with human intervention at the minimum level [1]. Besides that, automation of task management services is also to provide the conditions for a secure, equitable, unbiased distribution of water resources, for example, in inter-farm irrigation systems, ensuring quality and consistency in the supply of water while minimizing water losses [2]. It is possible to categorize automation into semi-automated and fully automated. Completely automated is when all operations, including making choices for certain conditions when turned on, are run and completed by the computer or device. Semi-automated, meanwhile, is when only up to a certain stage can the device or computer perform different tasks. Human interaction will be required to complete the task after reaching this stage, which is the focus of this research.

There are other sectors in the industry that can take advantage of the automation framework, as handling many activities in parallel is a possibility. The application for fully automated system in the water industry in particular would help increase the quality of the existing process. The water level in tanks in the plant can be managed by automatically increasing and decreasing the level of water according to a predefined conditions or fixed points [3]. In the water treatment facilities, for instance, the water level would be lower in the supply tank after a draft. More water would then be required, but not during the rainy seasons, in which there will be plenty of water. These parameters may be preset or programmed into the fully automatic system, thereby increasing the performance of the system. In this research, the laboratory setup utilized is a representation of a water distribution system, which will be run automatically using a programmable logic controller (PLC), as well as having the capability of control and monitoring from the supervisory control and data acquisition system (SCADA).

2 Automation by Programmable Logic Controller

In this research, to realize the automation process for the proposed device, a programmable logic controller (PLC) is utilized. From literature, PLC is a digital device which transforms an electromechanical process into a relay of signals/commands which forms into automation [4]. PLC is a robust and versatile controller that can automate the functioning of industrial equipment, processes, or even production lines. The use of PLC for industrial control had numerous advantages. Some of the benefit highlighted are ability to manage complex system in a cost-effective way, extendable to other systems and elevated reliability which able to operate for many years [1]. In this research, the CP1H-XA PLC OMRON is used as the main controller. OMRON CP1H PLC utilizes a peripheral USB port, pulse output for four axes, advanced power for high-precision positioning control, high-speed counters, differential phases for four axes, and easily handles multi-axis control with a single unit. The algorithm utilized are given as Algorithm 1. The laboratory scaled system is given in Fig. 1.

Referring to the algorithm and Fig. 1, the PLC receives analog inputs from the liquid-level sensor from the reservoir and the clear water tank. Then, the signal is sent to the VSD to regulate/control the speed of the motor, thus controlling the water level and manipulating the flow in the tanks. The level sensor used is a hydrostatic-level sensor from Milonetech eTape. The collected data then is sent to the PLC. The values from the flow rate sensors is then recorded to monitor the volume of water entering either the reservoir tank or the clear water tank. The values are then compared with the value from the sensor output, i.e., the set-point inserted in the ladder diagram. For example, when the level of the reservoir tank is between 0 and 10 cm, the ladder diagram will trigger a low signal; thus, the incoming motor speed is regulated to 50 Hz while the outgoing motor is 0 Hz. Then, the water level will slowly increase, and when it reaches a set-point level between 11 and 20 cm, the PLC ladder diagram

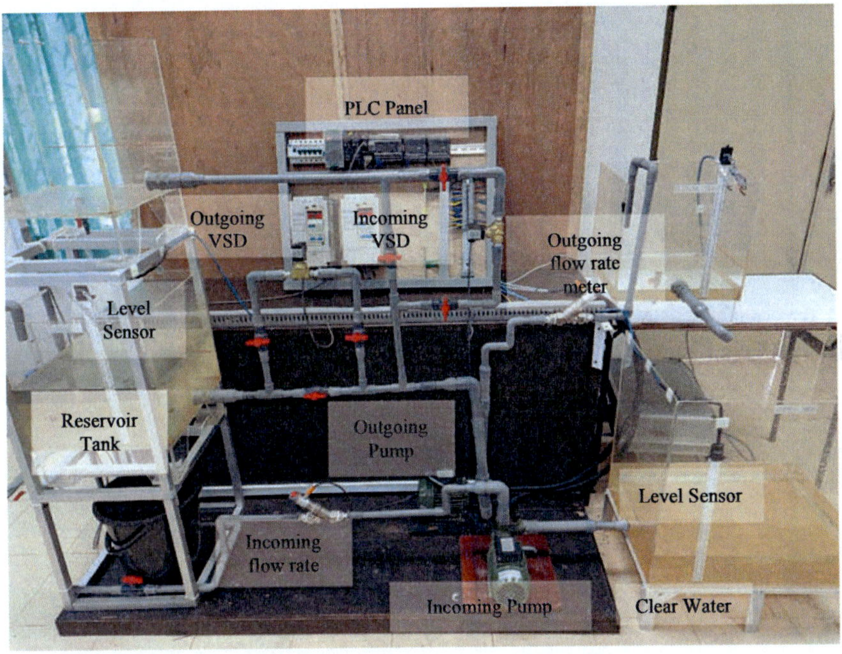

Fig. 1 Laboratory setup for the water distribution system

programming will trigger a medium signal, and the outgoing motor speed is adjusted to 30 Hz while the incoming motor speed remains the same. Similarly, when the level reaches a value between 21 and 25 cm, the program will trigger a high signal and incoming motor speed is adjusted to 30 Hz while the outgoing motor speed is set to 50 Hz. An emergency feature is also included in the system whereby when the water level reaches a level of 30 cm out of the sudden, the program will trigger a high-high signal, thus adjusting the incoming motor speed to 0 Hz while the outgoing motor speed is adjusted to 50 Hz, i.e., to quickly lower the water level of the tank in order to avoid spillage. This is a very important feature as water spill may damage the components nearby.

Algorithm 1

Step 1. Determine the set-point level L_A of the reservoir tank using trial and error.
Low level: $0\,\text{cm} \leq L_{A\_LL} \leq 10\,\text{cm}$
Medium level: $10\,\text{cm} \leq L_{A\_ML} \leq 20\,\text{cm}$
High level: $20\,\text{cm} \leq L_{A\_HL} \leq 25\,\text{cm}$
High-high level: $25\,\text{cm} \leq L_{A\_HL} \leq 30\,\text{cm}$.

Step 2. Determine the desired level for $L_A = L_D$ using trial and error.

$L_D = 25\,\text{cm}$

Step 3. Determine the incoming motor speed Hz_i according to the set-point level.
At low level, $Hz_i = 50\,Hz$
At medium level, $Hz_i = 50\,Hz$
At high level, $Hz_i = 25\,Hz$.

Step 4. Determine the outgoing motor speed Hz_o according to the set-point level.
At low level, $Hz_i = 0\,Hz$
At medium level, $Hz_i = 25\,Hz$
At high level, $Hz_i = 50\,Hz$.

Step 5. Trigger the overflow when L_A accidentally reaches high-high level.
$Hz_i = 0\,Hz$
$Hz_o = 50\,Hz$.

Step 6. Repeat the system while maintaining L_D.

3 Hardware Development

The laboratory setup is fabricated based on a real water distribution system [3] for the local water utility. The PLC program from the previous section is able to control the distribution of water from the clear water tank to the reservoir tank, automatically. The laboratory setup for the water distribution system is as given in Fig. 1. A PC workstation is connected to the PLC via a serial USB port communication.

From Fig. 1, initially, the set-point level of the reservoir tank, which represents the customer demand, is determined. Then, the speed of the pump is regulated by the VSD. In a real system, the clear water tank (CWT) represents the treated water tank out of a water treatment plant (WTP). The required volume of clean water can be represented by a set-point level in the reservoir tank. The distribution of water starts at the CWT, pumped out by the incoming motor through the water flow rate, into the reservoir tank. The level of the reservoir tank is captured by the level sensor. According to the PLC programming explained in the previous sections, the water level in the reservoir tank is maintained based on the set-point level inserted in the ladder programming and the pumps and VSDs are responsible in ensuring that. Since the operation of the incoming and outgoing pumps and VSDs are programmed according to the current level of the reservoir tank as well as the current speed of the incoming pump, an automated solution to the water distribution system is now provided. The incoming flow rate, i.e., the water flow into the reservoir tank, and the outgoing flow rate, i.e., the water flow out of the reservoir tank, as well as the level of the reservoir tank and the clear water tank, and the incoming and outgoing motor speed are monitored and logged. The process time is taken at an interval of 1 second. The system can be controlled, monitored, and logged through a supervisory control and data acquisition system (SCADA) via CX-Supervisor. Elaboration of the SCADA system is provided in the next section.

4 Supervisory Control and Data Acquisition System (SCADA)

Supervisory control and data acquisition (SCADA) is a combination of software and hardware components that allow real-time data process control, collection, and monitoring. Using human–machine interface (HMI), the SCADA system communicates directly with hardware devices such as pumps, motors, and valves [5]. A remote terminal unit (RTU) or programmable logic controller (PLC) performs all control actions in a device (PLC). Data acquisition starts at the stage of the RTU or PLC, where meter readings and status reports of the equipment are included [6–8]. Both the PLC and RTU have a similar function in translating the signal from input or output (I/O) devices in order to convert it to digital data and transfer it to the monitoring system to be processed. The data is then manipulated, allowing operators to track and manage the device remotely from a control room workstation [9]. In this research, an OMRON CX-Designer is used. The SCADA system developed for this research is given in Fig. 2.

From Fig. 2, the level for the reservoir tank (Tank A), the clear water tank (Supply tank), the water flow rate (Flow Rate A), and the motor speed can be monitored. From the data acquisition button on top of the page, the logged data can be observed, which are presented in Figs. 3, 4, and 5 for the water level of the tanks, the motor speed, and the water flow rate, respectively.

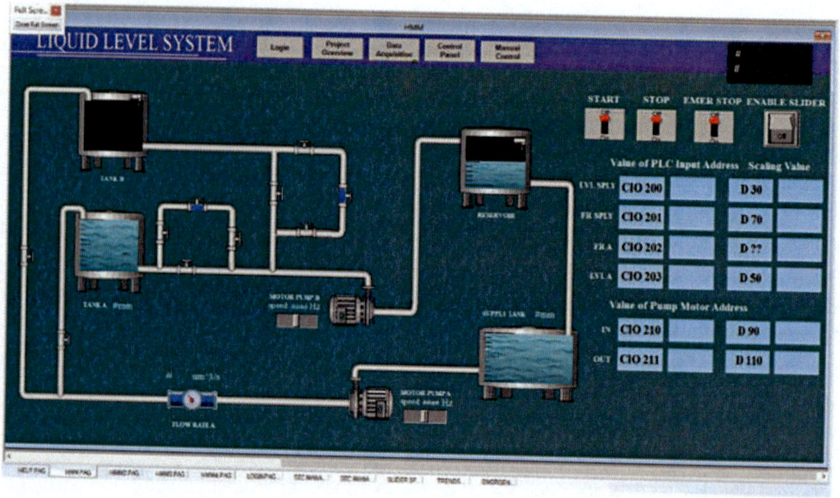

Fig. 2 SCADA system for the laboratory setup

From Fig. 4 in particular, it can be observed that the automation program designed through PLC is able to maintain the water level of the reservoir tank within a respectable limit. The next phase for this research is then to provide a mathematical model of the system followed by an optimized control algorithm to maintain the water level in the reservoir tank [10–12].

5 Conclusion

From the research conducted, it can be concluded that the proposed SCADA system is able to perform the automation process as intended. The system is able to run continuously without interruption until switched off, and the SCADA system is able to log all data in the Excel file as well as displaying the data logged graphically. From the data collected, the mathematical model of the system can be formulated, i.e. the next phase of the research.

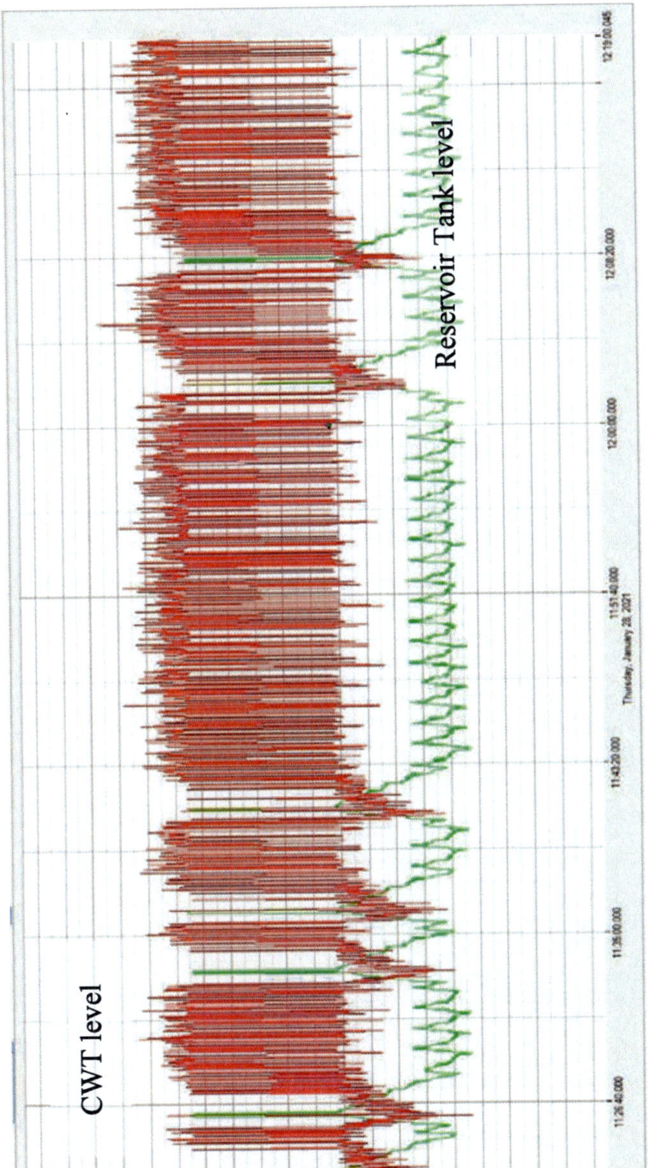

Fig. 3 Data trending for the water-level sensor

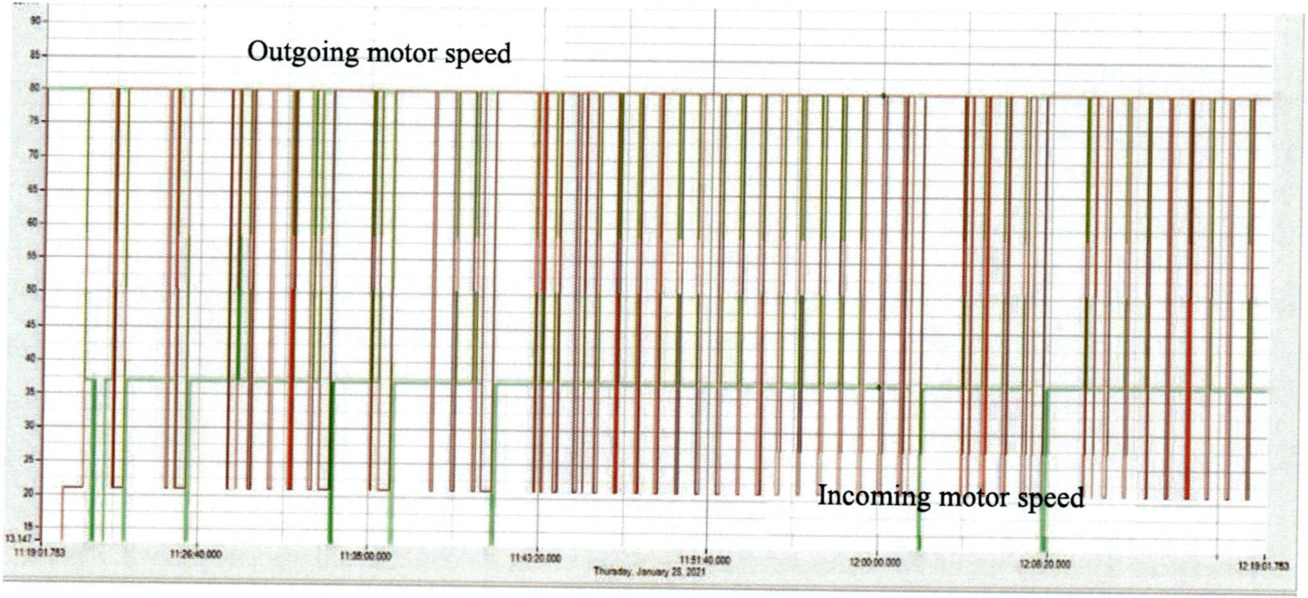

Fig. 4 Data trending for the incoming and outgoing motor speed

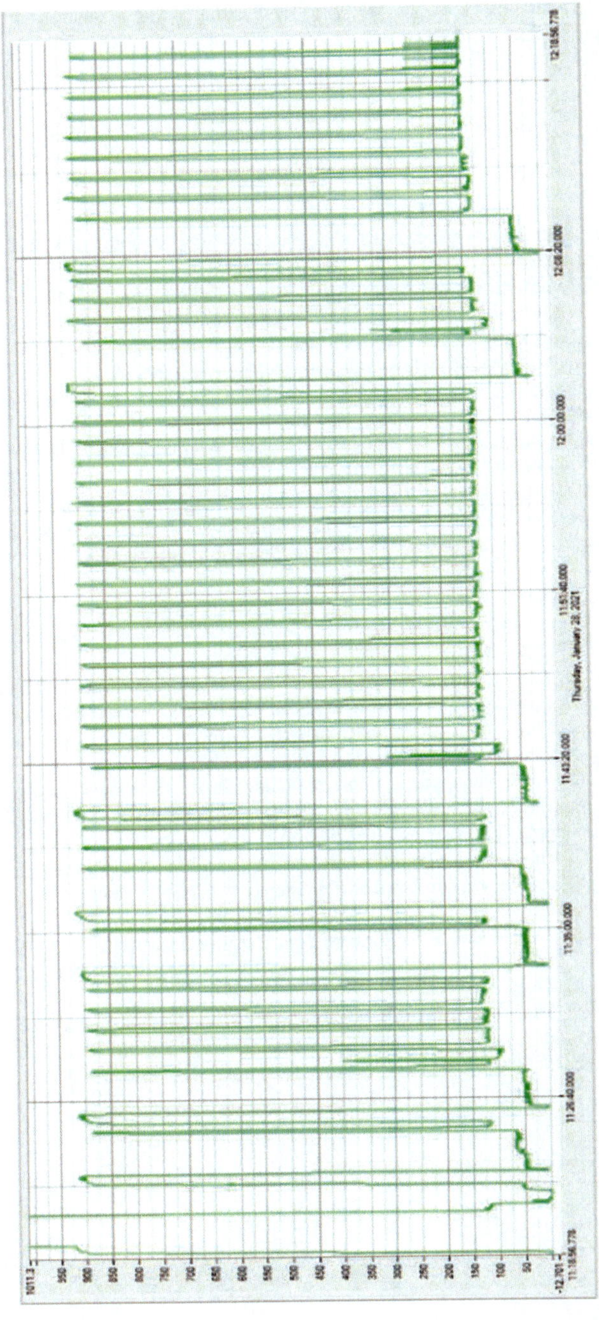

Fig. 5 Data trending for the incoming water flow rate sensor

Acknowledgements The authors would like to convey our gratitude to the PRGS-ICC grant from Universiti Teknologi Malaysia (Vote No: 4J399) as well as the continuous support from Ranhill SAJ Sdn Bhd for the realization of this research.

References

1. V. Soundhrya, Automation in water board using PLC with SCADA. Int. J. Adv. Res. Biol. Eng. Sci. Technol. **2** (2016)
2. I.F. Yurchenko, Automatization of water distribution control for irrigation. Int. J. Adv. Appl. Sci. **4**(2), 72–77 (2017). https://doi.org/10.21833/ijaas.2017.02.013
3. D. Jung, S. Lee, J.H. Kim, Robustness and water distribution system: state-of-the-art review. Water (Switzerland) **11**(5) (2019). https://doi.org/10.3390/w11050974
4. G. Alem, K. Vankdoth, Automatic fluid level control using programmable logic controller. Int. Res. J. Eng. Technol. **3** (2016)
5. N. Aphiratsakun, W. Suvittayakul, K. Saingam, *AU Water Level Process Control System* (IEEE, 2015). 978-1-4799-7961-5/15/$31.00 ©2015
6. A.A. Tkachev, Y.G. Ivanenko, V.V. Zarubin, I.V. Olgarenko, Automation of water distribution management during the reconstruction of main irrigation canals. IOP Conf. Ser. Mater. Sci. Eng. **537** (2019). Institute of Physics Publishing. https://doi.org/10.1088/1757-899X/537/3/032070
7. N. Quadar, A. Chehri, G. Jeon, A. Ahmad, Smart water distribution system based on IoT networks, a critical review, in *Smart Innovation, Systems and Technologies*, vol. 189 (Springer, 2021), pp. 293–303. https://doi.org/10.1007/978-981-15-5784-2_24
8. L.K. Narayanan, S. Sankaranarayanan, IoT enabled smart water distribution and underground pipe health monitoring architecture for smart cities, in *2019 IEEE 5th International Conference for Convergence in Technology, I2CT 2019* (Institute of Electrical and Electronics Engineers Inc., 2019). https://doi.org/10.1109/I2CT45611.2019.9033593
9. S. Wadekar, V. Vakare, R. Prajapati, S. Yadav, V. Yadav, Smart water management using IoT, in *2016 5th International Conference on Wireless Networks and Embedded Systems (WECON)* (2016). https://doi.org/10.1109/wecon.2016.7993425
10. S. Yazdekhasti, G. Vladeanu, C. Daly, *Evaluation of Artificial Intelligence Tool Performance for Predicting Water Pipe Failures* (American Society of Civil Engineers (ASCE), 2020), pp. 203–211. https://doi.org/10.1061/9780784483213.023
11. A. Gupta, N. Bokde, K. Kulat, Z.M. Yaseen, Nodal matrix analysis for optimal pressure-reducing valve localization in a water distribution system. Energies **13**(8) (2020). https://doi.org/10.3390/en13081878
12. B. Sadrfaridpour, H. Saeidi, J. Burke, K. Madathil, Y. Wang, Modeling and control of trust in human-robot collaborative manufacturing, in *Robust Intelligence and Trust in Autonomous Systems* (Springer US, 2016), pp. 115–142. https://doi.org/10.1007/978-1-4899-7668-0_7

Control of Thin McKibben Muscles in an Antagonistic Pair Configuration

Muhamad Hazwan Abdul Hafidz, Nor Mohd Haziq Norsahperi, Hong Win Soon, Mohd Najeb Jamaludin, and Ahmad Athif Mohd Faudzi

Abstract This paper proposes an antagonistic proportional integral (API) control strategy for thin McKibben muscles (TMM) in an antagonistic pair configuration. The study intends to explore a model independent control strategy for TMMs in an antagonistic pair assembly on a TMM platform which mimics the human finger at metacarpophalangeal (MCP) joint. Set-point tracking is achieved by eliminating dead zones due to muscle–tendon slack and using a mode selector for switching of flexion and extension muscles to provide smooth tracking. To test the effectiveness of the proposed method, experiments are carried out on a TMM platform and injected with step and sinusoidal input signals. Experiments conducted showed the TMM platform able to produce high accuracy and fast rise time by implementing the API control scheme. The control method can be very useful in other TMM applications requiring antagonistic muscle actuation.

Keywords Thin McKibben muscle · Antagonistic muscle control · Soft actuator

---

M. H. A. Hafidz · H. W. Soon · A. A. M. Faudzi (✉)
School of Electrical Engineering, Faculty of Engineering, Universiti Teknologi Malaysia, Johor Bahru, Malaysia
e-mail: athif@utm.my

N. M. H. Norsahperi
Department of Electrical and Electronics Engineering, Faculty of Engineering, Universiti Putra Malaysia, Seri Kembangan, Malaysia

M. N. Jamaludin
Sports Innovation and Technology Centre, Institute Human Centered Engineering, Universiti Teknologi Malaysia, Johor Bahru, Malaysia

A. A. M. Faudzi
Centre for Artificial Intelligence and Robotics, Universiti Teknologi Malaysia, Johor Bahru, Malaysia

A. Kumar et al. (eds.), *Computational Intelligence in Machine Learning*, Lecture Notes in Electrical Engineering 834, https://doi.org/10.1007/978-981-16-8484-5_13

1 Introduction

The human hand plays a vital role in our everyday lives. Activities of daily livings such as playing the piano, pinching objects, and sports such as tennis require high dexterity, accuracy and an acceptable response time to achieve these tasks as intended. The biomechanics of the hand and redundant mechanism of hand muscles play a vital role in achieving dexterous hand motions [1].

Although numerous research has been conducted on robotic hands, most hands neglect the need to preserve the biomechanics of the hands by mechanizing components such as the bones, muscles, ligaments, and tendons [2–4] which may introduce discrepancies. A few works have been carried out that closely replicate the human hand by using bones structure that determines the range of motion (ROM) at joints, use ligaments to bind the bones together and use tendons as transmission methods to produce motion [5]. Even though most of the parts closely replicates the human hand, the use of motors is still used which differs from the human hands in terms of compliancy and adds weight to the robotic hand. Although the use of electrical actuators is the most popular, the use artificial muscles is having been showing an increasing trend in research of artificial hand technology [6]. The human-like robotic hand not only replicates the hand in terms of the bones, ligaments, and tendons, but also incorporates the usage of pneumatic artificial muscles as actuators and has shown nearly similar sweeping motion compared to a cadaver motion [7].

TMMs were developed by [8] to solve problems of conventional McKibben muscles that are heavy and bulky which makes them difficult to mount in confined small spaces. TMMs have been used in robotic applications as the legged hexapod to produce walking motion [9]. The robot uses TMMs as actuators to produce forward and backward swing of legs and is assembled in antagonistic pairs. Soft manipulators have also been developed incorporating the use of multiple TMMs in antagonistic pairs as actuators [10, 11]. TMMs application which uses multiple TMM and antagonistic pair configuration has also been used in rehabilitation devices and human body parts [12]. Although aggressive research has been carried out on the use of TMMs in antagonistic pairs over the past few years, the area of control on TMMs in antagonistic configuration is understudied and requires attention. This study introduces a model-independent antagonistic proportional integral (API) controller for a TMM platform with antagonistic muscle pair configuration.

2 Thin McKibben Muscle Platform

The TMM platform aims to mimic the human hand from metacarpal bone to proximal bone. The platform in Fig. 1 consists of two links and 1 joint. The first link is connected to second link by hinge joint. Silicon sheets are placed at the hinge joint to replicate ligaments of the volar plate present in human finger joints. Two TMMs are attached on opposing side to the first link as the origin of Dyneema tendons, and Loctite 401

Fig. 1 Thin McKibben muscle platform in antagonistic pair configuration

is used to glue the tendons to the links. The other end of the TMMs is inserted at the second link using the same method. This is similar to the human hand where muscles are attached to bones by tendons at origin and insertion point. A tendon enters a tube at the volar plate ligament which is used to replicate tendon sheaths which binds the tendon to the bone surface. Two IMUs are used in the setup. The IMU data is sent to an Arduino Mega for joint angle feedback. Pulse width modulation (PWM) output is converted by PWM to analog voltage convertor to control the analog valves.

3 Control Strategy of Thin McKibben Muscle Platform

TMMs which are categorized as pneumatic artificial muscles (PAM) possess many advantages in the application of robotics such as compliance and lightweight. However due to the rubber tube elasticity, complex geometric structure of the braided fiber sleeve, vicious friction, etc., PAM systems suffer from nonlinearities, time-varying characteristics, and uncertainties [13]. Hence, the process of obtaining a perfect dynamic model for PAM systems is problematic [13–17]. Without access of a perfect model, model-independent controllers might be a solution to implement effective control for the TMM platform.

Proportional–integral–derivative (PID) controllers are widely used in industrial control systems due to the simplicity of the controller and reduced number of parameters to be tuned. It is one of the most utilized model-free control algorithms, consisting of a proportional term, integral term, and a derivative term. The controller can be mathematically formulated as Eq. (1) where $e(t)$ represents the error signal and $u(t)$ the control signal while the parameters to be tuned are K_p, T_i, and T_d.

$$U(t) = K_{\mathrm{p}}\left[e(t) + \frac{1}{T_{\mathrm{i}}}\int_{0}^{t} e(t)\mathrm{d}t + T_{\mathrm{d}}\frac{\mathrm{d}}{\mathrm{d}t}e(t)\right] \tag{1}$$

Although the control defined in Eq. (1) is commonly used to control different systems, there is a constant need of modification to be suited to different systems for the system to achieve better performance. In the case of the highly nonlinear TMM platform which possesses unsimilar dynamics for extension and flexion motion and also dead zone due to slack regions of the TMM actuator, alterations such as additional tuning parameters are incorporated to accomplish better tracking results.

Similar to our hand muscles at MCP joint, the TMM platform consists of an antagonistic TMM pair where the links inserted by muscles produce motion by contraction of one muscle while the other muscle is relaxed. When the muscle for flexion motion contracts, the antagonistic muscle which is relaxed needs to have a certain amount of slack in order for flexion to occur. Without the slack region, the flexion muscle is unable to pull the link for flexion. This slack region is achieved by fabricating additional length to the Dyneema tendon enough for flexion and extension to occur. However, the slack results in a dead zone where the control signal needs to exceed a certain level of contraction in order for motion to occur as shown in Fig. 2. The controller is modified to start actuation directly after flexion slack region, FS and extension slack region, ES as in Eqs. (2) and (3).

For implementing of an antagonistic pair control strategy, a mode selector $f, g \in \mathbb{R}$ is postured on flexion and extension mode appropriately. For the mode selector, $f \in [0, 1]$ is employed to form flexion action to decide whether the flexion muscle

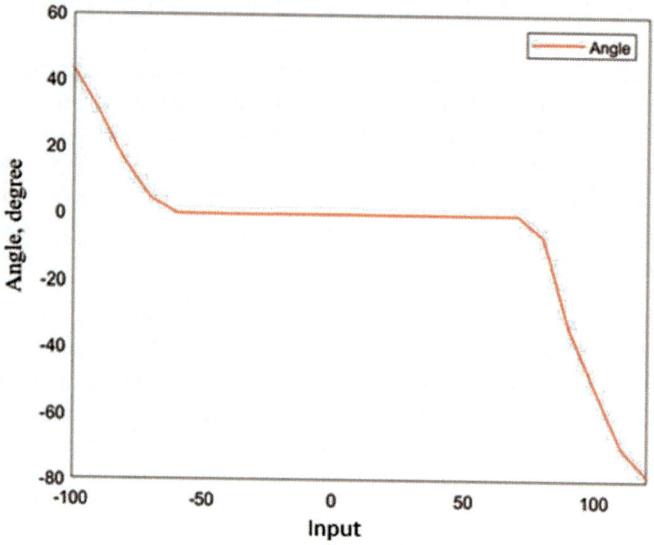

Fig. 2 Joint angle of TMM platform during different input values

should be contracting or relaxing. Flexion muscles need to contract when the set-point is smaller than the actual angle, hence $f = 1$ when $e < 0$ while $f = 0$ when $e > 0$. The opposing extension muscle is activated by the mode selector $g \in [0, 1]$ where extension muscle is actuated to obtain addition in angle where $g = 1$ when $e > 0$ and $g = 0$ when $e > 0$. Consequently, antagonistic actuation is obtained to form a contract and relaxed muscle pair without simultaneously activating both muscles at the same time. Separation of PID controller for flexion and extension also leads to additional parameters to tune which gives the designer more control over the system. The control signal formed is defined as

$$U_{\text{flex}}(t) = -f\left[K_{\text{pflex}}\left[e(t) + \frac{1}{T_{\text{iflex}}}\int_0^t e(t)\text{d}t + T_{\text{dflex}}\frac{\text{d}}{\text{d}t}e(t)\right]\right] + \text{FS} \quad (2)$$

$$U_{\text{ext}}(t) = g\left[K_{\text{pext}}\left[e(t) + \frac{1}{T_{\text{iext}}}\int_0^t e(t)\text{d}t + T_{\text{dext}}\frac{\text{d}}{\text{d}t}e(t)\right]\right] + \text{ES} \quad (3)$$

4 Experimental Results

This section presents the experimental studies to prove the overall efficiency of the proposed control strategy. Figure 3 shows the experiment SIMULINK diagram of the proposed control strategy run in real time. The four parameters K_{pflex}, K_{iflex}, K_{pext}, K_{iext} are initially determined by Ziegler–Nichols method and fine-tuned through extensive experimental trials to obtain API controller parameters as in. It should be noted that the experiments conducted in this work are executed with a sampling period of $\Delta T = 0.01$ s.

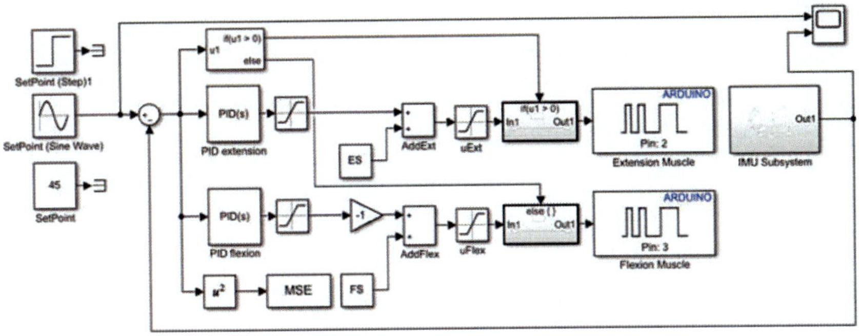

Fig. 3 Experiment SIMULINK model of TMM platform using API control

The API control strategy is evaluated firstly by introducing a step response from 0 to 45° and −45° to evaluate single muscle actuation of extension and flexion performance. The values of ES = 60 and FS = 70 are chosen for the slack parameters defined from Fig. 2, and the experiment was run with $\Delta T = 0.01$ s. The results shown in Figs. 4 and 5 show for both motions, and the API controller was able to achieve close to zero steady-state error with small rise time of below 0.5 s.

An input of a sinusoidal reference signal is with range of −45 to 45° at 0.5 rad/s. This would allow the muscles to work antagonistically by switching automatically to and from flexion and extension muscles according to the motion planned.

From the experimental results in Figs. 6 and 7, the responses produced via both controllers are able to track the sinusoidal signal to a certain degree of accuracy even though faced with challenges of disturbances due to nonlinear load caused by wires

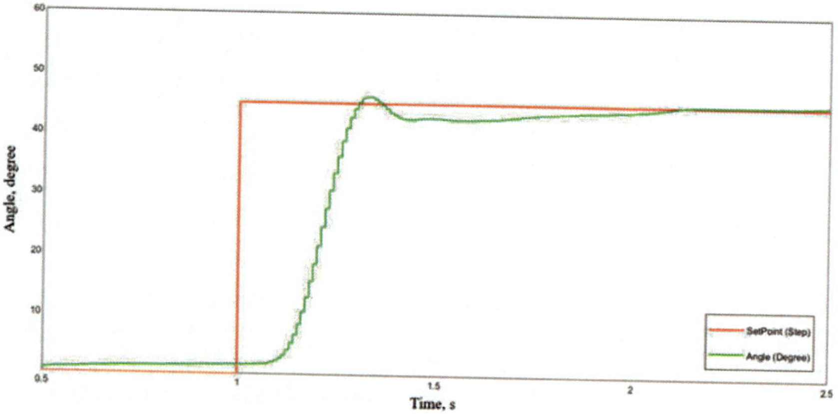

Fig. 4 Step response of TMM platform for extension

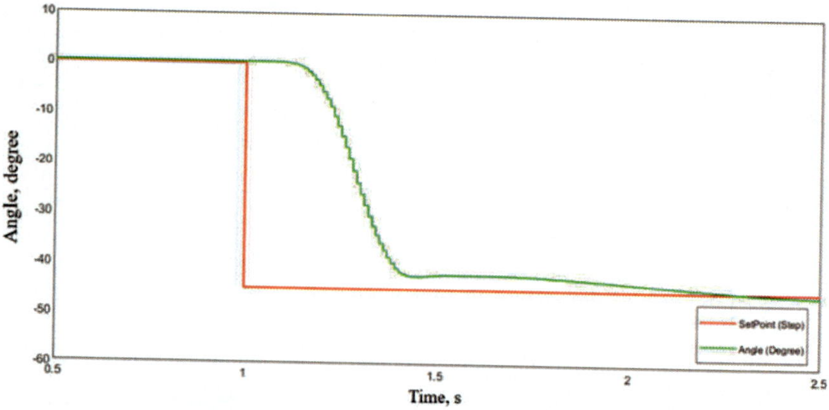

Fig. 5 Step response of TMM platform for flexion

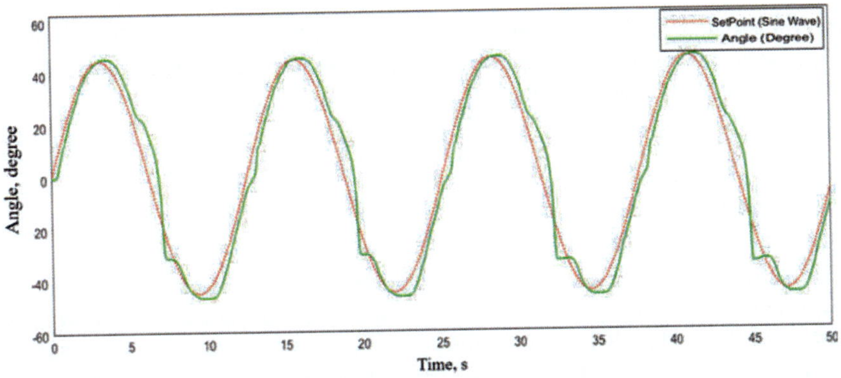

Fig. 6 Sinusoidal response of TMM platform using single PI controller

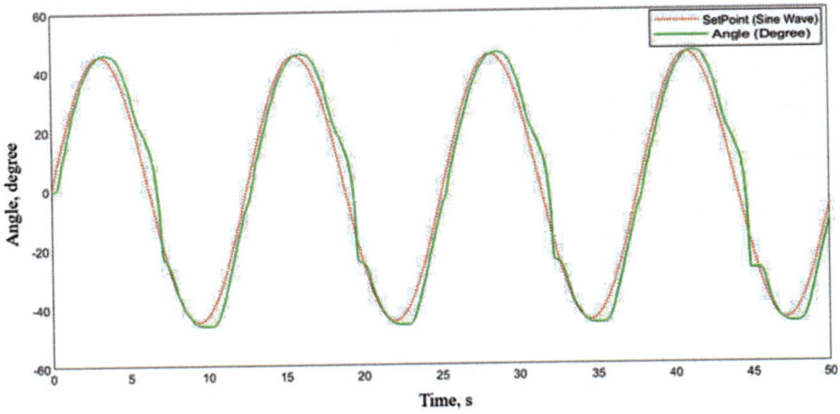

Fig. 7 Sinusoidal response of TMM platform using single API controller

attached to the IMU sensors and force caused by compression and expansion of the volar plate ligament attached to the joint. However, the API controller performed a much smoother and accurate performance compared to the single PID as in Table 1.

The main advantage of the TMM platform is the similarity of the platform to the human hand counterpart, thus allowing the implementation of control strategies without having to compensate in control software for the discrepancies produced by the difference in anatomy in humanoid systems. Implementation of other advanced

Table 1 Mean squared error

Controller	MSE
PI	47.53
Antagonistic PI	33.93

control methods might have better responses and accuracy for the TMM platform. This study intends to explore and implement novel control schemes for TMM in antagonistic pair focused on humanoid systems.

5 Conclusion

In this paper, a TMM platform is introduced which closely replicates mechanism of human index finger at MCP joint for flexion and extension motion. An API control strategy is also presented for controlling an antagonistic muscle pair. Influences of dead zones for the TMM platform have been compensated, and effectiveness of the proposed control strategy has been shown by angle tracking control experiments with fast response speed and acceptable control precision. An interesting future work to study would be to implement machine learning techniques to obtain better position tracking accuracy for a TMM antagonistic system.

Acknowledgements The authors would like to acknowledge the support provided by the Ministry of Higher Education Malaysia (MOHE) through support under the Fundamental Research Grant Scheme (FRGS/1/2019/TK04/UTM/02/41). The author also would like to thank Universiti Teknologi Malaysia (UTM) for work and facilities support.

References

1. A. Bicchi, Hands for dexterous manipulation and robust grasping: a difficult road toward simplicity. IEEE Trans. Robot. Autom. **16**(6), 652–662 (2000). https://doi.org/10.1109/70.897777
2. D. Lee, J. Park, S. Park, M. Baeg, KITECH-hand: a highly dexterous and modularized robotic hand. IEEE/ASME Trans. Mechatron. **22**(2), 876–887 (2017). https://doi.org/10.1109/TMECH.2016.2634602
3. H. Zhou, A. Mohammadi, D. Oetomo, G. Alici, A novel monolithic soft robotic thumb for an anthropomorphic prosthetic hand. IEEE Robot. Autom. Lett. **PP**, 1 (2019). https://doi.org/10.1109/LRA.2019.2892203
4. L.B. Bridgwater et al., The Robonant 2 hand—designed to do work with tools, in *2012 IEEE International Conference on Robotics and Automation* (2012), pp. 3425–3430. https://doi.org/10.1109/ICRA.2012.6224772
5. A.D. Deshpande, J. Ko, D. Fox, Y. Matsuoka, Anatomically correct testbed hand control: muscle and joint control strategies, in *Proceedings of IEEE International Conference on Robotics and Automation* (2009), pp. 4416–4422. https://doi.org/10.1109/ROBOT.2009.5152542
6. J. Szkopek, G. Redlarski, Artificial-hand technology—current state of knowledge in designing and forecasting changes. Appl. Sci. **9**(19) (2019). https://doi.org/10.3390/app9194090
7. A.A.M. Faudzi, J. Ooga, T. Goto, M. Takeichi, K. Suzumori, Index finger of a human-like robotic hand using thin soft muscles. IEEE Robot. Autom. Lett. **3**(1), 92–99 (2018). https://doi.org/10.1109/LRA.2017.2732059
8. S. Kurumaya, H. Nabae, G. Endo, K. Suzumori, Design of thin McKibben muscle and multi-filament structure. Sens. Actuators A Phys. **261**, 66–74 (2017). https://doi.org/10.1016/j.sna.2017.04.047

9. A.A.M. Faudzi, G. Endo, S. Kurumaya, K. Suzumori, Long-legged hexapod giacometti robot using thin soft McKibben actuator. IEEE Robot. Autom. Lett. **3**(1), 100–107 (2018). https://doi.org/10.1109/LRA.2017.2734244
10. Y. Peng et al., Development of continuum manipulator actuated by thin McKibben pneumatic artificial muscle. Mechatronics **60**, 56–65 (2019). https://doi.org/10.1016/j.mechatronics.2019.05.001
11. A.A. Faudzi, N.I. Azmi, M. Sayahkarajy, W.L. Xuan, K. Suzumori, Soft manipulator using thin McKibben actuator, in *IEEE/ASME International Conference on Advanced Intelligent Mechatronics, AIM*, vol. 2018, July 2018, pp. 334–339. https://doi.org/10.1109/AIM.2018.8452698
12. N. Endo, Y. Kizaki, N. Kamamichi, Flexible pneumatic bending actuator for a robotic tongue. J. Robot. Mechatron. **32**(5), 894–902 (2020). https://doi.org/10.20965/jrm.2020.p0894
13. Y. Chen, N. Sun, D. Liang, Y. Qin, Y. Fang, A neuroadaptive control method for pneumatic artificial muscle systems with hardware experiments. Mech. Syst. Signal Process. **146**, 106976 (2021). https://doi.org/10.1016/j.ymssp.2020.106976
14. Y. Yuan, Y. Yu, L. Guo, Nonlinear active disturbance rejection control for the pneumatic muscle actuators with discrete-time measurements. IEEE Trans. Ind. Electron. **66**(3), 2044–2053 (2019). https://doi.org/10.1109/TIE.2018.2838061
15. A. Nikkhah, C. Bradley, A. Sharif Ahmadian, Design, dynamic modeling, control and implementation of hydraulic artificial muscles in an antagonistic pair configuration. Mech. Mach. Theory **153**, 1–16 (2020). https://doi.org/10.1016/j.mechmachtheory.2020.104007
16. L. Zhao, H. Cheng, J. Zhang, Y. Xia, Adaptive control for a motion mechanism with pneumatic artificial muscles subject to dead-zones. Mech. Syst. Signal Process. **148**, 107155 (2021). https://doi.org/10.1016/j.ymssp.2020.107155
17. Q. Ai et al., High-order model-free adaptive iterative learning control of pneumatic artificial muscle with enhanced convergence. IEEE Trans. Ind. Electron. **67**(11), 9548–9559 (2020). https://doi.org/10.1109/TIE.2019.2952810

Defect Severity Classification of Complex Composites Using CWT and CNN

Lim Wilson, Anis Salwa Mohd Khairuddin, Uswah Khairuddin, and Bibi Intan Suraya Murat

Abstract Composite structures are prone to internal defects such as delamination. Due to this, it is vital to recognize internal flaws in composite materials accurately because there is possibility that these internal defects can severely degrade the composite structure's strength. This work aims to develop an intelligent complex composite defect severity classification which will contribute to efficient monitoring of composite structures during their service life. Firstly, the behavior of guided ultrasonic waves is processed and transformed into image database using continuous wavelet transform method. Then, a defect classification framework is proposed by using convolutional neural network to classify six types of defect sizes. A total of 798, 342, and 90 images are used for training, validation, and testing, respectively. The results present that the proposed system achieved approximately above 86% of precision and recall for all six defects classes.

Keywords Deep learning · Signal classification · Machine learning · Wavelet transform

1 Introduction

Complex composites are broadly applied in the aerospace and automotive industries because of their high strength-to-weight ratio, tolerance to fatigue and resistance to corrosion, flexible designs consequently leading to highly effective structures.

L. Wilson · A. S. M. Khairuddin
Department of Electrical Engineering, Faculty of Engineering, University of Malaya, Kuala Lumpur, Malaysia

U. Khairuddin (✉)
Department of Mechanical Precision Engineering, Malaysia-Japan International Institute of Technology, Universiti Teknologi Malaysia, Johor Bahru, Malaysia
e-mail: uswah.kl@utm.my

B. I. S. Murat
Faculty of Mechanical Engineering, Universiti Teknologi MARA, Shah Alam, Malaysia

A. Kumar et al. (eds.), *Computational Intelligence in Machine Learning*, Lecture Notes in Electrical Engineering 834, https://doi.org/10.1007/978-981-16-8484-5_14

However, composite structures are susceptible to internal defects such as delamination due to their layer-by-layer construction. Due to low-velocity impact damage, failure in composite structures leads to a significant maintenance concern because it can lead to damage that is hardly noticeable and difficult to perceive. Depending on the severity of the impact, matrix breakage or delamination may occur, reducing the load capacity of the structure. In composite structures, low-velocity impact can lead to hardly noticeable and difficult to perceive damage or delamination. Delamination developed at the initial stage of an impact event is therefore more dangerous to the integrity of the structure, as it could continue to spread to the entire structure [1].

Therefore, to detect and monitor damage of composite components, non-destructive testing (NDT) techniques, which can be deployed quickly and reliably, are vital to the continued growth of this sector. NDT is the examination of objects to distinguish defects, cracks, or other discontinuity without damaging the objects that we examine. Essentially, this examination is performed to guarantee that the materials we used are harmless and have not exceeded the damage tolerance [2]. Previous works have proposed intelligent defect classification for efficient monitoring of composite structures during their service life. Virupakshappa and Oruklu [3] proposed convolutional neural network (CNN) method to classify the flaws from ultrasonic signals. Satisfactory results have been obtained with the average classification accuracy of 90%, 93%, 95%, and 91% for flat bottom hole, rectangular void with greater width, side drilled hole, and rectangular void with greater length, respectively. Miranda et al. [4] proposed support vector machine (SVM) with CNN baseline model to classify the defect type and component type for defect classification on aircraft fuselage images. The average classification accuracy achieves 87.7% for hybrid method. Xu et al. [5] proposed CNN method for fatigue crack diagnosis (FCD) of aircraft structures. The minimum diagnosis accuracy achieved 86.64% with 1 mm diagnosis error. Clayton et al. [6] proposed artificial neural network (ANN) method for alkali–silica reaction (ASR) damaged classes to detect embedded defects. Although the model achieved classification accuracy above 95%, the model only can differentiate whether ASR is damaged or non-damaged (binary result).

Continuous wavelet transform (CWT) evaluates signal at different frequencies with different resolutions. CWT is considered to provide satisfactory time resolution but low-frequency resolution at high frequencies. Meanwhile, CWT provides satisfactory frequency resolution but poor time resolution at low frequencies. This kind of analysis is suitable if the signal is composed of high-frequency components for a short period and low-frequency components for a long period. This is generally encountered in practical applications, such as detecting complex composite defects [7]. The benefit of adopting wavelet scalogram over traditional short-time Fourier spectrogram is the flexibility in presenting time-varying characteristics with variable time–frequency resolution [8].

This work proposed an intelligent complex composite defect severity classification system by using CWT and CNN. Section 2 discusses on the signal preprocessing approach applied in this work while Sect. 3 discusses on the adoption of CNN method for classification purpose. The results and discussion are presented in Sect. 4. Finally, the work is concluded in Sect. 5.

2 Signal Preprocessing Using Continuous Wavelet Transform

The raw guided ultrasonic wave signal database used in this study includes 1200 samples for 6 classes (10, 20, 30, 40, 50, and 60 mm) or 200 samples for each defect size. The sample of signals is observed and analyzed by extracting the features using continuous wavelet transform (CWT). Wavelet scalogram is adopted to extract time–frequency features from the signals in order to detect faults [8, 9]. The scalogram is defined as:

$$\mathrm{SC}_x(t, a; \psi) = |\mathrm{CW}_x(t, a; \psi)|^2 \tag{1}$$

where CW_x is the continuous wavelet transform of x. The continuous wavelet transform is applied to project a signal x on a family of wavelets inferred from the mother wavelet by translation and dilations:

$$\mathrm{CW}_x(t, a; \psi) = \int_{-\infty}^{+\infty} x(\tau)\frac{1}{\sqrt{a}}\psi^*\left(\frac{\tau - t}{a}\right)\mathrm{d}\tau \tag{2}$$

where ψ^* is the conjugate of the complex of the wavelet ψ, a is the scale factor. The scale, a, is associated to frequency by the relationship $a = (\omega_c/\omega)$. ω_c is the center frequency of mother wavelet. The energy spread of ψ links to a Heisenberg box positioned at $(t, \omega_c/a)$ with the time interval proportional to a and the frequency interval proportional to $1/a$. The area of the box stays at constant value. In summary, the wavelet transform applies short windows at high frequencies and long windows at low frequencies. Thus, the frequency resolution becomes lesser while the time resolution increases as the analysis frequency rises.

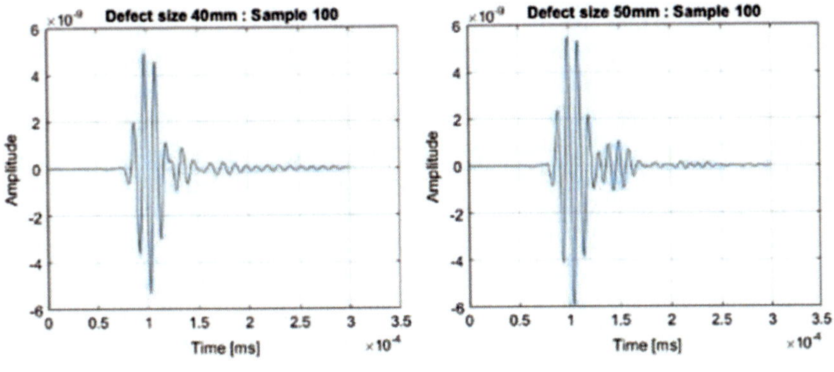

Fig. 1 Example of signal samples for different defect sizes

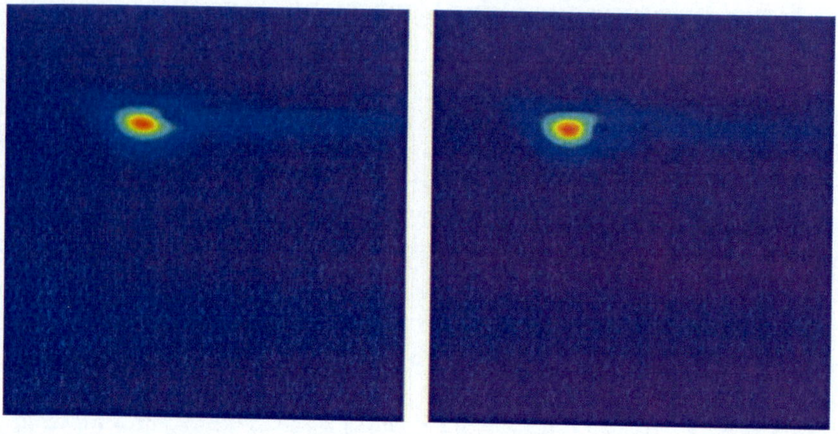

Fig. 2 Example of scalogram images using CWT method

Figure 1 shows the raw signal of each size with excitation frequency at 100 kHz. By observing each signal, the shape of the signals is quite similar before 1.5×10^{-4} ms but starting to differ after 1.5×10^{-4} ms as there is the presence of fluctuation and damping effects. The acoustic wave strength of the signal decreases as the distance of the emitter, and receiver increases due to the nature of sound wave. To extract the characteristics of each signal class, continuous wavelet transform (CWT) is used to develop 2D images. Figure 2 shows the scalogram images based on CWT method. The signals are presented in a circular shape image with different hue that represents the magnitude of the signal. These scalogram images are fed to CNN for the classification process.

3 Classification Using Convolutional Neural Network

The CNN model is adopted to extract discriminant features to the distinguish signals [10, 11]. Theoretically, in deep learning CNN models, each input image will go through a series of convolution layers with filters (kernels), pooling, fully connected layers (FC), and apply softmax function to distinguish an object. Convolution is the initial layer to extract features from an input image. Convolution conserves the connection between pixels by learning image features using small squares of input data. It is a mathematical operation that considers two inputs such as image matrix and a filter or kernel [12]. Meanwhile, the fully connected layer and classification layer of the model are modified to match the number of defect size classes which is 6. In this work, the selection of pretrained CNN between GoogleNet, ResNet-50, and Inception-ResNet-V2 is done based on the training and validation performance.

Table 1 Pretrained model training and validation performance

Pretrained model	Final validation accuracy (%)	Final validation loss (%)
GoogleNet	22.51	1.7917
ResNet-50	98.24	0.0479
Inception-ResNet-v2	99.12	0.0341

Table 2 Performance comparison in terms of number of epochs

Number of epochs	Final validation accuracy (%)	Final validation loss (%)
50	95.03	0.1923
100	99.12	0.0341
200	100.00	0.0023

4 Results and Discussion

A total of 798 and 342 images are used for training and validation, respectively. The model is tested with 90 sample images (15 images for each defect size). In this study, three models which are GoogleNet, ResNet-50, and Inception-ResNetV2 were trained with the image datasets, and the comparison results are presented in Table 1. The performance of the proposed system is evaluated in terms of precision, recall, and $F1$ score.

From the result above, Inception-ResNetV2 model is chosen since the validation accuracy is the highest. The model is then trained with different parameters in terms of number of epochs to identify the effect on validation accuracy. The results are shown in Table 2. From the training results shown, it can be seen that the highest validation accuracy of 100% is achieved when data is trained for 200 epochs.

The performance of the proposed system is assessed in terms of precision, recall, and $F1$ score as shown in Fig. 3. Generally, the precision, recall, and $F1$ score achieved for defect size 10, 20, 30, and 40 mm are approximately 86%. Meanwhile, the accuracy for defect size 50 and 60 mm is approximately 93%. This shows that it is easier to detect bigger defect size that is above 50 mm of size as compared to smaller defect size. The higher the percentage of precision, recall, and $F1$ score, the better the performance of the detection model.

5 Conclusion

In this paper, a feasible classification framework is proposed to classify defect severity using continuous wavelet transform and convolutional neural network. The wavelet scalogram is fed into the neural network by using Inception-ResNet-V2 pretrained model with 200 epochs and the final validation accuracy achieved is 100%. As for

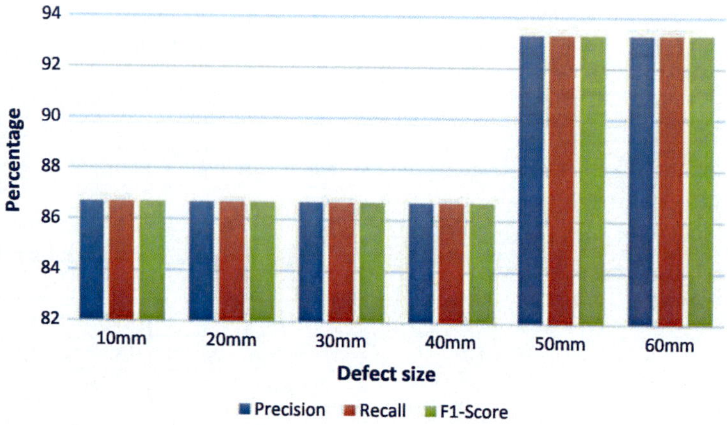

Fig. 3 Performance evaluation of the proposed system for various defect size

the testing phase, the precision, recall, and $F1$-score achieved approximately 86% for defect size of 10, 20, 30, and 40 mm while 93% for defect size 50 and 60 mm.

References

1. A.B. Hossein Towsyfyan, R. Boardman, T. Blumensath, Successes and challenges in non-destructive testing of aircraft composite structures. Chin. J. Aeronaut. **33**(3), 771–791 (2020). https://doi.org/10.1016/j.cja.2019.09.017
2. M. Kersemans, E. Verboven, J. Segers, S. Hedayatrasa, W.V. Paepegem, Non-destructive testing of composites by ultrasound, local defect resonance and thermography. Multidiscipl. Digit. Publ. Inst. Proc. **2**(8), 554 (2018)
3. K. Virupakshappa, E. Oruklu, Multi-class classification of defect types in ultrasonic NDT signals with convolutional neural networks, in *2019 IEEE International Ultrasonics Symposium (IUS)* (IEEE, 2019), pp. 1647–1650
4. J. Miranda, J. Veith, S. Larnier, A. Herbulot, M. Devy, Machine learning approaches for defect classification on aircraft fuselage images acquired by an UAV, in *Fourteenth International Conference on Quality Control by Artificial Vision*, vol. 11172 (International Society for Optics and Photonics, 2019), p. 1117208
5. L. Xu, S. Yuan, J. Chen, Y. Ren, Guided wave-convolutional neural network based fatigue crack diagnosis of aircraft structures. Sensors **19**(16), 3567 (2019)
6. D.A. Clayton, H. Santos-Villalobos, J. Baba, *Evaluation of Advanced Signal Processing Techniques to Improve Detection and Identification of Embedded Defects* (Oak Ridge National Lab. (ORNL), Oak Ridge, TN, 2016)
7. S. Barrios, D. Buldain, M.P. Comech, I. Gilbert, Partial discharge identification in MV switchgear using scalogram representations and convolutional AutoEncoder. IEEE Trans. Power Deliv. (2020)
8. X. Chen, Z. Feng, C. Zhao, Reassigned wavelet scalogram for gear fault detection under nonstationary operational conditions, in *2016 Prognostics and System Health Management Conference (PHM-Chengdu)* (IEEE, 2016), pp. 1–5

9. J.-A. Liu, K. Wei, J. Tan, J. Huang, X. Wu, A signal analysis method based on reassigned scalogram, in *2010 3rd International Congress on Image and Signal Processing*, vol. 6 (IEEE, 2010), pp. 2901–2904
10. H. Dai, F. Ge, Q. Li, W. Zhang, T. Liu, Optimize CNN model for FMRI signal classification via Adanet-based neural architecture search, in *2020 IEEE 17th International Symposium on Biomedical Imaging (ISBI)* (IEEE, 2020), pp. 1399–1403
11. S. Li, H. Feng, EEG signal classification method based on feature priority analysis and CNN, in *2019 International Conference on Communications, Information System and Computer Engineering (CISCE)* (IEEE, 2019), pp. 403–406
12. A. Khumaidi, E.M. Yuniarno, M.H. Purnomo, Welding defect classification based on convolution neural network (CNN) and Gaussian kernel, in *2017 International Seminar on Intelligent Technology and Its Applications (ISITIA)* (IEEE, 2017), pp. 261–265

Detection of Mobile Phone Usage While Driving Using Computer Vision and Deep Learning

Yeoh Mei Hwei, Usman Ullah Sheikh, and Nurul Ezaila Binti Alias

Abstract This work proposes a computer vision-based detection of the usage of mobile phone while driving. The majority of the existing works is based on in-car camera system mounted inside the vehicles, which is more suitable to be implemented as a driver assistance system. However, there are no existing datasets that are captured based on the camera system mounted outside the vehicles. Thus, the total amount of dataset acquired is limited for this work. Since there are no datasets available for this project, the images are captured and collected using Fujifilm XT 10 with XF 35 mm f2 lens at an overhead bridge nearby Spice Arena, Penang, Malaysia. There is a total of 1351 images captured and collected. The proposed system framework is developed based on Faster R-CNN with Inception-V2 architecture. It is based on detection evaluation metrics applied by COCO. In terms of overall detection accuracy, it obtained 88.71% for the accuracy.

Keywords Faster R-CNN · Inception-V2 · Object detection · Deep learning

1 Introduction

With the rapid subscriptions of the mobile phone throughout the world, distracted driving has become a critical topic in traffic safety. According to National Highway Traffic Safety Administration Analysis 2018, which carried out by National Safety Council, it stated that the total number of fatal distraction-affected traffic crashes has increased up to 2628 where 349 traffic crashes are caused by mobile phone usage

Y. M. Hwei
Device Development Group, Intel Microelectronics (M) Sdn. Bhd., Bayan Lepas, Pulau Pinang, Malaysia
e-mail: mei.hwei.yeoh@intel.com

U. U. Sheikh (✉) · N. E. B. Alias
School of Electrical Engineering, Faculty of Engineering, Universiti Teknologi Malaysia, Johor Bahru, Johor, Malaysia
e-mail: usman@utm.my

N. E. B. Alias
e-mail: ezaila@utm.my

A. Kumar et al. (eds.), *Computational Intelligence in Machine Learning*, Lecture Notes in Electrical Engineering 834, https://doi.org/10.1007/978-981-16-8484-5_15

while driving, which contributes about 13.3% of the total number of fatal distraction-affected traffic crashes [1]. The majority of the prior works is based on the in-car camera system mounted inside the vehicles. The captured input data images are not suitable to be used for this work, as it is more suitable for a driver assistance system. This is because it is not easy for the government to enforce a law for all drivers to install the camera monitor system into the driver assistance system, impossible to expect all the cars have the driver assistance system, which only available for highly advanced cars. Therefore, in order to monitor the driver behavioral on traffic, a driver monitoring system is required to be designed with a camera system mounted outside the vehicles. In this paper, we propose a camera-based automated image recognition system for detecting the driver using mobile phones for calling while driving. The first objective is to collect datasets for driving conditions of car drivers in Malaysia. The second objective is to design and develop a convolutional neural network-based approach deep learning system with high overall detection accuracy. The last objective is to provide performance evaluation between prior works with the proposed framework.

2 Research Methodology

2.1 *Image Acquisition*

The images are captured using Fujifilm XT 10 with XF 35 mm f2 lens at an overhead bridge nearby Spice Arena, Penang, Malaysia, due to the optimum height of the bridge. Most of the images are captured during the evening time, which is around 5–7.30 p.m. This is to avoid the captured images exposed to sunlight excessively during morning time and to avoid being blurred during night time. However, the other dominant challenge will be on the car window and windshield that are tinted. There are a total of 2340 images captured and collected. However, only 1351 images are applicable to be used as the dataset, which means that about 42% of the captured images are discarded for those unclear images and partial car images due to the factors of handshaking or fast car moving speed or excessive sunlight reflections. The captured images are saved in JPEG format with a resolution of 4896×3264.

2.2 *Data Preprocessing*

Since the limitation of this project is to detect a single object in an image, thus, the captured images with multiple objects have to be cropped into an image with a single object and resized to a resolution of 900×1000. In this project, the objects included are only cars. Thus, the cropped images are classified into two classes which are "Call" and "Not Call". For the cropped images with the driver's both hands holding on the steering wheel or without bending the driver's both hands toward their head, the cropped image is considered as a "Not Call" dataset. While, for the cropped images with bending of driver's hand toward their head, the cropped image is considered as a "Call" dataset. Most of the captured images are unable to present the full image for the driver's head, thus assuming of bending the driver's hand toward their head will be considered as an action for holding a mobile phone for calling while driving. Since the captured locations are in Malaysia, all the captured images are based on right-handed driving. In order for the system to detect left-handed driving, both types of datasets are flipped vertically to form a left-handed driver dataset.

2.3 Proposed Methodology

The mobile phone detection system is using Faster R-CNN with Inception-V2 architecture, configured for datasets processed. In terms of architecture, Faster R-CNN is the integration network for feature network, region proposal network, and detection network. For the feature network, it receives the resized input images of resolution 600 × 1024 pixels and sends to the convolutional model for generating the feature maps. In this case, Inception-V2 is applied as the pre-trained convolutional neural network for feature extractions. For the region proposal network, it is known as the first stage detector, where a region proposal with bounding boxes corresponded with objectness score is generated through the small network from the feature maps, ported RPN introduced in [2]. It is implemented on the last convolutional feature map of the convolution model through a 3 × 3 sliding window. The bounding box is also known as the region of interests (ROIs), which are generated by a mechanism with a predefined set for bounding boxes with different aspect ratio and scale called anchor boxes. Every position in the feature map is considered as an anchor. In this case, four scales and three aspect ratios are applied to each anchor, which means that there are a total of 12 anchor boxes are generated per pixel in the convolutional feature map. The scales set for each anchor are 0.25, 0.5, 1.0, 2.0, while the aspect ratio set for each anchor are 0.5, 1.0, 2.0. Thus, the total number of anchor box will be $W \times H \times k$, where W and H denote as weight and height of the convolutional feature map, respectively, and k denotes the number of anchor box per pixel. With the stride of 16 at height and width, respectively, there is a total of 2368 positions generated. Thus, the total number of anchor box will be 28,416. During the first stage of detection, the IOU threshold is set as 0.7. The anchor box with the highest IOU or with an IOU value higher than 0.7 with respect to any ground truth bounding box will be treated as foreground. At the same time, the anchor box with an IOU value lower than 0.3 with respect to all ground truth bounding boxes will be treated as background. The rest of the anchor box will be discarded during RPN training. However, during the marking of the anchor box, it is possible that multiple anchor boxes are taken as foreground on the same object. Thus, a reduction process on multiple anchor boxes that taken are as foreground on the same object has to be carried out, which is known as non-maximum suppression, NMS. All the anchor boxes will be sorted based on their confidence score, and the anchor box with the highest confidence score is selected. For those anchor boxes with a higher IOU value than the IOU value of the selected anchor box, it will be removed from the regional proposal list. In order to maintain the same number of regional proposals for each image, the maximum number of anchor boxes is set to be 300. As a result, it generates a total of 4×300 outputs in box regression layers, which represents four for coordinates of 300 anchor boxes and generates a total of 2×300 outputs in classification layers, which represent the objectness classification [3, 4]. In terms of detection network, it is known as the second stage "box classifier" detector, where the regional proposals are used to crop the features from the last convolutional feature map of the convolution model, resize all the proposal to the same size through ROI pooling layers thus, passing to fully connected layer for multi-class classification and bounding box regression. In the fully connected layer, softmax is used as the activation function.

3 Experimental Results

3.1 *Faster R-CNN Model Results*

The mobile phone detection system is proposed to train 20,000 steps with a batch size of 1. It means that only one image in the training datasets is trained for each step. Due to the properties of Faster R-CNN architecture, multiple image regions are generated from one image, based on different predefined aspect ratios and scales. There is a total of 1351 images in the training datasets, it means that all of the images are trained up to 14 times. The training process of 20,000 steps on a Dell Inspiron 15R laptop with CPU Intel Core i7 (4th Gen) 4500U at 1.8 GHz is about 64 h. The learning rate is set to be 0.0002 through the training process in order to avoid the divergence of the training model diverges due to unstable gradients.

During the first stage of the detection network, the objectness loss weight is set to be 1.0, and the localization loss weight is set to be 2.0, which contributes to RPN objectness loss and RPN localization loss. While, RPN objectness loss is a classification loss function in detecting the existence of an object in region proposals, whether classified as "object" or "non-object" which is trained to 8.4258×10^{-5}. RPN localization loss is a regression loss function in predicting a bounding box that contains an object which is trained down to 3.5107×10^{-4}. During the second stage of the detection network, the classification loss weight is set to be 1.0, and the localization loss weight is set to be 2.0, which contributes to box classifier classification loss and box classifier localization loss. Box classifier classification loss is a classification loss function for multiple class in ROI, either classified as "Call" or "Not Call", able to be trained down to 0.03829. The box classifier localization loss is a regression loss function between the predicted bounding box in ROI and the ground truth bounding box and trained to 0.02481. The total loss is the summation for four types of losses with a value of 0.06353. When the model successfully predicts the correct class on

Fig. 1 "Not Call" detected images (left 1, 2) and "Call" detected images (right 1, 2)

the testing datasets, either class for "Call" or class for "Not Call", it is known as true positive. While, when the model predicts the wrong class on the testing datasets, it is known as false positive. On the other hand, if the model fails to perform the detection when the ground truth object is present in the testing datasets, it is known as a false negative. In this work, all the testing datasets contain the ground truth object. Thus, true negative can be ignored. Most of the upper part of the windscreen for cars is tinted, which blocks us from capturing the images with the full face of the driver. This work assumes that bending of driver's left hand or right hand toward their head is considered as "Call". Besides that, the proposed system is limited to a single object in an image for car drivers using their mobile phone for calling or not calling only, and other distractions are not included. In terms of performance evaluation, the results are expressed in terms of mAP and mAR based on detection evaluation metrics COCO. The mAP for the IOU threshold of 0.50 is 97.75% for the model. It means that the model provides 97.75% accuracy in localizing the object along with the classes while the mAR with ten detections per image obtained 72.03% for the model. It means that the model provides 72.03% accuracy in predicting the positive class among all the ground truth objects along with the classes. With the IOU threshold set to 0.5 and the number of detections per image set to be 100, the total number of detected images for true positive is 707, the total number of detected images for false positive is 85, and the total number of detected images for a false negative is 5. Thus, the model provides 88.71% in terms of efficiency of the system (Fig. 1).

3.2 *Performance Evaluation Between Existing Works and Proposed Work*

For this section, the performance of [5, 6] is evaluated with the proposed work in terms of the dataset, proposed framework, contributions, and detection accuracy. The first evaluated prior work based on [5] proposed a framework on detecting the usage of mobile phone by the driver and the hands on the steering wheel through multiple scale Faster R-CNN (MSFRCNN) which is able to detect small objects in low resolution under various scales, poses, and environmental conditions. Based on the sample datasets presented in the article, it can be observed that the images from both of the resources are captured inside the vehicles. There are two types of datasets that have been applied, which are the front view for the car driver and the front view for the steering wheel. Based on the accuracy reported from this article, the accuracy for the usage of mobile phone by driver detection is 0.94; the accuracy for the presence of hands on the steering wheel is 0.93, and the accuracy for the number of hands on the steering wheel is 0.65. However, it is more suitable to be implemented as car driver assistance system. The second evaluated prior work based on [6] proposed a framework on detecting the usage of mobile phone by cyclists and bikers through spectral Faster R-CNN architecture by incorporating a deep actionlet network on training the deep action specific features based on posture for phoning actions only. There are two types of datasets that have been applied for deep actionlet network training: the front view and the side view for car driver ported from SHRP-2 face view videos and DRCAM database, respectively. Based on the accuracy reported, the accuracy for the usage of mobile phone by cyclist and bikers detection is 0.93. It is more suitable to be implemented as a cyclist and bikers monitoring system. In this work, the input datasets are ported from the captured datasets from the camera system mounted at the overhead bridge. All the images are captured from outside the vehicles with a front view of the car. It proposes a framework for detecting mobile phone usage by the driver through Faster R-CNN with Inception-V2 architecture by fine-tuning the training configuration parameters. With the proposed framework, it is able to detect the posture of single-car driver bending their left or right hand toward their head from outside of the vehicles, which is more suitable to be implemented as a driver monitoring system. Based on the performance results, the accuracy for the usage of mobile phone by driver detection is 0.88. To our best knowledge, it can be concluded that this work is the first work in developing a framework for car driver monitoring system with captured datasets from outside the vehicles.

4 Conclusion

This paper presented a convolutional neural network-based approach for the detection of mobile phone usage during driving using Faster R-CNN with Inception-V2 architecture. A total of 1351 images from the collected dataset were processed, and in terms of overall detection accuracy, it obtained 88.71%. For future work, adverse weather conditions must be considered, such as rainy weather, to be applicable for real-time implementation of the proposed algorithm.

References

1. Injury Facts, Distracted driving—injury facts (2019). [Online]. Available at: https://injuryfacts.nsc.org/motor-vehicle/motor-vehicle-safety-issues/distracted-driving/
2. S. Ren, K. He, R. Girshick, J. Sun, Faster R-CNN: towards real-time object detection with region proposal networks. IEEE Trans. Pattern Anal. Mach. Intell. **39**(6), 1137–1149 (2017)
3. R. Gandhi, R-CNN, fast R-CNN, faster R-CNN, YOLO—object detection algorithms (2018). [Online] Medium. Available at: https://towardsdatascience.com/r-cnn-fast-r-cnn-faster-r-cnn-yolo-object-detection-algorithms-36d53571365e
4. A. Khazri, Faster RCNN object detection (2019). [Online] Medium. Available at: https://towardsdatascience.com/faster-rcnn-object-detection-f865e5ed7fc4
5. T.H.N. Le, Y. Zheng, C. Zhu, K. Luu, M. Savvides, Multiple scale faster-RCNN approach to driver's cell-phone usage and hands on steering wheel detection, in *CVPR* (2016)
6. A. Ulhaq, J. He, Y. Zhang, Deep actionlet proposals for driver's behavior monitoring, in *2017 International Conference on Image and Vision Computing New Zealand (IVCNZ)*, Christchurch (2017), pp. 1–6. https://doi.org/10.1109/IVCNZ.2017.8402447

Industry Revolution 4.0 Knowledge Assessment in Malaysia

Sharifah Kamilah Syed Yusof, Usman Ullah Sheikh, Norjulia Mohamad Nordin, Naziha Ahmad Azli, Aida Ali, and Nurul Mu'azzah Abdul Latiff

Abstract The impacts of the Fourth Industrial Revolution (IR4.0) on our life is well noted. The future job market landscape will drastically change and needs to be addressed by the Higher Education Institutes (HEIs) for graduate employability. This paper presents engineering education for IR4.0 in the context of Malaysia. The initiatives taken by the Ministry of Education Malaysia (MoEM) in addressing the challenges of IR4.0 for the higher education sector is discussed. This work assessed the IR4.0 knowledge amongst industries, the adoption, the readiness and the skill sets required for future-ready engineers. This is followed by an overview on the approach taken by one of the public universities in Malaysia in addressing the needs of IR4.0 through its undergraduate engineering programmes.

Keywords Engineering education · Fourth Industrial Revolution · IR4.0

1 Introduction

The role of technology as a catalyst for continuous growth and revolution is undeniable. From the first to the third industrial revolutions, technology has been the pivotal foundation for the society's transformation. Disruptions in technology such as digitization, automation, Artificial Intelligence (AI) and the Internet of Things (IoT) are the drivers of another massive change—the Fourth Industrial Revolution (IR4.0), or more commonly known as IR4.0. The recent years have seen the impacts of IR4.0 cyber-physical systems on various aspects of our life, including manufacturing, business and education. The era of the IR4.0 not only alters the lifestyle of the people but also drastically changes the future scenario of job markets. Many tasks in business and manufacturing are known to be rapidly taken over by robots and

S. K. S. Yusof · U. U. Sheikh (✉) · N. M. Nordin · N. A. Azli · N. M. A. Latiff
School of Electrical Engineering, Faculty of Engineering, Universiti Teknologi Malaysia, Johor Bahru, Johor, Malaysia
e-mail: usman@utm.my

A. Ali
School of Computing, Faculty of Engineering, Universiti Teknologi Malaysia, Johor Bahru, Johor, Malaysia

A. Kumar et al. (eds.), *Computational Intelligence in Machine Learning*, Lecture Notes in Electrical Engineering 834, https://doi.org/10.1007/978-981-16-8484-5_16

intelligent machines. Thus, jobs that are currently available may cease to exist in the near future and replaced by new or yet to be created jobs. This calls for the need to reevaluate the industry's current job skills and future job skills, and also to prepare for the development of graduates at the Higher Education Institutes (HEIs) for future job skills, respectively [1]. The Malaysian Education Ministry has highlighted the planning to develop future-proof graduates with the right set of skills, abilities and humanistic values. The Malaysian Higher Education Framework 4.0 [2], highlighted the Future Ready Curriculum (FRC) as one of the main initiatives in preparing graduates of HEIs in Malaysia towards IR4.0. The Malaysia Qualification Framework (MQF) second Edition [3] was released in 2017, which addressed the academic programme learning outcomes as follows: (1) Knowledge and understanding, (2) Cognitive skills, (3) Functional work skills, (4) Personal and entrepreneurial skills and (5) Ethics and professionalism.

In response to the IR4.0, Industry4WRD, the national policy on Industry 4.0 was launched in 2018 by Malaysian Ministry of Trade and Industry (MITI). The policy provides a comprehensive digital transformation agenda for the manufacturing and related services sectors in Malaysia [4]. Adopting IR4.0 in Malaysia has undoubtedly implicated the learning, training and development roles. These roles are becoming more significant and challenging in the effort to yield knowledgeable and highly skilled human resource, especially in Malaysia. Universiti Teknologi Malaysia (UTM) is a partner in the Erasmus+ Capacity-building projects in the field of higher education (CBHE) project with the primary goal in Skill4.0 provision. Therefore, in this paper, we would like to share the findings from UTM team's survey study in relation to IR4.0 knowledge assessment in Malaysia. This survey involves participants from Malaysian companies. This paper presents the analysis of the study conducted from the responses gathered by UTM. This paper is organized as follows. In Sect. 2, an overview of the Erasmus+ project partnership is introduced. Section 3 describes the survey methodology followed by the results and discussion in Sect. 4, and the conclusion and the way forward in Sect. 5.

2 Building Skills 4.0 Through University and Enterprise Collaboration (SHYFTE)

Erasmus+ CBHE provides the opportunity for multilateral partnerships from eligible partner countries, mainly HEIs, to promote transnational cooperation through actions that include quality of higher education provision and its alignment with labour market needs [5]. The overall aims of SHYFTE Erasmus+ CBHE project are:

i. To support academic and administrative staff in Asian HEIs to design and implement new methodology and learning materials to enhance and improve the competences and skills related to IR4.0 that will enable HEIs organizations aligning their learning programme strategy both to the requirements of the regional industry and the global labour market.

ii. To build and strengthen links between HEIs and Industry 4.0, and promote the job market by minimizing the skills gap.
iii. To build Skills 4.0 LABs or Learning centres of excellence in Asian Partner's HEI to enable each of them to become the reference centre in its country and disseminate the results of the project nationally and regionally.

In order to fulfil the requirements of the IR4.0, SHYFTE project will jointly be carried out by HEIs and small and medium enterprises (SMEs) in Europe and Asia. The huge skills gap between HEIs and Industries cannot be eliminated by later stages of the chain. That is why HEIs must be developed in associations with industries and SMEs. The project partnership constitutes both EU and Asian partners. To provide a desirable job market platform, SHYFTE will provide the necessary and novel skill development strategy in both EU and Asia with the primary focus on four following research domains; (1) Industrial engineering and management, (2) Software engineering and big data analytics, (3) Wireless and networks analytics and (4) Artificial intelligence.

3 Industry Revolution 4.0 Knowledge Assessment Methodology

As mentioned earlier, IR4.0 has given a new impetus to education and industry sectors transformation with far greater needs to produce employees and graduates with relevant and meaningful skills. These have necessitated profound changes in the learning, training and development role provisions. These roles are becoming more significant and challenging in the effort to yield knowledgeable and highly skilled human resource relevant to IR4.0, especially in Malaysia. The IR4.0 knowledge assessment is very much linked to the demands of the human resource skills for IR4.0. Prior to the formation of the assessment tools, a literature study on the skills requirements are identified by the accountable partners and are formed as part of the items in the questionnaire survey. Notably, SMEs are the foundation of the growth of most countries' economy. There is still a lack of relevant discussion regarding the SMEs' development through the implementation of Industry 4.0 principles. Therefore, the target participants in this assessment are companies which include SMEs. In this paper, we are focussing on a survey conducted by the UTM team. In this paper, the analysis of the data gathered are presented. Strategically, from the assessment carried out from all the SHYFTE partners, it is anticipated the gaps in the IR4.0 adoption can be determined and will be referred to in the development of effective training learning materials.

4 Results and Discussions

The assessment survey is conducted by UTM with 95 Malaysia companies participation, consisting of national and multi-national companies. The main purpose of the survey is to evaluate the IR4.0 skills in the current job employer and future HEI graduates, hence, addressing the needs for Industry 4.0 revolution in the future undertakings. There are a total of 29 questions constructed by the SHYFTE. Figure 1 shows the distribution of respondents in the survey whereby manufacturing (30.5%), electric, gas and power supply companies (17.9%) and education (11.6%) makes up the top three respondents. In terms of capital, 52.6% of the respondents have capital of over RM 15 million, 31.6% between RM 300 k to RM 15 million and the remaining 15.8% with capital of lower than RM 300 k. 68.5% of the respondents employ over 100 employees, 12.7% employ between 20 and 99 employees, whilst 18.9% of the companies have less than 20 employees.

4.1 *Vision of Digital Transformation and IR4.0 Technology Importance Assessment*

One of the key aspects of identifying the readiness for IR4.0 is to evaluate and measure the vision of SMEs in terms of digital transformation. From the survey (Fig. 2), there is a worrying trend that ~12% have no vision to low vision for digital transformation, whilst ~36% have low to medium transformation. Moreover, 18.9% of the companies have not heard of the term Industry 4.0, showing the crucial need to educate SMEs. When enquired on evaluating the importance of IR4.0 technologies, SMEs in Malaysia mostly rate manufacturing technologies (e.g. robotics, IoT, wireless technology) with the highest importance, whilst fundamental technologies (e.g. nanotechnology) are rated as the least important. In terms of most important technology as assessed by the SMEs, the implementation of application of IoT is

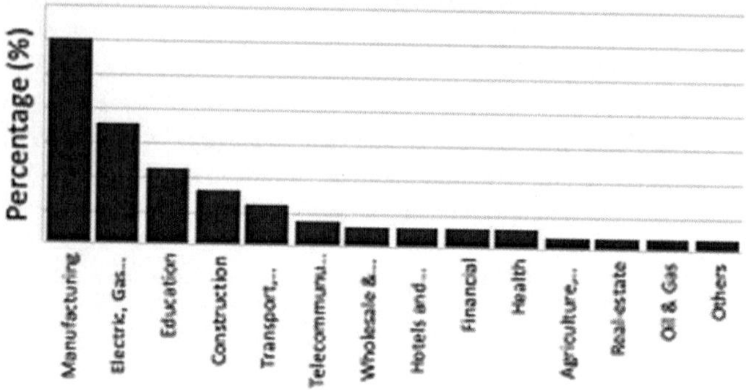

Fig. 1 Respondents in the survey conducted

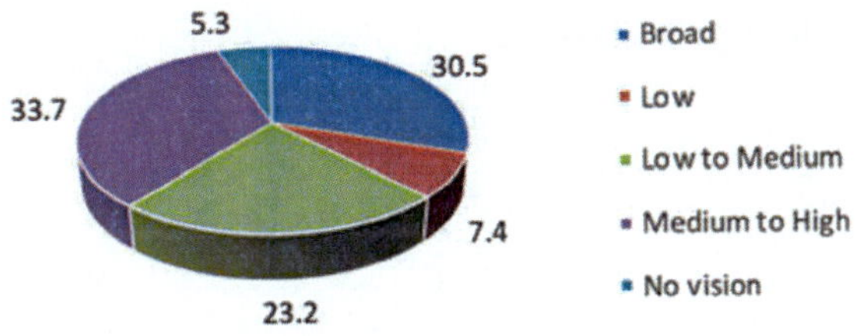

Fig. 2 Vision for digital transformation

Table 1 IR4.0 technology importance assessment by Malaysian SMEs

Technology	Importance
IoT	Advanced
Cloud Computing	Advanced
Wireless Technology	Advanced
Cyber Security	Advanced
Simulation Systems	Advanced
Robotics Automation	High
Collaborative Robotics	High
Virtual Augmented Reality	High
Cyber Physical Systems	High
Big Data/Data Science	High
Additive Manufacturing	High
Nanotechnology	Low
Smart materials	Low

most sought after, followed by cloud computing, wireless technology, cyber security and simulation systems. Table 1 summarizes the findings.

4.2 *IR4.0 Skills Assessment*

Having employees with the right IR4.0 skills set is crucial in propelling the company towards industry 4.0. Based on the findings, SMEs larger agree that the skills of the employees are greatly inadequate, and none of them assessed the skills as substantial (Table 2). Understanding the need for IR4.0, 45.4% agrees on the importance of deploying team of experts for adopting IR4.0 and 50.5% have plans to deploy team of experts in the near future. Although IR4 focuses on automation, artificial intelligence and technologies to accelerate manufacturing, the respondents emphasized that besides technology as the driving factor, SMEs greatly value skills such as decision making/problem solving, responsibility and adaptability and the ability to work in a group, thus reinforcing the fact that future-ready engineers must possess not only technology-related skills but also soft-skills.

Table 2 Skills assessment

Technology	Skill
Wireless Technology	**Adequate**
Cloud Computing	Inadequate
IoT	Inadequate
Cyber Security	Inadequate
Simulation Systems	Inadequate
Robotics Automation	Inadequate
Collaborative Robotics	Inadequate
Virtual Augmented Reality	Inadequate
Cyber Physical Systems	Inadequate
Big Data/Data Science	Inadequate
Additive Manufacturing	Inadequate
Nanotechnology	Inadequate
Smart materials	Inadequate

4.3 IR4 Establishment and Willingness to Adopt

One of the primary purposes of the survey conducted is to measure the establishment of IR4 and the willingness of enterprises to adopt IR4. In terms of an existing establishment, the current state of IR4 is 5.3% have no adoption, 45.3% have medium establishment and 43.2% have high implementation. What is encouraging is 62.1% are willing to implement IR4.0, 31.6% have a moderate interest in adopting IR4, whilst 6.3% are not willing.

4.4 Technology Adoption

IR4 requires various supporting technologies ranging from wireless technologies, data engineering and management and programming languages, to name a few. In the terms of wireless technologies, our research found that there is a good initial adoption of 5G whereby 10.3% respondents are already using it. IoT has been found to be quite matured in Malaysia as 48.3% are using IoT in their business. Data organization is key to big data and data analytics. Industry response shows, a lot of unstructured data is used but many large enterprises are already geared towards big data utilizing data warehouse and Hadoop. 37.8% respondents are currently deploying data engineering e.g. data acquisition (batch/real-time), pre-processing (offline/real-time) and batch processing. Whilst 15.8% currently need but not deploying. 37.8% respondents are actively using data analytics (e.g. descriptive, exploratory, causal, predictive, prescriptive), whilst 18.9% currently need. Automating industrial processes normally requires engineers to programme systems, and algorithms to perform data analysis, decision making and automation. Our survey found that industries in Malaysia favour open-sourced programming tools and languages. Most industries use at least one or more of the following languages in production; C++ (60.3%), Java (48.3%), Python (20.7%), R (12.1%) and Scala (10.3%).

When enquired about the adoption of artificial intelligence (AI), over 46.5% respondents do not use any AI algorithms or foresee no use for it (only 29.3% actively use it). However, owing to the fact of low AI adoption, 58.6%, 68.9%, 68.9% do not see the need for Python, Prolong or Lisp, respectively. The most actively used programming language for AI is Java (58.6%), C++ (55.2%) and Python (27.5%) (industries use at least one or more of the given programming languages).

In the next 5 years, respondents general are willing to spend on IT, production, marketing and sales and less on logistics. Industries agree that IR4.0 can be implemented in small businesses which will enable them to be more competitive and agile. However, industries need skills they do not have right now, and failing to cope up will mean possibility of being excluded from the market.

5 Conclusions

The status quo of the perception of IR4.0 from the sample companies in Malaysia have been identified. Most companies agreed on the benefits of IR4.0. Their willingness to adopt IR4.0 and acceptance in being agile and competitive are commendable. The support from the government-industry related entities is inevitable in making IR4.0 success a reality. In the case of Universiti Teknologi Malaysia, proper actions have been taken in addressing the needs of the IR4.0 through its academic programme, curricular transformation and training requirements. The training target participants would be current company employees, active HEI students and fresh graduates.

Acknowledgements The authors wish to express their gratitude to Ministry of Higher Education (MOHE), Research Management Centre (RMC) for the sponsorship, and Universiti Teknologi Malaysia for the financial support and advice for this project (vote number Q.J130000.3651.02M73).

References

1. Website of the World Economic Forum, https://www.weforum.org/
2. *Framing Malaysia Higher Education 4.0, Future-Proof Talents* (Department of Higher Education Malaysia, Ministry of Education Malaysia, 2018)
3. *Malaysia Qualification Framework (MQF)*, 2nd edn. (Malaysian Qualifications Agency, 2017)
4. *Industry4WRD* (Malaysian Ministry of International Trade & Industry, 2018)
5. Erasmus+ CBHE, https://ec.europa.eu/programmes/erasmus-plus/opportunities/organisations/innovation-good-practices/capacity-building-higher-education_en

Practical Approaches: User Stories Domain Prediction and Keyword Extraction for Insights

Vaishali Sharma, Khyati Ahlawat, and Shorya Sharma

Abstract This project aims to make a recommendation system for user stories domain prediction and extracting useful keywords and key phrases. It gives an insight to business analysts and helps them write their user stories in a particular well-defined manner. So that next time whenever a client gives a requirement to some new business analyst and at the time of writing user stories, they will know for what domain they are writing and what relevant information he/she needs to ask from the client. Moreover, we are extracting useful keywords and key phrases from user stories which can also be used as tagging and giving insights to them for their business purpose in future. A recommender system or recommendation system is a subdivision of information retrieval and information retrieval system used to predict the categorization or choices of a user on a particular item in general. The recommendation system is widely and primarily used in commercial areas. Later on, many of them are used in wide areas like Facebook, Netflix, Amazon Prime, YouTube, etc. As we have mentioned above, the aim of this project is to give an insight to business analysts to make their user stories fast and in an appropriate manner. So that whenever a client gives some requirements to a new business analyst, then he/she would have the ability to recognize that this user domain story would fall under which category and which domain they are talking about. We are going to make a recommendation system using different deep learning algorithms to analyze and evaluate the performance of different algorithms on it. After that, we are extracting important keywords and key phrases from it by implementing a natural language processing technique named "Named Entity Recognition" which further has multiple uses like tagging, maintaining stocks, helps in making insights of business, etc. We are trying to build an API for it.

Keywords Machine learning · Recommendation · Keyword extraction · Neural network · Natural language processing · User stories

---

V. Sharma (✉) · K. Ahlawat · S. Sharma
Department of CSE, Indira Gandhi Delhi Technical University for Women, Kashmere Gate, New Delhi, India

A. Kumar et al. (eds.), *Computational Intelligence in Machine Learning*, Lecture Notes in Electrical Engineering 834, https://doi.org/10.1007/978-981-16-8484-5_17

1 Introduction

1.1 Natural Language Processing

If we go back and look at the origins of languages, the basic idea to create languages was to simplify human communication. Everyone needs to communicate one way or another. Without languages, in place, society would be in chaos and will eventually collapse. All great inventions are a result of effective communication and the ways that we humans have used languages for the betterment of our own society. Our languages on Earth have rules and order and can be stored in a library for storage purposes.

Languages that are spoken by humans are termed as Natural Languages. It is our way of communicating. But what about computers? Well, NLP is a subdivision of AI and linguals dedicated to making machines understand the language of human languages. Natural language processing language, in this case, can be many which include English, Hindi, Chinese, etc. Also, we have to bear in mind a computer only understands machine-specific language; it is impossible for humans to speak in machine language. So we rather make the computer understand. The computer interprets our human language [4].

If we look back at the core of languages, languages are nothing but a collection of rules or even characters. Symbols are concatenated and are then used for displaying some information or even contradict thoughts or convey basic human ideas. The applications are varied. The rules put everything in order as to how languages are used and governed. They bring in order and not chaos. Libraries and books store the data.

We use natural language in almost each and every platform. We use it even on smartphones and computers. Encompasses everything related to computer understanding and generating natural languages. In simple terms, it is the process of making a computer a human who can contemplate each and every word said by a human mouth and respond accordingly. A natural language [3] is also known as the common language that is used by human beings to accomplish day-to-day tasks. Natural language processing is all about computers getting what we mean. It is for computers to understand our ways of communication. The language, in this case, can be many which include English, Hindi, Chinese, etc. Also, we have to bear in mind a computer only understands machine-specific language; it is impossible for humans to speak in machine language. So we rather make the computer understand. The computer interprets our human language.

If we look back at the core of languages, languages are nothing but a collection of rules or even characters. Symbols are concatenated and are then used for displaying some information or even contradict thoughts or convey basic human ideas. The applications are varied. The rules put everything in order as to how languages are used and governed. They bring in order and not chaos. Libraries and books store the data.

We use natural language in almost each and every platform. We use it even on smartphones and computers.

1.2 Natural Language Generation

Natural language generation on the other hand is a subclass of natural language processing (NLP) and is also known as the next generation. We should keep in mind that NLG [10] is the fastest-growing technology in the business world. The most common usage is to robotize reporting tasks and time-dependent data analysis.

Data was always required to be analyzed. In fact, that is why we collect data, so that we can analyze it and draw some conclusions. But the corporations are at huge pressure to reduce costs and meet customer needs along with reading the data and drawing conclusions. Hence, to automate and optimize the process, we make use of natural language generation. It is to combine the human mind and efficiency of a computer. For visual interfaces, we often feel drawn in. But with natural languages, some bullet points and words can do a lot more than the visual interface will ever be able to do. It is important to understand NLG a lot more than this. In big data, we are able to produce robotized results and that is when NLG comes into play. If it is only visual, there is definitely a need to increase the scope. It can combine both textual information and visual information.

1.3 Named Entity Recognition

It defines a streamlet of text and then further determines which entity in the document belongs to which nouns such as the name of place, country names of the person, or an object. Natural language processing transforms huge classes of text into a more formal manner in which it first defines the structures of a text because it will be easy for computer programs to understand and manipulate into natural language processing concepts. NER is also referred to as entity identification, chunking or extraction. Hence, from the very definition, NER is a form of NLP. NER is basically a two-step procedure.

a. A named entity is detected
b. Categorize the following entity.

Step one revolves around detecting a string or words that lead to the formation of an entity. Each word for example leads to a token. Example "The great falling" which is three tokens but represents one entity.

Step two revolves around common entity categories which are persons and organizations. We can create entities depending on the task that we want to perform accordingly. What is important is the training data. The more relevant the training data is, the better it is for the model to get trained. The more accomplished the model will be at completing tasks. Train your model for a different task, and it will not do the task of what was expected from the model. Once everything is defined and set, labeling data and creating a training dataset can be used to effectively train an algorithm that can eventually label text. With NER, quickly grouping texts is no big deal and the subject or theme of a body can be quickly understood at a glance.

2 Related Work

After reviewing, the complete literature survey on keywords extraction and named entity recognition techniques in the past few years. We analyzed that the problem of the user stories domain has not been resolved, and there has not been any research done on this issue. So this project aims to work on user domain stories and makes a kind of recommendation system based on user stories. Earlier work has been done on improving the parameters of the named entity recognition techniques such as precision and recall or try to improvise existing algorithms. Earlier research work has been done on text summarization for single documents and multiple documents also some of them developed few new models for improvising the accuracy of models as compared to previous existing models and techniques. Although few researches have been done on various application areas of keyword extraction technique, this problem has not been solved. Some research has been done on keyword extraction from single and individual documents, but the flaw was that research would not consider grammatical conventions. Also, few research has been done on making a model named Rake whose accuracy is more than the existing techniques like text rank algorithm, TF-IDF algorithm, etc., and various applications have also been solved which is keyword extraction from emails and newspapers.

3 Practical Implementation on Domain Classification and Keyword Extraction

Here, in this section, we have briefly mentioned the details, and the steps involved during this implementation of the project are described below.

3.1 Data Collection

The very first important task during this project was data collection. So, we had collected real-world data using Jira software. This project aims to work on a real-time dataset. This dataset has been collected with the help of experts, business analysts, and data consisting of user stories and their specific domain class in which they fall.

Our data had two columns, one of it was the user stories, and the second column was their domain. And the raw data that we had collected was not appropriate to directly use for interactive analysis. And the next step is data cleaning.

3.1.1 Data Cleaning/Data Wrangling

After the accomplishment of the data collection part, the data we had collected is messy, and we cannot proceed without cleaning that data. This step is very important in building any type of model in machine learning, deep learning, and natural language processing that after collecting data from the data source we cannot train our model on that. Typically, there is a need to transform and aggregate that data according to the model that we want to build. In this step, we remove some tuples from our dataset that are not necessary for the model, or we can say that they are some kind of outlier information in our data. This step is performed before we do an analysis of our data, and it is also called data preprocessing [2].

After preprocessing has been done, now the next step that comes under the data cleaning is data wrangling. Data wrangling is typically begun when we start doing interactive analysis and model building with our data. During this step, we can analyze our data that may change the shape of our data and extract useful attributes out of it. We can transform the dataset according to our views on the dataset. This step is very useful in providing insights for the dataset and generating a good interactive and analytical model.

3.1.2 Data Annotation

This is the process of adding metadata to our dataset. The metadata that we added it in the form of labels or tags and we can add metadata to any kind of structured and unstructured data. And this step comes under the training of the dataset. Data annotation is a crucial stage of data cleaning because based on annotated data our model predicts the outcomes of the pattern, labels, and tags. And in our dataset, it took 48 h to annotate the data, and after that, we can implement an algorithm on our dataset.

We had done data annotation using a tool named Doccano. Doccano [2] is an open-source tool available for data annotation. It provides data annotation in the form of labels, tags, text classification, and sequencing as represented in Fig. 1.

3.2 Algorithm Implementation

After accomplishing our data preprocessing part, we started implementing algorithms on our dataset.

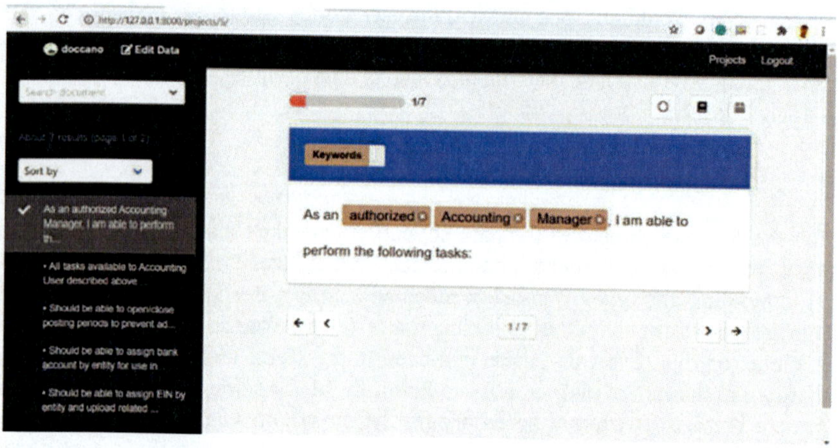

Fig. 1 Data annotation [4]

The LSTM and CNN models are both models used for implementation of different NLP which are trying to get convincing results. Model convolutional neural network removes high-level features using convolutional layers and layers of multiple integrations. The LSTM model is ready to find the most important conditions between word systems.

The first algorithm that we have implemented is LSTM which is a supervised deep learning algorithm.

1. Long short-term memory: LSTM algorithm is a better version of the feed-forward recurrent neural network model. It is a supervised deep learning algorithm. As the name of this algorithm explains itself, it works well on sequencing data as in Fig. 3. Sequencing data means that it can predict patterns based on the previous data that is stored in their memory. The accuracy of LSTM was only 34% on this dataset which is depicted in Fig. 2.
2. Convolution neural network: As the accuracy of the LSTM model is low and inefficient, now we are trying to implement another deep learning model [9] which is also known best for classification and sequencing of data. The working of CNN model is as represented in Fig. 4.

CNN is a deep learning structure often used for sequential record classification. Initially, CNN was designed for image processing and was successfully used for text-based content separation. In CNN's basic image processing, an image sensor is displayed with a set of characters of dimension $d \times d$. These layers of convolution are known as feature maps, and they can be integrated to provide multiple filters for input. A simple CNN with low hyperparameter tuning and static vector can achieve the best result on more than one benchmark.

When learning to practice CNN [6] on word embedding, maintaining tracks of the dimension of the matrices can be confusing. Max pooling is a commonly used pooling technique in the process of getting the maximum element where the top item

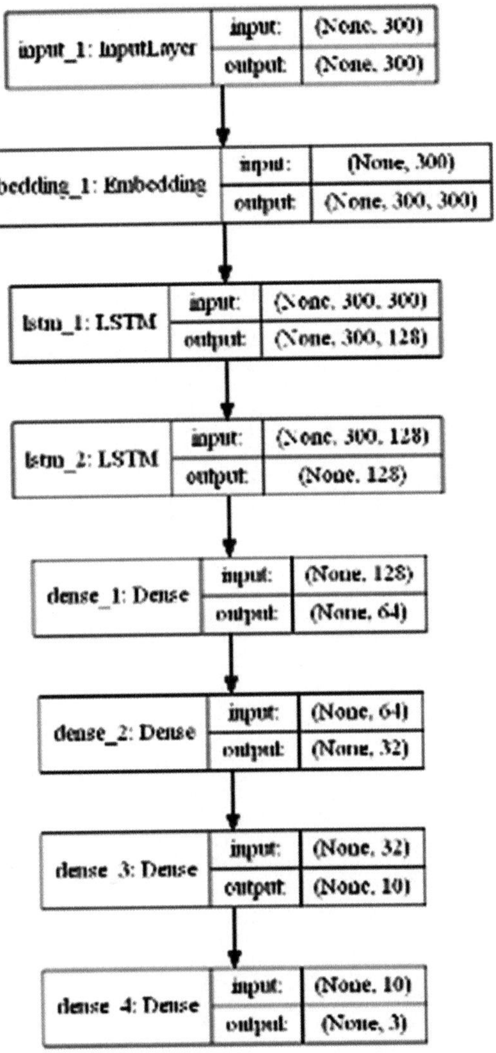

Fig. 2 LSTM graph model

is selected in the merge window. The maps are organized into a single column to feed all the aggregate pooled results from the stacked features which are included in the next layer. The last layers of convolution neural networks are usually completely connected [1].

A CNN is a deep learning technique which is also known as ConvNets [8]. It takes input as image, text, and assigning various parameters such as weight and bias to it. This will help in differentiating one image/object from the other. CNN is widely chosen for image processing, sentimental analysis, face recognition, X-ray analysis, cancer detection, and many more applications. The preprocessing of the data in a

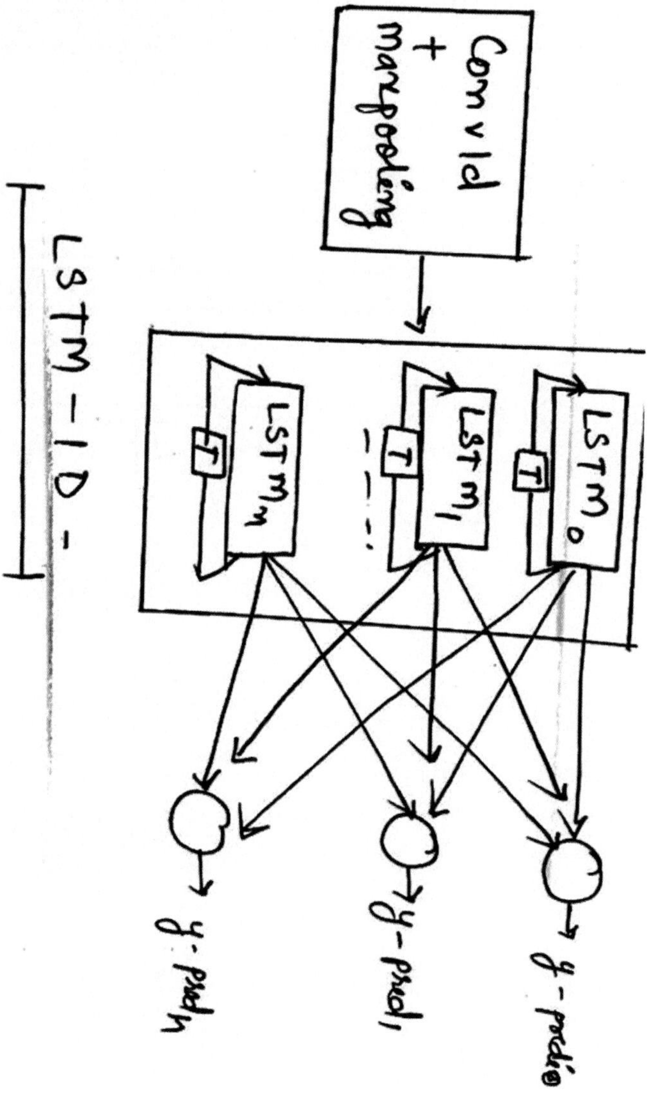

Fig. 3 LSTM 1D model [7]

convolution neural network is much low in comparison with other machine learning approaches.

CNN is a supervised feed-forward artificial neural network where connections are acyclic. It uses a multilayer perceptron design to perform the processing of the data. CNN can capture the contiguous and temporary dependencies of the filters of pictures. This will help in the better fitting of the image by reduction of various

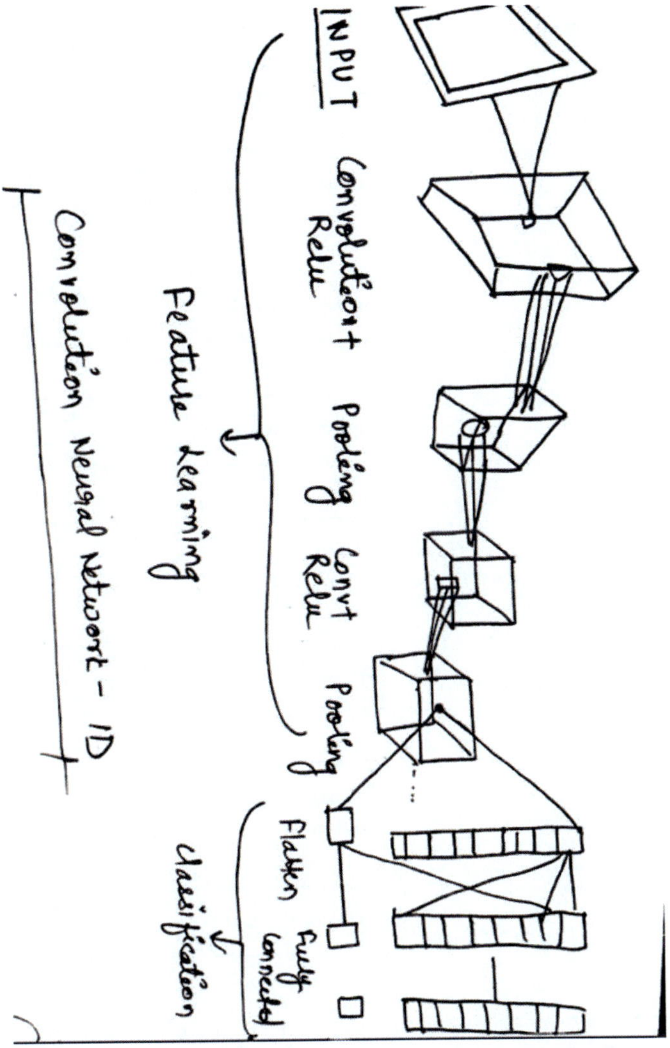

Fig. 4 Convolution neural network 1D model [8]

parameters and filters, and it will give surpassing results. The role of CNN is to reduce the preprocessing [5] of datasets without losing the vital attributes and features of the picture and classify it.

The accuracy of the CNN algorithm works well and efficiently on our dataset. Here, we attached the graph that we have obtained during implementation on our dataset.

Now, we successfully build the model with the help of CNN and LSTM deep learning algorithms, train them well, and test their accuracy. This is the part of

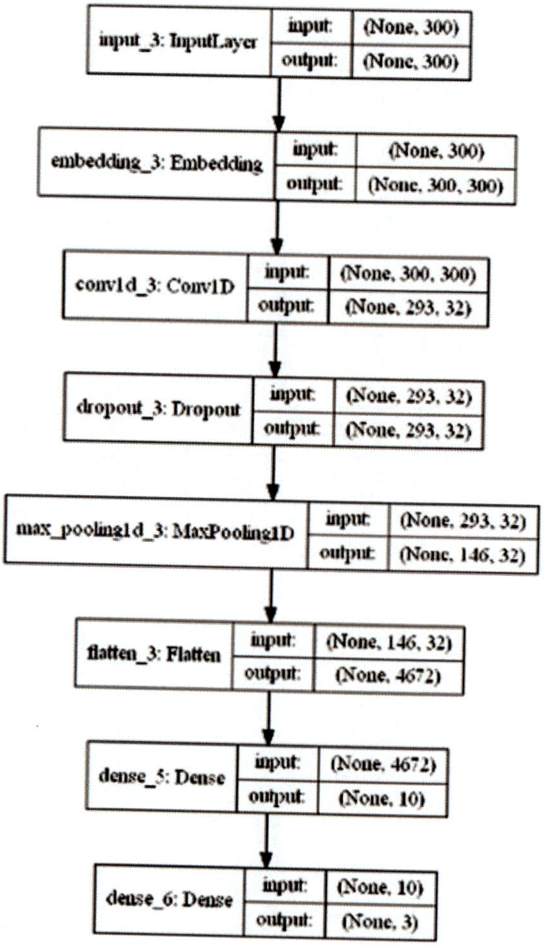

Fig. 5 CNN graph model [9]

making the recommendation model of our project. In the next part, we are trying to extract useful and relevant keywords from the dataset using keyword extraction named entity recognition. The accuracy of CNN [11] model is much better, and it worked well on this real-time dataset. The accuracy of CNN model was 93.4% as depicted in Fig. 5.

3.3 *Keyword Extraction*

Keyword analysis uses a text-based technique that automatically extracts the most used words from a text. It is to extract words from a text that are most used or are rather more important. It helps summarize the content and take out the most important word from that text. As suggested as our project deals with natural language processing [7], keyword analysis makes use of NLP, so that our human language or the search terms as you may say could be broken down and then studied by the machines. The application of keyword analysis is broad and is used in all forms of text be it business reports, on social media, forums, and reviews.

There are various approaches of keyword extraction based on which keyword can be extracted, and they have different principles on which they work upon. In our project, we used three algorithms of keyword extraction.

1. Text Rank Algorithm: Text rank in an unsupervised keyword extraction approach in which it first gathers all the text available in the document. In the next step, it tokenizes the text into sentences, and after that, it forms a vector embedding for each word and sentence present in the document. In the next step, we are trying to find a similarity between a word or a sentence vector and store it in a matrix as shown in Fig. 6. In the end, we generated a graph from the matrix representation.

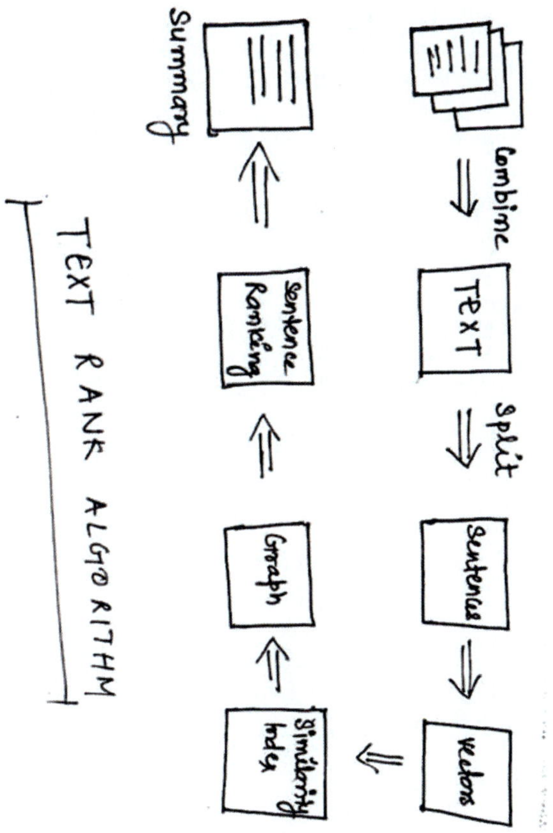

Fig. 6 Working of text rank algorithm [6]

2. NPR: NPR is another approach to finding out useful keywords from a text. But in this approach, first tokenizes the data into words and sentences and generates a pattern out of it. It works the same in the manner of the part of speech tagging approach generating a pattern and works on identifying keywords from a text.
3. SpaCy Custom Keyword Extraction: This was the third approach that we had followed while building our model. This was the approach where we had manually annotated our data. We had to carry an annotated file and download it in the form of JSON file format. And then, we trained our model on custom keywords and loaded it with our model. We had used the Doccano tool for the data annotation process.

3.4 API Building

The last step in our implementation details was of building an API for this model.

1. After successfully building an API, this was the appearance of our API window depicted in Fig. 7.
2. In the next step, we clicked on the keywords and domains, and clicked on the Try it Out Option which appeared on the right side of the API window, a next window came where we had to put the user story that we want to classify. The output of this is as shown in Fig. 8.

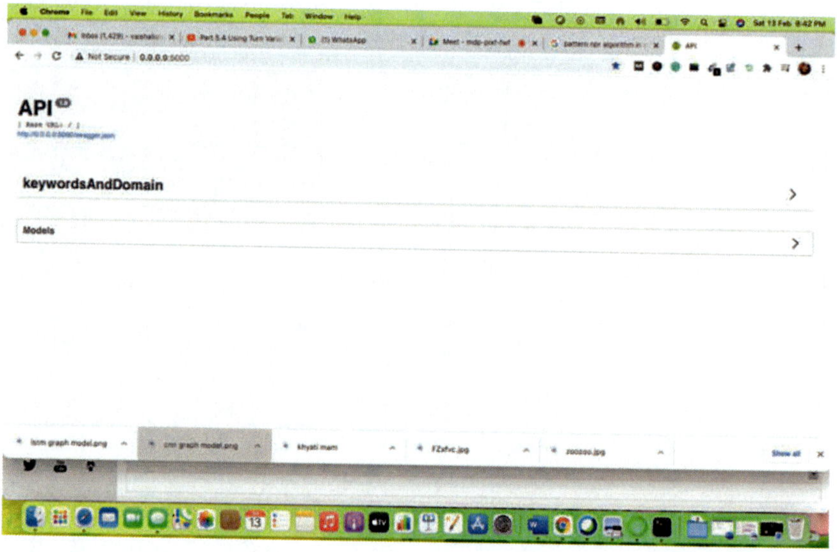

Fig. 7 API window

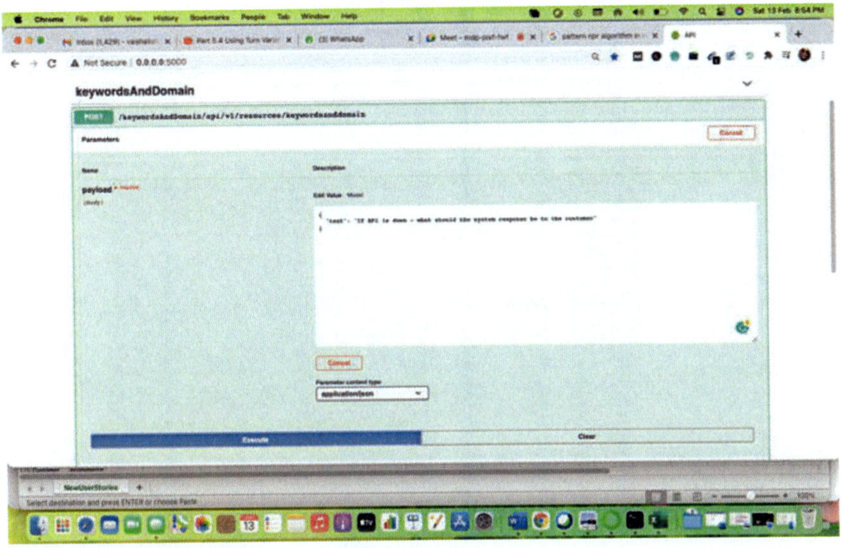

Fig. 8 Keyword and domain

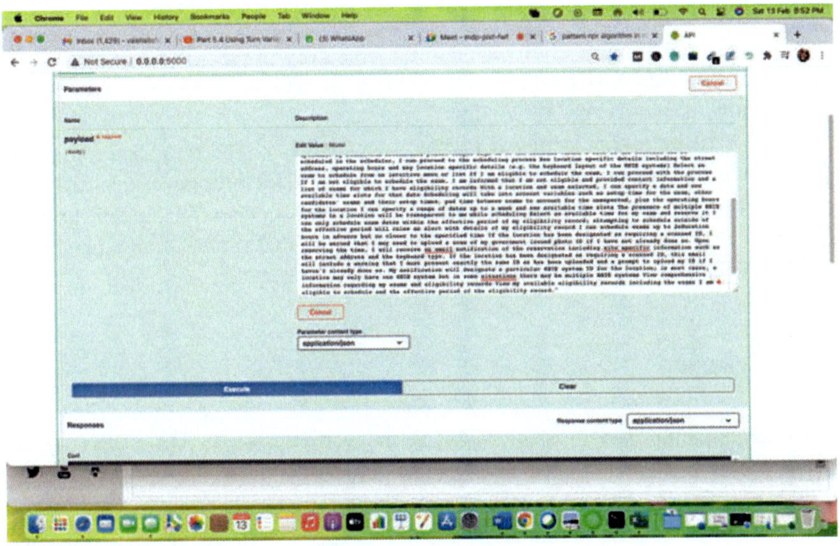

Fig. 9 Entering user story

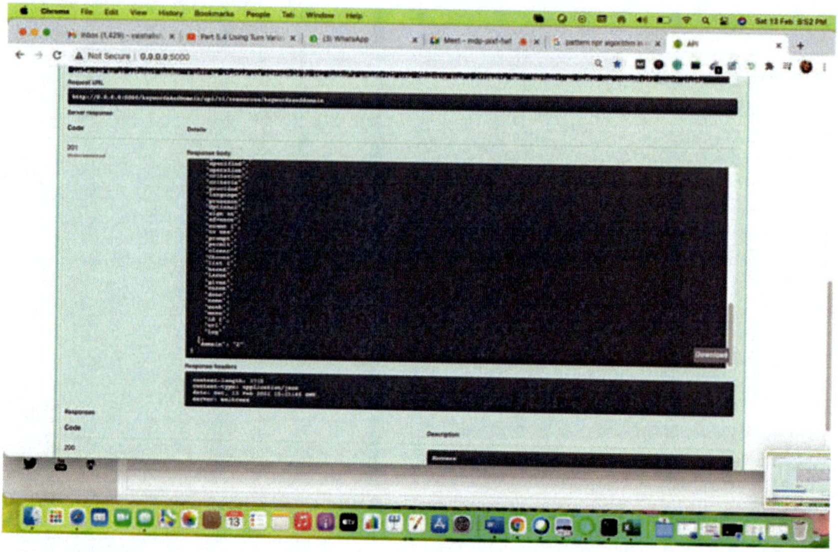

Fig. 10 Output of specifying domain

3. This was the window after we entered a user story in the API for classification of its correct domain in which it falls. Figure 9 depicted below is the output for entering user story in API window.
4. This was the output depicted in Fig. 10 which is of the classification of the domain of user story that we had entered. In our model, we have specified three domains, 0 is used for banking domain, 1 is used for education domain, and 2 is used for e-commerce domain. And the user story that we had entered is related to e-commerce, and it is correctly classified by our model.

4 Conclusion

A recommendation-based system based on the NER technique was developed to extract useful and relevant keywords and key phrases that would further recommend to us in the future. Also, this project will assist business analysts and society to save their precious time on manually extracting useful information. Also, we aim to develop an API for it. In this paper, we came up with a solution to the domain prediction problem in which we build a model that is capable of doing domain prediction as well as keyword extraction. This paper worked on building an API for user stories domain prediction and keyword extraction for insights. The model automatically reads user stories and predicts the specific domain in which they fall. It would help business analysts to classify the domain of their user stories, and this API will be really helpful for society and business analysts. To extend the functionality of domain prediction and keyword extraction technique, we tried to build API using both domain prediction and keyword extraction [12].

References

1. S. Beliga, *Keyword Extraction: A Review of Methods and Approaches* (Dep. Informatics, Univ. Rijeka, Rijeka, 2014), pp. 1–9
2. R. Campos, V. Mangaravite, A. Pasquali, A.M. Jorge, C. Nunes, A. Jatowt, A text feature based automatic keyword extraction method for single documents, in *Lecture Notes in Computer Science (Including Subseries Lecture Notes in Artificial Intelligence and Lecture Notes in Bioinformatics) LNCS*, vol. 10772 (2018), pp. 684–691
3. P.K. Das, M.K. Majumder, B.K. Kaushik, S.K. Manhas, in *Proceedings of the Second International Conference on Soft Computing for Problem Solving (SocProS 2012)*, 28–30 Dec 2012. Adv. Intell. Syst. Comput. **236**, 1117–1126 (2014)
4. P.W. Eklund, A. Hoang, A performance survey of public domain supervised machine learning algorithms. Aust. J. Intell. Inf. Syst. **9**(1), 1–47 (2002)
5. Y. Kim, J.H. Lee, S. Choi, J.M. Lee, J.H. Kim, J. Seok, H.J. Joo, Validation of deep learning natural language processing algorithm for keyword extraction from pathology reports in electronic health records. Sci. Rep. **10**, 1–9 (2020)
6. S. Lahiri, R. Mihalcea, P.H. Lai, Keyword extraction from emails. Nat. Lang. Eng. **23**, 295–317 (2017)
7. J. Li, A. Sun, J. Han, C. Li, A survey on deep learning for named entity recognition. IEEE Trans. Knowl. Data Eng. **XX**, 1 (2020)
8. J.R. Lin, Z.Z. Hu, J.L. Li, L.M. Chen, Understanding on-site inspection of construction projects based on keyword extraction and topic modeling. IEEE Access **8**, 198503–198517 (2020)
9. A. Onan, S. Korukoğlu, H. Bulut, Ensemble of keyword extraction methods and classifiers in text classification. Expert Syst. Appl. **57**, 232–247 (2016)
10. S. Rose, D. Engel, N. Cramer, W. Cowley, Automatic keyword extraction, in *Text Mining: Applications and Theory* (2010), 1–277
11. T. Sexton, M. Hodkiewicz, M.P. Brundage, T. Smoker, Benchmarking for keyword extraction methodologies in maintenance work orders, in *Proceedings of the Annual Conference of the Prognostics and Health Management Society PHM* (2018), pp. 1–10
12. J.R. Thomas, S.K. Bharti, K.S. Babu, Automatic keyword extraction for text summarization in e-newspapers, in *ACM International Conference Proceeding Series*, 25–26 Aug 2016

An Improved Content-Based Image Retrieval System for Tomato Leaf Disease Classification

Gurubelli Yogeswararao, R. Malmathanraj, and P. Palanisamy

Abstract In this research work, a novel content-based image retrieval (CBIR) system is developed to classify the tomato plant leaf diseases. The proposed CBIR system uses color, shape, and texture features of the tomato leaf to classify similar images. The HSV color histogram is used to extract color features and Fourier descriptors provide shape feature, in the form of the contour of the region of interest. In order to consider global texture features, a variant of local binary pattern (LBP) called completed LBP (CLBP) is utilized. Furthermore, feature fusion of all color, shape, and texture properties is done to increase accuracy. Based on this feature vector, classification of disease is done using a supervised learning technique called support vector machine (SVM). The analysis of different kernels like linear, RBF, polynomial, and hyper-parameter optimization showed that linear kernel is best suitable. In regards to classification, the mean accuracy of 97.3% is achieved in the linear SVM model by five-fold cross-validation.

Keywords Color histograms · Fourier descriptors · Local binary patterns · Support vector machine

1 Introduction

The agriculture industry is one of the most vital sectors in India, which hugely contributes to the economy of the nation. Also, it is one of the least touched upon sector with technology, where major part of the process is done physically by human intervention. As technology has developed over time, many agriculture components and processes have become automated to ensure speedy production of the highest quality products. Because of the increased demand in the agricultural industry, agri-

G. Yogeswararao (✉) · R. Malmathanraj · P. Palanisamy
Department of Electronics and Communication Engineering, National Institute of Technology Tiruchirappalli, Tiruchirappalli, India

A. Kumar et al. (eds.), *Computational Intelligence in Machine Learning*, Lecture Notes in Electrical Engineering 834, https://doi.org/10.1007/978-981-16-8484-5_18

culture product must be cultivated using an efficient process [1]. Due to the different climatic seasons and during the different growing stages of the tomato plant, several diseases and abnormal conditions may affect the plant resulting in considerable losses of production. Diseases and defects found in plants and crops have a significant impact on production in the agriculture industry and lead to substantial economic losses [2]. Early and accurate identification of these plant diseases is fundamental for various control mitigation strategies.

Tomato is one of India's most valued food crops. Due to these environment impairments, the tomato plant becomes very susceptible to diseases caused by fungi, bacteria, and viruses [3]. The image segmentation can help to classify the affected leaves from the non-affected one [4]. Content-based image retrieval (CBIR) is more reliable and effective technique to retrieve similar images, based on the query image using their visual features like color, shape, and texture. Another such application realm is agriculture, where CBIR can be used to detect diseases early in leaf, stem, and fruit [5]. Defect identification, feature extraction, feature fusion, training, and classification have the flow of steps implemented. For accurate defect segmentation, *K*-means clustering technique has been used to get the region of interest and to remove the background [6]. A new local color texture descriptor called local binary pattern for color images (LBPC) has been developed based on the concept of separating color pixels in an image using a hyper-plane in m number of dimensions [7]. The primary objectives in this research work are to develop a trained model which classifies whether the tomato plant leaf is healthy or not, identification of the exact disease and finally when a query image is given, similar images are classified based on color, shape, and texture from the database.

- In the case of classification, the goal is to extract the color, shape, and texture features from the image and to train a supervised machine learning model with multiclass SVM algorithm to classify and identify the diseases.
- As far as performance metrics are concerned, the aim is to increase the accuracy, precision and recall compared to the existing methods. The hyper-parameter tuning will be completed, to select the suitable model resulting in increased precision.
- The proposed automated disease identification system will help farmers by saving their plant, energy, money, and resources; further, this research work aims to eliminate the need to track naked eyes on leaves of tomato plants.

The remainder of the paper is organized as follows: The complete methodology is explained in Sect. 2. The implementation and classification based results and discussion are explained in Sect. 3. Finally, Sect. 4 concludes the research work.

2 Methodology

The proposed CBIR system is an improvisation of the work cited in the literature to identify tomato leaf diseases and to collect similar samples of the diseased leaf

(a) Bacterial Spot (b) Septorial spot (c) Tomato mosaic (d) Late blight

Fig. 1 Tomato leaf disease samples

from the database when a query leaf image is classified. For each leaf sample, color, texture, and shape features are initially obtained, and the feature vector is generated. For classification, 70% of the leaf samples is used for training the multiclass SVM model, and the remaining 30% is used for testing. The accuracy of the model is determined by the K-fold cross-validation of the trained model. The tomato leaf samples are taken from plant village database [8]; the symptoms are in the form of various visible patterns or holes which are shown on Fig. 1 which is the target to detect, observe, and classify.

2.1 *Pre-processing*

2.1.1 Resizing

The leaf samples taken into consideration are resized to standard dimensions to ensure uniformity. They resized image has a resolution of 800 pixels to 533 pixels in width.

2.1.2 Segmentation

A typical leaf image consists of regions of interest (ROI) along with background noise. It is necessary to select only the required information and suppress other undesirable components. This will result in lesser interference during feature extraction to ensure cleaner feature vectors.

The image segmentation is performed to select the ROI by *K*-means clustering method [9]. Based on the experimentation with the dataset and applying the algorithm to images present in data set, 'k' $= 2$ has given satisfactory results in selecting only region that contained leaf, while the background region has been clustered separately as shown in Fig. 2.

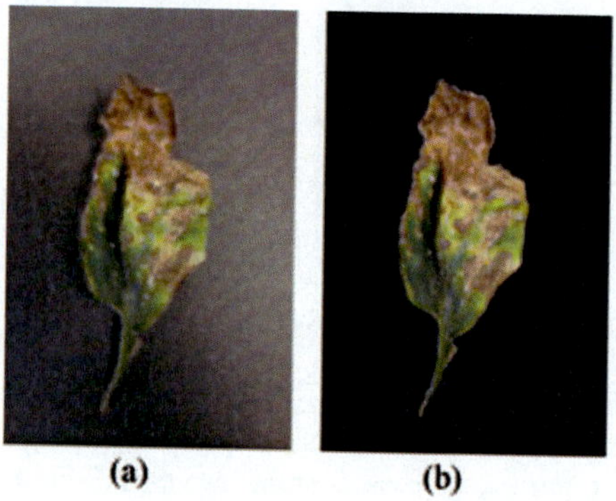

Fig. 2 **a** Original image, **b** segmented images

2.2 *Feature Extraction*

In the proposed methodology, color, shape, and texture features are extracted from leaf images. In texture features, LBP has some limitations such as they produce rather long histograms, which slows down the recognition speed especially on large-scale database. Furthermore, under certain circumstances, they miss the local structure as they fail to consider the effect of the center pixel. Also, they are sensitive to noise, and they have small spatial support area. As these are some concerns in the case of the classifying efficiency, in order to increase the efficiency, modified LBP called as complete local binary pattern has been used.

2.2.1 Completed Local Binary Pattern (CLBP)

In order to improve the performance of LBP, a new scheme considering global thresholding, named as CLBP has been proposed by [10]. Here, it finds two components: sign component and magnitude component. The magnitude component is calculated by the difference of central pixel and neighboring pixels. The difference is compared with the global mean and converted to an 8-bit value. The sign and magnitude component can be calculated as shown in the below equation:

$$S_P = S(W_P - W_C) \tag{1}$$

$$M_P = |W_P - W_C| \tag{2}$$

where S_P is sign component and M_P is magnitude component. S_P and M_P are used to find the CLBP_S and CLBP_M operators, respectively. CLBP_S is calculated as same as finding LBP (4), while in CLBP_M the local variance of the magnitude is calculated (6). These two components can be described mathematically as follows:

$$\text{CLBP\_}S_{\text{P,R}} = \sum_{P=0}^{P-1} 2^P S(W_\text{P} - W_\text{C}) \tag{3}$$

$$S_\text{P} = \begin{cases} 1, & W_\text{P} \geq W_\text{C} \\ 0, & W_\text{P} < W_\text{C} \end{cases} \tag{4}$$

$$\text{CLBP\_}M_{\text{P,R}} = \sum_{P=0}^{P-1} 2^P t(W_\text{P} - C) \tag{5}$$

$$t(M_\text{P}, C) = \begin{cases} 1, & |W_\text{P} - W_\text{C}| \geq C \\ 0, & |W_\text{P} - W_\text{C}| < C \end{cases} \tag{6}$$

where W_P and W_C are neighboring and central pixel, respectively, and C is the mean value of magnitude component in whole image. The limitation of LBP is that it does not take into account of the effect central pixel; in CLBP, it has been rectified by identifying the CLBP_C operator, which is calculated as below:

$$\text{CLBP\_}C = t(W_\text{C}, C_i) \tag{7}$$

where W_C is the center pixels' value, and C_i is the average gray level of the whole image. CLBP is applied on hue color plane in HSV color space also to include the color component and for grayscale image. Then, CLBP_S and CLBP_M components are joined together linearly, and feature vector is obtained. Moreover, dealing with reduced number of patterns becomes easier and more flexible than normal CLBP without losing information about the image primitives. This can be done by uniform rotation invariant method.

2.3 *Feature Combination*

The feature vectors obtained from color, shape, and texture properties are concatenated linearly to obtain a single feature vector for a particular leaf image in database. This process is repeated for every image sample in database to obtain a feature bank for classification and retrieval purposes.

2.4 *Disease Classification*

In this research work, a multiclass support vector machine (SVM) has been used to achieve the same. The support vector machine identifies the decision boundary that separates different classes and maximizes the margin.

2.4.1 Evaluation Metrics

Once a model has been finalized, the performance of the model has to be evaluated. The confusion matrix tests the performance of the classifier. The each predicted output class is plotted and counted against the actual class label. In this research work, precision, recall, and $F1$ score are computed based on the confusion matrix values.

3 Results

3.1 Implementation

The above mentioned methods have been implemented using Python 3.6 as the main programming language and used OpenCV, Numpy, Sklearn libraries for image processing and classification. Visualization of results have been carried out with the help of Matplotlib.

3.2 Classification

The proposed model has been trained with different kernels and hyper-parameters. From the simulations, it is found that the highest accuracy to be obtained by using linear kernel with C value as 1. The features obtained are linearly separable by great extent. The linear kernel SVM model has been used for further classification problems. Five-fold validated accuracy for various combinations of features has done to find the best possible combination is shown in Table 1.

The particular combination of HSV histogram, Fourier descriptors (FD), and CLBP of hue component in HSV color space has found to give the best possible results. This is may be due to the fact HSV color space considers actual color component and invariant of illumination differences. Also, CLBPs are better than normal LBPs because of the extra information provided by magnitude differences between central and neighboring pixels. The confusion matrix has plotted for a 70:30 train test split of 300 images in database using linear SVM. Figure 3 shows the confusion matrix with normalization obtained from linear SVM kernel. The precision, recall, and $F1$ score have been computed from confusion matrix and have shown in Table 2.

For each class, the training accuracy increases with increase in images used for training. This can be attributed to the fact that additional distinguishable features are discovered from new images. Also, the particular combination of HSV histogram, FD, and CLBP feature vectors have been found to perform better for classification.

Table 1 Validated accuracy

Features	5-fold validated accuracy
RGB hist	0.904
HSV hist	0.942
FD	0.608
Gray_LBP	0.880
Hue_LBP	0.897
Gray_CLBP	0.928
Hue_CLBP	0.917
RGB hist + FD + Gray_LBP	0.924
HSV hist + FD + Hue_LBP	0.957
RGB hist + FD + Gray_CLBP	0.964
HSV hist + FD + Hue_CLBP	**0.973**
RGB hist + HSV hist + FD + Gray_CLBP + Hue_CLBP	0.971

Bold represents the highest validation accuracy value as compared to existing features

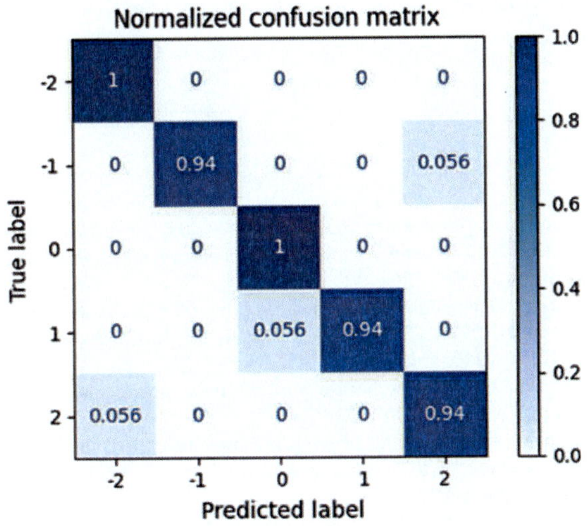

Fig. 3 Normalized confusion matrix

4 Conclusion

The proposed CBIR system is particularly implemented for agricultural uses. Four types of tomato leaf diseases are taken into account. *K*-means clustering and bilateral filtering are performed to segment the region of interest and remove the noise clearly. The individual performance of color, shape, and texture features is found out. It is analyzed that if the features are fused, there is a considerable increase

Table 2 Classification metrics

Class	Precision	Recall	$F1$ score	Support
Bacterial spot	0.947	1.000	0.973	18
Healthy	1.000	0.944	0.971	18
Late blight	0.947	1.000	0.973	18
Septorial spot	1.000	0.0944	0.971	18
Mosaic	0.944	0.944	0.914	18
Macro average	0.967	0.966	0.960	0.90

in the result. The training and classification model is implemented using support vector machines. The maximum mean classification accuracy of 97.3% is obtained by using a combination of HSV histogram, Fourier descriptors, and uniform hue CLBP features. In future, there is a scope for creating a mobile application which scans the leaf and identification and classification of disease. In the present research work, only four tomato leaf diseases are analyzed, the same can be explored further for other tomato leaf diseases too. Another area which can improve is replacing deep learning algorithms instead of similarity measurements metrics for the content-based image retrieval.

References

1. M. Jhuria, A. Kumar, R. Borse, Image processing for smart farming: detection of disease and fruit grading, in *2013 IEEE Second International Conference on Image Information Processing (ICIIP-2013)* (2013), pp. 521–526
2. S. Dubey, A.S. Jalal, Detection and classification of apple fruit diseases using complete local binary patterns, in *2012 Third International Conference on Computer and Communication Technology* (2012), pp. 346–351
3. S. Vetal, R.S. Khule, Tomato plant disease detection using image processing. Int. J. Adv. Res. Comput. Commun. Eng. **6**, 293–297 (2017)
4. S. Sai Satyanarayana Reddy, P. Chatterjee, Ch. Mamatha, Y. Vijaya Bhaskar Reddy, Use of image processing techniques to detect diseases in tomato leaves. Int. J. Civ. Eng. Technol. (IJCIET) **8**(12), 287–290 (2017)
5. J.K. Patil, R. Kumar, Analysis of content based image retrieval for plant leaf diseases using color, shape and texture features. Eng. Agric. Environ. Food **10**(2), 69–78 (2017)
6. S.R. Dubey, A.S. Jalal, Apple disease classification using color, texture and shape features from images. SIViP **10**, 819–826 (2016)
7. C. Singh, E. Walia, K.P. Kaur, Color texture description with novel local binary patterns for effective image retrieval. Pattern Recogn. **76**, 50–68 (2018)
8. G. Geetharamani, P.J. Arun, Identification of plant leaf diseases using a nine-layer deep convolutional neural network. Comput. Electr. Eng. **76**, 323–338 (2019)
9. G.B. Coleman, H.C. Andrews, Image segmentation by clustering. Proc. IEEE **67**(5), 773–785 (1979). https://doi.org/10.1109/PROC.1979.11327
10. Z. Guo, L. Zhang, D. Zhang, A completed modeling of local binary pattern operator for texture classification. IEEE Trans. Image Process. **19**(6), 1657–1663 (2010). https://doi.org/10.1109/TIP.2010.2044957

Plant Disease Identification Using Discrete Wavelet Transform and Convolutional Neural Network

Milind Choudhary, Banhi Sanyal, and Ramesh Kumar Mohapatra

Abstract Plant disease needs to be studied for detection and identification to minimize the losses incurred by these diseases. This paper discusses identification and detection of plant diseases through visual symptoms on the leaves of fruits and vegetables. Deep learning models have been integrated with feature extraction method, namely discrete wavelet analysis for the disease identification process. The algorithm developed has been used to pre-process the input image. Extraction of the necessary features is also done to reduce parameters of an image. These reduced parameters are used as input to deep CNN model to classify the images of disease-affected plant leaves. Transfer learning using Inception Resnet has been implemented as the classifier model. The average accuracy achieved on training and test dataset is 97.77% and 94.05%, respectively.

Keywords Plant disease identification · Deep learning · Bior 3.7 wavelet · Inception Resnet · Convolutional neural network (CNN)

1 Introduction

Crop diseases are the cause for wastage of approximately 10% of world wild crop yield [1]. India losses approximately 30% of its crops to various diseases [2]. Fruits and vegetables also face similar wastage due to diseases. Agriculture is one of the main contributors to economy in majority of the countries [3]. Furthermore, fruits and vegetables are also main source of nutrition, vitamins, and minerals. This necessitates further research to bring down the wastage of fruits and vegetables. Image processing is now being used in almost all fields. Plant disease identification is another field in which deep learning and image processing have been used. Still, there remains scope

M. Choudhary (✉) · B. Sanyal · R. K. Mohapatra
National Institute of Technology Rourkela, Rourkela, Odisha, India

B. Sanyal
e-mail: 517CS1009@nitrkl.ac.in

R. K. Mohapatra
e-mail: mohapatrark@nitrkl.ac.in

A. Kumar et al. (eds.), *Computational Intelligence in Machine Learning*, Lecture Notes in Electrical Engineering 834, https://doi.org/10.1007/978-981-16-8484-5_19

for further improvement in the plant disease detection, so that fruits of deep learning reaches to the mass.

A few studies have been found in literature, earlier, in this field of plant disease identification. These works provide results with which the current work can be compared.

Such works provide scope for improvement and also validate the databases which can be used. However, not much work was found encompassing the concepts of deep learning, namely CNN and transfer learning. Although, wavelet has been used as a feature extractor, but few works are available on combination of CNN and wavelet. The literature stated the challenges faced due to variety of database and similar structure leaves of different plants in the database. Earlier works also faced problems which they left for future works to be resolved. Using deeper networks and different feature extractors were some of the problems. Furthermore, specific problems like lack of pooling and batch normalization were also found.

The current study uses transfer learning using Inception Resnet [4] which provides a very deep model to understand the difference in such similar leaf structures. Feature extractor used is a wavelet-based one. Specifically, Bior 3.7 has been used. This combination of wavelet and deep CNN provides a very efficient model which is applied on images of plant leaves for detection of diseases.

This paper contains five sections. Literature survey is explained in Sect. 2. Section 3 describes the methodology involved in the creation of the model. Section 4 encompasses the results and discussion found on the test database and compares the current works with previous such works. Section 5 presents the conclusion and scope of further study.

2 Literature Review

Plant and crop diseases account for wastage of a huge amount of crop yield [1, 2]. Many countries in the world have agriculture-dependent economy [3]. To address such issue, the current study made use if the general repository of images of disease affected plant leaves [5]. Researchers have been applying various techniques to detect the diseases by analyzing images. Machine learning, deep learning, and feature extraction are some of such techniques.

It is a known fact that deep CNN can detect most of the differentiating factors in the various classes. The current study also uses a CNN model through transfer learning called Inception Resnet [4]. Inception Resnet contains skip connections like Resnet architecture [6]. These skip connections help to address the degradation problem [7]. Inception Resnet also uses the block structure of inception network [8]. This network includes normal activation functions as used in neural networks [9, 10]. Over fitting is avoided in deep CNNs with the help of drop out layers [11]. CNN is also robust at feature extraction; however, usage of feature extracting methods based on wavelet analysis is seen quite often in various domains [12, 13]. Wavelets are small wave like structures which are transformed to fit the image and reduce the number of input parameters through a techniques known as wavelet analysis [14, 15]. Wavelet analysis involves the resizing of the mother wavelet in length and height to fit the input wave [16, 17]. This study uses Bior 3.7 which has been observed to perform well on feature extraction [18]. Some works have used CNN, while some have used wavelet analysis coupled with machine learning algorithms for plant disease identification. Gabor wavelet with support vector machine (SVM) was seen in a couple of literature [19, 20] which gives an insight on usage of a combination of wavelet and machine learning techniques. Haar wavelet has also been used with K-nearest neighbors (K-NN) [21]. This work used an unsupervised learning method to group the diseases.

One of the initial works [22] in this domain used multiple linear regression (MLR) along with continuous wavelet transform (CWT). Apart from this combination method, deep CNN was also used to identify plant disease [11, 23]. Transfer learning using GoogLeNet [23] gives insight on using the latest of models and retraining it for the study purpose. The current study also uses transfer learning, and the model used is Inception Resnet, along with Bior 3.7 wavelet.

3 Methodology

3.1 Dataset

A set of 43,454 images for training and validation, and 10,849 images for testing are analyzed for the work. The images are distributed across 38 classes which contain healthy and disease inflicted leaves of various plants. The dataset for the image was obtained from plant disease database from Kaggle [5].

3.2 System Requirements

The plant disease identification system using preprocessing, feature extraction, and deep learning model utilizes both CPU and GPU. The exact specifications of the system are CPU—2 × 64 bit-Intel(R) Xeon(R) CPU @ 2.20 GHz, GPU—NVIDIA Tesla T4, RAM (CPU)—13.33 GB, VRAM (GPU)—15.079 GB, and disk space—69 GB.

3.3 Experimental Setup

Preprocessing The preprocessing steps involve resizing of the images from 256 × 256 to 150 × 150. It was necessary to use the images for feature extraction and CNN model. The resized images are converted into grayscale for feature extraction using wavelet.

Feature Extraction Using Wavelet Wavelets unlike sinusoidal waves are generally asymmetric and irregular. In wavelet analysis, the wavelet signal is scaled as well as shifted from the mother wavelet and is used for feature extraction [15].

This study uses discrete wavelet transform (DWT) for the reduction of input parameters from the images. DWT filters an image into four sub-bands. Multilevel decomposition is obtained by iterative subdivision of the sub-band which has the lowest frequency [16]. Our study uses the level 1 decomposition of image to reduce the number of input parameters in each image for the CNN. Four sub images produced using wavelet transform are vertical details, diagonal details, horizontal details, and approximation [17].

This study uses approximation image which has less number of parameters as compared to original image as input for the CNN. Bior 3.7 wavelet is used for the decomposition into approximation. Biorthogonal wavelet function is used since it has heavy noise suppression and edge preservation [24] as compared to haar and db4 wavelets. Biorthogonal wavelet has an odd symmetric structure, and this accounts for the usage of Bior 3.7 wavelet for the study. This odd symmetry helps in proper reduction of parameters without loss of essential parameters in the images of plant disease database.

Deep Convolutional Neural Network Model The study uses deep CNN for the plant disease identification process. After the input images are preprocessed and feature extraction using wavelet analysis has been done, the reduced number of input parameters is fed to the CNN. The identification of the disease was performed using transfer learning of Inception Resnet [4] model. The Inception Resnet model used was trained on the dataset as sufficient number of images was available. Inception Resnet uses a modified module as compared to Resnet model. This module is used along with a stem module. The usage of residual connection avoids the degradation problem [7]. It also reduces the training time.

4 Results and Discussion

Our study uses the level 1 decomposition of image to reduce the number of input parameters in each image for the CNN. No significant increase of accuracy is seen for level 2 decomposition, and the accuracy falls by using deeper level decomposition. The wavelet used was Bior 3.7, and it was seen that it performs better than haar and db4 wavelet. The model was run for 10 epochs with batch size of 32. 2713 images were used for validation. The following final accuracy and loss were observed for the training and validation set, and they are final training Accuracy—97.77%, final training loss—0.0724, final validation accuracy—98.57%, and final validation loss—0.0339. The accuracy and loss plots are shown in Fig. 1a and b. The test accuracy on

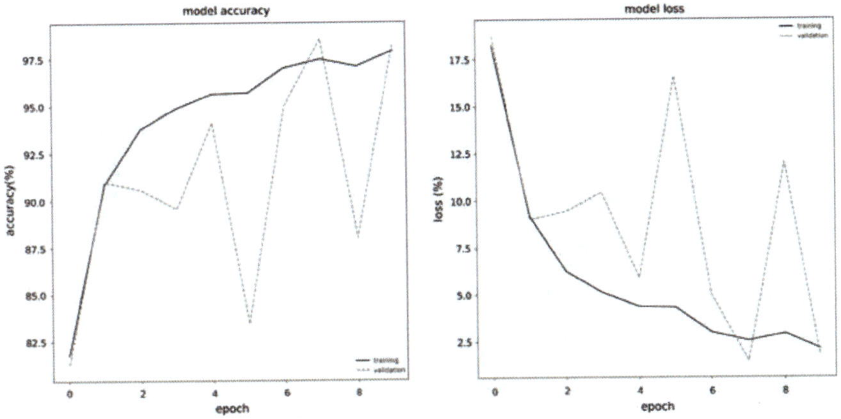

(a) Accuracy Curve (b) Loss Curve

Fig. 1 Plots on test dataset

a new distribution was observed to be 94.05%. Confusion matrix was calculated and plotted Fig. 2. The precision was found out to be 93.49%, recall was found out to be 91.85%, and the $F1$ score was 92.43%. These metrics were calculated on the test dataset. Figure 2 shows that the diseases associated with grapes are detected with negligible loss as compared to other plants.

4.1 Comparative Results

Table 1 compares the previous works of many authors. From the accuracy column, it is evident that the current work shows an increase of 4% accuracy as compared to the next best model which was SVM after extraction of features using Gabor wavelet. Current work which uses Inception Resnet performs the best as compared other works. Some of earlier work made use of CWT, DWT, and Gabor wavelet, but usage of Bior 3.7 for feature extraction in present work gave an accuracy of 94.05%.

5 Conclusion and Scope of Further Study

In this study, Bior 3.7 wavelet is used for feature extraction, and a deep CNN based model, namely Inception Resnet, is used for plant disease identification.

The results and the comparative results show that the methodology proposed is a potent one and can identify the plant diseases in the database with an accuracy of 94.05%. The methodology proposed gives better results as compared to other plant

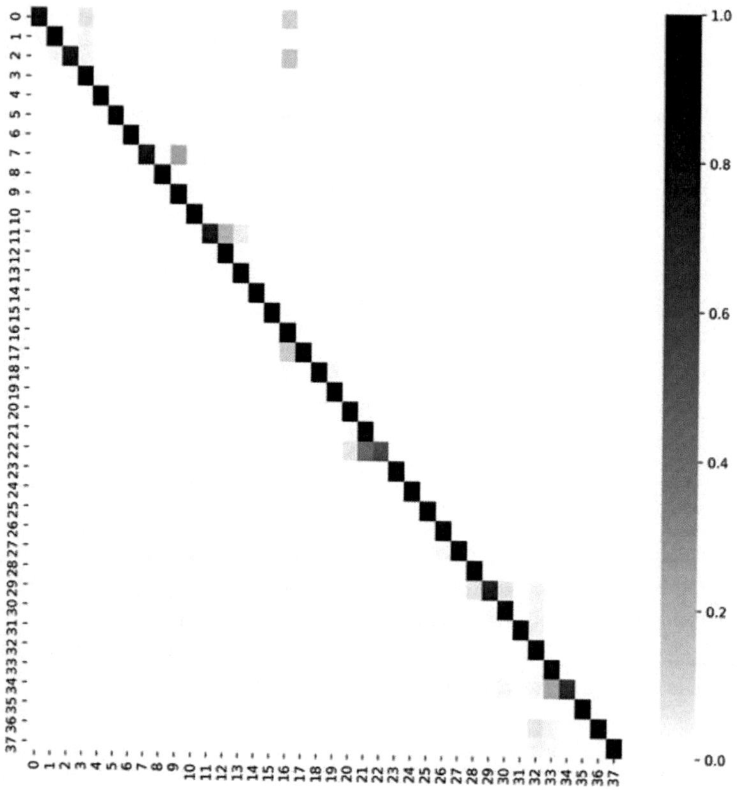

Fig. 2 Confusion matrix

disease identification systems. In future, this methodology can be applied to a wider variety of plants and crops with more number of diseases.

The current study gives insight on the usage of existing architecture and training it for a different database. Deep CNN has helped in effective localization of the differentiating factors in various healthy and disease affected plants. Subsequently, other architectures could be implemented to obtain even higher accuracy and faster models. Accuracy can also be improved by adding custom layers to the existing architecture. Since some misclassification is seen in tomato and its diseases as shown in Fig. 2, a larger database with more images in each type of disease and healthy leaf can also be used to improve the model.

Table 1 Comparison of accuracy of different models and feature extractors

Reference	Dataset	Feature extractor	Classifier	Accuracy (%)
Mishra et al. [11], 2019	Dataset created from plant village dataset, hosted at plant village disease classification challenge	NA	Deep neural network (DNN)	88.46
Prasad et al. [19], 2012	Self-prepared dataset	Gabor wavelet	SVM	89.00
Mousavi et al. [20], 2016	Public dataset labeled by plant pathologist	Gabor wavelet	SVM	90.04
Deshapande et al. [21], 2019	Agricultural fields of agricultural university, Dharwad	Haar wavelet	K-NN	85.00
Zhang et al. [22], 2012	Local Field, ASD Inc., Boulder, Colorado, USA	CWT	MLR	77.00
Mukherjee al. [23], 2017	Apple leaves from plant village dataset	NA	Transfer learning using GoogLeNet	85.04
Pujari et al. [25], 2014	Department of plant pathology, university of agricultural sciences, Dharwad	DWT	Probabilistic neural network (PNN)	86.48
Current study	**Plant disease dataset** [5]	**Bior 3.7 wavelet (DWT)**	**Transfer learning using Inception Resnet**	**94.05**

References

1. P. Christou, R.M. Twyman, The potential of genetically enhanced plants to address food insecurity. Nutr. Res. Rev. **17**(1), 23–42 (2004)
2. D. Sharma, D. Rao, A field study of pest of cauliflower cabbage and okra in some areas of Jaipur. Int. J. Life Sci. Biotechnol. Pharma Res **1**(2), 2250–3137 (2012)
3. J.M. Alstonand, P.G. Pardey, Agriculture in the global economy. J. Econ. Perspect. **28**(1), 121–146, February2014. [Online]. Available https://doi.org/10.1257/jep.28.1.121
4. C. Szegedy, S. Ioffe, V. Vanhoucke, A. Alemi, Inception-v4, inception-Resnet and the impact of residual connections on learning (2016)
5. S.R. Sharma, Plant disease dataset. https://www.kaggle.com/saroz014/plant-disease/version/1 (2018)
6. K. He, X. Zhang, S. Ren, J. Sun, Deep residual learning for image recognition (2015)
7. L.D. Nguyen, D. Lin, Z. Lin, J. Cao, Deep CNNs for microscopic image classification by exploiting transfer learning and feature concatenation, in *2018 IEEE International Symposium on Circuits and Systems (ISCAS)*, pp. 1–5 (2018)

8. C. Szegedy, V. Vanhoucke, S. Ioffe, J. Shlens, Z. Wojna, Rethinking the inception architecture for computer vision (2015)
9. A.F. Agarap, Deep learning using rectified linear units (relu). arXiv preprint arXiv:1803.08375 (2018)
10. R.A. Dunne, N.A. Campbell, On the pairing of the soft-max activation and cross-entropy penalty functions and the derivation of the soft-max activation function, in *Proc. 8th Aust. Conf. on the Neural Networks*, vol. 181 (Melbourne, Citeseer, 1997), p. 185
11. S. Mishra, R. Sachan, D. Rajpal, Deep convolutional neural network based detection system for real-time corn plant disease recognition. Proc. Comput. Sci. **167**, 2003–2010 (2020)
12. Z. Iqbal, M.A. Khan, M. Sharif, J.H. Shah, M.H. ur Rehman, K. Javed, An automated detection and classification of citrus plant diseases using image processing techniques: a review. Comput. Electron. Agric **153**, 12–32 (2018)
13. B.S. Prajapati, V.K. Dabhi, H.B. Prajapati, A survey on detection and classification of cotton leaf diseases, in *2016 International Conference on Electrical, Electronics, and Optimization Techniques (ICEEOT)* (IEEE, 2016), pp. 2499–2506
14. A. Jaiswal, J. Upadhyay, A. Somkuwar, Image de-noising and quality measurements by using filtering and wavelet based techniques. AEU-Int. J. Electron. Commun. **68**(8), 699–705 (2014)
15. R.K. Rai, T.R. Sontakke, Implementation of image de-noising using thresholding techniques. Int. J. Comput. Technol. Electron. Eng. (IJCTEE) **1**(2), 6–10 (2011)
16. B. Chinnarao, M. Madhavilatha, N. Pratap, Improved image de-noising algorithm using dual tree complex wavelet transform with new thresholding technique. IJEE 4(1), 135–139 (2012)
17. B. Ergen, Signal and image de-noising using wavelet transform, in *Advances in Wavelet Theory and Their Applications in Engineering, Physics and Technology*. IntechOpen (2012)
18. G. O. Gajbhiye and A. N Kamthane, "Automatic classification of glaucomatous images using wavelet and moment feature," in2015Annual IEEE India Conference (INDICON), 2015, pp. 1–5
19. S. Prasad, P. Kumar, R. Hazra, A. Kumar, Plant leaf disease detection using Gabor wavelet transform, in *International Conference on Swarm, Evolutionary, and Memetic Computing* (Springer, 2012), pp. 372–379
20. S. Mousavi, Z. Hanifeloo, P. Sumari, M.M. Arshad, Enhancing the diagnosis of corn pests using Gabor wavelet features and SVM classification (2016)
21. A.S. Deshapande, S.G. Giraddi, K. Karibasappa, S.D. Desai, Fungal disease detection in maize leaves using haar wavelet features, in *Information and Communication Technology for Intelligent Systems* (Springer, 2019), pp. 275–286
22. J.-c. Zhang, Y. Lin, J.-h. Wang, W.-j. Huang, L.-p. Chen, D.-y. Zhang, Spectroscopic leaf level detection of powdery mildew for winter wheat using continuous wavelet analysis. J. Integr. Agric. **11**(9), 1474–1484 (2012)
23. S. Mukherjee, P. Kumar, R. Saini, P.P. Roy, D.P. Dogra, B.-G. Kim, Plant disease identification using deep neural networks. J. Multimedia Inf. Syst. **4**(4), 233–238 (2017)
24. Y. Shen, Q. Liu, S. Lou, Y.-L. Hou, Wavelet-based total variation and non local similarity model for image de-noising. IEEE Signal Process Lett **24**(6), 877–881 (2017)
25. J.D. Pujari, R. Yakkundimath, A.S. Byadgi, Automatic fungal disease detection based on wavelet feature extraction and PCA analysing commercial crops. Int. J. Image Graph. Signal Process. **6**(1), 24–31 (2013)

Recognizing Art Style Automatically in Painting Using Convolutional Neural Network

Mahfuza Akter, Mst. Rasheda Akther, and Md. Khaliluzzaman

Abstract Recognition of art style in painting is an active and important research area in the computer visions field as art style recognition is not depending on the definite feature; however, extracting important in-definitive features from a painting image is a crucial fact. Moreover, using deep convolutional neural network (CNN) makes the model overfit and small CNN model makes the model underfit. To overcome these problems, in this work, a model is developed by us for art style recognition in painting based on the shallow convolutional neural network (SCNN). The proposed model is developed based on the two consecutive convolutional layers and single fully connected layer to extract painting features and recognize the art styles accordingly. For training and testing purposes in the art style recognition, we have used Wikipaintings dataset. We also used our created real-time dataset in this work. Our proposed model shows 60.37% accuracy which is the significant improvement over some current state of the arts.

Keywords Feature extraction · Overfit · Underfit · Art styles in paintings · Shallow convolutional network · Wikipaintings dataset

1 Introduction

In the current decay, art style recognition is a vital issue in the computer vision's field because of its growing demands in digital multimedia industries as well as academic area. As art styles, visual features are not defined in identifiable features way such as object or facial recognition, so, it is really challenging task to identify the art style in automatic way. The art styles classification is also difficult for non-representational art work [1].

Many researchers have done many research works to recognize the art style through various handcrafted features and machine learning approaches. From this research, they identify many problems to classify the art style. Such as, in [2],

M. Akter · Mst. R. Akther · Md. Khaliluzzaman (✉)
Department of Computer Science and Engineering, International Islamic University Chittagong (IIUC), Chittagong 4318, Bangladesh

A. Kumar et al. (eds.), *Computational Intelligence in Machine Learning*, Lecture Notes in Electrical Engineering 834, https://doi.org/10.1007/978-981-16-8484-5_20

authors examine the efficiency of various renown image features, i.e., special envelops, histogram-oriented gradients (HOGs), discriminative color names with various machine learning classification approaches, i.e., support vector machine (SVM), random forests. From this work, authors observe that it is hard to classify some art styles with these techniques. They also observe that adding more features, the model's performance is not improved.

In the recent years, to derive the critical visual features from an image, researchers focus on deep convolutional neural networks. It is a bio-inspired hierarchical multi-layered neural network which can capable to learn visual patterns immediately from pixels of image frames except any preprocessing steps [2, 3]. Many researchers apply DCNN models to identify the previous problems and recognize the art styles efficiently from different classes. Such as, in [4], authors observed the limitations of handcrafted feature and proposed a CNN model. Their CNN model is based on the AlexNet [5], and the model was trained with ImageNet. This model performs better with state of the art and increases the accuracy of art style classification to 54.5% for 25 classes.

Each art contains different features in various regions of paintings. In this respect, methods of feature extraction should be able to extract characteristics from the art's important parts. However, the extracting feature from a painting image is a big issue in the art styles painting classification. In this regard, many researchers develop many CNN models for art styles painting classifications. However, these models are sometime overfitted and underfitted because of large and small amount of convolutional layers in this network. To overcome these problems and develop a regularized CNN model, in this work, we have proposed a shallow convolutional neural network (SCNN). The proposed network contains two consecutive convolutional layers and one fully connected layer. In this work, Wikipaintings and real dataset are used to train as well as test the proposed shallow model.

The remaining part of this paper is presented as follows. In Sect. 2, recent state-of-the-art work is presented. In Sect. 3, proposed model is presented. In Sect. 4, experimental results are showed. The conclusion is given in Sect. 5.

2 Related Works

Recognition of art style is an important system, and several researchers have done many works on the recognition of art style because of its diverse applications. In an automatic way, identifying the style of an art is a difficult task. Some primitive challenges stand in the method of recognition of art style in painting. CNN model is a better and effective model for recognizing art style.

So many pieces of research observed that most of the systems designed for recognition of art style automatically were created on the features of handcrafted and used algorithms which are focused on deep learning to identify art styles. For example, in [6], authors proposed a convolutional neural network to classify painting images from challenging environments. Authors tested their designed model with the challenging test dataset which is collected from different art movements. In [7], an algorithm of multi-task learning was presented by them to learn a style-special dictionary representation. To exhibit the usefulness of their method, they introduced the dataset whose name is DART, which contains over 1500 pictures of arts representative of various styles.

An auto-encoder-based deep CNN model is proposed in [8]. Here, auto-encoder is used for the purpose of preprocessing. The encoded image is used to train a supervised model which is based on the CNN. The performance shows by the propose model is better than the state of the art works. It also shows that the deep neural network performs better to recognize the painting classification. Another CNN model is proposed in [9], where generated distorted painting image is used for training and validation process. They compared the proposed model with the handcrafted feature, i.e., SIFT method and shows the outperformance.

In [10], a deep CNN model is proposed by the authors to explain the perceptiveness of visual features to identify the artistic styles in painting. The authors also suggest a rigid binary representation of the painting. A style predicting approach is proposed in [4], where authors evaluate various image features to perform the task. This work exhibits several types of styles not previously considered in the literature, and they exhibited state-of-the-art results in the prediction of both style and aesthetic quality for multilayer networks.

In [11], authors investigate the different applicability's of metric for the learning approach as well as different applicability of visual features to estimate the learning similarities with the collection of painting. In this work, authors consider three basic concept, i.e., style, artist, and genre. A deep CNN-based fine art painting classification model is proposed in [1]. The authors divide the objective of the work in twofold. Firstly, authors develop an end-to-end DCNN model to examine the capability of CNN model. Finally, authors argue that the fine art painting classification is more challenging with respect to the other classification model such as face or object recognition.

In [12], an artificial system is introduced to generate artistic images through a deep neural network with high perceptual quality. The proposed system introduces neural system which creates an arbitrary image by separating and recombining image contents. This work shows a flow of algorithmic representation of how a person creates and perceives artistic image. The effect of very DCNN is estimated in [13]. The proposed network's complexity is very high because the network is very deep. However, the accuracy of the model is significantly improved. A deep residual neural network is introduced in [14] detecting artistic style painting. For that, the model is pre-trained by using the ImageNet. However, the model is underfitted.

3 The Proposed Framework

In this paper, an automatic art style recognition framework is established. Improving a deep neural network conjointly depends on adding a lot of layers or neurons, facilitating network gradient flow, notably for classification problems with a massive number of classes. As the art style recognition methods contain the small number of output classes as well as small number of datasets, in this paper we demonstrate a model with the small number of convolutional and fully connected layers, which show promising outcomes.

3.1 Model Explanation

In this thesis work, a CNN framework is proposed which contains two convolutional layers and one fully connected layer. In Fig. 1, the workflow diagram of our proposed model is presented. The proposed convolutional neural network architecture is presented in Fig. 2. Two convolutional layers are contained by the framework which extracts the important features of art's region. The framework contains one fully connected layer which recognizes art style.

In our proposed model, 32 filters consist in the first convolutional layer and filter's size is 3×3 and each has 3 channels. Same padding is used by us; that is, the input pictures are zero-padded so that the filters convolve over each input picture's pixel. In the convolution layer, we used an activation function that is ReLU. To improve model performance and stability, the batch normalization is used.

The size of max pooling filter is 2×2 that moves by a stride of 2. On the feature maps, it performs max pooling operation. We have used a dropout layer in the shake of dropping out the neurons which helps prevent this model from overfitting to the data which are trained. For the first convolution layer, dropout value has been set to 0.20.

64 filters of size 3×3 as well as each contain 3 channels the second convolutional layer is made. Here the SAME padding is also used. To the feature maps, the ReLU activation function is also applied. In the second layer to improve the model performance, again the batch normalization is used. The size of max pooling filter is 2×2 that moves by a stride of 2, and on the feature maps, it executes max pooling operation. For second convolution layer, the value 0.25 has been set as dropout value to overcome the problem of overfitting. After setting the dropout value of second convolutional layer, a flatten layer has been used.

The final pool layer's result is flattened into a victories type and succeeded by a fully connected layer which has 512 neurons. Then we have used fully connected layers for classifying the final features into the several classes in the output layer which are extracted from the preceding pooling layers and convolution layers. From features, the fully connected layers learn. We have applied batch normalization again. Then we used the dropout value 0.5. For activation function in dense layer, we used ReLU again.

Finally, the output layer consists of 6 nodes representing 6 classes. Then we used softmax activation function to classify the desired label in the output layer.

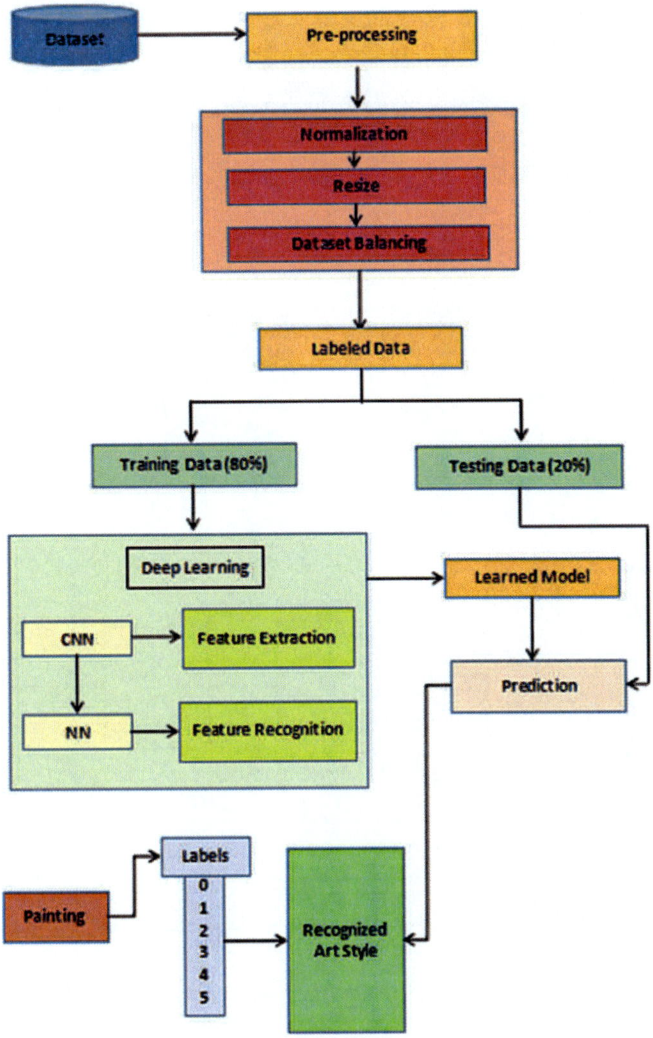

Fig. 1 Workflow diagram of the proposed framework

3.2 *Classification Loss Function*

We used a loss function for assessing the classification model's performance whose output is a probability value. In machine learning context, a log loss and cross-entropy loss can be considered equal. Performance of a model is measured by it, and a probability value among 0–1 is its output. For classification of binary, mathematical formula which is used to calculate cross-entropy loss is shown in (1).

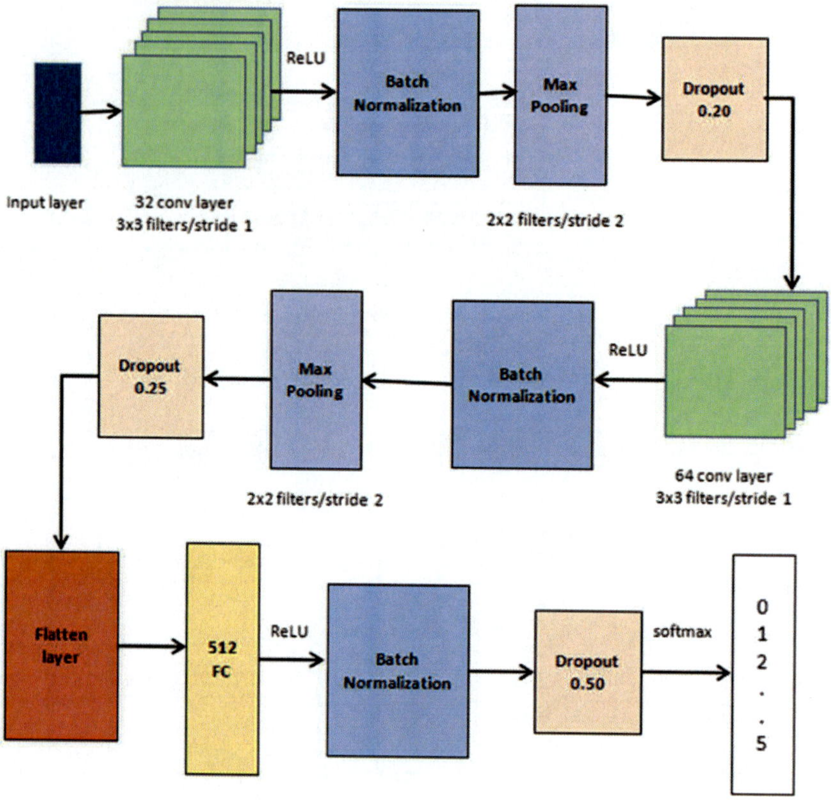

Fig. 2 Proposed art style recognition model

$$\text{Loss} = -(y(p)) + (1 - y)\log(1 - p) \tag{1}$$

Cross-entropy loss function is used when there are two or more label classes. In this work, we consider the categorical cross-entropy loss.

4 Experimental Results

In this section, we presented the detail review of experiments of our proposed model based on the recognition of art style datasets. We have showed our proposed model's performance on the Wikipaintings dataset, and some encouraging recent works compare the results. Several real datasets are also shown during the evaluation of the performance of our proposed model. By using the ideal network miniVGGNet [15], we train and test Wikipaintings dataset and real-time paintings. We have also shown the comparison of performance between the proposed model and miniVGGNet.

4.1 Datasets

We have used Wikipaintings dataset which is moderate in size. For training and testing of our model, we have used this dataset. Our dataset is divided into two parts. In the shake of training as well as testing purpose in the art style recognition, we have used 8000 images of Wikipaintings dataset. To train the model, 80% of images are used and rest 20% of dataset images are used for testing. We used 6 levels of Wikipaintings dataset. Examples of some images of Wikipaintings dataset and real dataset are presented in Figs. 3 and 4.

Fig. 3 Sample images from Wikipaintings dataset: **a** early renaissance, **b** high renaissance, **c** mannerism late renaissance, **d** Ukiyo-e, **e** minimalism, and **f** pop art

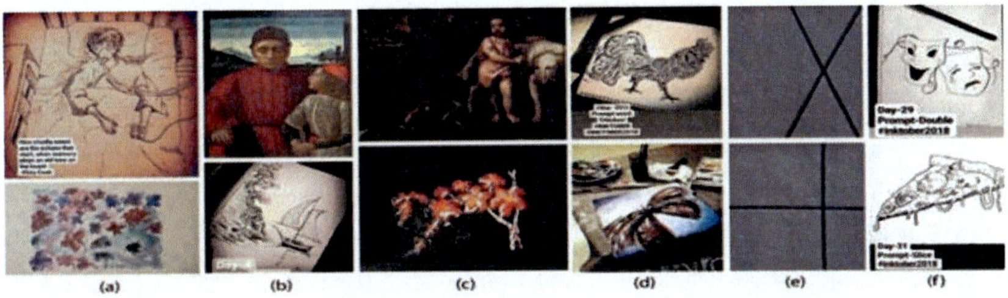

Fig. 4 Sample images from real dataset: **a** early renaissance, **b** high renaissance, **c** mannerism late renaissance, **d** Ukiyo-e, **e** minimalism, and **f** pop art

4.2 Experimental Analysis and Comparison

We drive the network on the Wikipaintings dataset's sub-set, justify the validation set, and investigate the accuracy through the test set. Before we get into the performance details of the model on various datasets, we discuss our training operation briefly. We trained each system per dataset in our simulations, and among such various models, we tried to keep the architecture and hyper-parameters identical. In this work, 40 epochs are used for training. For optimization, an empirical learning rate is used that is 0.001. The optimizer algorithm used in this method is Adam optimizer. As it is adaptive optimizer, it takes less reasonable time to train a model for Wikipaintings datasets. For the Wikipaintings dataset, we have used 20% of images for validation and rest of the images in the training which is 6400 images. We have used 1600 images for validation.

We evaluate the performance based on the different machine learning matrices such as accuracy, recall, precision, and $F1$-score. The proportion of total samples correctly recognized by the classifier is known as accuracy present in (2). The recall is the total positive samples that the classifier correctly predicted as positives present in (3). The total predicted positive samples by the classifier that are true positives are known as precision present in (4). $F1$-score measures a balanced average outcome by combining precision and recall present in (5). The above equations show how to calculate different machine learning matrices with true positive (TP), false positive (FP), true negative (TN), and false negative (FN), respectively.

$$\text{Accuracy} = \frac{\text{TP} + \text{TN}}{\text{TP} + \text{FP} + \text{TN} + \text{FN}} \tag{2}$$

$$\text{Recall} = \frac{\text{TP}}{\text{TP} + \text{FN}} \tag{3}$$

$$\text{Precision} = \frac{\text{TP}}{\text{TP} + \text{FP}} \tag{4}$$

$$F1 - \text{score} = \frac{2 * \text{precision} * \text{recall}}{\text{precision} + \text{recall}} \tag{5}$$

In Fig. 5a, curve of validation loss as well as training loss for miniVGGNet of Wikipaintings dataset is shown, and in Fig. 5b, curve of validation accuracy as well as training accuracy is presented. From the accuracy curve, we can see on the Wikipaintings dataset, miniVGGNet gives 41.87% accuracy. We can estimate the error in the validation from the loss curve. From the curves, it is observed that the loss both of training and testing are reduced constantly till 8 epochs; however, after that the loss is increasing rapidly. That means the model is going to overfit.

In Fig. 6, the evaluated result's confusion matrix is presented. How well the class labels are predicted can be calculated from the confusion matrix. In Table 1, the contribution of miniVGGNet on the prediction of each label is shown. The

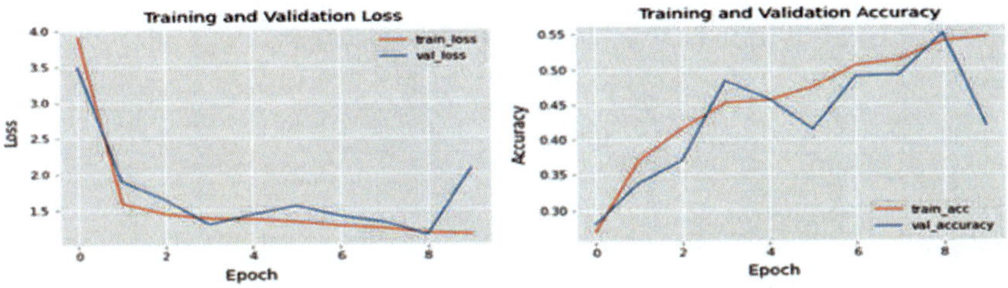

(a) (b)

Fig. 5 Loss and accuracy curve for miniVGGNet for Wikipaintings dataset: **a** training loss and validation loss curve, **b** training accuracy and validation accuracy curve

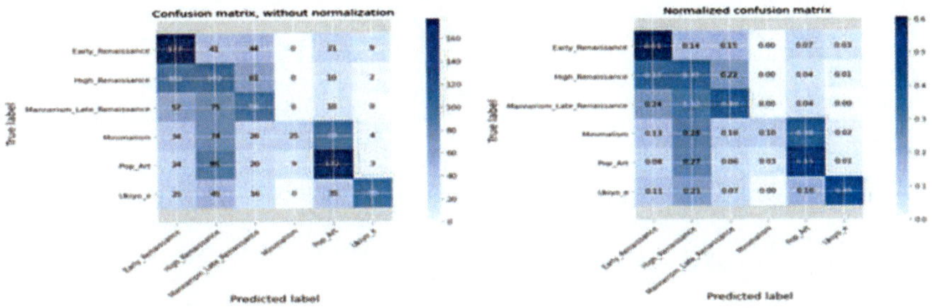

(a) (b)

Fig. 6 Confusion matrix for Wikipaintings dataset for miniVGGNet: **a** confusion matrix and **b** normalized confusion matrix

Table 1 Predicted score for each label by miniVGGNet for Wikipaintings dataset

Labels	Predicted score
Early renaissance	0.61
High renaissance	0.37
Mannerism late renaissance	0.40
Minimalism	0.10
Pop art	0.55
Ukiyo-e	0.45

classification report based on the Wikipaintings dataset applied by miniVGGNet is shown in Table 2.

In Fig. 7a, curve of training loss as well as validation loss for the Wikipaintings dataset for the proposed model is shown. In Fig. 7b, curve of training accuracy as well as validation accuracy is presented. The curve shows a smooth increase in accuracy. Accuracy remains constant from the 30th epoch. From the curves, it is observed that the loss both of training and testing are reduced constantly. That means the model is learning perfectly. On the overall evaluation of the loss and training curves, it is confirmed that the model is performed regularized. In Fig. 8, we have shown the

Table 2 Classification report for Wikipaintings dataset by miniVGGNet

Labels	Precision	Recall	$F1$-score	Support
Early renaissance	0.42	0.61	0.50	292
High renaissance	0.24	0.37	0.29	277
Mannerism late renaissance	0.37	0.40	0.38	238
Minimalism	0.74	0.10	0.17	261
Pop art	0.50	0.55	0.52	313
Ukiyo-e	0.84	0.45	0.59	219
Accuracy			0.42	1600
Macro average	0.52	0.41	0.41	1600
Weighted average	0.51	0.42	0.41	1600

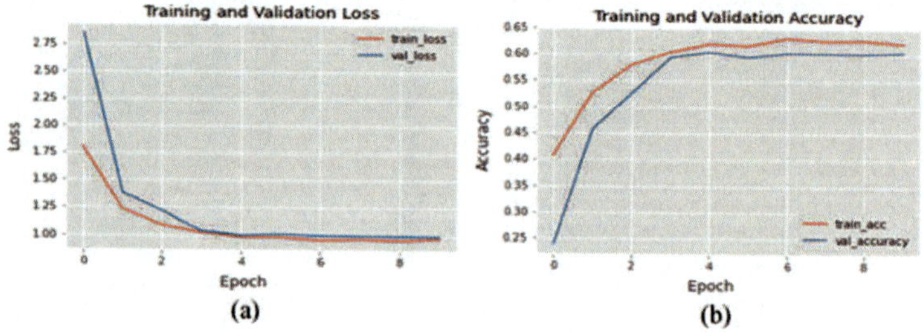

Fig. 7 Loss and accuracy curve of the proposed model for Wikipaintings datasets: **a** training loss and validation loss curve, **b** training accuracy and validation accuracy curve

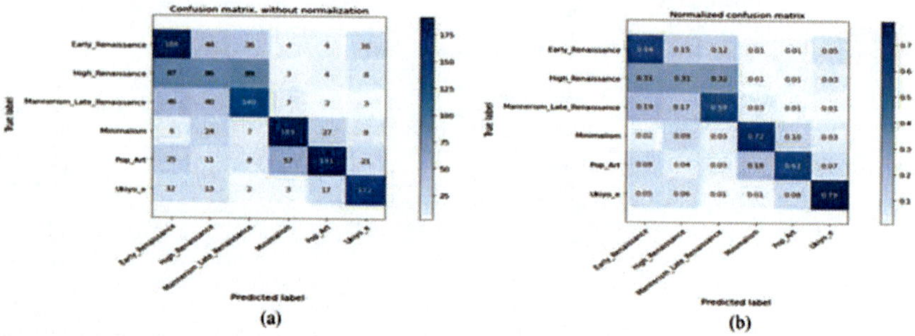

Fig. 8 Confusion matrix of Wikipaintings dataset for proposed model: **a** confusion matrix and **b** normalized confusion matrix

confusion matrix. The proposed network's accuracy on the Wikipaintings dataset is 60.37%. The proposed model of the Wikipaintings dataset is more than 18.50% better than the miniVGGNet. In Table 3, the predicted score for each label is pesented, where Table 4 presents the classification report for Wikipaintings dataset by proposed model.

Table 3 Predicted score for each label by proposed model

Labels	Predicted score
Early renaissance	0.64
High renaissance	0.31
Mannerism late renaissance	0.59
Minimalism	0.72
Pop art	0.61
Ukiyo-e	0.79

Table 4 Classification report for Wikipaintings dataset by proposed model

Labels	Precision	Recall	$F1$-score	Support
Early renaissance	0.52	0.64	0.57	292
High renaissance	0.39	0.31	0.35	277
Mannerism late renaissance	0.50	0.59	0.54	238
Minimalism	0.72	0.72	0.72	261
Pop art	0.78	0.61	0.68	313
Ukiyo-e	0.75	0.79	0.77	219
Accuracy			0.60	1600
Macro average	0.61	0.61	0.61	1600
Weighted average	0.61	0.60	0.61	1600

The model consequences for the Wikipaintings dataset are presented in Figs. 9 and 10 for random test images, respectively. Here, different art styles are recognized by the miniVGGNet and our proposed model competently. The corresponding confidence score of art style recognition is also shown in Figs. 9a and 10a, respectively. Then the recognition confidence scores in the bar chart of different art styles are presented in Figs. 9b and 10b, respectively.

Misclassified Art Style by miniVGGNet

Figure 11 shows some misclassification images by miniVGGNet. Confidence scores with recognition label are shown in Fig. 11a and b shows the respective confidence score in the bar chart.

Correctly Classified Art Style by Proposed Model Which Is Misclassified by miniVGGNet

Sample images of the Wikipaintings dataset which are predicted as wrong by miniVGGNet are tested by the proposed network. Predicted recognition label with confidence score is shown in Fig. 12a. Besides, Fig. 12b shows the respective confidence score in the bar chart. Prediction results of Fig. 12 show the accurate output as expected.

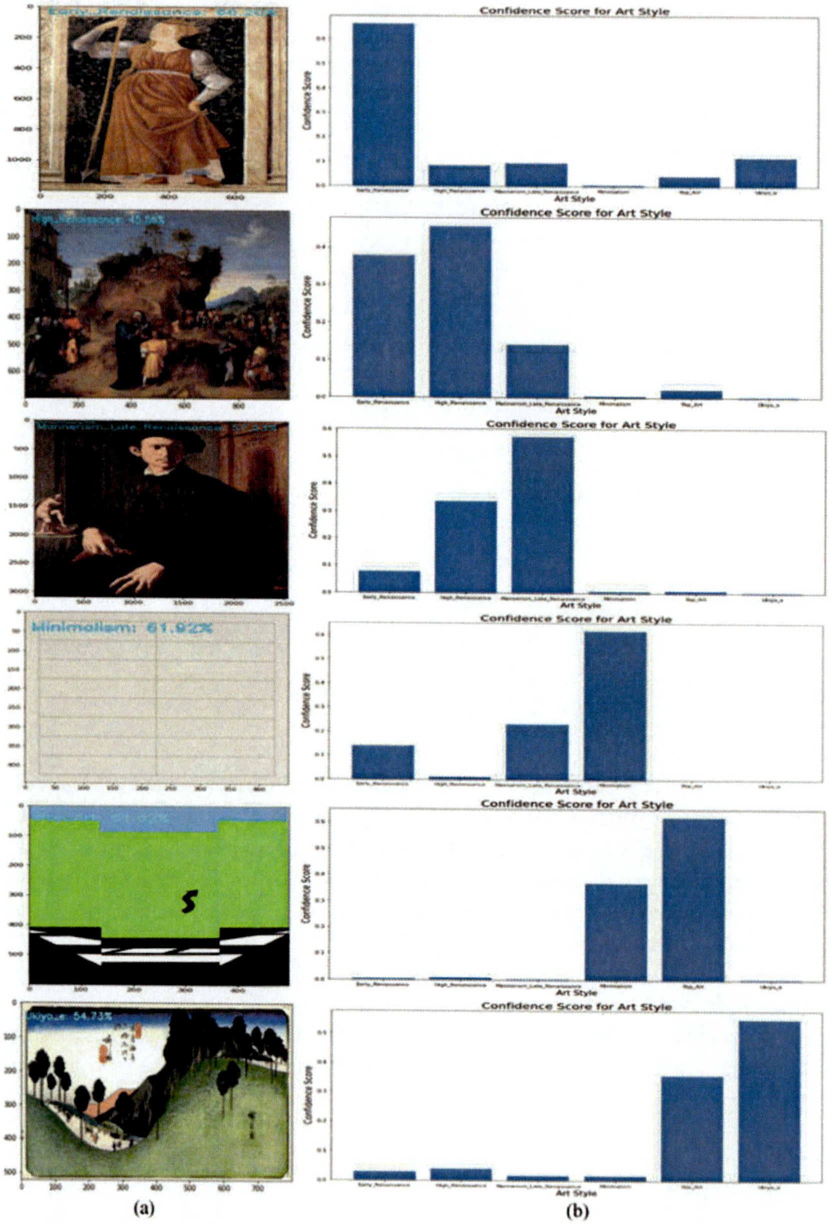

Fig. 9 Random experimental result of Wikipaintings dataset for miniVGGNet: **a** art style recognition with confidence score and **b** recognition confidence scores in the bar chart

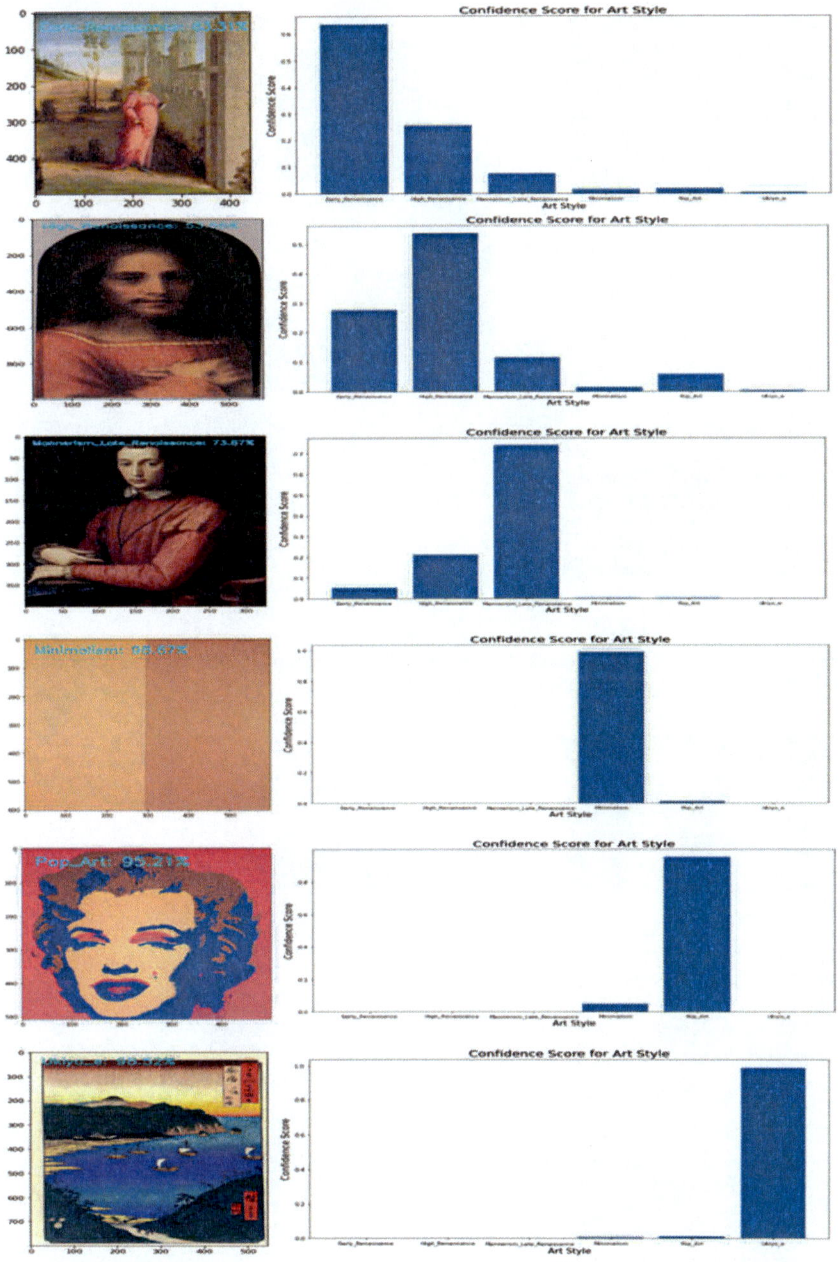

(a) (b)

Fig. 10 Random experimental result of Wikipaintings dataset for proposed model: **a** art style recognition with confidence score and **b** recognition confidence scores in the bar chart

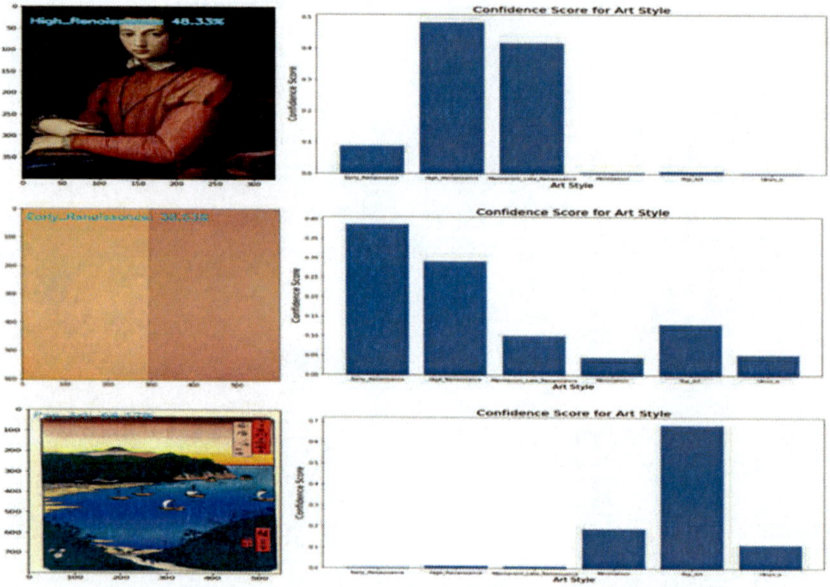

(a) (b)

Fig. 11 Some sample images which are wrong predicted by miniVGGNet, **a** confidence scores with recognition label and **b** confidence scores in the bar chart

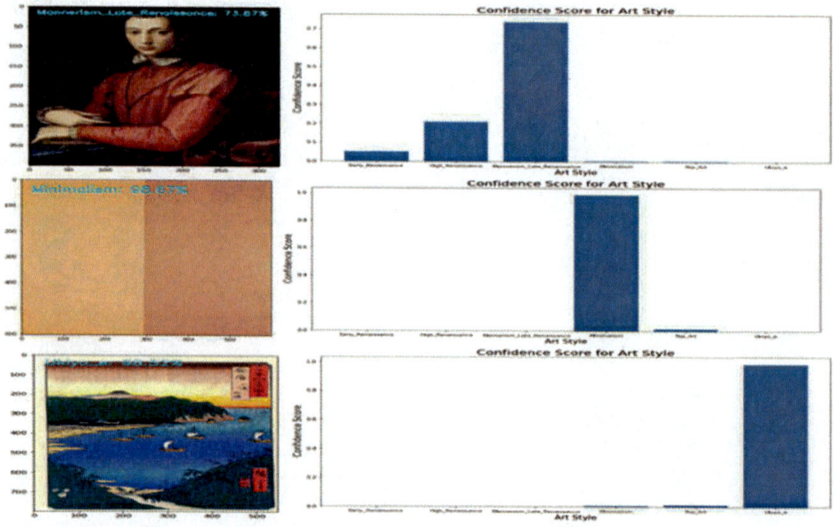

(a) (b)

Fig. 12 Correctly predicted by proposed model those are misclassified by miniVGGNet: **a** confidence scores with recognition label and **b** confidence scores in the bar chart

Table 5 Accuracy using different learning rate schedule

learning rate schedules	Accuracy (%)
Step-based	31.13
Linear	41.31
Polynomial (Proposed method)	**60.37**

The text is bold to indicate the best performance learning rate and method

Table 6 Classification accuracy on Wikipaintings dataset

Method	Accuracy (%)
The Proposed method	**60.37**
miniVGGNet	41.87
ResNet [14]	57.00

The text is bold to indicate the best performance learning rate and method

In our proposed model, we applied three types of learning rate schedule: step-based learning rate schedule, linear learning rate schedule as well as polynomial learning rate schedule. From Table 5, it is shown that the polynomial learning rate schedule shows the best accuracy. In this regard, in our work, we have used the polynomial learning rate schedule. When we used step-based learning rate schedules, we got 31.13% accuracy. When we used linear learning rate unless schedules, we got 41.31% accuracy. When we used polynomial learning rate schedules, we got 60.37% accuracy. Our model is compared with some existing state of the art that is miniVGGNet and ResNet. The comparison of our model with miniVGGNet and ResNet is shown in Table 6.

5 Conclusion

In this work, CNN based an art style recognition architecture is proposed. The proposed CNN model automatically detects several characteristics of individual paintings to support fast, correct, and reliable identification and recognition of an art style. We also have shown comparison between our proposed model and existing network, i.e., miniVGGNet and ResNet using Wikipaintings dataset. We also show some sample examples of real-time paintings. With the accuracy of 60.37%, our model which we have proposed performs efficiently on the Wikipaintings dataset using only one fully connected layer as well as two convolutional layers. In addition, the performance of miniVGGNet is also précised on the same dataset. In the future, we will try to provide better model that will extract the complicated painting features from the painting image and recognize the art style painting more accurately.

References

1. W.R. Tan, C.S. Chan, H.E. Aguirre, K. Tanaka, Ceci n'est pas une pipe: a deep convolutional network for fine-art paintings classification, in *2016 IEEE International Conference on Image Processing (ICIP)* (IEEE, 2016), pp. 3703–3707
2. Y. LeCun, K. Kavukcuoglu, C. Farabet, Convolutional networks and applications in vision, in *IEEE International Symposium on Circuits and Systems*, pp. 253–256 (2010)
3. A. Binder, W. Samek, G. Montavon, S. Bach, K.R. Müller, Analyzing and validating neural networks predictions, in *Proceedings of the ICML 2016 Workshop on Visualization for Deep Learning* (2016)
4. S. Karayev, M. Trentacoste, H. Han, A. Agarwala, T. Darrell, A. Hertzmann, H. Winnemoeller, Recognizing image style. arXiv preprint arXiv:1311.3715 (2013)
5. A. Krizhevsky, I. Sutskever, G.E. Hinton, Imagenet classification with deep convolutional neural networks. Adv. Neural. Inf. Process. Syst. **25**, 1097–1105 (2012)
6. M.O. Kelek, N. Calik, T. Yildirim, Painter classification over the novel art painting data set via the latest deep neural networks. Proc. Comput. Sci. **154**, 369–376 (2019)
7. G. Liu, Y. Yan, E. Ricci, Y. Yang, Y. Han, S. Winkler, N. Sebe, Inferring painting style with multi-task dictionary learning, in *IJCAI International Joint Conference on Artificial Intelligence* (2015)
8. O.E. David, N.S. Netanyahu, Deeppainter: Painter classification using deep convolutional autoencoders, in *International Conference on Artificial Neural Networks* (Springer, Cham, 2016), pp. 20–28
9. Y. Hong, J. Kim, Art painting identification using convolutional neural network. Int. J. Appl. Eng. Res. **12**(4), 532–539 (2017)
10. Y. Bar, N. Levy, L. Wolf, Classification of artistic styles using binarized features derived from a deep neural network, in *European Conference on Computer Vision* (Springer, Cham, 2014), pp. 71–84
11. B. Saleh, A. Elgammal, Large-scale classification of fine-art paintings: learning the right metric on the right feature. arXiv preprint arXiv:1505.00855 (2015)
12. L.A. Gatys, A.S. Ecker, M. Bethge, A neural algorithm of artistic style. arXiv preprint arXiv: 1508.06576 (2015)
13. K. Simonyan, A. Zisserman, Very deep convolutional networks for large-scale image recognition. arXiv preprint arXiv:1409.1556 (2014)
14. A. Lecoutre, B. Negrevergne, F. Yger, Recognizing art style automatically in painting with deep learning, in *Asian Conference on Machine Learning* (PMLR, 2017), pp. 327–342
15. A. Ismail, S.A. Ahmad, A.C. Soh, K. Hassan, H.H. Harith, Improving convolutional neural network (CNN) architecture (miniVGGNet) with batch normalization and learning rate decay factor for image classification. Int. J. Integr. Eng. **11**(4) (2019)

Depressive Bangla Text Detection from Social Media Post Using Different Data Mining Techniques

Farzana Tasnim, Sultana Umme Habiba, Nuren Nafisa, and Afsana Ahmed

Abstract Nowadays, social media have opened a door in front of us to share individual's expressions, emotions, and attitudes toward any incident. Sentiment analysis from different social media posts may help us to detect positive, negative, or emotional behavior toward society. Depressive text detection from social media posts is one of the most challenging parts of individual behavior or sentiment analysis. In this paper, different machine learning algorithms are used to detect depressive Bangla text from social media posts. Pre-processing steps like stemming, stop word removal, etc., are used to clean the collected data, and feature extraction techniques like count vectorization, TF-IDF, word embedding, etc., are applied to the collected dataset which consists of 6178 texts collected from social media posts. We have achieved the highest 97% classification accuracy using decision tree and 94% accuracy for bidirectional LSTM (deep learning model) to predict depressive text in Bangla language. Depressive text detection from social media posts will create an opportunity for psychologists to analyze sentiment from shared posts, reactions, and attitudes which may lessen the unwanted activities of the depressed people through diagnosis and taking them under treatment.

Keywords Depressive text · Bangla language · Deep learning · Social media post

1 Introduction

Mental disorders are becoming a great concern to the human being, depression in one of them. According to the statistics of WHO (World Health Organization) [1], almost 264 million of people are affected with this silent enemy. Almost 800,000 people commit suicide due to severe depression [1]. Loss of appetite, poor concentration on

F. Tasnim (✉) · N. Nafisa · A. Ahmed
Department of Computer Science and Engineering, International Islamic University Chittagong, Chittagong, Bangladesh

S. U. Habiba
Department of Computer Science and Engineering, Khulna University of Engineering and Technology, Khulna, Bangladesh

A. Kumar et al. (eds.), *Computational Intelligence in Machine Learning*, Lecture Notes in Electrical Engineering 834, https://doi.org/10.1007/978-981-16-8484-5_21

work, etc., are the primary effects off depression. In most of the underdeveloped and developing countries, people do not take depression so seriously, rather they feel shy to expose any mental disorder to go for treatment. Almost 76–85% people of low-income and middle-income countries remain untreated in case of mental disorders. Even in Bangladesh, 80% people do not go to psychologists to make a counselling for their problems [2].

At present world, people are getting involved in social media immensely to share and express their happiness, anxiety, sadness, curiosity, etc. So, social media has appeared as a source of data to analyze people's expression [3]. Most of the people in our country use Facebook to share their sentiment. If an automated system can detect depressive text from social media posts, this will be possible to predict depressed people within our network before they go under a malignant stage of depression.

In this paper, we have focused to build a trained model to detect depressive text from social media posts in Bangla language. This trained model will be able to predict if a text is depressive or not depressive. This work aims at reducing the number of unwanted behaviors like attempt to suicide, post-depression crime, etc., and to create opportunities for people working in the health sector. Detecting depressive text from social media post is a very challenging task [4]. First thing is that we have to select those factors which should be focused to analyze Facebook posts. These factors may include tag like "feeling Depressed," emoji, conversation in the comment section, etc. Again, Bangla Facebook status may contain unnecessary punctuations, misspelled word, English word, etc. These things should be taken under various preprocessing steps to achieve efficiency of the model. Bangla owns the seventh position in the most spoken language in the world. In recent years, researchers have shown interest to do this for Bangla language. In this paper, our motto is to collect a standard dataset from Bangla Facebook posts and predicting depressive text using both traditional and deep learning-based machine learning algorithms to do this challenging task.

2 Literature Review

A.H. Uddin et al. [5, 6] proposed to use long short-term memory recurrent neural network (LSTM-RNN) to do the task of depression analysis from social media post in Bangla. They collected Bangla tweets from Twitter to prepare the dataset for this task. This dataset contained 1968 tweets which was not large enough to perform well in deep learning model. To increase the performance of the model with this small dataset, they stratified the data in an order like that one depressive text was followed by another non-depressive text. They experimented with the dataset splitting 80% for training, 10% for validation, and 10% for testing the model. They tuned four hyper parameters of the model (LSTM size, batch size, number of epochs, and number of layers) to achieve maximum classification accuracy of the trained model. They evaluated the model and found 86.3% accuracy. Though Bangla tweets may require some preprocessing steps like stemming, stop word removal, etc., they did not do any of these processing. They did not mention any standard depending on which they stratified the data. Authors used the same dataset in another work [7] with gated recurrent neural network to predict depressive Bangla text. They also tuned the hyper parameters of the GRU model and achieved approximately 75% classification for this task.

Billah et al. [5] proposed to apply SGC classifier, multinomial Naïve Bayes, logistic regression, Nu SVC, and linear SVC to detect social media post if that was depressive or not. Since there is no benchmark dataset for depressive Bangla text detection, they also collected these data manually from Facebook posts. Psychologists generally diagnose patients through discussion with them, and they prefer some linguistic features that may help to detect depression among them. For example, depressed people usually use words like "me", "I", "myself", etc., which actually represent their self-centered thinking focusing on themselves rather than other people. Authors collected these type posts to enrich their dataset which consisted of 1000 text having depressive and not depressive texts. They applied several preprocessing steps to clean the data like punctuation removal, normalization, etc. They processed the data using tokenization also. They transformed the tweets in a set of feature vectors using Unigram and Emoticons. Bigram and Emoticons feature extraction did not prove well for this task.

They had achieved highest 77.9% classification accuracy for SGDC classifier. This work lacks in versatility of depression level because their dataset contains depressive text of those people who have made suicidal attempt. So, these depressive texts may not work well to detect mid-level clinical depression. Tweets of people having intense depression were included in their dataset.

Hassan et al. [8] proposed the workflow of analyzing sentiment from tweets to detect depression level of people who share their emotion and feelings through social media. They divided the tweets into sentences and then tokenized the words of those sentences to form a word vector for the training dataset. They removed the stop words. Stop word removal obviously clean the data for further processing. They applied different feature extraction techniques to ensure a vector of words as a feature vector. They applied N grams to detect co-occurring of words and syllables. POS tagging was applied to read words from text and to denote the parts of speech of the words. Negation was taken in concern to change the positivity or negativity of any word like not good, not bad, etc. They applied SVM, Naïve Bayes, and maximum entropy to classify depressive tweets where SVM showed highest 91% accuracy.

Islam et al. [9] proposed to use supervised learning to detect depression from social media posts. They analyzed various psycholinguistic features of the posts collected from social media, and this was done by using LIWC software. LIWC has more than 50 attributes, but they have used only 21 among of them. These 21 attributes were about linguistic information like verb, adverb, singular pronoun, etc.; emotional information like sadness, anxiety, positivity, etc.; and temporal information also. LIWC package analyzed the data using these factors and prepared data to train in supervised classifiers. They achieved highest 72% classification accuracy using SVM.

Aragón, López-Monroy et al. [10] proposed a new representation bag of sub-emotions.

3 Methodology

Bangla depressive text detection from social media posts require a standard dataset containing Bangla posts having both category like depressed and not depressed. Since we have no benchmark dataset in Bangla language for this work, we have collected dataset according to some rules. Data preprocessing and feature extraction techniques are applied before passing the dataset through training model. Punctuation removal, stop word removal, and stemming are done to clean the data. We have applied TF-IDF, count vectorizer, and word embedding feature extraction techniques in our work Fig. 1.

3.1 Dataset

Bangla Facebook status is the expected input data in our work to detect depressed people. We have collected 7000 data (Bangla text data) from Facebook where the number of depressed class is 3500 and the number of not depressed class is 3500. We have maintained consistency to categorize the dataset into two classes—depressed and not depressed. We have taken four facts in our concern to collect the data. These are: (i) Emotion expressed through emoji, (ii) Collecting status of those people who committed suicide and usually shared depressive posts in social media before their death, (iii) Collecting status of those people who share their posts with a tag "**feeling depressed**" and (iv) Collecting unambiguous data where class label is not confirmed only through reading the post. We have also

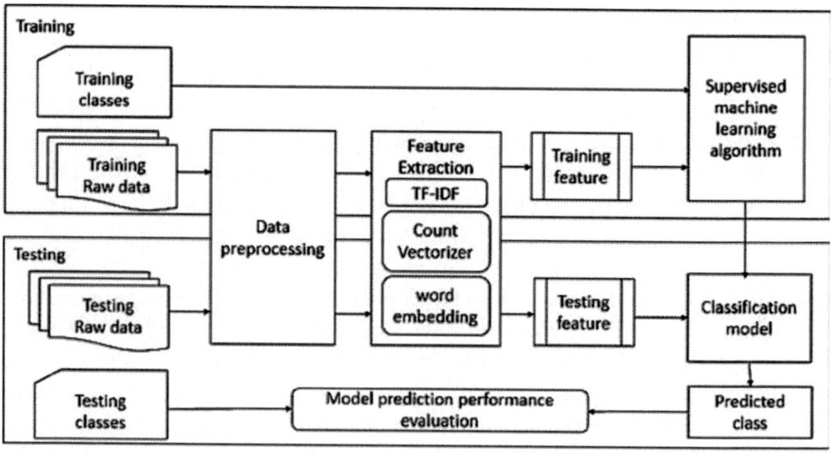

Fig. 1 Proposed methodology

Table 1 Examples of Bangla posts

Bangla Post	Class
পরিবার কি জিনিস প্রত্যেকটা মুহূর্ত বুঝতেসি....একা একটা বাসায় থেকে অফিস আসা যাওয়া, রান্না করা,কাপড় ধোয়া..	**Depressed**
রান্নাবান্নার কাজটা আমার কাছে একটা আর্টের মত লাগে ... এইখানে সব আছে ... ব্রেইন, স্কিল, এক্সপেরিয়েন্স আর ট্যালেন্ট !	**Not Depressed**

converted **emoji considering as a word** which may represent emotion more explicitly sometimes. At the time of labeling the data (which is confusing) as depressive or not depressive text, we have taken opinion of **20 people** within our friends and family and also **analyzed the comment section** of that post to label it, for example, মাঝে মাঝে নিজের নামে এমন কিছু কথা শুনি যা আমি নিজেও জানিনা. Since some Bangla word or punctuation using may create confusion to detect the category of the posts, we have analyzed the situation of the user through his Facebook profile and friends' comments, attitude, and response to their shared posts. Some examples of our collected dataset for this work are given in Table1.

3.2 *Text Preprocessing*

Bangla tweets from Facebook may contain noise, misspelling, unnecessary punctuation marks, etc. We have made some pre-processing steps to clean the data. These steps include punctuation removal, special character removal, stemming, etc.

Punctuation and Special Character Removal: Punctuations which are used to express emoji in the post are not removed. Only those punctuations (like '?', '/', '!") and special characters (like {@, #, $, %, —, „ ..) were eliminated from the dataset which played no role in the semantic part of the text, for example, "হুট করে পৃথিবীটা কেমন শূন্য শূন্য লাগে ... কেমন যেন একটা হাহাকার চারপাশে!!" After the removal of punctuations, we got like, "হুট করে পৃথিবীটা কেমন শূন্য শূন্য লাগে কেমন যেন একটা হাহাকার চারপাশে". Since these posts were in Bangla language, we removed English characters and digits from the posts also.

Tokenization: The sentences were tokenized for further processing of the data to be trained in the model.

Before tokenization: "বাসার নিচ থেকে মাঝে মাঝে অ্যাম্বুলেন্স যায় ... ফাঁকা রাস্তা দিয়ে হালকা সাইরেন বাজাতে বাজাতে ..."

After tokenization: <বাসার> <নিচ> <থেকে> <মাঝে> <মাঝে> <অ্যাম্বুলেন্স> <যায়> <ফাঁকা><রাস্তা><দিয়ে><হালকা><সাইরেন><বাজাতে><বাজাতে>

Stop Word Removal: Stop words in any language are those words adding no meaningful information in any sentence, and those can be ignored without hampering the original semantic of any sentence. In any text processing or sentiment analysis task, stop words are removed to decrease complexity and keep

Table 2 Example of stemming

Before stemming	After stemming
বাংলাদেশের	বাংলাদেশ
করছিলাম	কর
করছি	কর

the corpus free from unnecessary words. Some stop words in Bangla are like: {অতএব, অথচ, অথবা, আর,আরও, ই, কি, কিংবা, কিছু}, etc.

Stemming: Stemming is the process of turning a word into its stem or root word. Since our collected data consisted of Bangla social media posts, these might contain lots of misspelling. So, stemming is done to find the root word to avoid unambiguous meaning of the tweet Table 2.

3.3 *Feature Extraction Method*

Three feature extraction method are used to represent each tweet as a feature vector. Count vectorizer, TF-IDF, and word embedding methods are applied on our dataset.

Count Vectorizer

If the text data is represented in such a machine-readable form, where each word is shown as a vector is called vectorization [11]. Bag of words (BOW) model represents only the presence count of the words in a vector with a fixed size which contains only numbers. Count vectorizer encodes the data in a form where every word is given an index number (usually array index) and counts the occurrence of that word in the document. For example,

Vocabulary

আম্মু	আমার	গিয়েছিল	ছোট	জীবন	দুনিয়াটা	😢

Sample: আমার দুনিয়াটা অনেক ছোট 😢

Vectorization

আম্মু	আমার	গিয়েছিল	ছোট	জীবন	দুনিয়াটা	😢
0	1	0	1	0	1	1

TF-IDF Value

TF-IDF stands for "Term Frequency Inverse Document Frequency." It is one of most powerful feature extraction technique to represent word count in the documents or corpus. TF-IDF transforms text into number representation so that this can be fitted into machine learning algorithms.

$$\mathrm{TF}(t) = \frac{\text{(Number of times term } t \text{ appears in a document)}}{\text{(Total number of terms in the document)}}$$

$$\mathrm{IDF}(t) = \frac{\text{log\_ e(Total number of documents)}}{\text{(Number of documents with term } t \text{in it)}}$$

$$\mathrm{TF} - \mathrm{IDF} = \mathrm{TF}(t) * \mathrm{IDF}(t)$$

Word Embedding

Word embedding technique includes the individual word into a real-valued predefined vector space. It is more reliable than TF-IDF and count vectorizer in case of semantic representation. For example,

We have created a vocabulary where the size of the vocabulary was = 10,000.

One hot representation is the representation of the sentences using word indexes from vocabulary.

Sample: পৃথিবী সুস্থ হয়ে গেলে এমন এক ঝুম বৃষ্টি মনপ্রাণ উজাড় কর ভিজ

After pre-processing: পৃথিবী সুস্থ হয়ে গেলে এমন এক ঝুমবৃষ্টি মনপ্রাণ উজাড় কর ভিজ

One hot vector representation:

পৃথিবীটা সুস্থ হয়ে গেলে এমন এক ঝুম বৃষ্টিতে মনপ্রাণ উজাড় করে ভিজবো= [234, 356, 783, 1002,998, 3356, 8565, 9779, 2771, 712, 2003, 3870]

Vector representation after 40 length pre-padding: [0 0 0 … 2771 712 2003 3870]

3.4 *Supervised Learning and Classifier*

Since depressive text detection from social media posts is one of the most challenging tasks, it was not so easy to select appropriate machine learning algorithm to do this work. We have turned our work dataset into a large feature vector after all these pre-processing and feature extraction techniques. These data were experimented in both traditional machines learning algorithms and deep learning models. The experiment was done using multi-nominal NB, aggressive classifier, decision tree classifier, neural network, and linear support vector machine. We have also experimented in two deep learning model—bidirectional LSTM (BiLSTM) and gated recurrent units (GRU). We have used tenfold cross-validation for each of the model. Each *i*th fold is used a test dataset for *i*th training iteration. Remaining portion of the data is used as training data.

4 Result Analysis

We have experimented with the processed dataset using traditional machine learning algorithms and deep learning models. LTK (Natural Language Tool Kit) in Python was used to implement the work of stop word removal, stemming, tokenizer, etc. Different classifiers and feature extraction algorithms were implemented through scikit learn library. Feature extraction algorithms improve the classification performance immensely. In Fig. 2, the comparative study among the classifiers is given. It is showed that highest 97% classification accuracy is found for "Decision Tree" to detect depressive text from social media post. Between the two deep learning model, BiLSTM has shown better performance showing 94% classification accuracy. In Fig. 2, it is shown that accuracy got improvement in case of all classifiers after adding emoji with text data. For every classifier, accuracy for text + emoji is higher than accuracy for only text. In Table 3, precision and recall for the classifiers are shown for text and text + emoji cases. Decision tree and BiLSTM also showed highest precision and recall values.

T = true, P = positive, F = false, and N = negative.

$$\text{Accuracy} = \frac{\text{TP} + \text{TN}}{\text{TP} + \text{FP} + \text{TN} + \text{FN}}, \text{Precision} = \frac{\text{TP}}{\text{TP} + \text{FP}}, \text{Recall} = \frac{\text{TP}}{\text{FP} + \text{FN}}$$

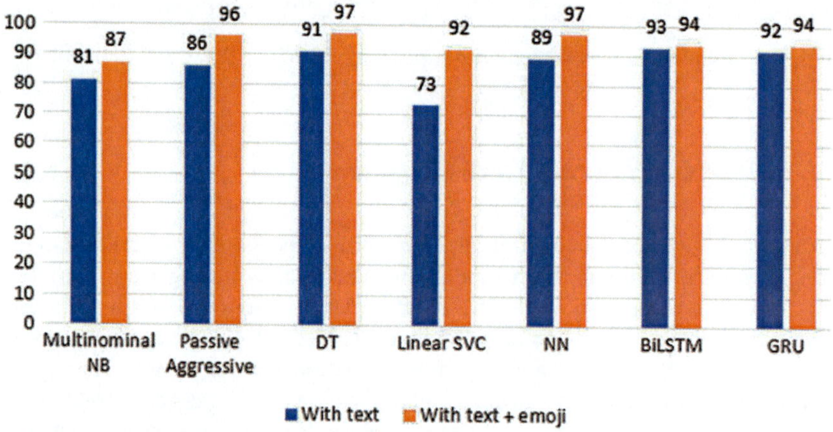

Fig. 2 Accuracy for depressive text detection

Table 3 Precision and recall for depressive text detection

Classifier name	Precision		Recall	
	Text	Text + emoji	Text	Text + emoji
Multi-nominal NB	81	87	81	87
Passive aggressive	87	97	87	97
DT	92	97	91	97
Linear SVC	74	92	74	92
NN	90	97	89	97
BiLSTM	93	94	93	94
GRU	93	94	93	94

$$F1\ \text{Score} = \frac{21}{\frac{1}{\text{precision}} + \frac{1}{\text{recall}}} = \frac{2 * \text{precision} * \text{recall}}{\text{precision} + \text{recall}}$$

In Fig. 3, the training accuracy and validation accuracy for BiLSTM model showed that both of them started to each a saturation at 6–7th epoch of training. Training accuracy indicates the performance of the model during training time on the training dataset. Validation accuracy measures how the trained model perform on the data beyond the training set. So, Fig. 3 indicates a good relation between training and validation accuracy. In Table 4, $F1$ score for the classifiers is given. It is showed that decision tree, neural network, BiLSTM, and GRU performed very well with respect to $F1$ score. The equations of the performance measuring parameters are given below:

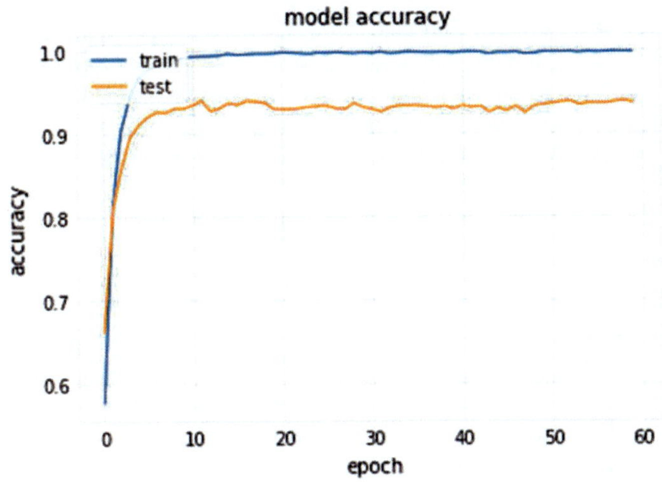

Fig. 3 Training and validation accuracy for BiLSTM

Table 4 $F1$ score for depressive text detection

Classifier name	With text		With text + emoji	
	Depressed	Not depressed	Depressed	Not depressed
Multinominal NB	81	81	87	87
Passive aggressive	86	87	96	96
DT	90	91	97	97
Linear SVC	72	75	92	92
NN	89	89	97	97
BiLSTM	93	93	94	94
GRU	92	92	94	94

Table 5 Comparison of proposed method with previous works

Ref	Methodology	Feature extraction	Language	Accuracy (%)
[3]	LSTM RNN	None	Bangla	86.3
[7]	KNN, SVM, DT	LIWC	English	73
[4]	SGDC, logistic regression	Unigram and emoticons with TF-IDF values	Bangla	74.57
[5]	GRU	None	Bangla	75.7
[6]	SVM, Naïve Bayes, ME	N gram, POS, negation checker	English	91
Proposed method	**DT, NN, BiLSTM, GRU**	**Count vectorizer, TF-IDF, word embedding**	**Bangla**	**97**

In Table 5, we have shown the comparison among our proposed methodology and previous works on depressive text detection (both in Bangla and English). Since the number of previous works on depressive text detection in Bangla language is very fewer, among [5–7] and our proposed methodology which work are done for Bangla language, our work has shown comparatively very good performance.

5 Conclusion

Depression detection from social media posts is one the most talked research issues in the world. But still there exist no benchmark dataset in Bangla language to do this task. In this paper, we applied different machine learning algorithms with feature extraction techniques to detect depressive text from social media posts. We experimented the work using with emoji and without emoji also. Emoji sometimes express sentiment more clearly in posts. In all classifiers, the classification performance improved using text with emoji. Three feature extraction techniques count vectorizer, TF-IDF, and word embedding prepared the collected data and transformed into a feature vector to be trainable in the classifier models. Decision tree showed highest classification accuracy 97% (with Text + emoji), and two deep learning models, BiLSTM and GRU, showed 94% classification accuracy. This work will be helpful to diagnose depressed people within our network. So that we may help them to take part in online counselling or prevent them doing any life-threatening occurrence owing to mental disorder.

References

1. World Health Organization. https://www.who.int/news-room/fact-sheets/detail/depression
2. What is the suicide rate among persons with depressive disorder (clinical depression)? https://www.medscape.com/answers/286759-14675/what-is-the-suicide-rate-among-persons-with-depressive-disorder-clinical-depression
3. A.U. Hassan et al., Sentiment analysis of social networking sites (SNS) data using machine learning approach for the measurement of depression, in *2017 International Conference on Information and Communication Technology Convergence (ICTC)* (IEEE, 2017)
4. M.R. Islam et al., Depression detection from social network data using machine learning techniques. Health Inf. Sci. Syst. **6**(1), 8 (2018)
5. A.H. Uddin, D. Bapery, A.S.M. Arif, Depression analysis from social media data in Bangla language using long short-term memory (LSTM) recurrent neural network technique, in *2019 International Conference on Computer, Communication, Chemical, Materials and Electronic Engineering (IC4ME2)* (IEEE, 2019)
6. S. Akhter, Social media bullying detection using machine learning on Bangla text, in *10th International Conference on Electrical and Computer Engineering (ICECE)* (IEEE, 2018)
7. A.N. Chy, M.H. Seddiqui, S. Das, Bangla news classification using naive Bayes classifier, in *16th Int'l Conf. Computer and Information Technology* (IEEE, 2014)
8. A.H. Uddin, D. Bapery, A.S.M. Arif, Depression analysis of Bangla social media data using gated recurrent neural network, in *2019 1st International Conference on Advances in Science, Engineering and Robotics Technology (ICASERT)* (IEEE, 2019)
9. M. Billah, E. Hassan, Depression detection from Bangla Facebook status using machine learning approach. Int. J. Comput. Appl. **975**, 8887
10. M.E. Aragón et al., Detecting depression in social media using fine-grained emotions, in *Proceedings of 2019 Conference of the North American Chapter of the Association for Computational Linguistics: Human Language Technologies*, vol. 1-Long and Short Papers (2019)
11. M.M. Tadesse et al., Detection of depression-related posts in Reddit social media forum. IEEE Access **7**, 44883–44893 (2019)

Implementation of Optimized PID Controllers in Real Time for Magnetic Levitation System

Soham Dey, Subrata Banerjee, and Jayati Dey

Abstract This paper aims to demonstrate a tuning method for the parameters of PID controller utilizing various modern meta-heuristic and swarm-intelligence-based optimization techniques for a laboratory scaled magnetic levitation system. Owing to the fact that classical controller tuning method failed to yield satisfactory result for a highly non-linear and unstable MAGLEV system; this work focuses on tuning of PID parameters employing a new artificial-intelligence-based optimization techniques, grey wolf optimization (GWO) and performance is compared with existing algorithms. The controller parameter optimization has been carried out by incorporating various error-based performance indices as objective functions. The proposed GWO-PID controller has proven its efficacy in terms of better transient specifications, optimum stability performance, accurate reference tracking capability, and improved convergence characteristic.

Keywords Evolutionary optimization · ITSE performance index · MAGLEV system · Optimized PID controller · Real-time implementation

1 Introduction

Magnetic levitation technique basically refers to a system, wherein a ferromagnetic object is levitated and suspended in the air with the help of electromagnetic forces without any visible means of mechanical or physical support [1]. Certain interesting features of MAGLEV system like: highly non-linear dynamic behavior, inherent instability, and free from mechanical frictional loss make this system suitable for application in many industrial applications. High speed MAGLEV trains [2], magnetically suspended wind turbines, magnetic bearings [3] are some of MAGLEV system's significant applications.

S. Dey (✉)
Dr. B. C. Roy Engineering College, Durgapur, India
e-mail: soham.dey@bcrec.ac.in

S. Banerjee · J. Dey
National Institute of Technology, Durgapur, India

A. Kumar et al. (eds.), *Computational Intelligence in Machine Learning*, Lecture Notes in Electrical Engineering 834, https://doi.org/10.1007/978-981-16-8484-5_22

Closed-loop stability of the MAGLEV system is achieved by applying proper controlling action. As found in the literature, several classical and modern control techniques have been applied to overcome the inherent instability problem. As in classical control approach cascaded lead-lag controller, lead-PI controller and PID controllers are mostly designed. As reported in [4], two control loops comprising of outer position control (lead controller) loop and inner current control (PI controller) loop can simultaneously stabilize the plant and improve performance. Robust linear control techniques such as μ-synthesis, Q-parameterization, H_2 and H_∞ control [5–7], have also been adopted by researchers to take care plant uncertainties and parameter variations for a wide range of operations.

Classical controller tuning procedure failed to provide prolific output as the system dealt with here is highly unstable and non-linear. As a result, researchers have opted for advanced tuning techniques including modern optimization-based parameter tuning [8]. Genetic algorithm-based parameter tuning is proposed in [9]. In [10], authors applied PSO to obtain optimum PID parameter values. Optimal PID controller tuning rule with set-point weighting is considered in [11].

The present work illustrates implementation of optimized PID controller for magnetic levitation system in software using MATALB/Simulink as well as in real time using MAGLEV hardware plant. The parameters of the PID controller are optimized considering various error-based performance criteria formulated as complex optimization problem. The real-time experimental results are in unison with the simulation work. The remaining paper is structured as follows: Sect. 2 gives a brief account of the physical and mathematical description of the system of interest. Section 3 provides proposed PID tuning method. Section 4 enumerates simulation and hardware results, and finally, Sect. 5 concludes the work.

2 Modeling of MAGLEV System

2.1 Brief Description of the MAGLEV Hardware Plant

The hardware setup used for experimentation in this work is manufactured by Feedback Instruments Ltd (UK). The main components of the MAGLEV setup are as follows: Electromagnetic coil, infrared light sensors, ferromagnetic ball (the object), analog/digital interface, and the controller. The magnetic field induced by the current in electromagnetic coil exerts an attractive force on the object which levitates it. The magnetic force is counterbalanced by the gravitational force. At the steady-state equilibrium condition, the object steadily floats in the magnetic field. The ball position is measured by the IR sensors, and the sensor output is used as feedback signal which in turn is used to produce required controlling force.

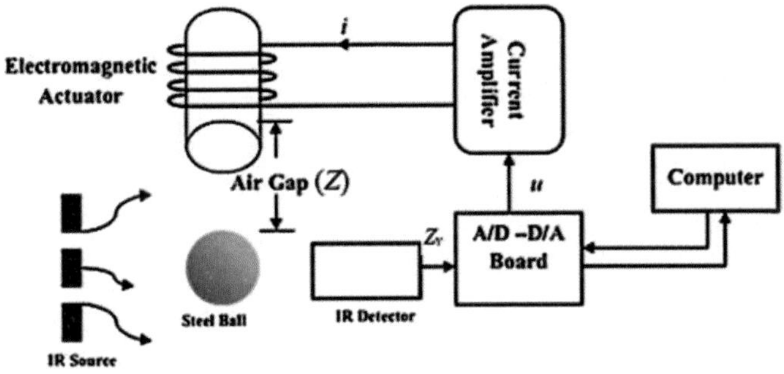

Fig. 1 Illustration of MAGLEV hardware setup

2.2 *Mathematical Model of the MAGLEV System*

The schematic diagram of MAGLV setup is shown in Fig. 1. Let's suppose, the ball having m is suspended in the magnetic field; the coil current and the ball position are denoted by i and z, respectively. Using Newtonian mechanics, the ball dynamics is formulated by the following differential equation:

$$F = m\ddot{z} = -c\left(\frac{i}{z}\right)^2 + mg \tag{1}$$

where F is the resultant force acting on the object; c is coil parameter constant, and g is acceleration due to gravity. In order to apply linear control algorithm, the non-linear.

Plant dynamics has to be linearized around a feasible equilibrium point (i_0, z_0). Let, Δi and Δz are very small deviations of current i and position z from equilibrium point (i_0, z_0), respectively; then:

$$i = i_0 + \Delta i \tag{2}$$

$$z = z_0 + \Delta z \tag{3}$$

Now, Eq. (1) is linearized by Taylor series expansion method:

$$\Delta\ddot{z} = -\left(\left.\frac{\partial f(z,i)}{\partial i}\right|_{z_0,i_0} \Delta i + \left.\frac{\partial f(z,i)}{\partial z}\right|_{z_0,i_0} \Delta z\right) \tag{4}$$

Here, $f(z, i) = {ci^2}/{mz^2}$; further doing simplification of the above equation:

$$\Delta\ddot{z} = -(K_i\Delta i - K_z\Delta z) \tag{5}$$

Table 1 Specifications of the MAGLEV

Components	Symbols	Values
Mass of the object (the ball)	m	0.02 kg
Input voltage to the controller	u	± 5 V
Equilibrium position of the ball	z_0	0.009 m
Equilibrium value of the coil current	i_0	0.8 A
Voltage to current gain	k_1	1.05 A/V
Feedback element (IR sensor) gain	k_2	143.48 V/m, −2.8 V

where $K_i = \frac{2g}{i_0}$ and $K_z = \frac{2g}{z_0}$. To obtain overall plant transfer function Laplace transform is applied on Eq. (5):

$$\frac{\Delta z(s)}{\Delta i(s)} = \frac{-K_i}{s^2 - K_z} \tag{6}$$

The parameter values of MAGLEV hardware setup are summarized in Table 1. Putting the parameter values in Eq. (6), the transfer function is obtained below:

$$\frac{\Delta z(s)}{\Delta u(s)} = \frac{-3694.78}{s^2 - 2180} \tag{7}$$

3 Optimized PID Design Algorithm

The most popularly used controller in process control action is the PID controller. Here, parameter tuning is accomplished using optimization algorithms, and objective functions are formulated by commonly used error-based performance indices: Integral absolute error (IAE), integral square error (ISE), and integral of time-weighted-square-error (ITSE). The mathematical expressions are shown below:

$$\text{IAE} = \int_0^\infty |e(t)|\mathrm{d}t;\quad \text{ISE} = \int_0^\infty e^2(t)\mathrm{d}t;\quad \text{and ITSE} = \int_0^\infty te^2(t)\mathrm{d}t$$

The implementation steps of proposed control algorithm are depicted as follows:

1. Initialize all the three controller parameter values: Proportional, integral, and derivative gain, respectively, by stability analysis-based design.
2. Run simulation to obtain the step response of the compensated system.
3. To achieve more precision in the result reduces time step by half and double the time range of the step response obtained in step 2.
4. Define performance criterion (objective function) according to requirements.

5. Execute optimization algorithm to obtain optimum controller parameter values that minimizes the performance criterion.
6. Reset the controller parameter values obtained in step 5.
7. Run the simulation again to get desired step response.

4 Simulation and Real-Time Results

The proposed optimized PID controller is simulated in MATLAB 8.3 software. A rigorous mathematical analysis and simulation study are conducted to tune the parameters of the PID controller. The simulation results are depicted in Figs. 2, 3 and 4. Table 2 enlists parameter values along with time domain specifications.

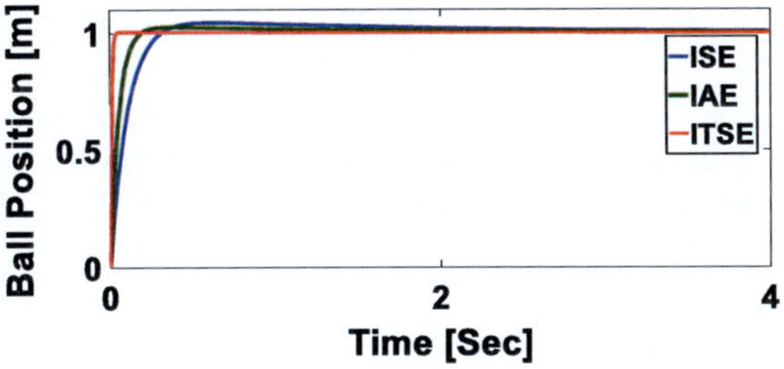

Fig. 2 Step response with GA-PID

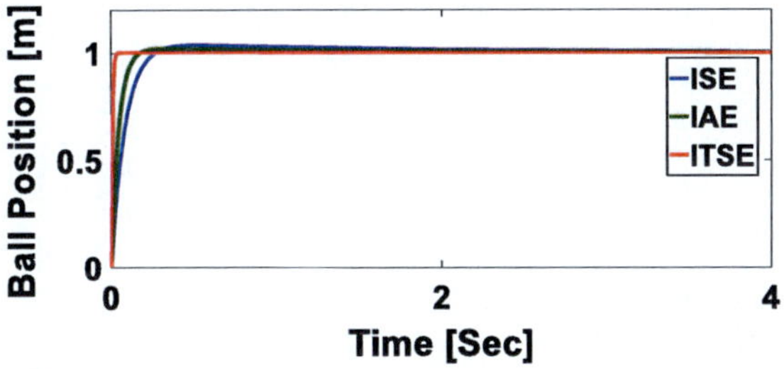

Fig. 3 Step response with PSO-PID

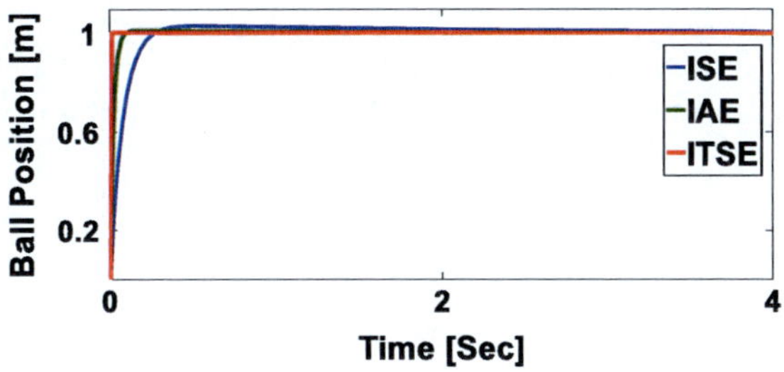

Fig. 4 Step response with GWO-PID

Table 2 Parameter values and performance specifications with optimized PID controllers

Controllers	Performance index	Controller parameter values			Time-domain specifications		
		K_p	K_i	K_d	Over shoot (%)	Rise time	Settling time
GA-PID	ISE	4.92	2.92	0.48	4.6	0.09	1.44
	IAE	5	3.04	0.25	2.4	0.05	1.01
	ITSE	4.95	3.05	0.128	1.4	0.02	0.16
PSO-PID	ISE	5	3	0.4	3	0.07	1.06
	IAE	5.21	2.92	0.23	2.1	0.037	0.93
	ITSE	5	3	0.245	1.1	0.02	0.1
GWO-PID	ISE	4.96	2.68	0.32	2.9	0.06	0.82
	IAE	3.38	2.88	0.10	2.02	0.02	0.65
	ITSE	4.78	3.50	0.284	1.5	0.008	0.04

Experimental results are obtained by putting controller parameter values in the real-time models installed in the PC with hardware–software interfacing. The real-time system output responses are captured in Figs. 5, 6 and 7.

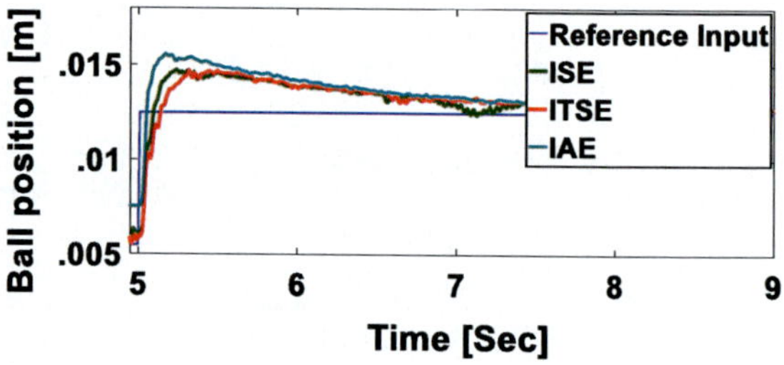

Fig. 5 Step response for GA-PID (Real time)

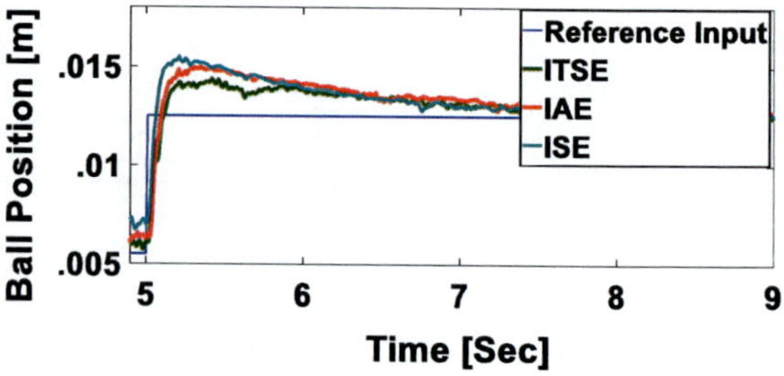

Fig. 6 Step response for PSO-PID (Real time)

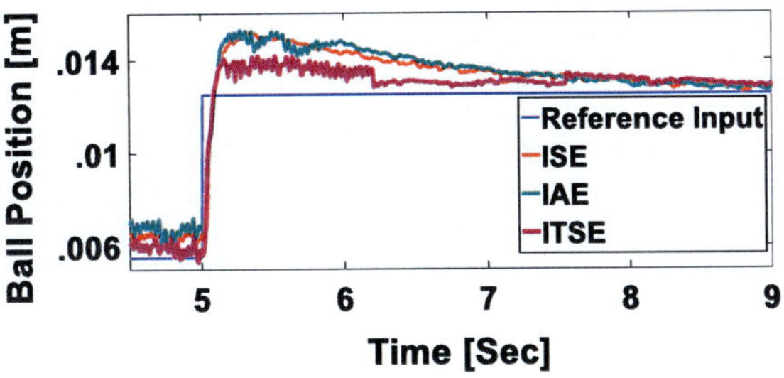

Fig. 7 Step response for GWO-PID (Real time)

5 Conclusions

In this paper, meta-heuristic algorithm-based optimal tuning of classical PID controller has successfully been exploited for real-time position control of a laboratory scale magnetic levitation system. An extensive simulation study has been carried out for the design of optimized PID controller. The simulation results show that, GWO-PID controller is able to provide better controller performance in terms of transient specifications. Maximum overshoot is significantly reduced to 1.5%, while speed of response (with rise time 0.008 s and settling time 0.04 s) has quite improved compared to other controllers. The designed optimized controllers have been implemented in real time with the laboratory scale experimental setup. It is seen that the experimental position responses are in unison with the simulation results. The GWO-PID controller utilizing ITSE performance index exhibits better position response than other controllers.

References

1. Y. Zhang, B. Xian, S. Ma, Continuous robust tracking control for magnetic levitation system with unidirectional input constraint. IEEE Trans. Ind. Electron. **62**(9), 5971–5980 (2015)
2. H.W. Lee, K.C. Kim, J. Lee, Review of maglev train technologies. IEEE Trans. Magn. **42**(7), 1917–1925 (2006)
3. A. Smirnov, N. Uzhegov, T. Sillanpää, J. Pyrhönen, O. Pyrhönen, High-speed electrical machine with active magnetic bearing system optimization. IEEE Trans. Ind. Electron. **64**(12), 9876–9885 (2017)
4. S. Banerjee, T.K.S. Kumar, J. Pal, D. Prasad, Controller design for large-gap control of electromagnetically levitated system by using an optimization technique. IEEE Trans. Control Syst. Technol. **16**(3), 408–415 (2008).
5. J. Lin, B. Tho, Analysis and μ-based controller design for an electromagnetic suspension system. IEEE Trans. Edu. **41**(2), 116–129 (1998)
6. A.M. Mohamed, F. Matsumura, T. Namerikawa, J.-H. Lee, Q-parameterization/μ-control of an electromagnetic suspension system, in *Proceedings of the 1997 IEEE International Conference on Control Applications (Connecticut)*, pp. 604–608 (1997)
7. A. Bittar, R.M. Sales, H_2 and H_∞-control for Maglev vehicles. IEEE Control Syst. Mag. **18**(4), 18–25 (1998)
8. S. Dey, J. Dey, S. Banerjee, Optimization algorithm based PID controller design for a magnetic levitation system, in *2020 IEEE Calcutta Conference (CALCON)*, pp. 258–262 (2020)

9. I. Ahmad, M. Shahzad, P. Palensky, Optimal PID control of magnetic levitation system using genetic algorithm, in *2014 IEEE International Energy Conference (ENERGYCON)*, (Cavtat, 2014), pp. 1429–1433
10. R. Wai, J. Lee, K. Chuang, Real-time PID control strategy for maglev transportation system via particle swarm optimization. IEEE Trans. Ind. Electron. **58**, 629–646 (2011)
11. M. Hast, T. Hägglund, Optimal proportional–integral–derivative set-point weighting and tuning rules for proportional set-point weights. IET Control Theory Appl. **9**(15), 2266–2272 (2015)

Anomalous Behavior Detection in Examination

Zaki Ahmed yadahalli and A. Parkavi

Abstract Monitoring a classroom full of students during examination is a tedious task, with much improvement in the technology there needs to be a system that can efficiently do the task of an invigilator during the examination that is to monitor the students for malpractice. The proposed system is expected to give an alert when some malpractice is detected by the system, malpractice here in the form of student's abnormal behavior, uncommon movements during examination. The feed given to the system is from the camera/CCTV installed in the classroom. The system analysis the live video feed from the classroom camera and evaluates the students behavior for malicious activities such as passing chits, making gestures at each others, looking here and there, shifting from their positions, copying, etc. In this system, we are not interested in facial recognition of students. The system only detects malicious behavior done by a particular student or multiple students and gives an alert as a box or any shape around that particular student/students who is involved in malpractice.

Keywords Gray scale · Anomalous · Computer vision · Frame

1 Introduction

Understanding human behavior through computer vision is both diverse and a complicated task. The examination invigilation task is rigorous task which is heavy workload on the invigilator and also not very efficient. In today's circumstances, where viruses like COVID, Ebola have restricted physical human interaction and pushed all academic activities to online mode there needs to be a way to incorporate these academic activities slowly into offline mode, one of these academic activities is conduction of examination. Conduction of examination in classroom with limited

Z. A. yadahalli (✉) · A. Parkavi
Department of Computer Science and Engineering, MSRIT, MSR Nagar, Bangalore, India
e-mail: 1ms19scs24@msrit.edu

A. Kumar et al. (eds.), *Computational Intelligence in Machine Learning*, Lecture Notes in Electrical Engineering 834, https://doi.org/10.1007/978-981-16-8484-5_23

number of students and exclusion of an invigilator whose task is to monitor students for malpractice by physically being present near the students is the requirement.

Here, we propose a system that would serve as a replacement for physical monitoring of student in the examination by an invigilator, the system is supposed to recognize individual students but not their identity (no facial recognition) and monitor them for their anomalous behavior. Whenever the system detects any unusual activity done by any student, it is required by the system to detect that particular student in the form of displaying some sort of shape around that particular student and give any specified type of alert.

There are quite a few challenges in building a video analytics system that analysis the video feed from a surveillance camera in an examination hall and overcomes the huge processing in the background [1], also there is nature of the examination hall which includes sufficient lighting and proper sitting pattern and arrangements. Other factors affecting the performance of such a system are the image depth and occlusion [2]. For instance, students sitting in the last benches are most probably not detected. Minor movements like writing in paper, drinking water should be ignored. It is essential that the software being built should rely on sequence of stages and not just on a single state to decide anomalous behavior.

2 Related Works

Researchers have converged their work relating to the topic being discussed in this paper, there are studies done by authors that focus their attention toward the task of preventing the malpractice occurring in examinations. One can observe methodologies being implemented like biometrics scans [3], emotion detection based on heartbeat, stress and anxiety analysis of students, detection of fraud using local binary patterns (LBP) method. Methodologies like fraud detection in hall tickets using pattern recognition, Ada boost methods [4], and ANN classifier for classification, etc.

Adaptive boost which is Ada boost is sensitive to noise and outliers in the data [5]. Multimodal biometric system which includes face, iris, and finger print detection is also developed for fraud detection. Computational stages and requirements are good in case of a multimodal system when compared with a traditional uni-model system. An important feature in this particular system is that it enhances the image quality of the images which have very low degree of complexity and clarity.

A variety of methodologies have been proposed for the extraction of features by making use of HMM (Hidden Markov model) and SVD (singular value decomposition) coefficient. Here, the author based on the priorities has considered to take seven states HMM [6]. For instance, while recognizing an id card, the image of the id card is captured. The features of the image are extracted, and then they are compared with the image which resides in the training database set. If both the images are similar, then no malpractice detected. Here instead of id card, it can be finger print scan or facially recognized image.

Another implemented methodology works with local binary pattern (LBP) [7]. In which weights are assigned to different regions in the image which improves the performance of the system drastically. This is usually concerned with the images where the contrast captured within the images varies based on different lighting conditions. Local binary pattern methodology focuses on the local features of the object and compares it with the object in the database which has the most similar local features.

3 Design of New Work

The input for the system is the video feed from the surveillance camera, from this video feed, the frames or images are extracted. Basically, all videos are made up of frames [8]. From these images, the initial image without students which is background image is constructed. Next are the usual frames with students, now we can consider a pair of images which are the current image and the other being background image. The difference of these two images is computed and if the difference is greater than the pre-defined threshold value then we increment the counter, based on how many times, the counter was incremented the decision will be made regarding anomalous activity being detected.

The background is static or say constant but the current image is not content, it is different based on the activity of the student in the video feed which is captured in the frames. Before this process of image subtraction, all the current images are required to change their color scale. There are various color scales provided but the common color scale which works efficiently in video based applications is gray scale. The reason to choose gray scale is that we don't really use the color attributes of the images/frames to analyze the images and detect the anomalous activity.

Initial steps involve establishing a connection to the camera from which the video feed is generated which servers as input to the system. It involves the following steps:

```
1.cv2.videocapture()
2.Infinite loop—read()
3.cv2.imshow()
```

First step specifies the camera from which the video feed is to be fetched. The infinite loop read is to used continue the video feed fetching from the specified camera. Below is the code snippet which shows the video feed being fetched from the specified camera, here mentioned as parameter 0. After specifying the so as to which camera is to be considered to fetch the video feed next step is reading the frames which are in the so fetched video feed. Last step is to mention the wait time to read the next frame from the feed.

```
camera = cv2.VideoCapture(0)        # start a connection to the camera
ret, frame = camera.read()          # read a frame
cv2.waitKey(300)                    # wait 5 minutes
```

There are a large number of frames in a single video, and analyzing these frames is the initial and most prominent task of the system. Hence, the pre-processing of these frames needs to be efficient and fast, so it is obvious to exclude the factors and attributes of the frames which are not required in analyzing the frames.

Reasons for converting the images to gray scale before processing are:

- Removal of noise, here color could be considered as noisy attribute.
- To improve code complexity, not having to deal with extra attributes of frames will give less complex code.
- To effectively visualize
- Improve execution speed
- Cv2.cvtcolor()

cv2.cvtcolor (src, code[,dst[,dstCn]])

An example is shown in Fig. 1 [9] so as to what difference is in the image after converting it to gray scale.

This is the syntax for changing the color scale of the frames obtained from the video feed. Here, cv2.cvtcolor is a python package to change color scale of an image, src is the source of the frames as in their location from where they have to be fetched. Code part in syntax specifies the required color scale to which the image needs to be converted. Dst and dstcn are optional fields. After achieving the conversion to gray scale as depicted in Fig. 1, next step is as discussed background subtraction. By making use of function which are morphological we track the student motion, for instance a student's head movement if it is not in a usual position. The steps

Fig. 1 Result after converting to gray scale (*Source* flickr.com/nasacommons)

involved in this process are segmentation, foreground/background analysis, feature extraction, tracking, and pre-processing.

```
import cv                                              #import cv2
path = r' C:\Users\Administrator\Desktop\image.png'    #path
src = cv2.imread(path)                                 # Reading an image in default mode
window_name = 'Image'                                  # Window name in which image is displayed
image = cv2.cvtColor(src, cv2.COLOR_BGR2GRAY )         # Using cv2.cvtColor() method, Using
                                                         cv2.COLOR_BGR2GRAY color space, conversion
                                                         code
cv2.imshow( window_name, image)                        # Displaying the image
```

Above is the code snippet for the conversion of color scale to gray scale.

System Architecture

As one can observe in Fig. 2 [10] that features are extracted and are subjected to the methods which are implemented and henceforth data is obtained from this analysis

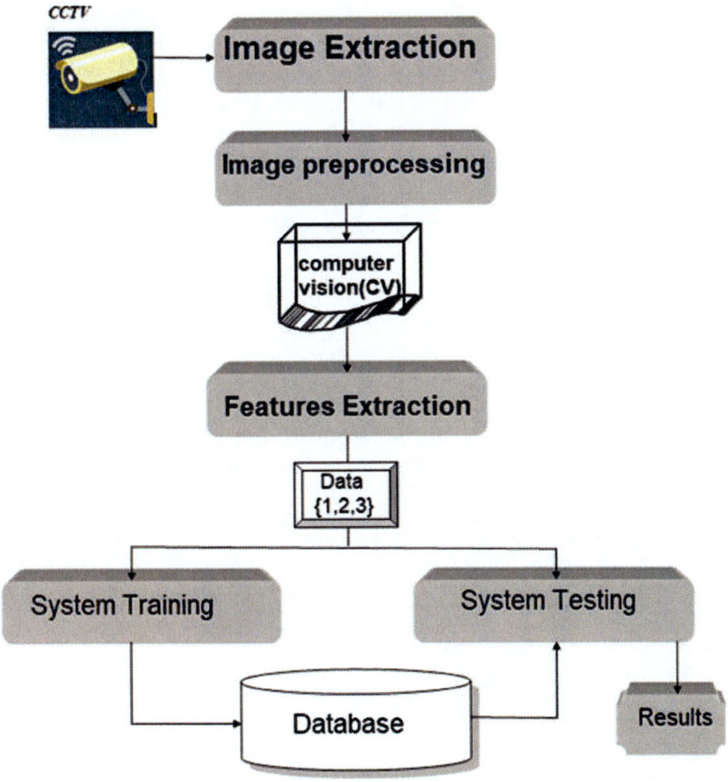

Fig. 2 System architecture (*Source* movement detection using image processing—IJSR)

process. Based on this, the system is trained with respect to common motion of students and anomalous behavior of the same students.

4 Conclusions

The performance of the system prominently depends on the quality of the video feed given to the system as input from the source as shown in Fig. 2 which is a CCTV. This requires the camera to be an efficient one that has quality video capturing properties. Also the system works only for static backgrounds and not for dynamic backgrounds, which is if the camera is in motion then there need to be a totally different set of approaches to be followed to get desired results. Performance of the system could be drastically be improved if there can be multiple high-quality video cameras are installed. Then, the system could be trained for different angles of students sitting position and patterns which would result in even more robust detection of malpractice by students in examinations.

If one is not limited in resources and can afford to install each camera for each individual student on their specific bench then even the eye motion detection of the students could be achieved. To make this system completely self-reliant, it could be incorporated with already implemented features like biometric, iris scans and electromagnetic pulse generator to detect electronic devices [11]. This is just the initial step for us to get accustomed to this new normal lifestyle that we are facing and hopefully come up with more solutions so as to how we can efficiently achieve the same work which was being done in offices now one can do from their respective homes.

References

1. S. Jaswanth Kumar, G. Spandan, N. Nitish Kumar, G.R. Mamatha, R. Roopesh, Online examination and malpractice detection. J. Xi'an Univ. Archit. Technol. (2020). ISSN No: 1006–7930
2. A.A. Ibrahim, G. Abosamra, M. Dahab, Real-time anomalous behavior detection of students in examination rooms using neural networks and Gaussian distribution. Int. J. Sci. Eng. Res. **9**(10) (2018)
3. O. Ojo, N.A Yekini, E.E. Aigbokhan, I. Onadokun, Deterring malpractice in a networked computer based examination using biometric control attendance register. Int. J. Adv. Netw. Appl. **10** (2019). ISSN: 0975–0290
4. K. Susheel Kumar, S. Prasad, V.B. Semwal, R.C. Tripathi, Real time face recognition using Ada-boost improved fast PCA algorithm. Int. J. Artif. Intell. Appl. **2**(3) (2011). https://doi.org/10.5121/ijaia.2011.2305
5. N. Yagnik Ankur, A.S. Singh, Outlier analysis using frequent pattern mining—a review. Int. J. Comput. Sci. Inf. Technol. **5** (2014). ISSN: 0975–9646
6. H. Miar-Naimi, P. Davari, A new fast and efficient HMM-based face recognition system using a 7-state HMM along with SVD coefficients. Iran. J. Electr. Electron. Eng. **4**(1–2) (2008)

7. S.G. Tejashwini, Fraud detection in examination using LBP method. Int. J. Latest Eng. Manage. Res. **2** (2017). ISSN: 2455–4847
8. S. Ahmed, N. Krishnnan, T. Ganta, G. JeyaKumar, A video analytics system for class room surveillance applications. IJRTE (2019). ISSN: 2277–3878
9. E.M. Collins, Date: 1998/10/30, Source: NASA/Johnson space center. https://www.flickr.com/photos/nasacommons/16504233985/
10. S.S. Kshirsagar, P.A. Ghonge, Movement detection using image processing. Int. J. Sci. Res. (IJSR) **4**(2), 2329–2333 (2015)
11. R. Mohanpriya, R. Indhumathi, L.K. Hema, Malpractice detection in examination hall using EMP. Int. J. Pure Appl. Math. **119**(16) (2018)

A Survey on DDoS Attacks from Compromised Devices to Enhance IoT Security

K. S. Niraja, Krishna Keerthi Chennam, and R. Madana Mohana

Abstract The Internet of Things had made IoT devices for cyber-attacks, and constant rise of concerns about IoT security is necessary as connectivity between different devices and objects is a key challenge. The traditional security approaches are not applicable because of size and computational power. There are many kinds of attacks like denial of service attacks, man in the middle attack, etc. Identification of non-authorized devices in the provision of security is very important to provide identification and authentication of different devices. Authenticating different objects and data is very complicated task especially from compromised devices. This paper presents a brief survey on distributed denial of service attacks.

Keywords IoT security · Cyber-attacks · Man in the middle attack · Distributed denial of service attack

1 Introduction

Denial of service attack (DoS) and distributed denial of service (DDoS) attack are the two major possible cyber-attacks in IoT networks where they can target the servers with overload network by utilizing the maximum bandwidth of the network. DoS is a type of attack where the hacker tries to send number of packets continually to the victim server, so that the packets may be overwhelmed in the network which may slow down the victim server and can gain access. Same way if the hacker tries to send packets from different systems in networks by accessing gain of those system by using different malware even the victims of those devices may not aware know as DDoS. The psychology motivations behind DoS and DDoS attacks vary based on the hacktivists do for political like protest against the victim, emotional like hate, revenge, etc., draw attention to potential security features failures, ideological like

K. S. Niraja (✉) · K. K. Chennam
Muffakhamjah College of Engineering, Banjara Hills, Hyderabad, Telangana 500034, India

R. Madana Mohana
Bharat Institute of Engineering and Technology, Ibrahimpatnam, Hyderabad, Telangana 501510, India

A. Kumar et al. (eds.), *Computational Intelligence in Machine Learning*, Lecture Notes in Electrical Engineering 834, https://doi.org/10.1007/978-981-16-8484-5_24

religious belief, commercial, financial, personal, and exploitation. If we see the recent DDoS attacks, out of five applications, four have been targeted using DDoS out of which approximately 40% on news, 25% on business, and 20% on financial services, and most of these attacks are been targeted from East Asia. Many industry experts consider IoT security as the first step in order to make IoT devices more acceptances as they carry sensitive data to enhance security against DDoS attacks.

2 Related Work

Nowadays, most of the researcher is concentrating on detection of malicious systems in the IoT environment because of the botnets attacks which are occurring in huge numbers. The CISCO survey says that mobile data traffic growth rate increases by more than 20% of total IP traffic by 2022, and there will be more possibility of DDoS attack in mere future also. Malicious activities can be detected on the network through intrusion detection in Snort [1] where they used detection system based on signatures. The pattern search method is updated against Snort to enhance the malicious activities by Aho Corasick [2]. When it comes to huge scale networks, it is very difficult to detect malicious activities in most reputed public intrusion detection systems, so Suricata [3] framed a multithreading structure using Raspberry Pi in order to detect scanning port attack. According to ZDNet article, the recent DDoS attack ever happened was been done by hijacked Web servers where though connectionless lightweight directory access protocol amplify the DDoS traffic by 56 to 70 times better than the initial size. Artificial-based neural network attack detection method is implemented to improve accuracy using Naïve Bayes [4]. Bring forward with new cyber-attack detection whose main target is different IoT devices with sinkhole attacks. To locate DDoS attack based on size of packet, Khater et al. [5] proposed multilayer perception with different datasets and achieved above 90% over other classifiers. Mitchell [6] applied rectified linear unit function over MLP for botnets detection.

Volume-based DDoS attacks is one of DDoS attack where the attacker tries to target the victim's network bandwidth with huge volumes of traffic by flooding with huge packets of data where, thereby, victims system may slow down with packet loss [7]. The next type of DDoS attack is protocol attacks where the victim's network layer is the target device which mainly focuses on firewalls and other server resources [8, 9]. Lastly, the application layer attack where the attackers try to target the victim's application layer by initiating different transaction request with fixed resources like available memory, disk space, etc. Detailed survey [10, 11] on these attacks will be discussed in the next section though we have to take into consideration that technophile may like to combine different types of DDoS attack.

3 Classification of DDoS Attacks

In the section above, we have seen the three types of possible attacks in DDoS attacks. In this section, we will discuss about types of attacks in each category. Volumetric attack we have [12] Memcached DDoS where the attacker tries to gain access to victim system by flooding the network. This attack can be relieved to disable UDP and to prevent IP spoofing. Network time protocol attack is volumetric reflect based where attacker supply amplified amount of traffic, so that victims are unable to access regular traffic. This attack can be relieved by NTP servers which should be reduced with source IP verification not allowing the packets to leave network. DNS protocol attack is where attacker holds open DNS resolvers. This attack can be relieved by lowering the open DNS resolvers. Simple service discovery protocol makes use of plug and play protocols to target victim which can be relieved by firewall [13]. Ping flood attack is where attacker echo-request packets are been overrun causing the victims system down. This attack can be relieved by disabling the ping flood.

The possible type of DDoS in protocol attacks are half open attack popularly known as SYN attack making victim's device heavily [14] SYN packets repeatedly. This attack can be relieved with SYN cookies and recycling half open TCP connection. Next possible attack is low and slow attack targeting the victims system slowly with less stream of traffic over network. This attack can be relieved by grading up the server day to day. ACK flood attack is where the attackers try to attack the victim's server with lot of junk data. This attack can be relieved ACK flooding detection tools. Quick UDP Internet connections protocol attack is transport layer protocol attack where the attacker sends malicious data to the victims system. This attack can be relieved by using strong TLS certificates.

Last but not the least attack is application layer attack, where attackers tries to target the top most layer in the OSI reference layer where the attack causes more damage compared to all the above attacks with less bandwidth. This attack can be relieved by analysis traffic is normal traffic or attack traffic with adapting various set of rules. In this section, we have done survey about all the possible attacks for DDOS attack and possible solutions to overcome such attacks in future.

4 Appliance of DDoS Attacks

In the above section, we have seen the possible attacks, whereas in this section, we have done survey on different tools available for DDoS attacks. We can define tools as collections of programs to debug the error in the networks coming to the DDoS attack. Low orbit ion which is an open-source tool is widely used DDoS attack tool to target small entrepreneur. High orbit ion tool is also an open source tool to target the victim's network causing vast traffic making shut down the resource. R U Dead Yet tool is where the attacker targets the victim's server with too much volume of requests [15]. Slowloris tool is where the attacker targets the victims system with too many HTTP connections.

In order to prevent DDoS attack, we need to be aware of typical inbound traffic in our network, maintain more bandwidth which likely to need more filters to adopt to initiate router to drop packets when overflow in the network. Some of the system analysts say that writing code must be slow because of faster code more chances for hackers to target the victims system. Nowadays, due to availability of public cloud, data is been dumped into it where more possibility of DDoS attacks. Big entrepreneur maintains IT operations people, so there may be chance of more attacks, so we have to get rid of them. Every resource in the system must be automatically secure in the network.

5 Artificial Intelligence in Identification of DDoS Attacks

AI plays a vital role in every field; hence, some of the prediction algorithms can be vastly used to identify the malicious traffic in the network. Threshold detection algorithms need to be refined with deep learning and machine learning algorithms to early detection of DDoS attacks [16, 17]. In future, more number of devices is being existence with 5G with more chances of threat possibility. Hence, smart mode detection has to be refined with artificial-based networks. Auto-encoder and decoder have to be refined much in the network to keep target of all malicious activities carried out with active participation of all devices in the networks. Hence, more filters are needed to be deployed in the network to keep track of all the malicious activity in daily bases.

6 Conclusion

In this paper, we sum up with different types of possible DDoS attacks to reduce the spread of victims system. Description about layers is involved for DDoS attacks and its importance with protocols. We conclude it by knowing different evaluations done previously.

References

1. A.R. Baker, J. Esler, *Snort IDS, IPS Toolkit; 30 Corporate, Dr.* (Elsevier Inc., Burlington, 2007). ISBN 9783540449119
2. A.V. Aho, M.J. Corasick, Efficient string matching: an aid to bibliographic search. Commun. ACM **18**, 333–340 (1975)
3. OISF, *Suricata User Guide* (Open Information Security Foundation, Boston, 2019)

4. C. Cervantes, D. Poplade, M. Nogueira, A. Santos, Detection of sinkhole attacks for supporting secure routing on 6LoWPAN for Internet of Things, in *Proceedings of the 2015 IFIP/IEEE International Symposium on Integrated Network Management (IM)*, Ottawa, ON, Canada, 11–15 May 2015, pp. 606–611
5. B.S. Khater, A. Wahid, B. Abdul, M. Yamani, I. Bin, M.A. Hussain, A.A. Ibrahim, A lightweight perceptron-based intrusion detection system for fog computing. Appl. Sci. **9**, 178 (2019)
6. T.M. Mitchell, *Machine Learning* (McGraw-Hill, New York, 1997). ISBN 978-0-07-042807-2
7. C. Kolias, G. Kambourakis, A. Stavrou, J. Voas, DDoS in the IoT: Mirai and Other Botnets. Computer **50**(7), 80–84 (2017)
8. R. Hallman, J. Bryan, G. Palavicini, J. Divita, J. Romero-Mariona, IoDDoS the Internet of distributed denial of service attacks—a case study of the Mirai malware and IoT-based Botnets, in *Proceedings of the 2nd International Conference on Internet of Things, Big Data and Security—Volume 1: IoTBDS* (SciTePress, 2017), pp. 47–58
9. M. Ozcelik, N. Chalabianloo, G. Gur, Software-defined edge defense against IoT-based DDoS, in *2017 IEEE International Conference on Computer and Information Technology (CIT)* (IEEE, 2017), pp. 308–313
10. D.H. Summerville, K.M. Zach, Y. Chen, Ultra-lightweight deep packet anomaly detection for Internet of Things devices, in *2015 IEEE 34th International Performance Computing and Communications Conference, IPCCC 2015* (2016)
11. Y.M.P. Pa, S. Suzuki, K. Yoshioka, T. Matsumoto, T. Kasama, C. Rossow, IoTPOT: a novel honeypot for revealing current IoT threats. J. Inf. Proces. **24**(3), 522–533 (2016)
12. H. Sedjelmaci, S.M. Senouci, M. Al-Bahri, A lightweight anomaly detection technique for low-resource IoT devices: a game-theoretic methodology, in *2016 IEEE International Conference on Communications (ICC)* (IEEE, 2016), pp. 1–6
13. H. Bostani, M. Sheikhan, Hybrid of anomaly-based and specification-based IDS for Internet of Things using unsupervised OPF based on MapReduce approach. Comput. Commun. (2017)
14. I. Butun, B. Kantarci, M. Erol-Kantarci, Anomaly detection and privacy preservation in cloud-centric Internet of Things, in *2015 IEEE International Conference on Communication Workshop (ICCW)* (IEEE, 2015), pp. 2610–2615
15. D. Midi, A. Rullo, A. Mudgerikar, E. Bertino, "Kalis—a system for knowledge-driven adaptable intrusion detection for the Internet of Things, in *2017 IEEE 37th International Conference on Distributed Computing Systems (ICDCS)* (IEEE, 2017), pp. 656–666
16. B.B. Zarpelo, R.S. Miani, C.T. Kawakani, S.C. de Alvarenga, A survey of intrusion detection in Internet of Things. J. Netw. Comput. Appl. **84**, 25–37 (2017)
17. A. Tuor, S. Kaplan, B. Hutchinson, N. Nichols, S. Robinson, Deep learning for unsupervised insider threat detection in structured cybersecurity data streams, in *Artificial Intelligence for Cybersecurity Workshop at AAAI* (2017)

A Study of Glaucoma Diagnosis Using Brain–Computer Interface Technology

R. Geethalakshmi, R. Vani, and Meenalosini Vimal Cruz

Abstract Brain–computer interface (BCI) is technology emerged over decade's plays a vital role in medical diagnosis. Glaucoma is a disease that mainly affects the optic nerve of the eye causing permanent vision loss in elder people. Usually, visual dysfunctions due to glaucoma occur at a slow rate. The people over the age of 45 years need to be diagnosed for the presence of glaucoma since glaucoma causes peripheral vision loss. Standard automated perimetry (SAP) is the existing diagnosis by physician where the patient needs to focus the flickering light stimuli over a period of time. Whenever the light stimuli occur patient needs to press the button, physician note down the response based on the button press. This provides inaccurate result due to patient's lack of knowledge about the experiment, fatigue, anxiety or distraction. This paper discusses steady-state visual evoked potential (SSVEP)-based glaucoma diagnosis for accurate prediction of the disease. In SSVEP, the patient needs to focus on the flickering stimuli, based on applied stimuli brain signal generated which can be detected using electroencephalography. Visual response map is drawn based on the detected brain signal. Analysing the visual response map, early detection of glaucoma diagnosis is possible glaucoma diagnosis using BCI provides automatic result with higher accuracy.

Keywords Standard automated perimetry (SAP) · Steady-state visual evoked potential (SSVEP) · Glaucoma · Brain–computer interface (BCI)

R. Geethalakshmi · R. Vani
Department of ECE, SRM Institute of Science and Technology, Ramapuram, Chennai, India
e-mail: gr7830@srmist.edu.in

R. Vani
e-mail: vanir@srmist.edu.in

M. V. Cruz (✉)
Georgia Southern University, Statesboro, GA, USA
e-mail: mvimalcruz@georgiasouthern.edu

A. Kumar et al. (eds.), *Computational Intelligence in Machine Learning*, Lecture Notes in Electrical Engineering 834, https://doi.org/10.1007/978-981-16-8484-5_25

1 Introduction

Glaucoma is disease that mainly affects the optical nerve creating permanent vision loss. Intraocular pressure inside the retina creates damage to the optic nerve. The analysis of glaucoma involves the three things: one is measuring intraocular pressure, second is the status of optic nerve, and finally if there is intraocular pressure then we need to analysis reason for that. If there is intraocular pressure with no optic nerve damage, it is said to ocular hypertension. Intraocular pressure with optic nerve damage is said to be glaucoma. If there is any damage or problem in the optic nerve, then the signal received by the brain also gets affected. Thus, by analysing the brain signal (electroencephalogram) according to the applied visual simulation is interest of this paper.

Visual evoked potential is visual stimuli applied to the eyes, and it may be flickering light or checkerboard. According to the applied visual stimuli, an electrical potential gets developed in the brain [1]. This VEP waveform can be detected by using EEG. Brain signal or EEG is approximately equal to the applied stimuli. The detected EEG is plotted using visual response map, on analysing visual response map level of optic nerve damage or glaucoma of detected. VEP gives higher accuracy than analysis retinal fundus image analysis. Analysing fundus image has the following drawbacks: blood vessel segmentation is tedious, accuracy is less because of bifurcation of blood vessel, artery and vein differentiation are difficult. Thus, SSVEP-based glaucoma detection is discussed in this paper.

Brain–computer interface (BCI) is the mechanism that creates a communication medium between brain and computer through brain waves. For example, electroencephalogram (EEG) signal is detected from the brain control the outside world device like wheelchair, robot arm and biometric. These brain signal can be used in diagnosing disease like glaucoma and epilepsy. The BCI-based glaucoma detection involves the detection of brain wave generated through visual stimuli applied [2]. The structure of the paper includes Sect. 2 existing work and its drawback, Sect. 3 emerging technology in visual field test, Sect. 4 how SSVEP-based BCI best for glaucoma diagnosis and Sect. 5 conclusion.

2 Related Works

Initially, glaucoma diagnosis is by means of analysing blood vessel in retinal fundus image. Other scope of diagnosing the disease made the way to visual evoked potential-based glaucoma detection. Historically standard automated perimetry (SAP) is used, where visual stimuli are applied at varying intensity and frequency, at time of recognizing the stimuli, the patient needs to press the button. The physician or ophthalmologist manually needs to record the response of button by the patient. In this case, the accurate diagnosis is not possible due to the following reasons. [3] The patient may not press the button due to anxiety, fatigue, distraction or lack of knowledge of the experiment conducted. We need an accurate diagnosis since glaucoma creates peripheral vision loss that cannot be recovered. SAP-based diagnosis of glaucoma is not accurate; we focus on steady-state visual evoked potential (SSVEP) for accurate detection of glaucoma.

Finding central and peripheral vision loss in children are very difficult since they cannot able say what is going on in them. SSVEP-based central and peripheral visual defect identification by using two different frequencies, flickering checkerboard is discussed [4], where the 6 and 7.5 Hz flickering checkerboard is displayed to the 52 participant of age between 5 and 34 years. EEG follows 10–20 electrode configuration in which electrode located at O1, O2 and Oz are the electrode focussed for visual analysis. Above all electrode provide similar response, since central visual response is relatively high in Oz location, we concentrate in that, likewise O1 and O2 provide high peripheral visual function. According to Hebert et al. [4], there is no significant difference of visual response between healthy adult and children, and also, there is no difference in response of sex.

BCI-based glaucoma paves a way for portable visual field test. Portable device for testing glaucoma includes virtual reality goggle along with the electrode in occipital lobe. Duke ophthalmologist found that virtual reality headset and electrode mounted in occipital lobe absolutely detects the brain waves generated. Medeiros et al. developed nGoggle, a portable device to diagnosis the vision loss due to glaucoma using brain waves. They used [5, 6] Samsung Gear VR goggle with six electrodes Oz, O3, O4, Pz, left and right mastoids with the reference at Fz, dual-band 602.11 a/b/g/n Wi-Fi and Bluetooth 4.0 radios and dual-core processor. The flicker checkerboard of black and white at a frequency of 8 to 11.8 Hz is applied, and EEG signal detected is plotted in visual response map.

There are two types of visual field test used widely humphrey visual field test and Goldmann perimetry test. Goldmann perimetry is the standard visual field test where the target is moved over a distance by the technician, the patient needs to press the button while seeing that target without moving the head and eyes. The physician notice the response and note it down. In humphrey visual field test, patient needs to focus on a fixed central light, several frequencies light flicker on the side of visual field, on identifying the flickering light stimuli the patient needs to press the button; accordingly, the response is noted. [7] The examination of both visual field test is conducted on 42 eyes of 20 participants along with the trained technician. The result stated that 60% of the patient feels comfortable with the Goldmann perimetry in that 50% already gone through the test. While 70% of technician feels humphrey visual field test is comfortable because it is fully automated. Humphrey visual field test is prevalent in USA in glaucoma diagnosis.

3 Emerging Technology in Visual Field Test

Traditional visual field test equipment has the following limitation; they are bulky; they need technician to control the operation, lack of portability, subjective analysis, time-consuming. Some of the emerging technologies [8] to overcome these limitations are discussed.

3.1 *Cambridge Glaucoma Visual Function Test*

A computer-based objective analysis of glaucoma is proposed in Cambridge University, named Cambridge glaucoma visual function test abbreviated as CGVFT. The test involves detecting visual function based on the repeated task, and here the collection of 63 images was displayed through large projector with 120 degree tilted horizontally. Initially, the patient starts looking at the central, spinning yellow star shown in Fig. 1 to evaluate the peripheral visual function. This design helps in diagnosing the peripheral vision loss due to the glaucoma. The task is repeated several times to get an accurate response.

Fig. 1 Cambridge glaucoma visual function test

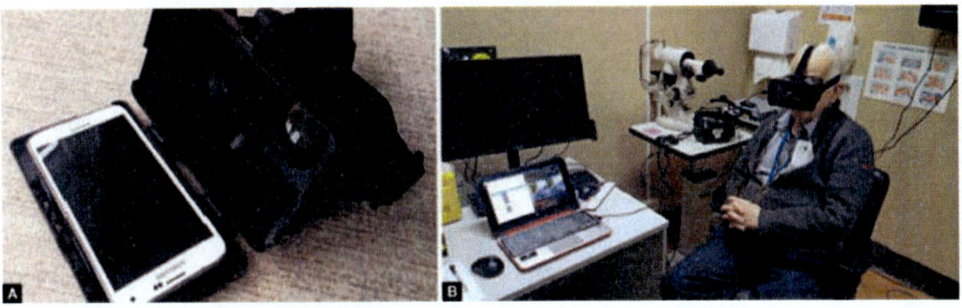

Fig. 2 Setup of virtual reality-based glaucoma diagnosis

3.2 Visual Function Test Using Virtual Reality Goggle

Virtual reality goggle also plays a role in diagnosing glaucoma; the smart phone is inserted into the virtual reality (VR) headset. Smart phone displays twice one for each eye. Virtual reality headset provides 360 degree view. Thus, image is visible to the participant in 360 degree. In this analysis, the participant is asked to look around the image, and the head movement of the participant is detected by the smart phone as shown in Fig. 2. If the participant changes the viewing angle or moves head, then the possibility of vision loss is diagnosed. The accuracy of VR-based virtual function is question. The experiment involves three different tasks: stationary objects, moving ball and driving. In which stationary object produces better result than the moving ball and driving. On comparing, the result of CGVFT and VR-based VFT, CGVFT provides better accuracy since it has fixation point to analysis peripheral vision function.

3.3 Perimetry Analysis Using Tablet

Tablet can also be used to analysis vision loss due to glaucoma. In tablet-based perimetry analysis is highly efficient compared with the above method, since it uses the protocol same as traditional SAP. The tablet needs to install with the software that displays the flickering light starts from the central point towards the periphery. The participant needs to touch the light stimuli displayed on the tablet screen as shown in Fig. 3. The installed software has in build computer voice instructing what the participant has to do in further. Finally, the 24-2 visual field response is obtained based on the participant touch the tablet. 24-2 visual field response is plotted same as humphrey visual test. The efficiency of this technique is comparatively high.

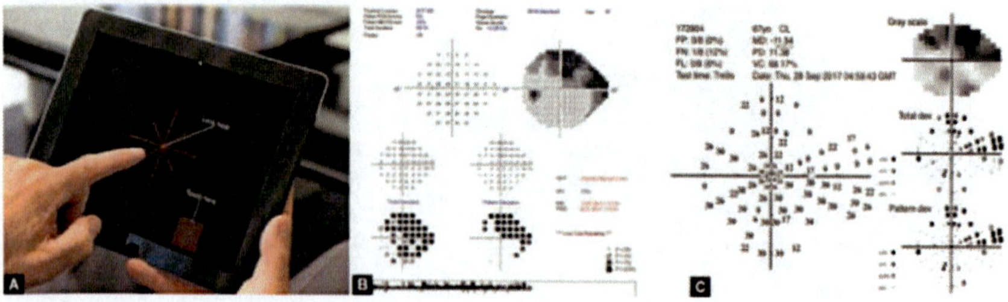

Fig. 3 Tablet-based perimetry analysis of glaucoma. **a** User view. Visual field response using **b** humphrey perimetry, **c** Melbourne rapid field from tablet

3.4 *Tablet Perimetry with Gaze Tracking*

Gaze tracker along with tablet-based perimetry is studied, where gaze tracker is attached to the tablet in this method one advantage is that the participant no need to press button or touch the screen the gaze tracker monitor the movement of the eyes. It gives accurate result comparing with the tablet without gaze tracker. This technique has higher research scope.

3.5 *Visual Field Test Using Robot*

The latest technology humanoid or robot can also be entertained in diagnosis of glaucoma which replaces the need of technician. The experiment is made as same as the traditional SAP. The patient is asked to pay attention to the flickering light stimuli starts from the central fixing point. McKendrick performed experiment on visual field tests under four circumstances based on feedback or instruction: one is optometrist providing feedback, second humanoid robot instructing the participant, third thing is computer voice control and fourth without instruction. All the result produced is similar with no improvement. Robots are the emerging technology, if there is any change in design and development of robot further improvement in glaucoma diagnosis can be achieved.

4 How SSVEP-Based BCI Best for Glaucoma Diagnosis

Traditional glaucoma diagnosis involves standard automated perimetry (SAP) in which varying frequency light stimuli or checkerboard are displayed on the screen. The participant needs to press the button whenever he/she views the stimuli. This traditional method is subjective assessment since the participants need to press the button. They may not press at the right time in spite of vision loss, and there are

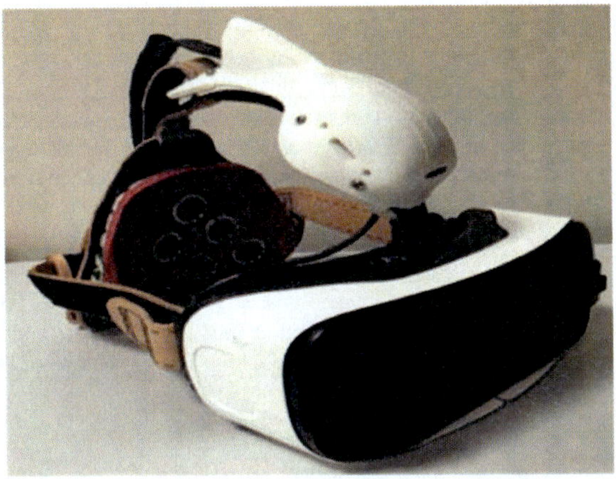

Fig. 4 ngoggle developed by duke opthalmologist

some reason like the lack of knowledge about the examination, anxiety, fatigue due to which they difficult in press the button. Thus, automatic diagnosis paves the interest of study. SSVEP-based BCI combines the SSVEP and brain waves for diagnosing the glaucomatous vision loss. The visual stimuli-like flickering light or flickering checkerboard is applied at different frequencies. Whenever a light stimulus is detected by the eyes corresponding brain waves will be generated in the occipital by detecting the brain waves we can automatically find visual response of the participant.

Duke university ophthalmologist [6–10] conducted SSVEP-based glaucoma analysis using ngoggle a virtual reality headset with six electrode placed at the occipital lobe of the brain. The visual stimuli is projected into the ngoggle through smart phone as shown in Fig. 4. Each and every time the light flickers in the ngoggle the corresponding brain signal is measured using software. On analysing the visual response map, the darker region shows the dim vision or vision loss, whereas the light region denotes the normal vision. Let us see the advantage of SSVEP with BCI-based glaucoma it provides portability of the diagnosis, it provides automatic diagnosis, it provides accurate result compared with emerging techniques discussed above, it provides ease of computation. Table 1 summarizes the various glaucoma diagnosis procedure emerging in this field. From the comparison, it highlights the SSVEP-based BCI is better compared with other technology in diagnosing glaucoma.

5 Conclusion

The various technologies in glaucoma diagnosis have their own advantage and disadvantage with the traditional SAP-based visual field test in glaucoma. The drawbacks of SAP involve the need for specialized technician to diagnosis the disease, whereas other discussed emerging technique like smartphone, tablet, computer paves more better result than traditional one. Since most of the developed countries suffer from

Table 1 Comparison of various emerging brain–computer interface technology

Technique	Implementation	Accuracy	Limitation
Cambridge glaucoma visual function test	Patient needs to look at the central, spinning yellow star in the image to evaluate the peripheral visual function	Task needs to be repeated to get accurate result	Eye-tracking is not implemented
Visual function test using virtual reality goggle	Image is visible to the participant in 360 degree, and the participant needs to look around the image. If the participant changes the viewing angle or moves head, then the possibility of vision loss is diagnosed	Diagnosis Procedure is easy to produce 50% accurate result	Semiautomatic
Perimetry analysis using tablet	The participant needs to touch the light stimuli displayed on the tablet screen	Gives better accuracy than virtual reality goggle	Semiautomatic
Tablet perimetry with gaze tracking	Similar to the perimetry analysis, but include gaze tracker to check eye movement	Better than perimetry analysis without gaze tracker	Semiautomatic
Visual field test using robot	Humanoid or robot perform the function of human	No improvement in accuracy compared with human	Performance is less

glaucoma a challenging disease to diagnosis. Almost 50% of the vision loss due to glaucoma is not yet diagnosed, since most of the affected people are willing to meet ophthalmologist due to their lack of knowledge about severity of the disease. At the same time, SAP-based diagnosis needs periodic check-up and diagnosis which is difficult for rural people. Thus, SSVEP-based BCI in diagnosing glaucoma overcome these difficulties. It provides accurate result as same as traditional SAP with all difficulties improved. Based on the analysis, the work can be extended for obtaining better accuracy.

References

1. A. Namburu, S. Chakkaravathy, M.V. Cruz, Single valued triangular neutrosophic fuzzy c-Means for MR brain image, in *Recent Advances in Computer Based Systems, Processes and application*. https://doi.org/10.1201/9781003043980-11
2. A. Namburu, S. Chakkaravathy, M.V. Cruz, H. Seetha, Single valued triangular neutrosophic fuzzy c-means for MR brain image segmentation, in *International Conference on Data Science, Machine Learning and Statistics-2019 (DMS-2019)*
3. R. Vani, D. Subitha, Health monitoring wearable device using Internet of Things. Eur. J. Mol. Clin. Med. **7**(03) (2020)
4. N. Hébert-Lalonde, D.S. Lionel Carmant, M.L. Marie-Sylvie Roy, D. Saint-Amour, A frequency-tagging electrophysiological method to identify central and peripheral visual field deficits. Documenta Ophthalmologica **129**(1), 17–26 (2014)
5. F.A. Medeiros, J.K. Zao, Y. Wang, M. Nakanishi, Y.-P. Lin, A. Diniz-Filho, T.-P. Jung, The nGoggle: a portable brain-based method for assessment of visual function deficits in glaucoma. Invest. Ophthalmol. Vis. Sci. **57**(12), 3940–3940 (2016)
6. M. Nakanishi, Y.-T. Wang, T.-P. Jung, J.K. Zao, Y.-Y. Chien, A. Diniz-Filho, F.B. Daga, Y.-P. Lin, Y. Wang, F.A. Medeiros, Detecting glaucoma with a portable brain-computer interface for objective assessment of visual function loss. JAMA Ophthalmol. **135**(6), 550–557 (2017)
7. S.C. Stock, A. Armengol-Urpi, B. Kovács, H. Maier, M. Gerdes, W Stork, S.E. Sarma, A system approach for closed-loop assessment of neuro-visual function based on convolutional neural network analysis of EEG signals, in *Neurophotonics*, vol. 11360 (International Society for Optics and Photonics, 2020), p. 1136008
8. S.E. Skalicky, Y.X.K. George, Novel means of clinical visual function testing among glaucoma patients, including virtual reality. J. Curr. Glaucoma Pract. **13**(3), 83 (2019)
9. J. Tang, M. Xu, Z. Liu, J. Qiao, S. Liu, S. Chen, T.-P. Jung, D. Ming, A brain-computer interface based on multifocal SSVEPs detected by inter-task-related component analysis. IEEE Access **8**, 138539–138550 (2020)
10. H.J. Khok, V.T.C. Koh, C. Guan, Deep multi-task learning for SSVEP detection and visual response mapping, in *2020 IEEE International Conference on Systems, Man, and Cybernetics (SMC)* (IEEE, 2020), pp. 1280–1285

Seafloor Habitat Mapping Using Machine Learning and Underwater Acoustic Sonar

Rozaimi Che Hasan, Najhan Md. Said, and Idham Khalil

Abstract The need for detailed spatial map of marine habitats is increasingly important and demanding in managing and preserving marine biodiversity. This study integrates machine learning technique with the in-situ dataset and underwater acoustic mapping data to produce habitat classification maps. For the acoustic data, high-spatial resolution bathymetry and backscatter data were acquired using Kongsberg EM2040C multibeam sonar echosounder. A set of derivative layers were computed as follows (from bathymetry); slope, aspect, rugosity, Benthic Position Index (BPI) (broad and fine scale BPI), whilst from backscatter layers were; Hue, Saturation and Intensity (HSI), Grey Level Texture Co-Occurrence (GLCM) layers (homogeneity, entropy, correlation and mean). Layers from inversion of acoustic properties using Angular Range Analysis (ARA) were also produced such as characterization (sediment class), phi, fluid factor, gradient, index of impendence, intercept, mean far, mean near, mean outer, mean total, roughness and volume homogeneity. Habitat classification map was derived using Random Forests decision trees. Full coverage benthic habitat maps at 1 m spatial resolution were successfully constructed which explained the spatial distribution of coral, fine sand and coarse sand. Bathymetry (and its derivatives) and GLCM mean were identified as the important variables to predict these habitats. This study demonstrated the contribution of machine learning technique to be integrated with underwater sonar data for seafloor habitat mapping.

R. Che Hasan (✉)
Razak Faculty of Technology and Informatics, Universiti Teknologi Malaysia, Jalan Sultan Yahya Petra, 54100 Kuala Lumpur, Malaysia
e-mail: rozaimi.kl@utm.my

N. Md. Said
National Hydrographic Centre, Bandar Armada Putra, Pulau Indah, 42009 Port Klang, Selangor, Malaysia
e-mail: najhan@navy.mil.my

I. Khalil
Institute of Oceanography and Environment (INOS), Universiti Malaysia Terengganu, Kuala Nerus, 21030 Terengganu, Malaysia
e-mail: idham@umt.edu.my

A. Kumar et al. (eds.), *Computational Intelligence in Machine Learning*, Lecture Notes in Electrical Engineering 834, https://doi.org/10.1007/978-981-16-8484-5_26

Keywords Acoustic sonar · Multibeam sonar · Bathymetry · Backscatter · Random Forest decision trees · Habitat maps

1 Introduction

Seabed mapping had been widely used to determine the fisheries habitat, characterize the deep-water coral reefs and recognize the marine biodiversity for conservation [1–3] and the high resolution of seafloor map had been increasingly in demand due to the effective planning and management of the marine ecosystems and resources. Adoption of Marine Protected Areas (MPAs) was one of the keys for keeping our marine biodiversity protected from the effect of human activities, understanding the distribution of endangered habitat, and marines, and as an industry for the fishery and tourism activities. For this, spatial habitat maps were needed to assist marine managers in developing a strategic conservation plan.

This study attempts to produce the habitat maps for Redang Marine Park located in Terengganu, Malaysia by using the integration of acoustic technique (i.e. multibeam echosounder (MBES)), in-situ ground truth dataset and machine learning approach. The outcome will be a habitat map (or classification map) which presented the spatial distribution of coral and sediment types (coarse and fine sand), at Redang Marine Park. The contribution of each variable used in the machine learning will also be assessed using the machine learning approach.

2 Methodology

2.1 *Study Area*

This study was carried out at Redang Marine Park. It is a part of the Redang archipelago, which also comprises the islands of *Pulau Lima, Pulau Paku Besar, Pulau Paku Kecil, Pulau Kerengga Kecil, Pulau Kerengga Besar, Pulau Ekor Tebu, Pulau Ling* and *Pulau Pinang*. The island is located approximately 24 nautical miles or 45 km into the South China Sea, off the east coast of Terengganu mainland. The island measures about 6 km wide and 7 km long. Apart from being a popular tourists destination spot, the island is also gazetted as a Marine Protected Area (MPA) and houses a vast variety of marine species and organisms. This area is also a landing and nesting site for turtles [4].

2.2 *Data Collection*

Field survey for acoustical mapping using multibeam echosounder (MBES) was conducted with the collaboration from the National Hydrographic Centre (Pusat Hidrografi Nasional, PHN) from 6 to 25th of April 2019 within 2 nautical miles from the coastline of *Pulau Redang* on a local fishing vessel. The survey was completed using the Kongsberg Maritime series EM 2040C MBES. The MBES was chosen since it could produce high-resolution data and had a wide frequency, ranging from 200 to 400 kHz. Moreover, the device also produced short pulse lengths with large bandwidth and had an angular coverage up to 170°. Daily sound velocity and tidal data were recorded to remove the effect of sound velocity and tidal effect. The acoustic survey had recorded two types of data; the depth data and intensity returns data (i.e. backscatter). Field survey for ground truthing was conducted from 8 to 10th July of 2019 along the coastal areas of *Pulau Redang* by using remote underwater video observation.

2.3 *Data Processing*

The raw multibeam bathymetric dataset was then processed using CARIS HIPS and SIPS software to remove any depth spikes and outliers. The depth data were rasterised into gridded bathymetry surface at 1-m pixel resolution. Apart from the bathymetry, the benthic seascape derivatives were also produced from the original bathymetric surface. Bathymetric derivatives were analyzed in ArcGIS using the geoprocessing tool Benthic Terrain Modeller (BTM). The derivatives used were the Fine Scale and Broad Scale Bathymetric Position Index (BPI), Aspect and Slope. Backscatter data were processed using FMGT (FM Geocoder Toolbox) to obtain backscatter mosaic (1-m pixel resolution). FMGT applied corrections in terms of georeferencing each beam data to the correct position and performed radiometric corrections to reduce the intensity values in the raw data to correct the source level, two-way transmission losses, insonified areas and finally, applied the Angular Varied-Gain (AVG) correction. The final mosaic was then exported to raster the format at 1-m pixel resolution. From the backscatter mosaic, texture analysis using the Grey-level Co-occurrence Matrices (GLCM) method was used to produce the texture layers such as, GLCM correlation, GLCM entropy, GLCM homogeneity and GLCM mean. In addition to this, Hue, Saturation and Intensity (HSI) layers (which originally used for radar images) were also derived from the backscatter mosaic, resulted in three HSI layers (Red, Green and Blue bands). Apart from backscatter mosaic, this study also used the sediment characterization or predicted sediment class that was generated using Angular Range Analysis (ARA) in FMGT [5, 6].

2.4 *Machine Learning*

The habitat map of Redang Marine Park was completed using the machine learning classification technique; Random Forest (RF) decision trees. RF had been widely used in terrestrial remote sensing, but recent studies had successfully applied this technique in benthic habitat mapping [7–13]. Compared to normal decision trees (such as CART–Classification and Regression Trees), RF is an ensemble method that generated many trees at each node (i.e. forest) and each individual tree will be voted a class [14]. Random selection of features performed by RF could also decrease the use of high-correlated variables. RF could also quantify the variables importance therefore it would be useful in knowing which variables (i.e. layers) have the highest contribution towards the classification process. RF classification was performed using open source software EnMAP-Box [15], accessible from http://www.enmap.org/enmapbox.html. The training data used to construct the RF model were extracted from the ground truth data using four main classes (fine sand, coarse sand, coral and algae).

3 Results and Discussion

The classification process produced habitat map with three classes (coral, fine sand and coarse sand), where most of the areas were dominated by coarse sand (Fig. 1).

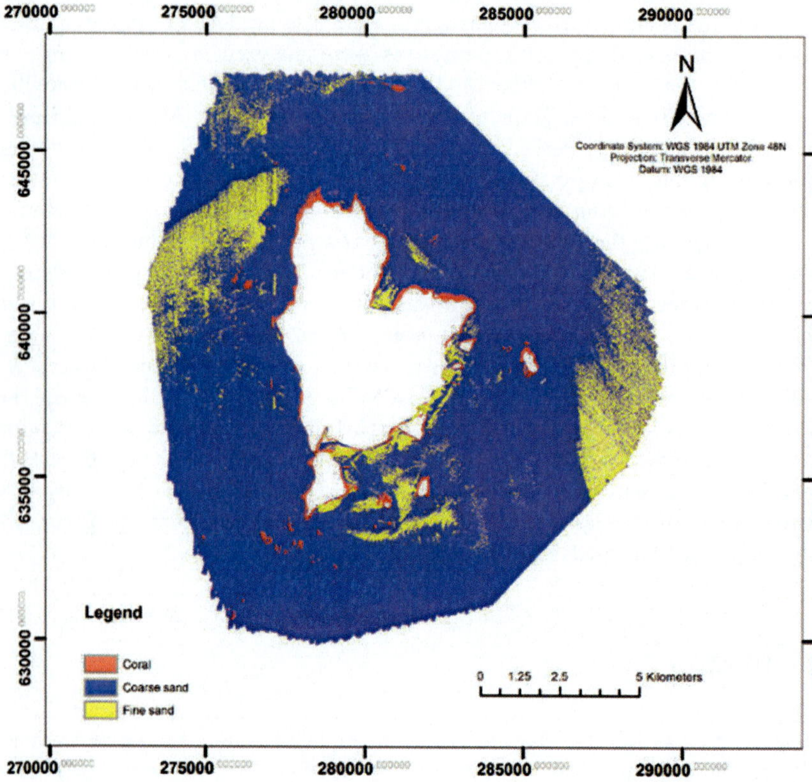

Fig. 1 Habitat classification map for Redang Marine Park using three classes (coral, coarse sand and fine sand)

Fine sand could be found in shallower areas close to the coastline especially at the southern part of Redang Island. Corals were mainly distributed around Redang Island, especially in the northern region. In terms of variable importance, most of the bathymetry and its derivatives especially slope, BPI (fine and broad scale), and aspect appeared to be important for this classification. This suggested that water depth played the main factor in predicting the spatial distribution of the coral, coarse sand and fine sand. In contrast, the variables derived from backscatter were less important except for GLCM Mean. GLCM Mean was found to be more important as compared to the original backscatter mosaic, either in the decibel values format, or the 8-bit greyscale backscatter mosaic. It showed that the mean grey values using GLCM technique had the capability to differentiate different sediment types (fine and coarse sands) as well as delineating coral from these sediments. This also suggested that the texture analysis from backscatter mosaic was useful for sediment characterization (Fig. 2).

For validation of the classification map produced in this study, previous sediment sampling as in Sokiman et al. [16] was used which comprising four types of sediment class: coarse sand, fine sand, medium sand and very coarse sand. The distribution of sediment sampling can be seen in Fig. 3 where in general, fine sand were located close to the coast and coarse sand/medium sand could be found as the distance is further away from the coastal area. Coarse sand from classification map was shown to be agreed with the coarse sand and very coarse sand from the sediment sampling point. The same comments can be made between fine sand (classification map) and the sampling points (medium and fine sand). However, there were also some misclassifications between coral (classification map) and coarse sand (sampling point).

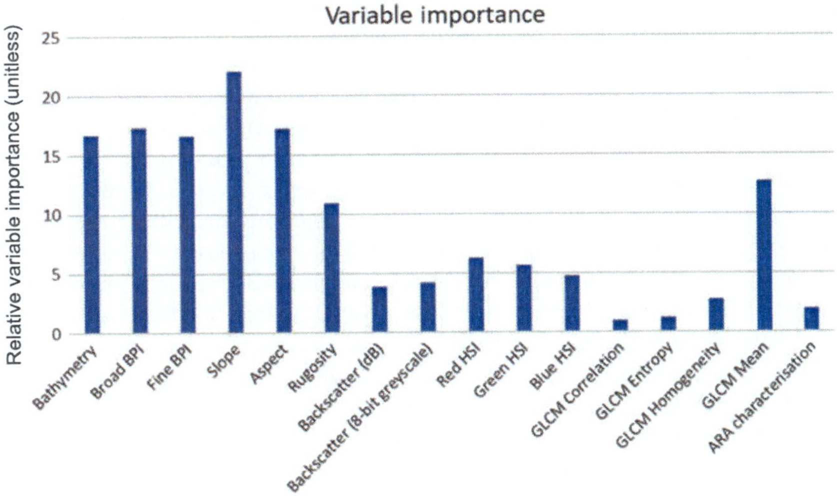

Fig. 2 Variable importance from Random Forest decision trees

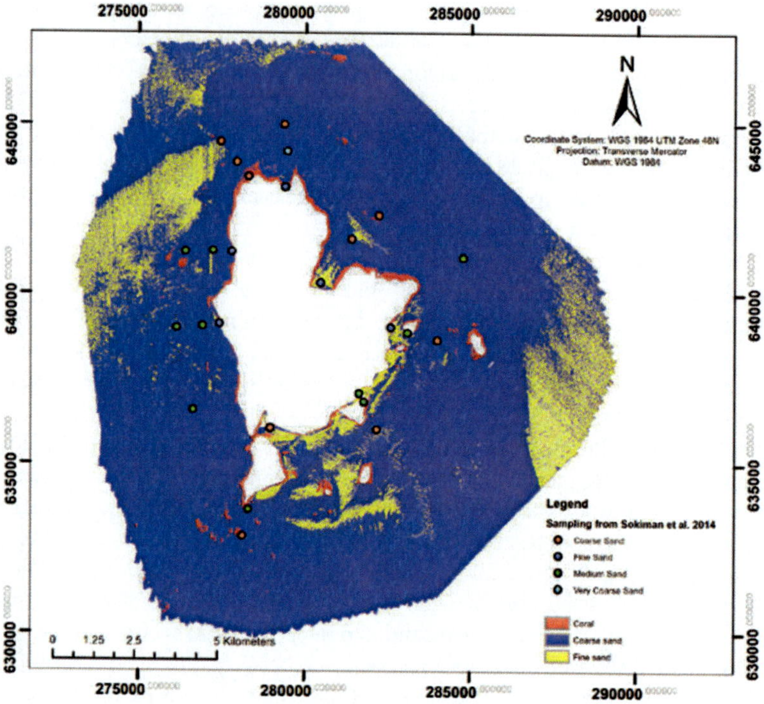

Fig. 3 Comparison of classification map with sediment sampling from previous study in Sokiman et al. [16]

4 Conclusion

This study had successfully produced detailed, spatially explicit and high-resolution marine habitat maps for Redang Marine Park. To our best knowledge, this was the first marine habitat map of Redang Marine Park ever produced using the high spatial resolution (1 m) acoustic mapping data coupled with the in-situ data by utilizing machine learning approach. The advantage of this technique is the ability to produce detailed habitat map without the need to obtain substantial ground truth dataset. Existing study on benthic habitat mapping had been focussing on sampling the data collection on selected locations (i.e. point sampling), and therefore difficult to explain in detail the spatial distributions and abundances of a specific marine habitat. With the habitat maps and habitat suitability derived in this study, marine managers and authorities will have additional and useful information especially on the coral distributions in Redang Marine Park.

Acknowledgements The authors would like to thank all the team members who involve with the data collection process and the Malaysia Department of Fisheries for providing support for this study.

References

1. A. Copeland, E. Edinger, R. Devillers, T. Bell, P. LeBlanc, J. Wroblewski, Marine habitat mapping in support of Marine Protected Area management in a subarctic fjord: Gilbert Bay, Labrador, Canada. J. Coast. Conserv. **17**(2), 225–237 (2013)
2. V.E. Kostylev, B.J. Todd, G.B. Fader, R. Courtney, G.D. Cameron, R.A. Pickrill, Benthic habitat mapping on the Scotian Shelf based on multibeam bathymetry, surficial geology and sea floor photographs. Mar. Ecol. Prog. Ser. **219**, 121–137 (2001)
3. J. Roberts, C. Brown, D. Long, C. Bates, Acoustic mapping using a multibeam echosounder reveals cold-water coral reefs and surrounding habitats. Coral Reefs **24**(4), 654–669 (2005)
4. T.A. Stewart, D.T. Booth, M.U. Rusli, Influence of sand grain size and nest microenvironment on incubation success, hatchling morphology and locomotion performance of green turtles (Chelonia mydas) at the Chagar Hutang Turtle Sanctuary, Redang Island, Malaysia. Aust. J. Zool. **66**(6), 356–368 (2018). https://doi.org/10.1071/ZO19025
5. L. Fonseca, C. Brown, B. Calder, L. Mayer, Y. Rzhanov, Angular range analysis of acoustic themes from Stanton Banks Ireland: a link between visual interpretation and multibeam echosounder angular signatures. Appl. Acoust. **70**(10), 1298–1304 (2009)
6. L. Fonseca, L. Mayer, Remote estimation of surficial seafloor properties through the application Angular Range Analysis to multibeam sonar data. Mar. Geophys. Res. **28**, 119–126 (2007)
7. B. Misiuk, M. Diesing, A. Aitken, C.J. Brown, E.N. Edinger, T. Bell, A spatially explicit comparison of quantitative and categorical modelling approaches for mapping seabed sediments using Random Forest. Geosciences **9**(6), 254 (2019)
8. R. Che Hasan, D. Ierodiaconou, L. Laurenson, A. Schimel, Integrating multibeam backscatter angular response, mosaic and bathymetry data for benthic habitat mapping. PLoS ONE **9**(5), e97339 (2014). https://doi.org/10.1371/journal.pone.0097339
9. Z. Huang, S.L. Nichol, J.P.W. Siwabessy, J. Daniell, B.P. Brooke, Predictive modelling of seabed sediment parameters using multibeam acoustic data: a case study on the Carnarvon Shelf, Western Australia. Int. J. Geogr. Inf. Sci. **26**(2), 283–307 (2012)
10. P.J. Mitchell, A.-L. Downie, M. Diesing, How good is my map? A tool for semi-automated thematic mapping and spatially explicit confidence assessment. Environ. Model. Softw. **108**, 111–122 (2018). https://doi.org/10.1016/j.envsoft.2018.07.014
11. M. Diesing, S.L. Green, D. Stephens, R.M. Lark, H.A. Stewart, D. Dove, Mapping seabed sediments: comparison of manual, geostatistical, object-based image analysis and machine learning approaches. Cont. Shelf Res. **84**, 107–119 (2014). https://doi.org/10.1016/j.csr.2014.05.004
12. D. Stephens, M. Diesing, A comparison of supervised classification methods for the prediction of substrate type using multibeam acoustic and legacy grain-size data. PLoS ONE **9**(4), e93950 (2014). https://doi.org/10.1371/journal.pone.0093950
13. D. Ierodiaconou, A.C.G. Schimel, D. Kennedy, J. Monk, G. Gaylard, M. Young, M. Diesing, A. Rattray, Combining pixel and object based image analysis of ultra-high resolution multibeam bathymetry and backscatter for habitat mapping in shallow marine waters. Marine Geophys. Res. **39**(1), 271–288 (2018). https://doi.org/10.1007/s11001-017-9338-z
14. L. Breiman, Random forests. Mach. Learn. **45**(1), 5–32 (2001)
15. S. Van der Linden, A. Rabe, M. Held, B. Jakimow, P.J. Leitão, A. Okujeni, M. Schwieder, S. Suess, P. Hostert, The EnMAP-Box—a toolbox and application programming interface for EnMAP data processing. Remote Sens. **7**(9), 11249–11266 (2015)
16. M.S. Sokiman, S.N.F.S. Nooh, N.A. Abdullah, Sedimentology and geochemistry of Redang Island, Terengganu. Paper presented at the National Geoscience Conference, 13–14 June 2014, Kuala Lumpur, 13–14 June 2014

Smart Irrigation for Enhanced Agriculture Production Using IoT

G. Saranyadevi, R. Vani, A. Jeshron Sonali, and S. Minoviya

Abstract In developing countries like India, where agriculture plays a major role a smart sensor for controlling the environmental changes in areas where crop fields are grown is necessary. For improving the productivity of the crop using less quantity of water as well as to reduce the complexity this Raspberry Pi based internet of things (IoT) irrigation system has been developed. On whole day, these environmental changes are controlled using IoT through simple message transfer protocol (SMTP) and message queue telemetry transport protocol (MQTT) having temperature sensor, moisture sensor, gas sensor and LDR where information from all these sensors estimates the need of water to the fields. Message queue telemetry transport protocol (MQTT) having high-speed data transmission both the old data and the current data collected by the sensors are sent to Message queue telemetry transport client. To analyze the weather conditions and to increase the yield of crops precision agriculture with cloud computing using water fertilizers has been implemented.

Keywords Smart irrigation · Protocol · IoT

1 Introduction

The main factor for the agriculture is to predict the weather, climatic changes and also mainly to predict that when the atmosphere conditions are going to change. Here, both the automated irrigation system and the IoT is employed for the irrigation monitoring. It will be helpful for us when to irrigate the crop or field at what conditions will the crop be able to sustain and also the other environment and also the other changes. To monitor the weather and climatic conditions which are varying with the time and space, so the use of these electrical devices in the agriculture has been

G. Saranyadevi (✉) · R. Vani · A. Jeshron Sonali · S. Minoviya
Department of ECE, SRM Institute of Science and Technology, Ramapuram, Chennai, Tamilnadu, India
e-mail: saranyag4@srmist.edu.in

R. Vani
e-mail: vanir@srmist.edu.in

A. Kumar et al. (eds.), *Computational Intelligence in Machine Learning*, Lecture Notes in Electrical Engineering 834, https://doi.org/10.1007/978-981-16-8484-5_27

increasing rapidly. So the use of these WSN technology is applied to the irrigation systems, and it automatically removes high risk to the cable line due to the automatic operations of the machines [1]. Due to the high cost of the WSN system modules, we are not using these type of technology and also it requires high energy for the data transmission between the client and server. So we will use the IoT and with the help of the IoT we are employing the several systems in the different crop fields by naming it as the clients and this clients, which is getting reported every time to the server about the status of the crop, climatic changes of the temperature, humidity, and also the moisture of the certain place. So that the watering and also the when the pesticides can be served based on the conditions on the field. The paper is organized as follows:

Section 2 describes the literature survey. Section 3 explains the proposed system in which the environmental changes are controlled using IoT through simple message transfer protocol (SMTP) and message queue telemetry transport protocol (MQTT). Section 4 mentions about the hardware description. In Sect. 5, results are discussed. Finally, conclusion in Sect. 6.

2 Literature Survey

Wireless sensor network and GPRS module are used in automatic irrigation system as shown in Fig. 1. A wireless sensor network in which the required parameter sensors are placed in the root zone of the crops. To transmit the information received through sensors, it has gate way unit and trigger acculators [2]. To know the required amount

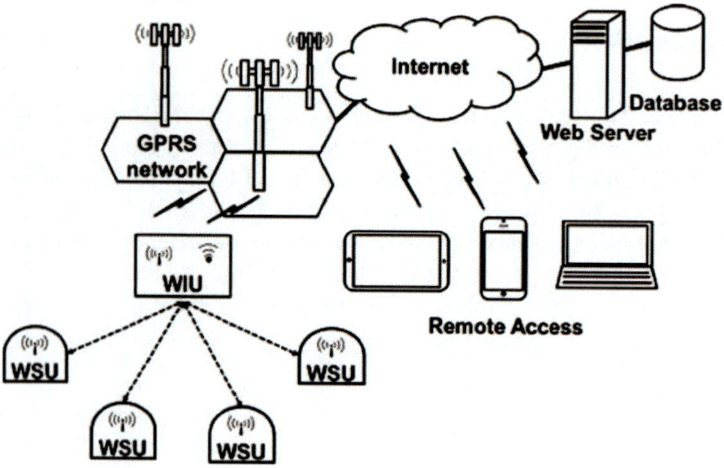

Fig. 1 GPRS module in automated irrigation system

of water starting values of soil moisture and temperature are taken into consideration and algorithm is developed. It uses photovoltaic panel and cellular internet communication link for data transference and irrigation check. First and foremost, it uses expensive and complex algorithms to get the required information. Secondly, it doesn't help to control the environmental changes and low-communication speed.

2.1 *Smart Phone Irrigation*

Taking agricultural crops into consideration, irrigation sensor was designed. The camera captures the images of the soil and roots of the crop to estimate the requirement of water. To estimate the wet and dry area of the crop using antireflective glass window an android app was developed so that this app activates the device to capture the images. A Wi-Fi connection is also needed to control the irrigation as in Fig. 2. This model addresses a lot of issues like having a smart phone with a Wi-Fi connection resulting in battery issues of phone, poor signal and time consuming.

2.2 *Wireless Sensor Network (WSN)*

To get the best crop conduction, increase the productivity WSN is used as in Fig. 3. Many environmental parameters are sensed using sensors which continuously sends the data about water shortage. The soil conditions, need of water and climatic changes are known by using a Bluetooth device [3, 4]. This Bluetooth is used to transfer the information from sender to the receiver. Though this model is of low cost and gives the output results, it cannot recollect the old data, and it is less secure weather forecasting cloud technologies are not used.

2.3 *IoT in Agriculture*

IoT has the ability to rule the world in many industries, smart cities and especially when it comes to agricultural sector it has a great wallop. The conventional based farming approach is too difficult when compared to smart farming based IoT [5, 6]. It has great advantages like less water usage, control over environment and updated input treatments. It makes stock man and farmers to get the useful data, making the farmers understand the process by installing the necessary smart ways to increase the productivity as in Fig. 4.

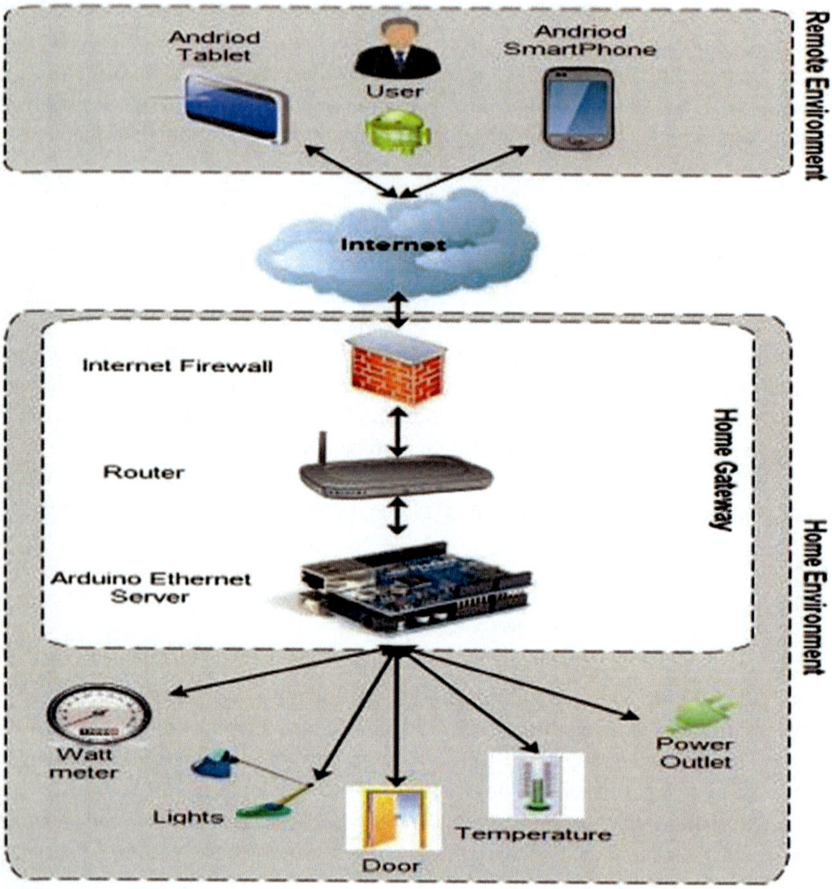

Fig. 2 Smart phone irrigation

2.4 Drone Applications of IoT in Agriculture

It is one if the best application of today's growing technology. It handles crop health, irrigation, crop records, crop spraying and soil field analysis. Drones get the data by insight monitoring and observations [7]. To select a ground resolution farmer, enter the data so that plant height measurement, yield prediction and so on can be identified. It is used by many organizations and one of the most recognized application of IoT agricultural division. It utilizes various sensors, control systems, robotics, automated hardware, free vehicles [8]. It adopts high-speed internet, mobile devices and low-cost satellites are few technological trends. The chance of crops being damaged will be reduced if accuracy of weather is high. It is used for food production and safety. Agriculture IoT not only concentrates on increasing the productivity of the crop and control of weather conditions but also takes responsibility about food security

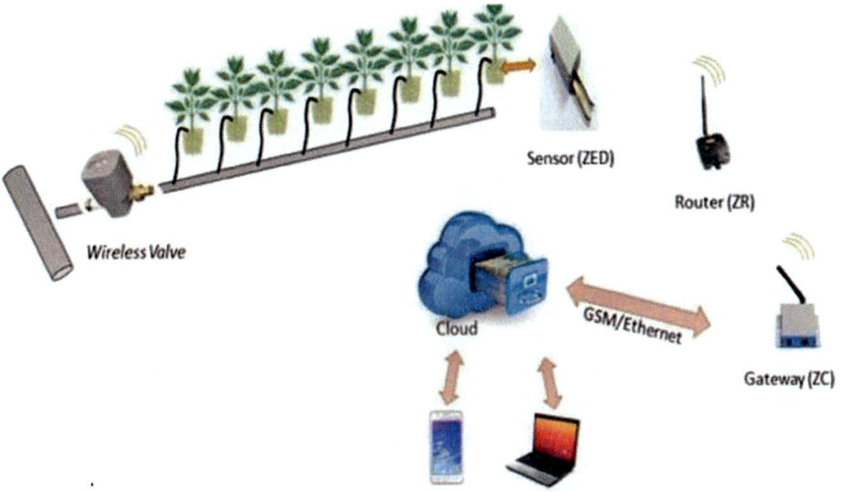

Fig. 3 Wireless sensor network

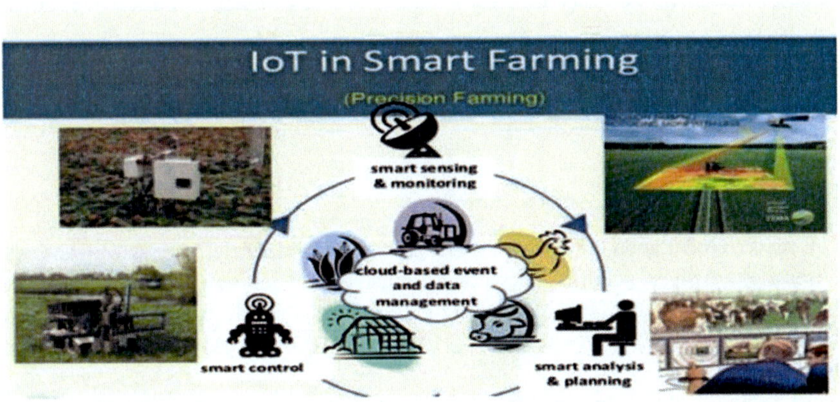

Fig. 4 IoT in agriculture

like storage transportation, etc. Shipping time, storage temperature and cloud-based records are monitored by this system.

3 Proposed System

In the present system, we are using the crop field algorithm. So the resulting data is having the high accuracy about the environment, and also mainly we are using the message queuing telemetry transport (MQTT) instead of using the Wireless Sensor

Network (WSN) is MQTT is very much faster than the WSN. We are using this message queuing telemetry transport to transfer all the information about the climatic conditions are processed automatically, like if there is any possibility of rain in the weather cloud, then the climatic conditions and also the upcoming possibilities of the rain data log and also the current image of the environment will also be sent to the users mail.

3.1 Message Queuing Telemetry Transport (MQTT)

Message queuing telemetry transport is a feather weight ISO standard publisher-subscriber messaging protocol specially designed for the internet of things. Internet of Things is creating the massive machine to machine network so that all of the devices like sensors, systems and also the actuators can connect to and communicate on the internet and with they need a communications protocol so that they can understand each other one of those protocols is message queuing telemetry transport. Message queuing telemetry transport is a TCP/IP protocol based on subscribe and publish messaging protocol designed for light weight machine-to-machine communications.

MQTT has built the QoS to deliver the message.
QoS 0: Once in a time "fire and forget," if the device is online, it sends the message, and it gets delivered or if the device is offline it can't be delivered it will be thrown away.
QoS 1: Atleast once.
QoS 2: Exactly once, 2 phase commit.

Both the QoS 1 and QoS 2 have the effectively guaranteed delivery, while the QoS 2 uses the two phase summit guaranteeing exactly once delivery and also it can give the end-to-end protocol level but not within the stacks on each side of the connection these QoS 1 and QoS 2 are queued for clients when the user is also online but have no timed out.

To work the message queuing telemetry transport, it has three stages they are called as the message queuing broker. It is like a gateway receiver or server and allows the clients to send and also to receive the short messages from the broker only if they have subscribed to any topic like weather, temperature and also the humidity. Here in the below diagram, the client A is the publisher and the remaining client B and C are the subscribers whom receives the information which is sent by the sensor known as the client A.

And also, these two B and C clients subscribe to the particular topic and gets published to the broker all the three will have the TCP connection with the broker as in Fig. 5. The sensor or the publisher which is called as the client A updates the topic and publishes the information about the topic to the MQ broker. These clients have subscribed to the topic so when the new piece of information gets published to the topic they then get the information from the MQ broker.

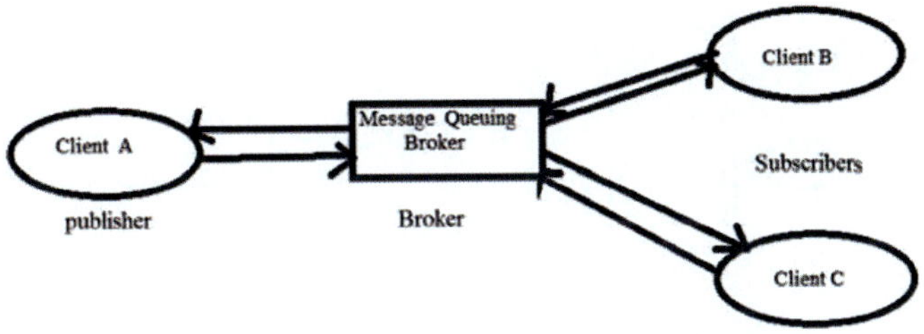

Fig. 5 Message queuing telemetry transport

3.2 Sensor/Publisher

The publisher has the information about the topic, and also this publisher has the data about the topic which the subscribers have been subscribed earlier. It is connected to the broker via TCP connections on all the sides whenever the new data comes, it will be sent to the broker. And also the publishers and also the subscribers are not connected directly through the TCP/IP connection. And this MQTT publishers allow username and password for the authentication and these clients can directly log on via password. These passwords are sent by the MQTT via the plain text or in encrypted form. If no data was sent, then the publisher will be in the off mode.

The subscriber receives the filtered messages every message is being filtered by the broker which is the middle of the publisher and it also knows that who will send the message and also who will receive the message. The subscriber is the application based. The broker filter the messages based on the subject and the content based and also the type based filtering by the broker. A user can also subscribe to the one topics or more topics. When subscribed to the one or more than one topics two characters can be used. MQTT supports disconnect and last will and testament messages in this they are two types: namely hash character—multi-level wildcard and plus character—single-level wildcard MQTT.

Clients which disconnect intentionally use disconnect message as in Fig. 6. MQTT broker will automatically publish last will and the testament messages on the behalf of the clients with unintentionally terminated messages. It also supports the retained messages which are automatically delivered when a client subscribes to the topic.

The topics which we select in the MQTT are similar to the files in a system using the forward slash(/). In message queuing telemetry transport, there is no formal structure to select the topic and a publisher is free to choose, it's own topic names and also the structure but it should consists atleast on character to be valid. There is no topics created by the broker. All topics are created by a subscriber and also the publisher client, are not permanent. A topic only exists only if a subscriber has already subscribed to it before only or a broker has restore the last messages stored for that topic. This is the example for that E.g.: School/Room/Sensor.

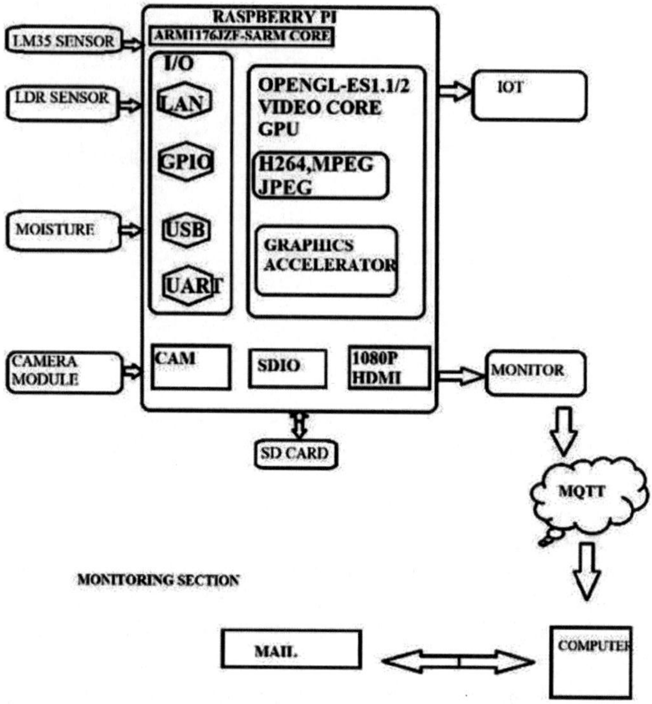

Fig. 6 Block diagram of MQTT subscriber/client

4 Hardware Description

Raspberry pi 3 begins as a toy computer and ends up being a real desktop PC having an upgraded ARMv7 multicore processor and gigabytes RAM. In general, Raspberry pi has mode B and B+, mode A and A+, Pi zero. This Raspberry pi mode B and B+ are the simplest and cheapest but B+ has a bit more improved design of embedded applications. Raspberry pi zero has input or output, general purpose input or output (GPIO) and Bluetooth capability. They mostly operate at a frequency of 800–1000 MHz, in extreme case they can be operated upto 1500 MHz. In this model neither USB nor ethernet ports are changed.

GPIO are the general purpose input and output pins which are used to turn off and to turn the devices like LED which will be present in the Raspberry pi. These pins allow you to connect and also to talk to the hardware modules that supports the protocol. UART is the universal asynchronous receiver and transmitter. It receives several bytes of the data from the GPIO pins but it transmits only bits of the data in the sequential order. In the asynchronous transmission of the data, it allows the bits to be sent or transmit without the sender but it will send the signal like the clock signal to the receiver side.

UART can be used between the two devices only like simplex means data transmitted on one side only and also the half duplex in which the data can be transmitted in the either direction or the Full duplex. MPC3008 is a high performed and low power 10-bit analog-to-digital converter (ADC).

In general to know the hotness as well as the coldness of an object, temperature sensor is used. In this LM 35 temperature sensor, the output is almost equal to the temperature. It doesn't require any amplified output rather provides higher output than thermocouples. It is sometimes used as a power supply, battery, etc. To measure the amount of water in the soil moisture sensor is used. Indirectly either by using electrical resistance or neutron interactions water volume is measured.

Light dependent resistor (LDR) is a variable light control resistor. As the light intensity increases, the resistance of LDR decreases. It can be used in circuits with light and dark activation switching and also in some light sensitive devices.

5 Results

Figure 7 shows that the sensor parameters and also the weather conditions like temperature, moisture, humidity and also amount of light which is receiving. It compares the data log with the already existing weather reporting conditions. It also finds whether there is any possibility of the rainfall then it will intimates to the user through the registered mail. It will also predict the parameters like thunder storms, etc., as in Fig. 7.

These sensor values are being sent to the clients using the cloud and MQTT protocol. These parameters can be checked by the client through the registered mail.

6 Conclusion

In this system, the irrigation system becomes more simple and easy with the quick transmission of the weather parameters by using the MQTT protocol and sending this data through cloud to the cloud who has been subscribed. The main advantage of using these protocol is when the client is not available then the parameters will be sent through the mail. When the client is available with the network then he can able to see the data parameters and also he can compare the present data with the existing data which has been already sent. So that the farmer will be able to know the changes in the atmosphere every day. This can also be implemented by using the computer-based algorithm and also the data can be processed without any further complexity. With the other specifications, sensors are used to make the process better.

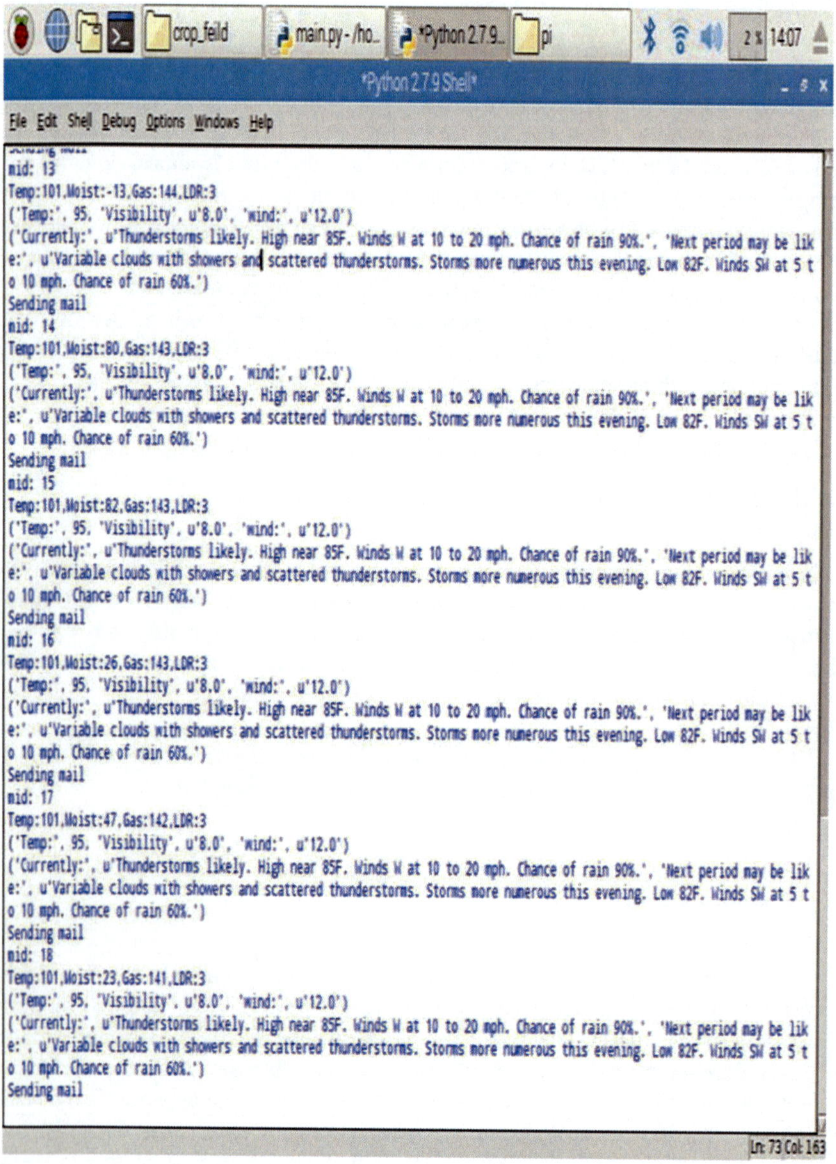

Fig. 7 Message queuing telemetry transport conversation/data log

References

1. J. Gutiérrez, J.F. Villa-Medina, A. Nieto-Garibay, M.Á. Porta-Gándara, Automated irrigation system using a wireless sensor network and GPRS module. IEEE Trans. Instrum. Measur. **63**(1), 166–176 (2013)
2. S.A. Nikolidakis, D. Kandris, D.D. Vergados, C. Douligeris, Energy efficient automated control of irrigation in agriculture by using wireless sensor networks. Comput. Electron. Agric. **113**, 154–163 (2015)
3. S.R. Nandurkar, V.R. Thool, R.C. Thool, Design and development of precision agriculture system using wireless sensor network, in *2014 First International Conference on Automation, Control, Energy and Systems (ACES)* (IEEE, 2014), pp. 1–6
4. R. Vani, N. Thendral, J.C. Kavitha, N.P.G. Bhavani, Intelligent traffic control system with priority to emergency vehicles, in *IOP Conference Series: Materials Science and Engineering*, vol. 455, no. 1 (IOP Publishing, 2018), p. 012023
5. S.K. Mandal, A. Maity, Precision farming for small agricultural farm: Indian scenario. J. Exp. Agricul. Int. 200–217 (2013)
6. D. Subitha, R. Vani, A. Raja, S. Balasubramani, A. Manjunathan, K. Sahile, Development of novel nano-silver-based antenna for green agriculture: Adv. Mater. Sci. Eng. (2021)
7. R. Hussain, J.L. Sahgal, M. Riyaj, Control of irrigation automatically by using wireless sensor network (2013)
8. R. Angeline, D. Subitha, R. Vani, Automation of transportation process using IoT for enhanced production. J. Green Eng. **10**(11), 11031–11039 (2020)

Design and Development of Medicine Retrieval Robot for Pharmaceutical Application

V. G. Pratheep, T. Tamilarasi, K. Ravichandran, A. Shanmugam, S. K. Thangarasu, and A. Prenitha

Abstract Now a day's pharmacy can be found in every major villages and cities. It became an integral part of our life. There is at least one person in every household who takes medicine on a daily basis. This paper focused on reducing the workload of the pharmacist by analyzing the prescription. Examining the handwriting of doctors is done by replacing the notepad with pressure pads. It records their script, which is decoded using handwriting recognition software; this information is then processed with the help of Arduino Mega (ATMega 2560) microcontroller hardware. After authentication from the microcontroller, an actuator goes to the defined position where the medicine is located to either store or retrieve the medicine depending on the application. Stepper motors and lead screws are used to retrieve the medicines.

Keywords Doctors script reading · Automation · Pharmacy · Handwriting input

1 Introduction

Pharmacies are the important part of our day-to-day life. The number of people affected by new diseases keeps on increasing. For this purpose, the researchers doing investigation in developing new drugs and vaccines to battle these newly developed diseases. According to the World Health Organization (WHO), at least 2.6 million people are well-trained pharmacists or pharmaceutical personals across the world. All the hospitals have trained professionals in their pharmaceutical section. The problem arises at the community pharmacies who belongs to any hospital or clinic. Medicines are stored and dispensed with the help of retail storefront in community pharmacies. It is used as a place where most pharmacists practice the profession of pharmacy. Nevertheless, in India, most of the pharmacies of community pharmacies and the

V. G. Pratheep (✉) · T. Tamilarasi · A. Shanmugam · S. K. Thangarasu
Department of Mechatronics Engineering, Kongu Engineering College, Perundurai, Tamilnadu 638060, India

K. Ravichandran · A. Prenitha
Department of Automobile Engineering, Kongu Polytechnic College, Perundurai, Tamilnadu 638060, India

A. Kumar et al. (eds.), *Computational Intelligence in Machine Learning*, Lecture Notes in Electrical Engineering 834, https://doi.org/10.1007/978-981-16-8484-5_28

person who working in the pharmacy is mostly untrained. According to Times of Indian, medical errors is one of the top 10 killers in world, as 15% of the patients as misdiagnosed. There are deaths outside the hospital due to the consumption of drugs in wrong dosage. In India, it is extreme as there are more pharmacies have untrained pharmacists. Some pharmacies provide their services without any prescription.

In 2015, the 12.5 million United States people were abused by uploads, and it was predicted that 24,861 people died because of over dosage of prescription. Across the country, 44,000 people died because of drug overdoses in 2013, and it increases double the time in 1999. More than 95% of deaths are due to the drugs-alprazolam (e.g., Xanax) and diazepam (e.g., Valium) involved with other drugs and with an average of more than two additional drugs. In drug overdose deaths involving diazepam, only 3% mentioned no concurrent drug use, 22% mentioned one other drug, 62% mentioned two to four other drugs and 13% mentioned five or more other drugs [1].

2 Existing System

Medicine cabinet as shown in Fig. 1 is widely used in pharmacies. In these cabinets, the medicines are stored in plastic containers. The boxes are placed in an alphabetical order so that it is easy to access by the person working there. Each row and column has its own specific number, which is stored in computer. The position of the drug can be determined using these numbers [2].

In addition to this indicator like LED (Light Emitting Diode) or buzzers are provided at the end of the boxes so that it can be accessed easily. At the same time, there are some drawbacks such as only one medicine at a time can be processed and multiple medicines results in confusion due to the multiple indications caused by the LED. In some cases, it is difficult to see the indication because of the low lighting of the room.

Fig. 1 Medicine cabinet

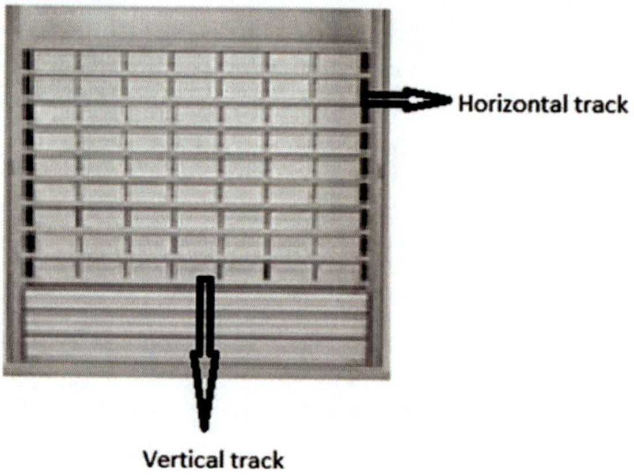

3 Proposed System

The proposed method is to further optimize the process by analyzing the prescription and automatically store the retrieving the prescribed medicines and drugs. This reduces the workload on the pharmacist as well as the error due to misinterpretation of the prescribed drugs. The inventory management of pharmacy can be improved by using the concepts Internet of Things (IoT).

The significant part of this research is handwriting recognition, which is done using pressure pad sensor [3–5] (in our case, it is smart phone). The prescription pad in doctor's table is replaced with pressure pads. Consequently, the medicines that are prescribed are transferred in the electronic form to the processor [1, 2, 6]. The data is sent to the processor via GSM module. This information is processed using Arduino mega (At mega 2560) which is the heart of the development. The Arduino processes the received information. If the prescribed medicine matches the name of the medicine in the program, it then calls the subprogram where the location of the drug is stored. This location data mostly consists of the timing intervals for the motor supply. It is transferred to driver IC's in the form of pluses which operates and slides the arm along the aluminum frame. This movable arm has an end effector likely a pneumatic piston, which is operated by using electrically actuated solenoid valve, which controls the direction in which the cylinder actuates. At the end of the piston, a suction cup is provided which operates at a pressure of 1.5 bar. The pressure is maintained at this value in order not to tear the foil of medicine during grasping process. After grasping the medicine, the end effector is returned to the home position where the medicine is dropped for further process. The pharmacist then cuts the medicine and delivers it along with bill and prescription. In order to store the medicine at its original place, it is then placed at the same position where the piston is used for this purpose. The detailed CAD model of the storage and retrieval set up is shown in Fig. 2.

4 Design Methodology

The setup is broadly divided in to three major parts; one is the mechanical setup involving the frame, lead screws, bearings, and sliding arm. While the second part is the electrical circuit-involving microcontroller, relays, driver IC and the third part is pneumatic system involving pneumatic cylinders and suction cup. The electrical part checks the data on the prescribed medicine and sends the location. Then, the mechanical and pneumatic part come into play for retrieving the medicine [7–9].

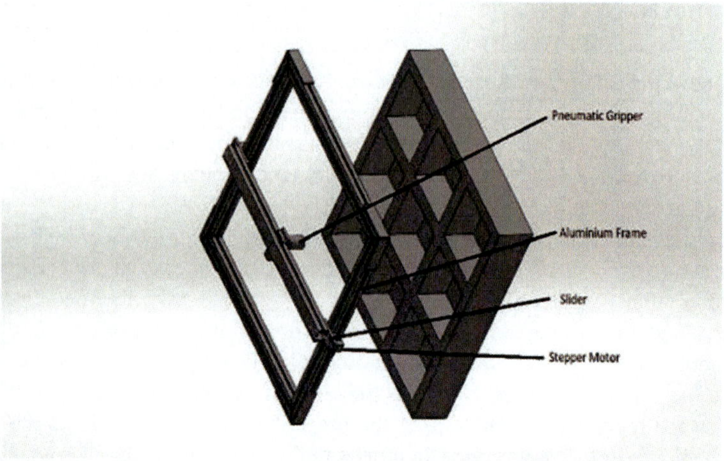

Fig. 2 CAD model storage and retrieval mechanism

4.1 Design of Electrical System

The electrical section as shown in Fig. 3 consists of microcontroller circuit, relay circuit, and driver circuit. The driver circuits are placed subsequent to their respective motor. Whereas the relay circuit is placed afterward to the pneumatic cylinder. The Arduino drives the two motors used for vertical motion in a synchronizing motion; this is achieved by providing, parallel outputs from Arduino.

Fig. 3 Electrical set up for the pharmaceutical robot

The electrical setup operates at 5 V except for the stepper motor and the solenoid valve. This 5 V is supplied by the Arduino mega, which is connected to the computer. Stepper motor and solenoid valve are connected with 12 V, which is drawn from the adapter.

4.2 Design of Mechanical System

The mechanical setup consists of aluminum frame, bearings, lead screws, and stepper motor. The aluminum frame has slots in it where the bearings and the stepper motor are attached. The vertical motion of the sliding arm is achieved by using bearing and lead screws and the horizontal motion is done using slider rack and pinion system [7–9]. This slider consists of stepper motor, driver IC and the pneumatic cylinder as shown in Fig. 4. It is important to placement the cylinder because of the change in center of gravity during extension and retraction of cylinder. Direction of the cylinder is controlled by using solenoid valve. At the end of the arm, a suction cup is placed. It must be operated at a pressure of 0.5 bar, and a valve is used to maintain at this value.

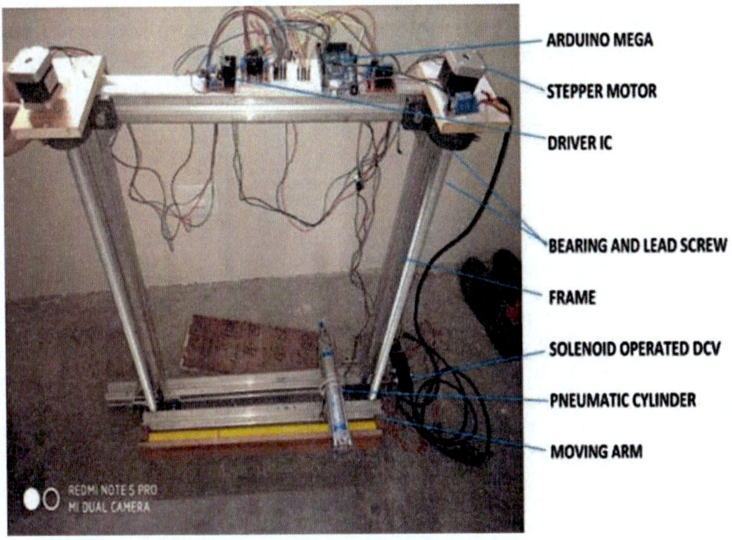

Fig. 4 Mechanical set up for the pharmaceutical robot

4.3 Feasibility Study

The proposed model is economically feasible, as all the components are easily available and the pharmacist can well connect the processor to his existing database. It is economically feasible where the price of the frame and motors is less than 15,000 rupees. Since the proposed setup adding the existing database in the computer to the processor and there is no complex mechanical setup needed so that the pharmacist can easily assemble the entire setup. For maintaining and servicing purposes, the frame can be disassembled and reassembled easily. In addition to this, database can also be easily updated. By using IOT, the whole process can be controlled from the clinic or the hospital itself. A normal person can quickly upload the database of the available medicines, and it can be readily updated.

5 Conclusion

The medicine retrieval robot for pharmaceutical applications greatly reduces the workload on the pharmacists by reducing the effort required to analyze the prescription and storing and retrieval of medicines. In addition, the errors of giving the wrong drug are greatly reduced. This system can also be improved by implementing IOT concept. The process of converting the developed prototype model into an inventory management for pharmaceutical storage facilities can be an effective method.

References

1. J. Brandt, S. Alessi-Severini, A. Singer, C. Leong, Novel measures of benzodiazepine and Z-Drug utilisation trends in a canadian provincial adult population (2001–2016). J. Popul. Ther. Clin. Pharmacol. **26**(1), 22–38 (2019)
2. A.R. Kernodle, C.K. Frail, S.A. Gernant, K.S. Pater et al., Patients' experiences using a brief screening tool for medication-related problems in a community pharmacy setting: a qualitative study. J. Pharm. Pract. **30**(1), 49–57 (2017)
3. K. Wang, B. Babenko, S. Belongie, End-to-end scene text recognition, in *IEEE In Computer Vision Conference (ICCV)* (IEEE Press, New York, 2011), pp. 1457–1464
4. Character recognition using two- stage convolutional neural network, in *16th International Conference on Frontiers in Handwriting Recognition (ICFHR)* (IEEE Press, New York, 2018), pp. 205–210
5. Z. Li, L. Jin, S. Lai, Rotation-free online handwritten Chinese (2018)
6. C.J. Makdessi, C. Day, B.B. Chaar, Challenges faced with opioid prescriptions in the community setting—Australian pharmacists' perspectives. Res. Soc. Admin. Pharm. (2019)
7. V.G. Pratheep, E.B. Priyanka, S. Thangavel, K. Heenalisha, M. Ariya Manickam, A.P. Logaram, Automatic body posture corrector for spinal cord patients, in *Materials, Design, and Manufacturing for Sustainable Environment*, ed. by S. Mohan, S. Shankar, G. Rajeshkumar. Lecture Notes in Mechanical Engineering. Springer, Singapore (2021)

8. V.G. Pratheep, M. Chinnathambi, E.B. Priyanka, P. Ponmurugan, P. Thiagarajan, Design and Analysis of six DOF robotic manipulator, in *IOP Conference Series: Materials Science and Engineering*, vol. 1055. p. 012005 (2021)
9. V.G. Pratheep, T. Tamilarasi, K. Dhileephan, A.J. Antony, A. Heeraj, M. Akash, Design and fabrication of multitasking transit robot, in *IOP Conference Series: Materials Science and Engineering*, vol. 1057, p. 012034 (2021)

Machine Learning Concept in Smart Water Purifier and Dispenser for Averting Spread of Infections

Arshad Mohammed, Mahabbob Shaik, and G. Sailaja

Abstract Recent pandemic made everyone health-conscious and more particularly about hygienist in food and drink water. The general water purifier and dispenser only control the purity in one way or the another but not according to individual user requirement. Moreover, the available water dispensers are not self-officiated to dispense the water without a manual touch of the user. It is worthwhile to note that based on the user's age and gender the purity or quality of the water has to be consumed to assist their metabolism and growth. Hence, in this dossier, a novel water purification system is presented wherein the microcontroller identifies the user age and gender by executing relevant machine learning algorithm and select the appropriate filter configuration to dispense desired water quality. Further, the pandemic has made us learn about social distancing and disinfection, this is applicable for water dispensers also. Hence, a dispensing system which dispenses the water based on the tumbler shape/size which is accomplished by executing a relevant machine learning algorithm.

Keywords Machine learning · Image processing

1 Introduction

The Internet of things or IoT implicits to a method that has interrelated devices which include people and also animals, besides unique identifiers, that can transport information with no individual communication during a set up. It is anticipated to achieve additional 80 billion appliances by 2022. The few related systems coming under this are smart houses, health care and transportation. We can see that few devices which cannot operate according to environmental conditions, they can be replaced with the smart devices which can work according to the concept they have designed for contextual [1, 2]. It is observed that when sensors are attached to the smarter objects, it will become more smart and sensible and will give new chance in the supervision process [3]. Few devices which can be worn on the body, and are

A. Mohammed · M. Shaik (✉) · G. Sailaja
EED, Muffakham Jah College of Engineering and Technology, Hyderabad, India

A. Kumar et al. (eds.), *Computational Intelligence in Machine Learning*, Lecture Notes in Electrical Engineering 834, https://doi.org/10.1007/978-981-16-8484-5_29

connected to the sensors are helped by smart devices and smart watches. When we observe these smart watches, which are available in the market are mostly connected to sensors which help in detecting the heart rate, temperature, workout time and steps calculation. Some smart watches which are currently available in the market, are connected with sensors also receive and send messages, calls, etc. These sensors along with smart devices has made big data science a trending technology in the society. Data got from sensors and these gadgets gives information mining its new bits of knowledge. As the human body comprises of 50–65% water [4].

As we as a whole know, it is actually quite significant for the people to devour water. Any place they work or in any circumstance, there is an interest to fulfill the thirst of water. The framework which is recommended in this paper achieves the solicitation of individuals who are having achieves the solicitation of individuals who are approaching the water dispenser and need to be careful about the overall infection of COVID-19. Embracing to this, it will decrease the human exertion in contacting just as decrease of contention in the territory. Along these lines, more astute water allocator approves particular users to get the cautions and updates with regards to when the distributor ought to be topped off. Additionally, for all the clients who will have the admittance to the specific floor, will get the alarms identified with the utilization of water dependent on the temperature. This paper propose the sensor subordinate keen water container's underlying outcomes. We will introduce the framework architecture, methodology and execution of current framework.

As of now, numerous profoundly evolved thoughts are arranged in the plan of advance programmed control techniques for the nature of water [5]. In this view, methods of profound learning are being used as a component to help the creation administrators settling on their choice to get extra clear arrangements from unrefined estimations of the boundaries appeared in the instruments. The fundamental thought behind this is that the determined estimations of the nature of water, which are in an orderly fashion, are to be set up into very much judged and reasonable data to bring out fundamental restorative activities continuously. This additionally permits one to use the unpracticed or new administrators in making. Typically, the water quality which is connected with the numbers, now, a huge subjective understanding framework is suggested. This subjective upheld translation empowers simpler restorative activities on water quality underway utilizing man-made consciousness calculations.

2 Literature Review

The current examination subject in quality control of water, is the acknowledgment of sporadic physicochemical properties of drinking water with more prominent accuracy, subsequently web based controlling and rectifying water quality will turn into a harder work. A constant arrangement of water quality observing and anomaly acknowledgment, in view of multiclass uphold vector machines (MSVM) recommended that suits RO creation plants. The prescribed framework is reasonable to give easy to understand data to creation controlling administrators to settle on choices and perform usable adjustments.

Muharemi et al. [6] expressed numerous techniques for distinguishing varieties or irregularities occurring in time arrangement information of nature of water. Their work too talked about and arranged answers for beat certain issues while overseeing information of time arrangement. They utilized calculated relapse, direct separate examination (LDA), uphold vector machines (SVM), profound neural organization (DNN), counterfeit neural organization (ANN), long transient memory (LSTM), and intermittent neural organization (RNN) to apply extra time arrangement information and performed assessment utilizing F score.

Ahmed et al. [7] has indicated that the utilization of IoT frameworks in water quality observing for constant and AI techniques like ANN and SVM to anticipate WQI, and recognize unpredictable occasions, explicitly, for deliberate contamination to allow continuous contamination identification and furthermore control activity. The proposed strategy by the creators, likewise applied relapse, connection examination, various leveled grouping and so on, to get familiar with different bearings which incorporate learning method to help dynamic.

Liu et al. focused on a forecast model of the nature of water which required information of higher caliber. They built and worked a keen arrangement of water quality checking which depended on the IoT, which made enormous information at a predominant speed thus made water quality information troublesome. They attempted to plan and set up a drinking-water quality time arrangement expectation model to figure the water quality large information by methods for cutting edge profound learning (DL) technique by methods for LSTM. They obtained information of drinking-water quality from the programmed water quality observing station of Guazhou water source of the Yangtze river in Yangzhou.

Karthick et al. zeroed in on planning a framework to ascertain boundaries, expressly, broke down oxygen, chloride, smelling salts, nitrate, hardness and turbidity to choose the water quality in well-water by utilizing IOT-based AI for water quality observing framework. They acquired spacetime information that will advance the simple revelation of constant examples to get familiar with the effect of pollution on the water quality in customary water bodies.

SeanWei Xiang Ong, MBBS Yian Kim Tan, PhD in their work depicted that there was enormous natural contamination by 1 SARS-CoV-2, COVID-19 admitted to the emergency unit at Evergreen Hospital persistent with mellow upper respiratory lot inclusion. Latrine bowl and sink tests were additionally certain, noteworthy that viral shedding in excreta could be a potential course of infection transmission. Post cleaning tests were negative, suggesting that current disinfecting activities are satisfactory. Extensive ecological defilement done by the patients with SARS-CoV-2 through respiratory drops and fecal issue, proposes that the climate as a likely vehicle of transmission and supports the need for severe recognition to water cleanliness and furthermore natural mindfulness.

Dongki Yang [8], has audited that the gathered data of the previous SARS-CoV and MERS-CoV contaminations and infers to help us and comprehend the recently showed up SARS-CoV-2 disease. Also, it deciphers the improvements in developing

COVID-19 immunizations and therapeutics through existing methodologies. In conclusion, the applicable methodology of nanotechnology is clarified in detail, and antibodies and restorative medications created dependent on nanomedicine, which is as of now going through clinical preliminaries, have introduced the possibility to become imaginative options for defeating COVID-19.

3 Challenges

In the present era, water has been added as a commodity in the wall street which led most of the inventors to look into the water quality and its dispencing aspects. Though water is abundant in nature the drinkable-water is very rare and this made the water to be logged in to wall street as a commodity. Under such circumstances, the main focus of most of the inventors and investors turned on water quality and wastage elimination. The water quality is such an aspect which cannot be undermined as different people requires a different quality of water such as children's do not need Alkaline water and in some cases, users needed zero TDS/minerals in water.

Though several inventions were made to get a good quality of water by deploying several filtration techniques. These techniques vary from reverse osmosis to UV filtrations and it is evident that they were successful in doing so.

4 Methodology

The present work is directed toward the integration of microcontroller 101a with camera 101b to make Central Imaging System 101. Whose primary function is to identify the users and their age by deploying relevant machine learning algorithm and it is directed toward the integration of microcontroller 101a with camera 101b for detecting the size and predicting the quantity of water to be dispensed by using an image detection algorithm. The microcontroller 101a is in integration with Water Filtering Units 102a and High-Pressure Motor 102b to constitute Water Quality Adjusting System 102.

The present subject matter is directed toward the use of relevant MLA in microcontroller 102a for detecting users and their age and issue an appropriate command to the Water Quality Adjusting System 102. The Water Quality Adjusting System 102 whose primary function is to filter the water in terms of, sludges, TDS, PH, Alkaline and adding minerals.

The deployment of Central Imaging System 101 for identifying the tumbler/glass/container size using relevant machine learning algorithms (MLA) and also to predict the exact amount of water to be dispensed. The execution of relevant MLA is to identify the user age and predict the exact quality of water to be needed and make relevant command to Water Quality Adjusting System 102 for necessary adjustments.

5 Description of the Model

Referring to Fig. 1 is a diagram depicting the 100 Machine Learning-Based Smart Water Purifier and Dispenser consisting of Central Imaging System 101, Water Quality Adjusting System (WQAS) 102 and Water Dispensing Adjusting System (WDAS) 103. Wherein the Central Imaging System (CIS) 101 comprising of microcontroller 101a and camera 101b, which identifies the user by deploying relevant machine learning algorithm (MLA). In coordination, microcontroller 101a identifies the user along with his/her age and by deploying relevant MLA determines the quality of water needed by the user and adjust the water quality using Water Quality Adjusting System 102.

Further to it, the microcontroller 101a in combination with camera 101b identifies the size and shape of the tumbler or container held by the user by executing relevant MLA on the images captured by the camera 101b. Once the size and shape are determined, the relevant MLA asses the time for which the Water Dispensing Adjusting System (WDAS) 103 should operate. Once this data is fed to the microcontroller 101a, it sends a command to the Water Dispensing Adjusting System (WDAS) 103 and starts dispensing.

Further to it, the microcontroller 101a captures the images of the water dispensing in the tumbler and execute relevant MLA to determine the status of water filled in the container. Once the tumbler is full or the predicted/estimated time to fill the tumbler is over, the microcontroller 101a stops the Water Dispensing Adjusting System (WDAS) 103.

In accordance with a non-limiting exemplary embodiment of the present subject matter, Fig. 2 is a depiction of Component Layout of the Machine Learning-Based Smart Water Purifier and Dispenser. The central commanding post is hosted by an advanced microcontroller 101a which is capable of not only controlling, coordinating

Fig. 1 Machine learning-based water purifier and dispenser

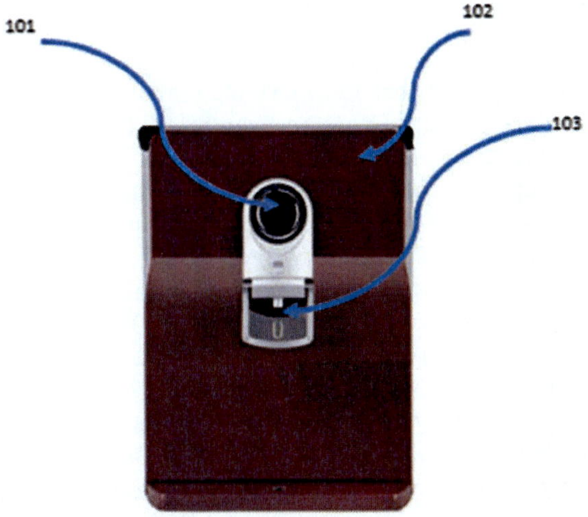

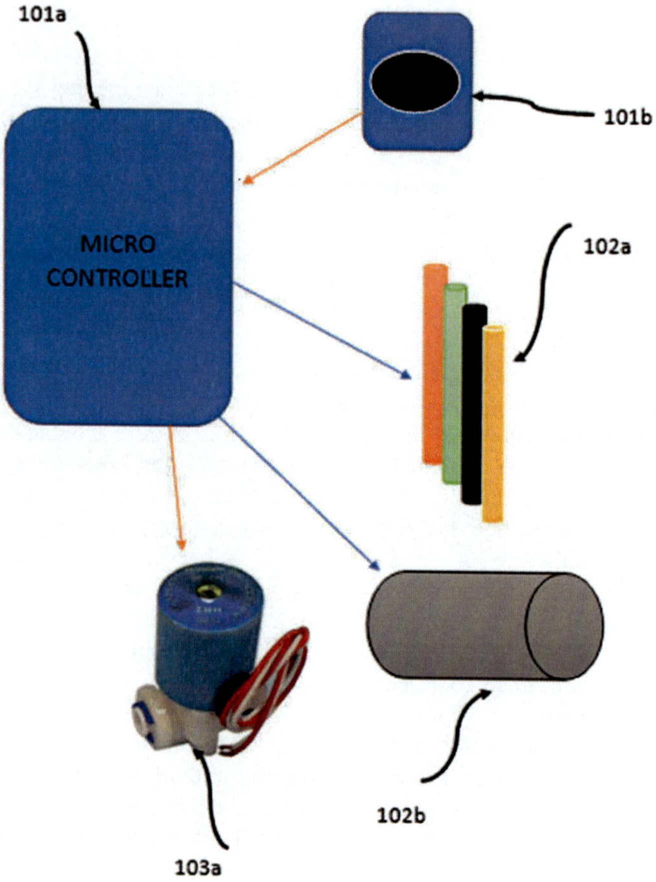

Fig. 2 Component layout of the system

the components but also executing artificial intelligence algorithms which may be a relevant machine learning algorithms (MLA). The microcontroller 101a acquires the images by using a live video feed from the camera module 101b and based on the output of relevant machine learning algorithms (MLA) command the Water Quality Adjusting Unit 102. Wherein the Water Quality Adjusting System 102 comprising of water filters 102a which may be carbon, graphite, reverse osmosis, mineral, copper, alkaline composition type. Once the type of filters to be used to get the desired water quality is selected then, microcontroller 101a divert the water by operating the high-pressure water pump so that water only passes through those water filters which are just needed.

Further to it, microcontroller 101a is integrated with water regulating valve 103a to form water dispensing unit 103. The water regulating valve 103a is a solenoid-operated valve which opens and closes to regulate the water flow. The operation depends on the command given by microcontroller which ascertains the valve operation based on the predicted time to fill the tumbler or determined level of water dispensed into the tumbler.

Following is a non-limiting exemplary embodiment of the present subject matter, as shown in Fig. 1, which is a 300 Process Executed In SMART WATER PURIFIER AND DISPENSER. The process starts at step 301, wherein the microcontroller 101a acquires user images through Camera 101b. In step 302, microcontroller 101a execute machine learning algorithm (MLA) 1 on image received and ascertain user and their age. In Step 303, microcontroller 101a verify the user identity to load the age of the user and if not found then it saves the user face and age. Further, it executes machine learning algorithm-2 (MLA-1) to ascertain the quality of water to be dispensed as per their age.

Further in step 304, based on the quality of water required, microcontroller 101a selects the multiple FILTER CARTRIDGES to get the desired quality of water and activated those filter values only such that water flows through those water filters only. In subsequent step 305, microcontroller 101a execute machine learning algorithm-3 (MLA-3) to ascertain the tumbler size and thereby estimate the quantity of water to be dispensed. Based on the quantity of water to be dispensed further, the microcontroller 101a selects the time for which the valve to be operated to dispense the said quantity of water. In step 306, microcontroller 101a commands the Motor 102b to pump feed water through preselected FILTER CARTRIDGES 102a. Step 307, microcontroller 101a sends a command to the dispensing unit to operate for a specific time in which tumbler/ bottle fills up. In step 308, microcontroller 101a executes MLA 4 on the live video feed to ascertain that tumbler is filled or not and once it is filled it commands the pump/dispensing unit to shut.

6 Conclusions

The Machine Learning-Based Smart Water Purifier And Dispenser, identifies the users and their age by deploying relevant machine learning algorithm; and wherein Water Quality Adjusting System 102 comprises of a plurality of water filter that may be carbon filter, alkaline or RO type; and Water Dispensing Adjusting System 103 consists of a valve controller integrated with high-pressure motor controlled by the microcontroller, and the microcontroller 101a executes relevant machine learning algorithm on camera data and ascertains users and their ages. The microcontroller 101a executes relevant machine learning algorithm on tumbler dimensions and thereby ascertain the time interval for which the valve controller is to be operated.

References

1. K.S. Chun, A.B. Sanders, R. Adaimi, N. Streeper, D.E. Conroy, E. Thomaz, Towards a generalizable method for detecting fluid intake with wrist-mounted sensors and adaptive segmentation, in *Proceedings of the 24th International Conference on Intelligent User Interfaces* (2019). https://doi.org/10.1145/3301275.3302315
2. O. Amft, D. Bannach, G. Pirkl, M. Kreil, P. Lukowicz, Towardsa wearable sensing-based assessment of fluid intake (2010)
3. ElProCus—Electronic Projects for Engineering Students. Heat Sensor Circuit and Its Working Principle (2017). [online] Available at: https://www.elprocus.com/heatsensor-circuit-and-working-operation/. Accessed 9 Oct 2017
4. M.C. Chiu, S.P. Chang, Y.C. Chang, H.H. Chu, C.C.H. Chen, F.-H. Hsiao, J.-C. Ko, *Playful bottle: A Mobile Social Persuasion System to Motivate Healthy Water Intake*, New York, NY, USA (2009), pp. 185–194
5. M.R. Zemouri, Contribution à la surveillance des systèmes de production à l'aide des réseaux de neuronesdynamique. Application à la e-maintenece (Contribution to the monitoring of production systems using dynamic neural networks: Application to e-maintenance). Ph.D. Thesis, Université de Franche-Comté, France 2003
6. F. Muharemi, D. Logofătu, F. Leon, Machine learning approaches for anomaly detection of water quality on a realworld data set. J. Inf. Telecommun. **3**(3), 294–307 (2019)
7. U. Ahmed, R. Mumtaz, H. Anwar, S. Mumtaz, A.M. Qamar, Water quality monitoring: from conventional to emerging technologies. Water Supply **20**(1), 28–45 (2020)
8. D. Yang, Int. J. Nanomed. **16**, 623–649. https://doi.org/10.2147/IJN.S296383. eCollection 2021

Regressors with Anova-Reduced Features for Tariff Rate Prediction Using Machine Learning

M. Shyamala Devi, Ganta Umamaheswar Reddy, Balla Teja Swaroop, and Rachaputi Kiran Kumar

Abstract Tariff is the collection of extra charges that is charged for the transportation of goods and materials from one place to another. The extra amount collected as tariff completely differs based on the factors like period, route, petrol cost, road quality, luggage type, demand, employee salary, seasonal impacts, and non-seasonal impacts. Due to the floating changes in the tariff rate, it remains a challenging task to predict the tariff rate for the industry sectors. With this overview, in order to overcome this problem, we are using machine learning techniques for tariff rate prediction. The classification of mushroom edible classes are achieved in four ways. Firstly, the dataset is preprocessed with feature scaling and missing values. Secondly, raw dataset is fitted to all the linear and ensembling regressors to analyze the performance of the metrics before and after feature scaling. Thirdly, the Anova test is applied to verify the features with PR(>*F*) < 0.05 that highly influence the target tariff rate. Fourth, the dimensionality reduced dataset after Anova test is fitted to all the linear and ensembling regressors to analyze the performance of the metrics before and after feature scaling. Fifth, the performance metrics used for analysis are EVS, MSE, MAE, and Rscore. The implementation is done using Python language under Spyder platform with Anaconda navigator. Experimental results show that the gradient boosting regressors tend to retain 0.915 RScore before and after feature scaling. The attributes seasonal impact and wttons having the PR as 0.952 and 0.253, respectively, and the above features are eliminated from the raw dataset. The gradient boosting regressors retain RScore as 0.916 for the Anova-reduced dataset before and after feature scaling.

M. Shyamala Devi (✉) · G. U. Reddy · B. T. Swaroop · R. K. Kumar
Computer Science & Engineering, Vel Tech Rangarajan Dr, Sagunthala R&D Institute of Science and Technology, Chennai, India
e-mail: shyamaladevim@veltech.edu.in

G. U. Reddy
e-mail: vtu10061@veltechuniv.edu.in

B. T. Swaroop
e-mail: vtu10073@veltechuniv.edu.in

R. K. Kumar
e-mail: vtu10006@veltechuniv.edu.in

A. Kumar et al. (eds.), *Computational Intelligence in Machine Learning*, Lecture Notes in Electrical Engineering 834, https://doi.org/10.1007/978-981-16-8484-5_30

Keywords Machine learning · Regression · Anova · And feature scaling

1 Introduction

The industry need to deliver their good and services to various parts of the country. The total revenue of the business depends on various factors that includes the tariff amount that they spend for their transportation of goods and services. If the industry is able to predict the future tariff rate for their goods, then, they can plan their financial expenses accordingly. So that they can take the necessary actions and they can make their business run in profitable way. With this analytical view, this project focuses on prediction of tariff rate using various regression algorithms.

2 Literature Review

In this paper, analysis of the regression model is done to predict the cost model, and the raw dataset is filtered with the help of linear regression, lasso regression, ridge regression, and elastic net regression. The ideal prediction line is plotted with the values of regression models [1]. This paper predicts the tariff rate for truck load by using feature extraction, data acquiring, training the model, model deploying and live testing. To get higher accuracy, residual values are handled [2]. This paper estimates benchmarking less than truckload market rates. This approach mainly categorizes in tangible and intangible cost toward tariff rate. It formulates multiple regression in order to estimate LTL price and is implemented by regression techniques [3]. This paper attempts to find optimal residential load control with price prediction in real-time electricity pricing environment. The automation systems are utilized completely, and the price prediction problem introduces and evaluated on a week basis [4].

This paper finds the residential electricity tariff switching based on customer response elasticity. They have used clustering and meter readings and creates homogenous group of households thereby finding the mean using k-means algorithm [5]. This paper predicts the dynamic tariff rate with smart grid using CGP evolved artificial neural networks. In this one, electricity forecasting method is defined that is involved with smart grid thereby projecting the beneficial aspects to the user [6]. This paper deals with efficient buyer groups with prediction-of-use electricity tariffs. It introduces prediction price of electricity by defining the POU tariffs using DP algorithm [7]. This paper finds the framework for determining tariff rate prediction at coordinating aggregators of plug in electric vehicles. It defines a framework to define ahead prediction. It uses tariff and plug in vehicles. POU strategy is distributed effectively [8]. This paper predicts the traffic flow with big data using SAE model that is comprised of encoders. Encoders act as building blocks to create a deep network and is effective as it delivers traffic flow in best way [9]. This paper predicts

Table 1 Dataset features

Features	Range of values
1. Road	National highway = nh, state highway = sh
2. Seasonal impact	Yes, no
3. Demand	High = h, normal = n, less = s, low = w
4. Wttons	Numerical values = $0 - \infty$
5. Labor	Numerical values = $0 - \infty$
6. Toll	Numerical values = $0 - \infty$
7. Distance	Numerical values = $0 - \infty$
8. Fuel cost	Numerical values = $0 - \infty$
9. Miscellaneous	Numerical values = $0 - \infty$
10. Tariff	numerical values = $0 - \infty$

price movement using convolutional neural network and is combined with the model and other techniques [10]. This paper predicts the property prices using machine learning; SVM input is mapped on m-dimensional to construct a linear model with zero correlation [11]. This paper is used to detect the air pollution using machine learning that measures quality index of air like particulate matter, ground level, sulfur dioxide, carbon dioxide [12]. This paper predicts the stock price using regression analysis using auto-regressive model to define current value where the correlation finds related stocks [13]. This paper evaluates the salary for the IT job market with few high-dimensional samples. Embedded feature selection is done with regularization methods [14]. This paper predicts the transformation of crypto-currency into a global phenomenon on bitcoin that deals with cryptographic cash [15].

3 Overall Proposed Architecture

3.1 Dataset Preparation

The tariff rate dataset from UCI machine repository is used. The dataset description is shown below in Table 1.

3.2 Proposed System

The overall workflow is shown in Fig. 1. The paper contributions are given below:

(i) Firstly, the dataset is preprocessed with feature scaling and missing values.

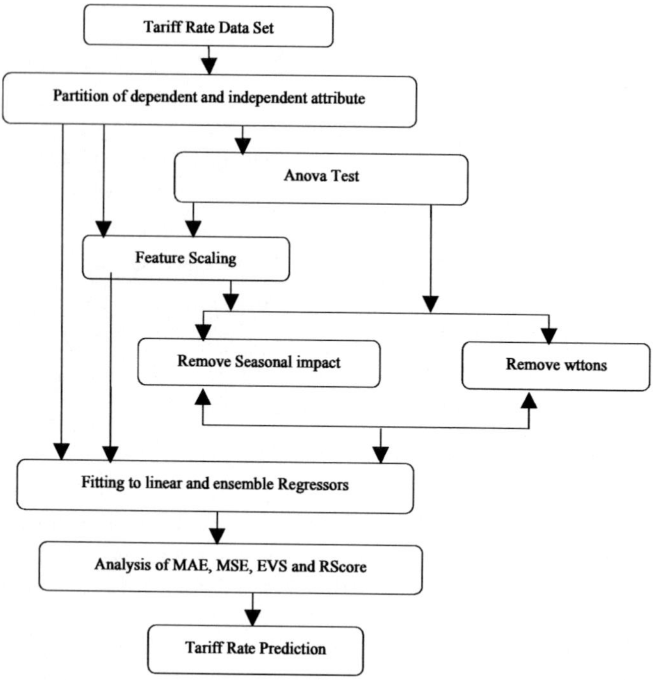

Fig. 1 Overall architecture flow

(ii) Secondly, raw dataset is fitted to all the linear and ensembling regressors to analyze the performance of the metrics before and after feature scaling.
(iii) Thirdly, the Anova test is applied to verify the features with PR(>F) < 0.05 that highly influence the target tariff rate.
(iv) Fourth, the dimensionality reduced dataset after Anova test is fitted to all the linear and ensembling regressors to analyze the performance of the metrics before and after feature scaling.
(v) Fifth, the performance metrics used for analysis are EVS, MSE, MAE, and RScore.

4 Results and Discussion

4.1 Implementation Setup and Anova Test

The tariff rate dataset extracted from the UCI machine learning repository is used for implementation. The dataset consists of the features, namely (road, seasonal impact, demand, wttons, labor, toll, distance, fuel cost, miscellaneous, and tariff). The code is implemented with Python under Anaconda navigator with Spyder IDE.

Table 2 Anova test analysis with the dataset features

Features	sum_sq	df	*F*	PR(>*F*)
Road	1.02499e+09	1	61.265255	0.00096
Seasonal impact	7.09149e+08	1	0.0034932	0.95291
Demand	5.22699e+08	1	28.2837	2.11E−07
Wttons	2.64304e+08	1	1.307809	0.253745
Labor	2.48763e+09	1	13.82165	1.43E−36
Toll	4.95164e+09	1	16.24166	3.55E−12
Distance	5.14438e+08	1	27.79339	2.38E−07
Fuel cost	4.63114e+08	1	11.79856	1.02E−10

The correlation of all the attributes toward the target is done and dimensionality reduction is applied. The dataset is splitted with 80:20 for training and testing dataset. Anova test is applied to dataset features and results shows that the attributes seasonal impact and wttons having the PR as 0.952 and 0.253, respectively, having value of PR(>*F*) > 0.05 and does not contribute to target as in Table 2. Anova test is used to determining the importance of features by comparing both the null and alternate hypothesis. If the P value of the feature associated with the F statistic is less than 0.05, then the existence of that attribute highly influence the target.

4.2 Regressor Analysis Before and After Feature Scaling

The raw dataset is fitted to all the regressors like Linear, Ridge, RidgeCV, ElasticNet, ElasticNetCV, Lars, LarsCV, Lasso, LassoCV, LassoLars, LassoLarsCV, LassoLarsIC, BayesianRidge, Polynomial, Linear Support Vector, Nu Support Vector, Decision Tree, Extra Tree, AdaBoost, Bagging, GradientBoosting, RandomForest, and Partial Least Square regressors with and without the presence of feature scaling. The performance analysis is done and is shown in Tables 3 and 4.

4.3 Anova Analysis Before and After Feature Scaling

The raw dataset is applied with the Anova test, and the features like seasonal impact and wttons are removed from the dataset. The Anova-reduced dataset is fitted to all the above mentioned regressors with and without the presence of feature scaling. The performance analysis for Anova-reduced dataset is done and is in Tables 5 and 6.

Table 3 Performance metrics of raw dataset before feature scaling

Regressors	EVS	MSE	MAE	R2SCORE
Linear Regression	0.878215	3,439,756	263.7683	0.876123
Ridge Regression	0.878475	3,432,406	263.4699	0.876388
RidgeCV Regression	0.880947	3,362,629	260.6849	0.878901
ElasticNet Regression	0.905799	2,660,730	233.1792	0.904178
ElasticNetCV	0.878215	3,439,756	263.7683	0.876123
Lars Regression	0.814072	5,251,790	323.2451	0.810866
LarsCV Regression	0.883009	3,304,454	258.348	0.880996
Lasso Regression	0.963268	1,044,796	270.2237	0.962373
LassoCV Regression	0.661147	9,570,700	440.8538	0.655328
LassoLars Regression	0.774165	6,378,845	357.2396	0.770277
LassoLarsCV	0.488464	14,430,751	727.2042	0.480302
LassoLarsIC	0.879929	3,391,380	261.8242	0.877865
BayesianRidge	0.878438	3,433,493	263.6519	0.876349
Polynomial	0.995046	139,980	49.32284	0.994959
Linear Support Vector	0.973948	737,499.5	136.0928	0.97344
Nu Support Vector	0.002013	30,811,476	4647.443	−0.10962
Decision Tree	0.898478	2,824,970	655.8103	0.898264
Extra Tree Regression	0.882195	3,287,443	728.431	0.881608
AdaBoost Regression	0.87945	3,349,106	981.8095	0.879388
Bagging Regression	0.010522	28,607,052	3536.757	−0.03023
GradientBoosting	0.915922	2,340,364	467.755	0.915716
RandomForest	0.838915	4,474,660	1109.475	0.838853
Partial Least Square	−3.40579	1.24E + 08	1872.932	−3.47631

5 Conclusion

This paper explores the efficiency of testing Anova analysis to the tariff rate dataset by removing the seasonal impact and wttons features from the dataset. Feature analysis of the dataset is done by exploring the correlation between the features. The raw dataset and Anova-reduced dataset are subjected to regressors to analyze the performance. Experimental results show that the gradient boosting regressors tend to retain 0.915 RScore before and after feature scaling. The attributes seasonal impact and wttons having the PR as 0.952 and 0.253, respectively, and the above features are eliminated from the raw dataset. The gradient boosting regressors retain RScore as 0.916 for the Anova-reduced dataset before and after feature scaling.

Table 4 Performance metrics of raw dataset after feature scaling

Regressors	EVS	MSE	MAE	R2SCORE
Linear Regression	0.898478	2,824,970	655.8103	0.898264
Ridge Regression	0.882195	3,287,443	728.431	0.881608
RidgeCV Regression	0.881731	3,285,030	959.1597	0.881695
ElasticNet Regression	0.006233	28,709,085	3544.119	−0.03391
ElasticNetCV	0.916043	2,336,997	467.5732	0.915837
Lars Regression	0.838915	4,474,660	1109.475	0.838853
LarsCV Regression	−3.40579	1.24E+08	1872.932	−3.47631
Lasso Regression	0.898478	2,824,970	655.8103	0.898264
LassoCV Regression	0.882195	3,287,443	728.431	0.881608
LassoLars Regression	0.881731	3,285,030	959.1597	0.881695
LassoLarsCV	0.006233	28,709,085	3544.119	−0.03391
LassoLarsIC	0.916043	2,336,997	467.5732	0.915837
BayesianRidge	0.838915	4,474,660	1109.475	0.838853
Polynomial	−3.40579	1.24E+08	1872.932	−3.47631
Linear Support Vector	0.898478	2,824,970	655.8103	0.898264
Nu Support Vector	0.882195	3,287,443	728.431	0.881608
Decision Tree	0.881731	3,285,030	959.1597	0.881695
Extra Tree Regression	0.006233	28,709,085	3544.119	−0.03391
AdaBoost Regression	0.916043	2,336,997	467.5732	0.915837
Bagging Regression	0.838915	4,474,660	1109.475	0.838853
GradientBoosting	−3.40579	1.24E+08	1872.932	−3.47631
RandomForest	0.898478	2,824,970	655.8103	0.898264
Partial Least Square	0.882195	3,287,443	728.431	0.881608

Table 5 Performance metrics of Anova-reduced dataset before feature scaling

Regressors	EVS	MSE	MAE	R2SCORE
Linear Regression	0.974081	732,447.8	134.9083	0.973622
Ridge Regression	0.974079	732,494.2	134.903	0.97362
RidgeCV Regression	0.974069	732,775.9	134.8533	0.97361
ElasticNet Regression	0.974043	733,423.9	134.4595	0.973587
ElasticNetCV	0.974081	732,447.8	134.9083	0.973622
Lars Regression	0.973238	756,334.8	141.4328	0.972762
LarsCV Regression	0.974045	733,421.1	134.7385	0.973587
Lasso Regression	0.963268	1,044,796	270.2237	0.962373
LassoCV Regression	0.972739	770,583.2	148.2824	0.972249
LassoLars Regression	0.973238	756,334.8	141.4328	0.972762
LassoLarsCV	0.95693	1,214,985	382.6386	0.956244
LassoLarsIC	0.97404	733,448.1	134.3087	0.973586
BayesianRidge	0.974045	733,451.8	134.8892	0.973586
Polynomial	0.969772	854,051.1	121.5073	0.969243
Linear Support Vector	0.9742	730,399.7	135.5991	0.973696
Nu Support Vector	0.002011	30,811,525	4647.442	−0.10962
Decision Tree	0.898891	2,818,037	621.9655	0.898513
Extra Tree Regression	0.909052	2,527,482	626.6207	0.908977
AdaBoost Regression	0.897042	2,859,573	925.5766	0.897017
Bagging Regression	0.010653	28,604,681	3536.314	−0.03015
GradientBoosting	0.91699	2,309,801	449.9083	0.916817
RandomForest	0.838803	4,478,079	1107.627	0.83873
Partial Least Square	0.964052	1,006,137	486.222	0.963766

Table 6 Performance metrics of Anova-reduced dataset after feature scaling

Regressors	EVS	MSE	MAE	R2SCORE
Linear Regression	0.974081	732,447.8	134.9083	0.973622
Ridge Regression	0.974079	732,494.2	134.903	0.97362
RidgeCV Regression	0.974069	732,775.9	134.8533	0.97361
ElasticNet Regression	0.974043	733,423.9	134.4595	0.973587
ElasticNetCV	0.974081	732,447.8	134.9083	0.973622
Lars Regression	0.973238	756,334.8	141.4328	0.972762
LarsCV Regression	0.974045	733,421.1	134.7385	0.973587
Lasso Regression	0.963268	1,044,796	270.2237	0.962373
LassoCV Regression	0.972739	770,583.2	148.2824	0.972249
LassoLars Regression	0.973238	756,334.8	141.4328	0.972762
LassoLarsCV	0.95693	1,214,985	382.6386	0.956244
LassoLarsIC	0.97404	733,448.1	134.3087	0.973586
BayesianRidge	0.974045	733,451.8	134.8892	0.973586
Polynomial	0.969772	854,051.1	121.5073	0.969243
Linear Support Vector	0.9742	730,399.7	135.5991	0.973696
Nu Support Vector	0.002011	30,811,525	4647.442	−0.10962
Decision Tree	0.898891	2,818,037	621.9655	0.898513
Extra Tree Regression	0.909052	2,527,482	626.6207	0.908977
AdaBoost Regression	0.891771	3,010,649	918.3653	0.891577
Bagging Regression	0.007764	28,673,020	3539.771	−0.03261
GradientBoosting	0.91699	2,309,801	449.9083	0.916817
RandomForest	0.838803	4,478,079	1107.627	0.83873
Partial Least Square	0.964052	1,006,137	486.222	0.963766

References

1. K. Sree Ram Murthy, K. Kranthi Kumar, D. Sai krishna, D. Sai Prasad, M.R. Dara, G. Divya, Prediction of tariff rate. Int. J. Adv. Sci. Technol. 07(05) (2020)
2. V. Arulkumar, S. Agalya, S. Gokul Chakkaravarthi, E. Gokula Priya, Tariff rate prediction for truckload based on predictive analysis using machine learning. Int. J. Adv. Sci. Technol. **29**(9s), 5477–5482 (2020)
3. E. Ozkaya, P. Keskinocak, V. Roshan Joseph, R. Weight, Estimating and benchmarking less-than-truckload market rates. Transp Res 667–682 (2019)
4. A. Hamed, M. Rad, A.L. Garcia, Optimal residential load control with price prediction in real-time electricity pricing environment. IEEE Trans. Smart Grid **1**(2) (2010)
5. S. Davarzani, I. Pisica, A. Gareth, *Assessment of Residential Electricity Tariff Switching Based on Customer Response Elasticity* (Institute of Energy Futures, Smart Power Networks Brunel University London, 2016)
6. G.M. Khan, R. Arshad, N.M. Khan, Efficient prediction of dynamic tariff in smart grid using CGP evolved Artificial Neural Networks, in *Proceedings of 16th IEEE International Conference on Machine Learning and Applications* (2017)

7. Valentin Robu., Meritxell Vinyals., Alex Roger., Nicholas Jennings, R.: Efficient buyer groups with prediction-of-use electricity tariffs. IEEE Trans. Smart Grid (2017)
8. E. Gustavo Coria, M. Angel Sanchez, Ameena S. Al-Sumaiti, A. Guiseppe Ratta, R. Sergio Rivera, A. Andres Romero, *A Framework for Determining a Prediction-of-Use Tariff Aimed at Coordinating Aggregators of Plug-In Electric Vehicles* (MDPI, Basel, Switzerland, 2019)
9. L. Yisheng, Y. Duan, W. Kang, Z. Li., F.-Y. Wang, Traffic flow prediction with big data: a deep learning approach. IEEE Trans. Intell. Transp. Syst. **16**(2) (2015)
10. C. Yang, J. Zhai, G. Tao, Deep learning for price movement prediction using convolutional neural network and long short-term memory. Math. Probl. Eng. **2020** (2020)
11. W.K.O. Ho, B.-S. Tang, S.W. Wong, Predicting property prices with machine learning algorithms. J. Prop. Res. (2020). https://doi.org/10.1080/09599916.2020.1832558
12. Vineeta, A. Bhat, S. Asha, P. Mishra, Machine learning based prediction system for detecting air pollution. Int. J. Eng. Res. Technol. **8**(09) (2019)
13. P.K. Sahoo, K. Charlapally, Stock price prediction using regression analysis. Int. J. Sci. Eng. Res. **6**(3) (2015)
14. I. Martin, A. Mariello, R. Battiti, J.A. Hernandez, Salary prediction in the IT job market with few high-dimensional samples. Int. J. Comput. Intell. Syst. **11**, 1192–1209 (2018)
15. A. Manimuthu, V. Raja Sreedharan, D. Marwaha, G. Rejikumar, Transformation of crypto currency into a global phenomenon. IEEE Eng. Manag. Rev. **47**(1) (2019)

A Theoretical Survey of Wireless Body Area Networks and Their Feasibility

G. Balaraju, Milindkumar, H. V. Ramachandra, and Harish Patil

Abstract The wireless body area networks (WBANs) are widely used in the various environments for the efficient detection of the body parts and internal organs. This is most widely used in the medical and medicine-oriented departments for the treatment-wise orientation. Apart from it, this is also used in other environments. This paper states the theoretical survey of the uses, issues, and technological aspects of the wireless body area networks with their proper explanations. The WBAN is stated with the future enhancement also in the below sections.

Keywords WBAN · Medicine · Patient homecare · Routing WBAN · Sensing and monitoring

1 Introduction

As social insurance costs are quickly expanding as the population, a patient is needed to screen wellbeing status anyplace both all through the medical clinic. This interest and the progression in innovation in versatile electronic gadgets, remote correspondence, versatile players, and sensors as prompted the advancement of WBANs. A WBAN is a system with an extraordinary reason configuration to work consequently can independently interface and collaborate with different therapeutic machines and sensors; they can be present inside as well as outside of the human body.

Separated from cost decrease and adaptability uses of WBAN in social insurance will offer huge preferences, for example, versatility of patients since convenient checking gadgets and sensors are being utilized, and besides, WBAN utilizes area

G. Balaraju (✉)
Department of CSE, Jain University, Bengaluru, India

Milindkumar
Govt Minority's Morarji Desai Girls Residential PU College, Koppal, India

H. V. Ramachandra
Department of CSE, CMR University, Bengaluru, India

H. Patil
Department of CSE, ISBM, Pune, India

A. Kumar et al. (eds.), *Computational Intelligence in Machine Learning*, Lecture Notes in Electrical Engineering 834, https://doi.org/10.1007/978-981-16-8484-5_31

autonomous checking gadgets; there is no checking gadgets present in contemporary electronic checking frameworks.

The fundamental motivation behind this research work is to exhibit a very thorough and succinct overview on WBAN and its different applications inside the human services enterprises.

2 Sensors

Sensing and monitoring of computerized gadgets that a person can carry on their body and depend on remote innovation are as of now in presence. There can be as savvy material or gauzes which permits the consistent observing of blood glucose levels, circulatory strain, etc. Wearable biosensors are very clear gadgets that survives the confinements of mobile innovation and furthermore give a quick furthermore, enthusiastic reaction needed for observing patients, and it is also functional in inaccessible region. We will begin with the officially existing sensors; fundamentally, we have a couple of different type of sensors—wearable sensors and the embedded sensor. These sensors are been plans regarding the reality that the result must be improved, while the expense and massiveness of the sensor are diminished. This present reality physiological information extracted from these sensors is being sent remotely to the Web which is at that point sent to the specialist. It aides in the enhancement of the prosperity of the patient's health by making the medicinal advance more good and simpler in the accumulation and examination of the tolerant information.

2.1 Smart Shirt

It resembles a wearable mother board. Essentially, it is a material produced using brilliant texture for the most part used to permit remote physiological observing of different indispensable indications of the wearer, for example, breath rate, pulse, temperature, movement, and stance. Vivo Metrics is asserted that Life Shirt was the primary industrially accessible shrewd shirt, and recorded ECG, breath utilizing inductance plethysmography, accelerometer with discretionary module heartbeat oximetry, GSR, amplifier blood weight, and electronic journal catch. Data from the shirt might be put away locally, or exchange to the wearer's specialist, mentor, or through an individual server.

2.2 Ring Sensor

It is a heartbeat oximetry sensor that helps in ceaseless observing of oxygen immersion and pulse in a very clear way. It is a ring formed gadget that is worn for significant lots serenely. The ring sensor is comprised of detecting gadgets like photo-diode, transmitter, LEDs, and any other chips. Such sensors are comprised of an opt electric parts which bolsters blood oxygen immersion non-invasively, consistently and a long haul observing of the patient's blood vessel blood volume waveforms. This framework gives a consistent checking of the patient; it gives very essential data which helps in preventive conclusions. The ring sensor is structure, so it is easy to wear all day or all the time; it is totally remote. The uprooting of vessel dividers happens that is estimated to distinguish the adjustments in the volumes of blood when this beat of weight goes through the vessels. This procedure is finished by photo-electric strategy in which the photoconductors and resistors were utilized; at that point, the measure of blood will increase in the finger when blood is compelled to the furthest points.

2.3 Motion Detection

Motion can be measured by sensors such as an accelerometer; it tends to be utilized to decide the introduction of certain part of the body without development consequently. The nature of movement and sort of exercises an individual takes part in can be distinguished by a framework which consolidates a gyrator. It is a sensor that measures the precise speed; it can likewise be utilized to decide the introduction of a moving body.

2.4 Electroencephalography [1]

It is essentially the chronicle of electrical movement of the cerebrum. In clinical environment, electroencephalography is essentially the chronicle of the cerebrum's unconstrained electrical action over some stretch of time, more often than not 28–43 min. Diagnostic applications mostly center around the unearthly substance of electroencephalography; it is essentially the neural motions those are seen in electroencephalography signals. Electroencephalography estimates the changes voltage coming about because of ionic current streams around the neuron of the mind. Directly, mobile electroencephalography chronicles have been appeared to have incredible incentive in the determination of epilepsy and in the checking of understanding reaction to treatment. A great part of the data obtained from AEEG chronicles may not be obtained amid a daily practice, 20-min electroencephalography test.

2.5 *Implantable Sensors*

An implantable sensor sends electrical signals into the mind or spinal string for the most part; it is used to the treat the Parkinson's, obstinate epilepsy, and interminable agony types of sicknesses. Spinal line incitement, in the most straightforward structure, made up of animating cathodes, embedded more often than not in the lower stomach region, directing wires serves to interface the anodes to the generator. Spinal string trigger is utilized generally in treatment at the point when flopped back medical procedure disorder happens and complex territorial torment disorder.

3 WBAN Architectures

In this area, common framework engineering of a remote body zone arrange is given regard to its physical area, different system topology, applications, and explicit system plan. A run of the mill WBAN framework engineering is comprised of three primary segments; these are as follows:

(i) Wireless body area networks.
(ii) Individual server.
(iii) Restorative separate for wellbeing observing system (MSHM) in this segment sensors that are wearable being appended to the patient's body, these sensors will detect the fundamental changes in soundness of the patients, just like electrocardiogram sensors, circulatory strain sensor, and so forth and giving a steady criticism to the human services checking framework. Coming up next are the most widely recognized primary pieces of the medicinal sensors.

3.1 *Radio Trans-collector*

A radio trans-collector transmits and get detected physiological information, and it also helps in interchanges between different hubs remotely.

3.2 *Memory*

Serves to temporarily store the information that was detected by a restorative sensor.

3.3 Microcontroller/Microprocessor

These can control the usefulness of the part in the sensor hub and furthermore utilized nearby information preparing.

3.4 Power Supply

Supply of power is considered main in the restorative sensor as it can, without much of a stretch influence the transportability of the medicinal sensor. The restorative sensor batteries are ordinarily utilized, these perform well as long as a while despite the fact that a solitary sensor can detect and process numerous physiological signs simultaneously. A solitary battery control supply can control numerous sensors simultaneously such the electromyogram which is utilized for muscles checking exercises, an electro cardio sensor can likewise be utilized in the meantime to screen the heart exercises and so on. These different sensors collaborate with one another with the assistance of a nearby radio recurrence, while ZigBee is utilized as the primary correspondence convention to speak with the separate. A Medical Super Sensor is utilized to gather numerous tests of the detected information from different WBAN sensors. Therefore, it decreases the measure of information that will be transmitted by the BSN's since the Medical Super Sensor shift through all the repetitive information. This thus improves the general transfer speed use just as a decrease in the power utilization of the BSs.

3.5 Individual Server

In this area, the individual server utilizes a correspondence convention with ZigBee to interface the WBAN hubs. This setup is actualized on an "Intelligent Personal Digital Assistant (IPDA)." This can keep the validation data of the patients and is arranged with the medicinal server Internet protocol addresses so as to associate the medicinal administrations. It gathers physiological imperative signs from WBANs, whenever if any an unexpected clinical changes in the ongoing persistent wellbeing conditions, for example, changes in oxygen immersion, cardiovascular signs, and so on the transmission of basic information will be organizes lastly.

Also, the IPDA can play out the errand of breaking down the physiological information shrewdly and can likewise complete a nearby thinking to decide client's wellbeing status and give criticism through an easy to understand and intuitive graphical UI in view of information got from medical super sensor. A 3-G correspondence convention and other long range correspondences conventions, for example, GPRS and WWAN are utilized to interface individual servers and the third level together.

So as to improve the general nature of administration for information transmission utilizing intelligent personal digital assistant, regarding dormancy, band-width, and power utilization an isolated administration dependent on two frameworks are appeared.

3.6 Medical Server for Healthcare Monitoring (MSHM)

MSHM obtains the data from individual servers, and it is the foundation of the whole engineering. It is arranged at a social insurance focuses where therapeutic administrations are given like a center, emergency clinic, and so on. MSHM maintains the electronic medicinal records (EMRs) of enlisted patients that is available by various restorative staff in the restorative focus; this incorporates authorities, specialists, and general specialists from their workplaces in the restorative focus through the Web.

In addition, the MSHM additionally gives directions to the patient, for example, recommended prescriptions, practices, and so on.

4 WBAN Network Design Issues

Information transmission unwavering quality and inertness are essential in WBANs which collect non-basic and basic information from the different piece of the human body [2]. The unwavering quality and inertness of WBANs will fundamentally rely upon the structure of the MAC layer and its physical plan. This layer decides the system proficiency and usage issues which fundamentally decide a framework and working expenses of WBANs. The structure of this layer additionally decides the control utilization of WBANs which is a vital plan issue. The physical layer likewise decides the unwavering quality of the WBAN all the while.

4.1 Power Efficiency

Power efficiency is dependably a vital operational issue in WBANs. The power efficiency in WBANs can be upgraded by the PHY (physical) and the MAC layer forms." Macintosh layer presents a lot more elevated amount of intensity sparing by utilizing a few strategies, for example, bundle transmission booking and channel get to systems it executes the utilization a shrewd flagging strategies and an ideal bundle structure. "By choosing suitable balance and coding systems, the PHY layer can increment the likelihood of fruitful transmissions. Overall delays and the power dissipations of WBANs can be decreased through a higher data transmission likelihood.

4.2 *Reliable Design*

The reliable model of WBAN is straightforwardly corresponding to the data transmission delay and the packets loss probabilities. These probabilities are impacted by the BERs of the MAC layer transmission and that of the channel. By utilizing a versatile adaptive modulation and coding strategies which suites the channel conditions in which the transmission happens the PHY layer of a WBAN can diminish the viable BERs of a transmission link. Anyway, the BER can be decreased by executing an advanced forward correcting error. The utilization of this procedure requires transmission of extra bits, but these increment the power spending plan of the WBANs because of the transmission of additional bits. The condition of a system can likewise influence the reliability and power budgets of the WBANs.

4.3 *Scalability*

Scalability is exceptionally fundamental for a patient observing framework, for example, WBAN in light of the fact that it is calm frequently important to change the number of hubs and gather diverse physiological information from the persistent body. At the point, when a WBAN is adaptable, it is simple for human services staffs to include or expel a few hubs without influencing the whole WBAN activity. Since the PHY layers are fixed, the adaptability of WBAN is to a great extent subject to MAC layer; this MAC layer assumes an indispensable job in keeping up unwavering quality under factor transmission and traffic conditions; MAC layer additionally serves to keep up a decent nature of administration.

5 WBAN Routing

Visiting of organized allocation due to its postural flexible dynamicity of sensors that are installed, sensors intensity of low transmitting capacity, misfortune with high expansion on human body, and less dependency from start to finish extending from sink directly from source remains the main characteristics of wireless body area network (WBAN) which drives infrastructure of routing resolution essential. [3] Researched conduction of link layer associated with WBANs having operating frequency 2.4 GHz resulting in the observation that follows:

(i) PDR is definitely affected by environment. More than 70% of networks in a lab settings exhibits PDR 90% or more; almost 50% of networks shows PDR of more than 90% even in open setting (on the rooftop).

(ii) PDR expands in an area which has low multipath by increased control of transmission.

(iii) Increase in average parcel conveyance proportion (PDR) increases with the increment in controlled transmission. The designer eventually came to conclusion that direct symmetry is best in situations with improving surface of intelligent (extra multiple path). While handling with BANs direction, basically, there are two ways to achieve it. In the first technique is to design a directing layer on Macintosh layer, where interface features are estimated which rely on chosen parameters which are considered in computation and other is executing steering capabilities with layers of MAC, with the approach of cross-layer architecture.

The core technique has been observed in [4] which the designer have projected a possible bundle directing resolution, probabilistic routing with postural link cost (PRPLC), in which stochastic network cost is used. Hence, projected convention is in accordance with postural network cost plan using a period changing expenditure that is calculated for each connection that relies on area in accessibility that are examples of connections. Here, resolution uses postural connection expenditure to find out possible transmitting parcels of data.

Youngsters are enlightened by each hub related to their turns for transmitting their respective data. Movement of data occurs one after another in form of cycles: control cycle and data cycle. Information related to the request in transmission is provided to each hub in control cycle. Whenever all hubs receives their control validations, beginning of data cycles takes place. In cycle of information, each plot of datum constitute of main two sections: period of holding and an information period. Time frame of information provides ample amount of space to help hubs for joining a tree. There is a possibility that this can provide dynamic support to system which results in separation of hubs because of postural portability too. Designers' talks exceedingly regarding strength effectiveness of the computation, depending upon the topology organized. As the hubs are required to exhaust energy because of intuning of gear and catching during the control cycle, understandably the tree assumes an important tasks in regulating the convention in strength production.

6 WBANS Security Protocols

It is required of WBAN to execute security tasks with assurance regarding patient's therapeutic information trustworthiness, privacy, and security. While referring to issues concerning to security in WBAN, it should make sure that the movability of Health Insurance and Accountability Act of 1996 (HIPAA) 1996 be maintained.

Secrecy: All the system should be capable to make sure that credibility mystery of message exchanges within the hubs. Information integrity: It is anticipated to maintain modifying of data navigation by using the correspondence ways in all hubs, moreover reply assault to be predicted without delays. Protection: Sensitive information regarding the patient should not be divulge to unauthorized entities.

Data-related medicinal detail remains aloof in most sensitive types of personnel data.

Accessibility: As such kind of system shows exceptionally sensitive, imperious and credibly information related to life-sparing data, it becomes most important that continuous accessibility of system assets to be performed.

Validation: This remains essential to make WBAN more powerful and versatile to validate hubs arrangements and similarly avoid organized bargain.

Different developing security approaches in WBANs are as shown in the following continuation.

6.1 IEEE 802.15.4 Security

Some of defensive case scenarios are realized in IEEE 802.15.4 [5]. It is mainly divided into two important modes: unbound mode and verified mode. No security suite had been selected in unbound mode, while remaining characterized based on security steps they possesses. Following is the IEEE 802.15.4 Security Suites.

6.2 Encryption of Equipment

With off-the-rack ZigBee stage, encryption of equipment is realized in WBAN still every sensor hub equipment lacks equipment encryption supporting. With the use of ChipCon 2420 ZigBee, consistent RF transceiver equipment encryption can be executed. With AES encryption utilizing 128-piece keys for security, CC2420 can easily execute IEEE 802.15.4 tasks. Such security tasks constitute of counter (CTR) mode unscrambling and encryption, CBC-MAC authentication and CCM encryption in addition to confirmation.

6.3 Small Sec

Small Sec remains mainstream in remote sensor network, assortment of custom equipment has also been executed. The product design is based on security that executes encryption of link-layer protocol. It is an essential part of authority Tiny OS operating systems. With a gathering key, regular Small Sec encodes the bundle of information to the sensor hubs and computes a message confirmation code (MAC) for entire parcel that includes header. This gathering key is distributed system wise and physically changed into hubs preceding organization. A lone purpose of defenselessness is shown by these network wide keys. Guarantees against hub catch are not provided by Minor Sec.

6.4 *Cryptography of Elliptic Curve*

Reasonable substitution for open key cryptography in remote sensor systems is presented itself by elliptic bend cryptography (ECC) [13]. The important and main explanation regarding this is its very small size of key and decreased marks and its likewise fast computation. Even though ECC has been efficiently realized in some of varieties, it has not yet reached on top decision related to WBAN. This result is because of its vitality requirements which still remains higher than the frameworks symmetric. In accordance with this, recommendation from many of the researchers is that ECC must be executed only for security-touchy and inconsistent tasks, for instance, key foundation while the main setup of codes or system refreshes.

7 Conclusion

Although in field of medicinal applications and medicinal services, remote innovation is as yet early wireless body area network (WBAN) that ensures to change it into main underlining part in locally confirmed, remotely controlled, patient restoration. It does not costs much and give comparatively better services to human with much more benefits and advantages to each and every one. Business products that are designed and produced efficiently by few organizations to escort and fathom advantageous future of human services existing problems. WBAN provides an unending regulation of patient wellbeing and health; it helps in supporting patients thereby satisfying them as it will definitely allow patients to perform daily tasks life without any hassles and sure of medical caretaking, instead of confining themselves close to specific medical care administrations such as medical clinic centers and intensive care units.

References

1. G.V. Crosby, C.A. Chin, T. Ghosh, R. Murimi, Wireless body area networks in mHealth, in *Mobile Health*, ed. by S. Adibi. Springer Series in Bio-/Neuroinformatics, vol. 5 (Springer, Cham, 2015). https://doi.org/10.1007/978-3-319-12817-7_37
2. R. Chauhan, R. Malik, Research on intelligent routing metric for wireless BAN: a survey. Int. J. Eng. Res. Technol. (IJERT), NCIETM-2017 **5**(11)
3. Q. Fang, S.-Y. Lee, H. Permana, K. Ghorbani, I. Cosic, Developing a wireless implantable bodysensor network in MICS Band. IEEE Trans Inf Technol Biomed **15**(4):567–576 (2011)
4. S. Josephine Selvarani, Online health monitoring system using Zigbee. Int. J. Comput. Sci. Eng. (IJCSE) **3**(4), 1578–1583 (2011)
5. M. Sridhar, et al., Wireless body area networks: requirements, characteristics, design consideration, and challenges, in *Incorporating the Internet of Things in Healthcare Applications and Wearable Devices*, ed. by P.B. Pankajavalli, G.S. Karthick (IGI Global, 2020), pp. 67–85. https://doi.org/10.4018/978-1-7998-1090-2.ch004
6. A. Pantelopoulos, N.G. Bourbakis, A Survey on wearable sensor-based systems for health monitoring and prognosis. IEEE Trans. Syst. Cybern. Part C Appl. Rev. **40**(1), 1–12 (2010)

7. C.-C. Lin, R.-G. Leeb, C.-C. Hsiaoc, A pervasive health monitoring service system based on ubiquitous network technology. Int. J. Med. Inf. **77**, 461–469 (2008)
8. C. Otto, A. Milenković, C. Sanders, E. Jovanov, System architecture of a wireless body area sensor network for ubiquitous health monitoring. J. Mobile Multimed. **1**(4):307–326 (2006)
9. E. Jovanov, Wireless technology and system integration in body area networks for M-Health applications, in *Engineering in Medicine and Biology Society, 2005. IEEEEMBS 2005, Conference Publications*, pp. 7158–7160, 17–18 Jan 2006
10. Royal College of Physicians, National early warning score (NEWS): standardising the assessment of acute illness severity in the NHS. Report of a working party. London, RCP, pp. 1–29 (2012)
11. R.S.H. Istepanian, E. Jovanov, Y.T. Zhang, Guest editorial introduction to the special section M-Health: beyond seamless mobility and global wireless health-care connectivity. IEEE Trans. Inf. Technol. Biomed. **8**(4), 405–414
12. A. Milenkovic, C. Otto, E. Jovanov, Wireless sensor networks for personal health monitoring: issues and an implementation. Comput. Commun. **29**, 2521–2533

Flower Pollination Algorithm for Convolutional Neural Network Training in Vibration Classification

Md. Fadil Md. Esa, Noorfa Haszlinna Mustaffa, Nor Haizan Mohamed Radzi, and Roselina Sallehuddin

Abstract A convolutional neural network (CNN) is among the branches in deep learning (DL) which is a new field of research in machine learning. This model artificially mimics the visual cortex to learn the representation of visuals from low to high levels of abstraction and representation to mining data patterns such as image, sound, text, and signal. Despite the proven CNN advantages in various applications, training this model is challenging especially in complex scenarios. Some methods to optimize CNN training have been proposed, such as stochastic and meta-heuristic algorithms. In this paper, we propose a flower pollination algorithm (FPA) to train CNN. A CWRU bearing dataset is used to ensure the accuracy and efficiency of the proposed method. Moreover, we also compare our proposed method with the original CNN. The results show that the proposed method needs to be refined to achieve the required performance of CNN.

Keywords Deep learning · Optimization · Flower pollination algorithm · Convolution neural network

1 Introduction

Recently, deep learning (DL) has received increased attention from researchers due to rapid advances in computational technology and its ability to outperform other approaches. DL can learn directly from the raw data, and the feature extraction is performed inside the training process. With these advantages, DL is helpful to solve

Md. F. Md. Esa (✉) · N. H. Mustaffa · N. H. M. Radzi · R. Sallehuddin
School of Computing, Faculty of Engineering, Universiti Teknologi Malaysia 81310, UTM Johor Bahru, Johor, Malaysia

N. H. Mustaffa
e-mail: noorfa@utm.my

N. H. M. Radzi
e-mail: haizan@utm.my

R. Sallehuddin
e-mail: roselina@utm.my

A. Kumar et al. (eds.), *Computational Intelligence in Machine Learning*, Lecture Notes in Electrical Engineering 834, https://doi.org/10.1007/978-981-16-8484-5_32

the classification and feature visualization in many complex problems [1]. DL models include deep neural networks (DNNs), convolutional neural networks (CNNs), recurrent neural network (RNNs), Boltzmann machine (RBMs), deep belief networks (DBNs), regularized auto-encoders, and deep Boltzmann machines (DBMs). CNN has shown an excellent performance in image and sound recognition compared with other DL models.

Although DL has an excellent performance in learning tasks in many applications, the training process is very challenging [2]. Training is a process to search the optimized parameter combination based on supplied data into the DL model. Commonly, DL is structured by many hyper-parameter settings and a large number of parameters to be trained. Additionally, with large-scale data, the training phase suffers from a high computational cost and takes more time [3].

There are many successful methods in DL model training such as stochastic gradient descent (SGD), adaptive moment estimation (ADAM), and meta-heuristics. SGD is the widely popular optimization method used in ML [4]. The advantages of the SGD method are its simplicity and fast execution. However, this method involves several hyper-parameter tunings to achieve the optimal parameters for the model. Meta-heuristics have been proven through examining their simplicity, flexibility, deprivation-free mechanisms, and ability to avoid local optimum [5] and to solve complex nonlinear problems. They outperform classical optimization algorithms when dimensions and study size are increased [6]. Meta-heuristics use a set of natural strategies to enhance the efficiency of heuristic procedures such as evolutionary strategy, social behavior, and information transition [7].

Examples of the nature-inspired meta-heuristic-based algorithms are genetic algorithm (GA) from the evolution process of genetics, particle swarm optimization (PSO) from the movement of organisms, ant colony optimization algorithm (ACO) from ant movement during foraging for food, artificial bee colony algorithm (ABC) from the foraging behavior of a honey bee swarm, and flower pollination algorithm (FPA) from the pollination behavior of flowering plants [8]. Meta-heuristic approaches that have been used to train CNN include simulated annealing (SA), differential evolution (DE), and harmony search algorithm (HS) [8], GACNN [9], hybrid gravitational search algorithm [10], and PSO [11]. There are many variants of FPA that have been proposed to solve the various optimization problems including for training a shallow neural network.

Thus, this paper proposes an FPA to train a CNN model in order to optimize the accuracy of bearing fault classification for a CWEU dataset. The goal of the paper is to evaluate FPA's performance and study its properties in CNN training.

The remainder of this paper is organized as follows; Sect. 2 discusses the CNN architecture and optimizer. Then, Sect. 3 illustrates a basic FPA algorithm, and Sect. 4 describes the proposed method. Section 5 presents the experimental results and some discussion. The conclusion is presented in Sect. 6.

2 CNN Architecture

CNN is a multilayer artificial neural network (ANN) that can comprise several types of layers and ending with a classification layer [12]. The structure can be designed in stacked or parallel convolutional block. Principally, each block contains three basic operation types of layers which are the filter, activation, and pooling layers. The last layer is fully connected in a form of multilayer perceptron. Figure 1 shows the basic layer of CNN architecture for one-dimensional data.

The filter or sometimes called the kernel layer is the small size of the input region to combine one neuron in the first hidden layer. It is arranged by sliding throughout the whole input layer either in an overlapped window or not. The output of this process is called a feature map and works as the input layer for the next layer. The activation layer creates a complex mapping between the input and output layers by using mathematical equations such as rectified linear unit (ReLU). Meanwhile, the pooling layer is performed to compress feature representation. Commonly, pooling involves simple statistical functions such as minimum, maximum, or average functions. At the last layer or fully connected layer, all feature maps from the earlier layers will be combined together to complete the CNN architecture [1]. A feature of the structure including size, number, and type function is called as a hyper-parameter of CNN.

2.1 CNN Optimizer

The iterative approach is usually used to train a network to get optimal parameters within the allocated time. The function of the optimizer is to search for combinations

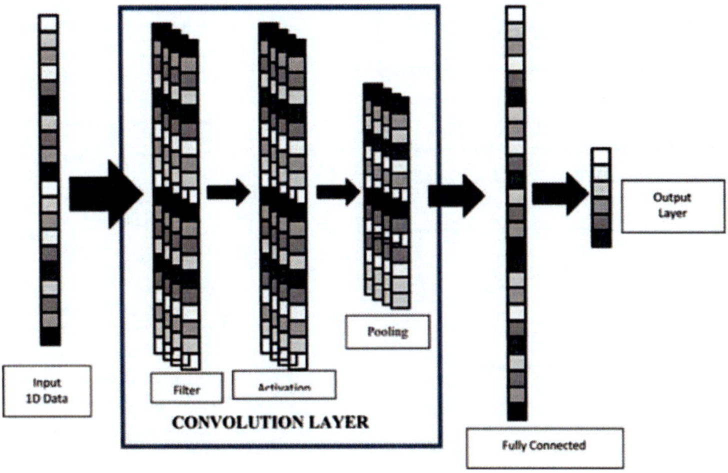

Fig. 1 Basic structure of convolution neural network for 1D data

of parameters, and the same time attempts to optimize the lost function in CNN. Therefore, the objective performance can be converged at global optimum or near optimum of the non-convex plane.

3 Flower Pollination Algorithm

The flower pollination algorithm (FPA) is categorized as a nature-based meta-heuristic algorithm. This algorithm was pioneered by Yang and inspired by the pollination behavior of flowering plants [13]. The idea of this algorithm is to simulate the four pollination processes of a flower in which it searches for the optimal flower combination to maintain the species. Pollination is the process where the male gamete (pollen) is transferred to the stigma, so it can fertilize the female gamete [14].

Yang formulated the four concepts in pollination which are abiotic, biotic, flower constancy, and switching control. The first is abiotic which is the pollination process that not involves the pollinators but other forces of nature such as gravity and diffusion in water or wind. Meanwhile, pollination using pollinators such as birds, insects, and mammals is called biotic. Since this type of pollination can be performed over a longer range, it is considered as global pollination. Flower constancy is a natural process in reproducing a new plant and at the same time maintaining its identity in the same species of plant involving a minimum of two flowers. Lastly, switching is to control the probability of abiotic or biotic processes at one time [13]. Figure 2 shows the flowchart of FPA by Yang.

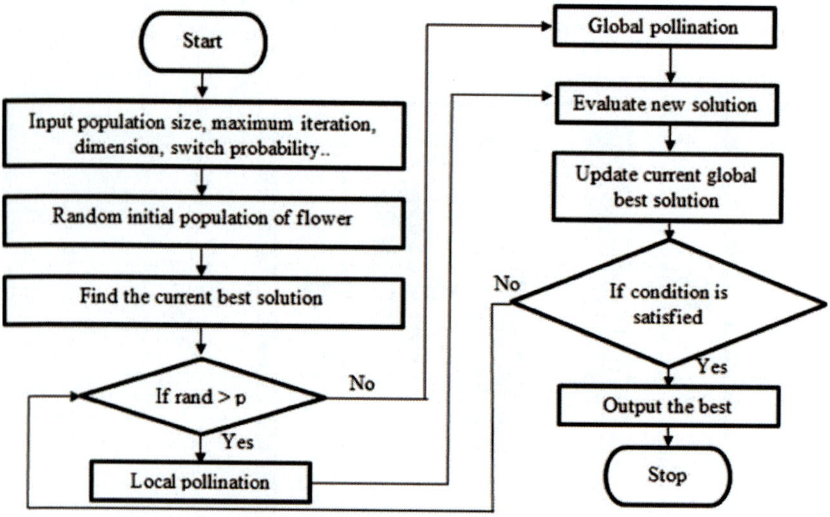

Fig. 2 Flow chart of flower pollination algorithm [13]

4 Proposed Method

In this paper, the flower pollination algorithm is applied to optimize the parameters of CNN. The flowchart of the proposed method called CNN with FPA (CNNFPA) is illustrated in Fig. 3.

Data from the Case Western Reserve University (CWRU) Bearing Data Center [15] is used in this study to validate the proposed method. The dataset is a 1D vibration signal for bearing fault classification problem consisting of ten conditions in four motor loads, two frequencies, three speeds, and four bearing sizes. This study selects the dataset with properties as follows: 48 kHz frequency, 1 hp load, 1772 rpm, and ten vibration signals that represent nine defects and a good condition bearing.

The fixed window technique with a 2048 sized window is implemented which produces 200 sub-signals for each signal. All the generated signals are then divided into three parts with ratios of 70% training and 15% validation. Meanwhile, testing datasets are collected at the original raw data of 15%. Moreover, these settings are run 20 times each.

The architecture of this proposed method refers to a simple five-layer CNN structure with three block convolution layers with ten classed as the output layer. The objective function used is categorical canopy entropy (CCE). The CNN structure and training parameter settings in the experiments are as given in Table 1.

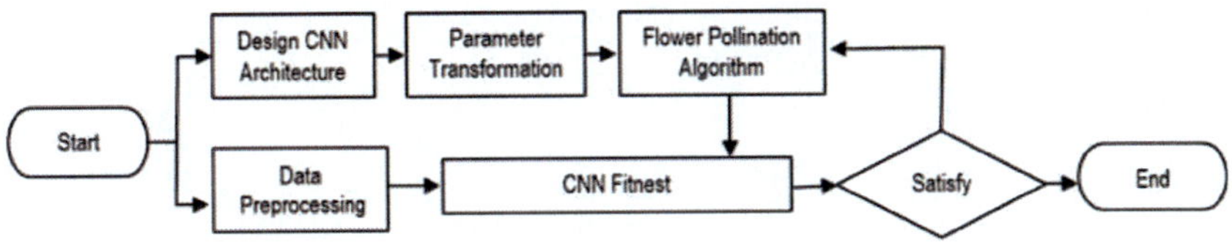

Fig. 3 Flowchart of proposed method (CNNFPA)

Table 1 Structures of the CNN, training, and variables parameters

Layer	Type	Variables and dimensions	Training parameters
1	Convolution	FN = 10; FS = 256;	• ADAM • Minibatch size = 20 • Epochs = 20
2	Pooling	S = 2	
3	Convolution	FN = 15; FS = 32;	
4	Pooling	S = 2	
5	Convolution	FN = 15; FS = 16; B = 15	
6	Fully	ReLU activation function	
7	Softmax	10 outputs	

ADAM Adaptive moment; *No* overlapping filter; *FN* filter number; *FS* filter size; *S* down-sampling rate

5 Simulation and Results

The primary concern of the simulation in this paper is to explore the performance of the CNN training algorithm using FPA. The objective is to minimize the error and maximize the accuracy of classification problem in an acceptable time consumption.

There are three experiments conducted for the CNN and CNNFPA methods using CWRU datasets. The first is CNN using the ADAM method (CNN). The second is CNNFPA by different population sizes, and the third experiment is CNNFPA with different ranges of switch control probabilities. The experimental result for the various population sizes of CNNFPA compared with CNN is given in Table 2.

In this experiment, ADAM is applied to train CNN using 20 epochs. Meanwhile, the CNNFPA uses 200 iterations, $p = 0.2$, lambda $= 0.1$, and gamma $= 0.1$. The results show that CNNFPA is not able to compete with the performance of CNN using the ADAM method even with the maximum number of iterations. The computation time also increases with the increment in population size. Hence, the experiments show that the usage of different population sizes significantly influences the performance of CNNFPA in terms of accuracy and computation time.

Table 2 Comparison of CNN performance result with various CNNFPA population sizes

Algorithm	CNN	CNNFPA				
Population size	N/A	3	5	10	20	30
Accuracy %	1	0.235	0.11	0.11	0.095	0.094
Time (*H:M:S*)	0:1:17	1:44:59	2:47:07	5:36:29	11:17:38	16:50:29

Table 3 Comparison of CNNFPA performance with varied population sizes

Algorithm	CNNFPA		
Switch control (*P*)	Pop size	Accuracy %	Time (H:M)
0.2	3	0.235	1:40
	5	0.11	2:47
	10	0.11	5:36
0.4	3	0.13	1:40
	5	0.29	2:49
	10	0.13	5:39
0.6	3	0.08	1:43
	5	0.11	2:55
	10	0.1	5:44
0.8	3	0.086	1:44
	5	0.11	2:50
	10	0.1	5:40

Furthermore, the experimental result of CNNFPA with different switch controls (P) is given in Table 3. This experiment uses parameter settings of 200 iterations, lambda $= 0.1$, gamma $= 0.1$, and population size $= 3, 5$, and 10.

The result given in Table 3 indicates that generally there are no significant performance changes between different switch controls with the same population size. A different result occurs at $P = 0.4$ and population size $= 5$ due to the random initial value of FPA.

In general, the experimental results show that the accuracy of CNN with ADAM outperforms the CNNFPA method. CNNFPA with a small population size is better than other CNNFPA settings. Moreover, there is no significant difference in accuracy and computation time with different switch control parameter settings. We assume that several factors have contributed in this scenario such as the high-dimensional problem like CNN and the complexity of FPA computation.

6 Conclusion

The objective of our proposed method is to optimize CNN using the flower pollination algorithm. Even though there is an increase in computation time, the classification accuracy from the proposed method is lower than CNN with ADAM for all variations of iterations. This proposed method could potentially be employed if the high dimensionality and computation complexity problems are solved. For future studies, the CNNFPA can be tested with other DL models and compared with other metaheuristic algorithms for DL model such as recurrent neural network or deep belief network.

Acknowledgements This work was supported by Universiti Teknologi Malaysia under the UTM Encouragement Research Grant (UTMER Vot No. 18J50).

References

1. K. Yu, L. Jia, Y. Chen, W. Xu, Deep learning: yesterday, today, and tomorrow. J. Comput. Res. Dev. **50**(9), 1799–1804 (2013)
2. P.O. Glauner, Comparison of training methods for deep neural networks. M.S. thesis (2015)
3. M. Frank, D. Drikakis, V. Charissis, Machine-learning methods for computational science and engineering. Computation **8**(1), 15 (2020)
4. D. Saad, Online algorithms and stochastic approximations. Online Learn. **5**, 6–3 (1998)
5. S.M.J. Jalali, S. Ahmadian, A. Khosravi, S. Mirjalili, M.R. Mahmoudi, S. Nahavandi, Neuroevolution-based autonomous robot navigation: a comparative study. Cogn. Syst. Res. **62**, 35–43 (2020)
6. E. Rashedi, H. Nezamabadi-Pour, S. Saryazdi, GSA: a gravitational search algorithm. Inf. Sci. **179**(13), 2232–2248 (2009)
7. X.S. Yang, Nature-inspired metaheuristic algorithms (Luniver Press, 2010)

8. L.M. Rere, M.I. Fanany, A.M. Arymurthy, Metaheuristic algorithms for convolution neural network. Comput. Intell. Neurosci. **2016** (2016)
9. P. Esfahanian, M. Akhavan, Gacnn: Training deep convolutional neural networks with genetic algorithm (2019). arXiv preprint arXiv:1909.13354
10. K. Kang, C. Bae, H.W.F. Yeung, Y.Y. Chung, A hybrid gravitational search algorithm with swarm intelligence and deep convolutional feature for object tracking optimization. Appl. Soft Comput. **66**, 319–329 (2018)
11. M.H. Khalifa, M. Ammar, W. Ouarda, A.M. Alimi particle swarm optimization for deep learning of convolution neural network, in *2017 Sudan Conference on Computer Science and Information Technology (SCCSIT)*. IEEE, pp. 1–5 (2017)
12. A. Lopez-Rincon, A. Tonda, M. Elati, O. Schwander, B. Piwowarski, P. Gallinari, Evolutionary optimization of convolutional neural networks for cancer miRNA biomarkers classification. Appl. Soft. Comput. **65**, 91–100 (2018)
13. X.-S. Yang, Flower pollination algorithm for global optimization. Unconvent. Comput. Nat. Comput. **7445**, 240–249 (2012)
14. R. Frankel, E. Galun, Pollination Mechanisms, Reproduction and Plant Breeding, vol. 2 (Springer Science & Business Media, 2012)
15. X. Lou, K.A. Loparo, Bearing fault diagnosis based on wavelet transform and fuzzy inference. Mech. Syst. Signal Process. **18**, 1077–1095 (2004)

Online Logistics Booking System Utilizing Artificial Intelligence

Hisham Mohamad Malik Ghazal, Shafishuhaza Sahlan, Hazlina Selamat, and Fatimah Sham Ismail

Abstract In this paper, the application of genetic algorithm (GA) in linear and non-linear dynamic modeling, as well as the development of an alternative GA-based model structure selection algorithm is presented. As a benchmark for the proposed algorithm, orthogonal least square (OLS), a gradient descent method, is utilized. To reduce problems of premature convergence in simple GA (SGA), a model structure selection based on modified genetic algorithm (MGA) is proposed. The effect of different combinations of MGA operators on the performance of the developed model was studied, and the effectiveness and shortcomings of MGA were highlighted. Results were compared between SGA, MGA, and OLS. From the results obtained, it was observed that with similar number of dynamic terms, MGA performs better than SGA, in terms of exploring potential solution, meanwhile outperformed the OLS algorithm in terms of selected number of terms and predictive accuracy. In addition, for fine-tuning, the use of local search of memetic algorithm (MA) with MGA is also proposed and investigated. From the results obtained, it can be observed that MA can produce an adequate which satisfy the model validation tests with significant advantages over OLS, SGA, and MGA methods.

Keywords Genetic algorithm · Orthogonal least square · Modified genetic algorithm · Simple genetic algorithm · Artificial intelligence

1 Introduction

In recent times, whenever innovation or the future of technology is mentioned, the first term that comes to mind is artificial intelligence (AI). Artificial intelligence is a broad branch of computer science concerned with programming machines in a way to simulate human thinking and mimic their actions. There are several types of artificial intelligence where instead of the machine being programmed on what to think or simulate, the machines would observe, analyze, and learn from the information and

H. M. M. Ghazal · S. Sahlan (✉) · H. Selamat · F. S. Ismail
Faculty of Engineering, School of Electrical Engineering, Universiti Teknologi Malaysia, UTM Skudai, 81310 Johor, Malaysia
e-mail: shafis@utm.my

A. Kumar et al. (eds.), *Computational Intelligence in Machine Learning*, Lecture Notes in Electrical Engineering 834, https://doi.org/10.1007/978-981-16-8484-5_33

mistakes just the same way the human brain would. This innovative technology is affecting consumer products and has taken us to significant breakthroughs in physics and healthcare as well as altered industries as diverse as manufacturing, finance, and retail. We could be decades away from generalized AI, where a machine can do anything a human brain could, but AI in its current state is an important and essential part of our world.

This paper focuses on logistics operation [1], which is the operation of planning and executing an efficient shipment and storage of cargo from an origin point to a desired destination. Logistics operations in its traditional methods are a lengthy and tedious process. Conventionally, to discuss the available options for logistic bookings, phone calls between users and staffs managing the schedules are required as well as getting in contact with the designated deliverymen. The current process, however, is difficult to optimize in meeting customers' requirements in a timely and cost-effective method. The technology involved is the methods of data maintenance and how to coordinate the work processes [2]. To improve the performance of the system and the experience for the user through the conventional methods is intractable and time consuming which means, it is not effective to achieve optimum time and cost efficiency. This is where data management and AI algorithms [1, 3] come into hand, for the aid in such complex processes is essential to ease the process.

Therefore, in this paper, to manage logistic bookings, a Website is developed, utilized as a medium between users and deliverymen by manipulating datasets within the database, along with developing an AI algorithm for simulating and calculating booking rates, eliminating the need of a middle-man staff to handle the bookings, and simplifying the booking process for the user [4]. The motivation of this paper is to explore the complex branch of AI through its application to an online logistics booking system, improving the existing system.

2 Artificial Intelligence

Artificial intelligence follows the concept that human intelligence could be executed in a way which a machine could easily mimic and execute tasks, ranging from straightforward to the ones which are even more complex. The aims of AI incorporate learning, reasoning, and perception [5, 6]. AI can be categorized into two categorize which are the weak and the strong AI systems. The weak AI is a system designed to carry out one specific job such as video games like chess or even handling seven personal assistants such as Apple's Siri and Google assistant. Meanwhile, the strong AI system carries out more complex tasks that tend to be considered as humanlike such as programmed to control situations where they may be needed to problem solve without the intervention of human. An example of such systems could be seen in self-driving cars or in hospital operating rooms. In this research, the use of Python language is utilized. The three common types of AI algorithms include.

1. Supervised Learning: This algorithm consists of an outcome variable (dependent variable) which is to be predicted from a given set of predictors (independent variables). Some examples of supervised learning are linear regression, decision tree, random forest, and KNN.
2. Unsupervised Learning: This algorithm does not have an outcome variable to predict or estimate. It is utilized for clustering population in different groups. Some examples of unsupervised learning are Apriori algorithm and K-means.
3. Reinforcement Learning: In this algorithm, the machine is trained to make specific decisions. It works in a way where the machine is exposed to an environment, where it trains itself continually using trial and error. This machine learns from past experience and tries to capture the best possible knowledge to make accurate business decisions. An example of reinforced learning is Markov decision process.

2.1 Python

Due to its simplicity, Python is reviewed as the top tier language in AI. Among the apparent advantages are Python is used friendly and easy to learn, cross-platform is supported, there exist a huge community support, they are extremely powerful and of course, it is an open-source language.

Python accommodates a vast quantity of libraries. Numerous of those are designed for artificial intelligence and machine learning. Python can be used in MATLAB. The advantages of using Python over MATLAB would be that the code is more compact and simpler to read. Unlike MATLAB's one-base indexing, Python utilizes zero-based indexing. The object-oriented programming in Python is much more flexible as compared to MATLAB. Python is free and open-source, and the developers of language encourage users to give in suggestions for the software Python also offers a greater set of options in graphics package and toolsets. All in all, Python has been rapidly obtaining momentum as being the preferred programming language for beginner users working on AI project, hence utilized in this project.

3 Web Development

Python Web development is one of the main use cases of Python language. It is applicable for the backend development only, meaning that it will connect to the database, and send or retrieve data from the server. It builds the logic behind the procedures that happen on the backend of an application. Meanwhile, conventionally, Web application's frontend is constructed using HTML or JavaScript. An integrated development environment (IDE) is a software which provides comprehensive facilities to computer programmers for software development. Visual Studio Code by Microsoft is one of the most popular IDEs used for Python Web development.

A Web framework is a code library which fastens and eases the Web development process by providing customary patterns for constructing reliable, scalable, and maintainable Web applications. Frameworks supply functionality in their code or through extensions to carry out customary operations which are required in order to run a Web application. Some examples of these operations could be URL routing, input form handling and validation, database connections configuration, and persistent data manipulations through an object-relational mapper (ORM), session storage and retrieval, and much more. Python offers a range of open-source Web frameworks for its users to suit their needs. Django and flask are the most highly favored Python Web development frameworks by the Python community [5, 7, 8].

3.1 Methodology

Linear regression is utilized to proximate real values such as costs, number of phone calls, or business's total sales, based on continuous variables. The algorithm works by constructing a relationship between independent and dependent variables by fitting a best line, also known as regression line which is represented by a linear equation. In the Web development, a multiple linear regression AI algorithm is designed using a dataset of 40 different logistic booking information. Each row of data contains five independent variables which are height, width, length, weight, and distance, in order to find the one dependent variable and price. The booking information dataset utilized for this project is given in Fig. 1.

To explain the processes of each used in the system, a use case diagram is utilized, as shown in Fig. 2.

The use case diagram is composed of four actors, which are anonymous unregistered used, user, deliverymen, and admin, with the use case description given in Table 1.

3.2 Webpage

In this section, the utilization of Python coding for the logistic Website is explained and demonstrated. As a start, the user is redirected to main Webpage for the proposed automated system in order to sign up or login to continue.

The sign-up page is constructed using HTML frontend and Python backend. Once landed on this page, users need to choose either *customer* or *deliverymen*, as shown in Fig. 3 with the function source code in Fig. 4. At this page, Python code is utilized. For the rest of the steps, the Python source code is not provided, due to limitation of space

Following the sign-up process, the user needs to log in. When logged in as customer, the user will be directed to the customer dashboard in Fig. 5.

Booking	Date	Height	Width	Length	Weight	Distance	Price
1	1/1/2020	8	25	35	2	10	8
2	13/1/2020	16	35	45	10	100	40
3	26/1/2020	4	15	25	1.5	10	6.5
4	5/2/2020	6	20	30	2.5	30	12
5	16/2/2020	8	25	35	3	20	11
6	4/3/2020	16	25	45	8	10	20
7	6/3/2020	8	35	35	2	35	13
8	11/3/2020	4	15	25	2	45	14.5
9	27/3/2020	8	25	35	2.5	60	20
10	2/4/2020	8	20	35	3.5	15	12
11	6/4/2020	8	20	35	3	25	12
12	9/4/2020	8	25	35	3	70	25
13	15/4/2020	8	20	35	2	80	10
14	29/4/2020	16	35	45	4	95	17
15	3/5/2020	8	35	30	5	30	20
16	8/5/2020	16	25	45	5	30	20
17	11/5/2020	16	35	45	8	25	30
18	18/5/2020	8	20	45	3	45	10
19	23/5/2020	8	20	45	2.5	30	15
20	1/6/2020	4	15	25	2.5	15	10
21	5/6/2020	16	35	45	3	10	9
22	15/6/2020	6	20	30	3	10	9
23	24/6/2020	8	25	35	4	10	11
24	30/6/2020	6	25	30	3.5	65	20
25	1/7/2020	16	35	45	2.5	90	20
26	15/7/2020	8	20	35	5	60	28
27	18/7/2020	6	20	30	4.5	45	19
28	21/7/2020	16	25	25	9	30	30
29	29/7/2020	16	35	45	8.5	30	29
30	1/8/2020	8	20	35	6	30	20
31	15/8/2020	4	15	25	3	35	14
32	20/9/2020	16	25	25	3.5	60	22
33	23/9/2020	16	35	45	8	65	36
34	11/10/2020	6	20	30	4	15	13
35	30/10/2020	8	20	35	3	80	13
36	2/11/2020	16	25	35	7	75	37
37	11/11/2020	4	15	20	2	30	11
38	5/12/2020	16	35	35	5.5	25	22
39	27/12/2020	8	25	35	4.5	20	19
40	1/1/2021	16	25	45	6	20	20.5

Fig. 1 Bookings information dataset utilized for this project

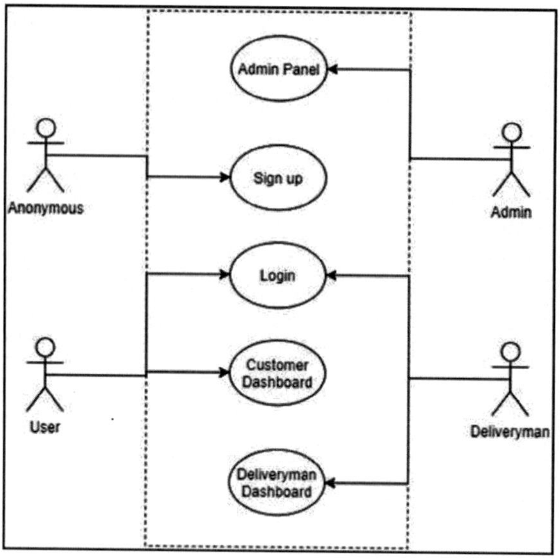

Fig. 2 Use case diagram for the research project

Table 1 Use case description for the project

Use case	Description
Admin panel	This use case is only used by the admin. It provides access to the Web application's database, where the information of registered users and booking orders is found
Sign up	This use case is only used by an anonymous. It provides the facilities for an unregistered user to register and create an account for the Web application
Login	This use case is used by both users and deliverymen to login to the Web application
Customer dashboard	This use case is used to navigate through user's options. Where users have the ability to place an order, or check ongoing order or check order history
Deliveryman dashboard	This use case is used to navigate through deliverymen options. Where deliverymen have the ability to pick up an order from the order list or to deliver the picked order

Similar as the sign-up page, the login page and the customer dashboard are run using Python code. When users entered wrong credentials, the code detected it and the Web App displays a message "*invalid credentials,*" and if the user is an existing user, their details will be displayed on the dashboard.

When placing order into the system, the Python code was programmed that if an ongoing order has not yet completed and attempts to place new orders will not be

Fig. 3 Sign-up page

successful, which is displayed by an error message. Once a deliveryman receives and picks up the customer's order, the deliveryman's information is then displayed on the ongoing page with his or her details for the customer.

Once the delivery has been completed, the customer's ongoing page will be cleared hence a new order can be placed.

On the other hand, when logged in as *deliverymen*, a different dashboard will appear, as shown in Fig. 6.

On the deliveryman dashboard, the list of deliveries is listed. The deliveryman is able to navigate through the dashboard, browsing through the available options, checking the order list, either they want to pick an order, or complete a delivery at the picked order page. The system is programmed such that the deliveryman is not able to pick another order until the current delivery is delivered. Once a delivery is completed, a message will appear on the dashboard, hence notifying the deliveryman. If the deliveryman attempts to pick up a completed order, an error message will be displayed.

4 AI Algorithm Evaluated

In this section, utilization of the proposed AI algorithm to estimate the optimum pricing for the delivery service is evaluated. To do this, 80% of the dataset is utilized for training meanwhile the remaining 20% is used for testing. The average error is calculated using root mean square error (RMS) is 4.14043, which indicates that the AI algorithm has an average error margin of positive or negative RMSE values.

```python
def signup(request):
    if request.method == 'POST':
        name = request.POST['name']
        phone = request.POST['phone'].replace(" ","")
        password1 = request.POST['password1'].replace(" ","")
        password2 = request.POST['password2'].replace(" ","")
        usertype = request.POST['usertype'].replace(" ","")
        email = request.POST['email'].replace(" ","")
        address = request.POST['address']

        register = Register()

        if str(password1) == str(password2):
            register.name = name
            register.phone = phone
            register.password = password1
            register.usertype = usertype
            register.email = email
            register.address = address
            register.save()

            return redirect("/login")
        else:
            return redirect("/signup")

    return render(request, "signup.html")
```

Fig. 4 Python source code

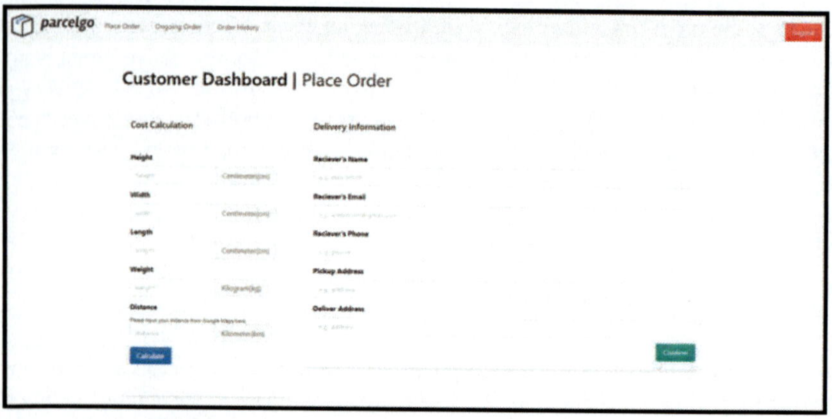

Fig. 5 Customer dashboard

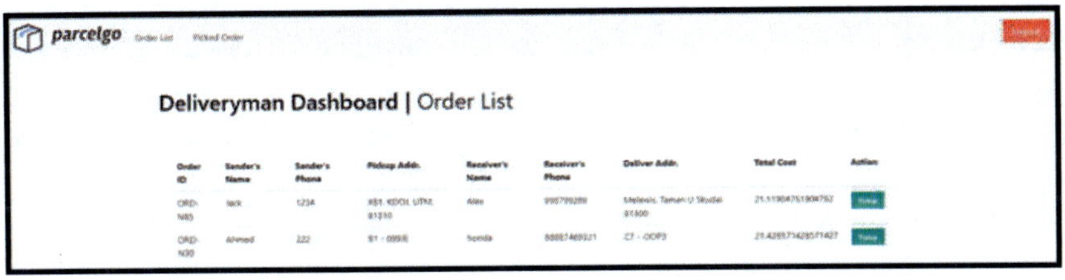

Fig. 6 Deliveryman dashboard

	Booking	Height	Width	Length	Weight	Distance	Price
count	40.000000	40.000000	40.000000	40.000000	40.000000	40.000000	40.000000
mean	20.500000	10.050000	24.750000	35.375000	4.262500	39.625000	18.212500
std	11.690452	4.607408	6.788943	7.458217	2.238897	26.296814	8.394051
min	1.000000	4.000000	15.000000	20.000000	1.500000	10.000000	6.500000
25%	10.750000	7.500000	20.000000	30.000000	2.500000	20.000000	11.750000
50%	20.500000	8.000000	25.000000	35.000000	3.500000	30.000000	18.000000
75%	30.250000	16.000000	27.500000	45.000000	5.125000	60.000000	20.875000
max	40.000000	16.000000	35.000000	45.000000	10.000000	100.000000	40.000000

Fig. 7 Finding the actual price mean value

Comparing the RMSE value to the price mean valued at 18.2125, as shown in Fig. 7, it gives context that the price estimate RMSE is around 20% of the actual price mean. This error margin percentage is not a devastating value yet not an optimum value.

In addition to the customer and the deliveryman database, access by admin to the Webserver is also available, along with admin dashboard. In the dashboard, there are registered user list, registered user entity, booking order list, booking order entity as well as updated delivery status.

5 Conclusion

In this paper, AI algorithm, Python in particular, is utilized to assist in logistic booking and management. The Web application serves as a medium between users and deliverymen, supervised by an admin, constructed by Python as the backend on manipulating the datasets in the database. The utilization of multiple linear regression supervised learning algorithm shown its effectiveness in this project which signifies further utilization in a variety of application.

Acknowledgements The authors would like to convey their deepest gratitude to CAIRO UTM for the realization of this paper.

References

1. A. Pasumpon Pandian, Artificial intelligence application in smart warehousing environment for automated logistics. J. Artif. Intell. Capsule Netw. **01**(02), 63–72 (2019)
2. A.S. Tomar, M. Singh, G. Sharma, K.V. Arya, Traffic Management using Logistic Regression with Fuzzy Logic. Procedia Comput. Sci. **132**, 451–560 (2018). https://doi.org/10.1016/j.procs.2018.05.159
3. M. Christopher, *Logistics & Supply Chain Management*, 4th edn (Prentice Hall Financial Times, 2011)
4. P. Charoenporn, Smart logistic system by IOT technology, in *Proceedings of the 6th International Conference on Information and Education Technology*, pp. 149–153 (2018). https://doi.org/10.1145/3178158.3178186
5. G. Kumar, H. Saini, Novel noncommutative cryptography scheme using extra special group. Secur. Commun. Netw. **2017**(5), 1–21 (2017)
6. H. Kaur, H. Nori, S. Jenkins, R. Caruana, H. Wallach, J.W. Vaughan, Interpreting interpretability: understanding data scientists' use of interpretability tools for machine learning, in *Proceedings of the 2020 CHI Conference on Human Factors in Computing Science* (2020), pp. 1–14. https://doi.org/10.1145/3313831.3376219
7. W. Lee, Python® machine learning, in *Python® Machine Learning*, 1st edn. (Wiley, 2019). https://doi.org/10.1002/9781119557500
8. G. Kumar, H. Saini, Secure and robust telemedicine using ECC on Radix-8 with formal verification. Int. J. Inf. Secur. Privacy **12**(1), 13–28 (2018)

Fuzzy Logic Control for Quadrotor Micro-aerial Vehicle Altitude and Attitude Stabilization

Aminurrashid Noordin, Mohd Ariffanan Mohd Basri, and Zaharuddin Mohamed

Abstract This paper presents a fuzzy logic controller (FLC) for altitude and attitude stabilization of a quadrotor micro-aerial vehicle (MAV). The MAV is a small vehicle that weighs less than 0.1 kg; therefore, a slight perturbation will affect its performance. Hence, for the actuated dynamics, roll (ϕ), pitch (θ), yaw (ψ), and altitude (z) stabilization, a FLC scheme is proposed. A 3×3 heuristics rules is used with error and derivative error as the inputs. In addition, five memberships function is created for the output comprise of triangles type and sigmoidal type. In this Mamdani-model, centroid is chosen as the defuzzification process. All individual gains for the FLC are tuned manually until achieving the desired responses. The Newton–Euler model of quadrotor is simulated using Simulink with a slight force perturbation which is applied on the altitude (z) to investigate the system performances. The simulation result shows that the flight control scheme provides good performance in the presence of perturbation.

Keywords UAV · Quadrotor · Fuzzy logic control · Micro-aerial vehicle · Perturbation

1 Introduction

Quadrotor has four rotors arranged symmetrically, with two diagonal motors (1 and 2) running in the same direction and the others (3 and 4) running in the opposite direction to eliminate anti-torque. With ability to perform vertical take-off and landing (VTOL) and omni-directional flying, quadrotor applications become more diverse, including civil, commercial, meteorological observation, agricultural, and

A. Noordin · M. A. M. Basri (✉) · Z. Mohamed
Faculty of Engineering, School of Electrical Engineering, Universiti Teknologi Malaysia, UTM Johor Bahru, 81310 Johor, Malaysia
e-mail: ariffanan@fke.utm.my

A. Noordin
Faculty of Electrical and Electronic Engineering Technology, Universiti Teknikal Malaysia Melaka, Hang Tuah Jaya, 76100 Durian Tunggal, Melaka, Malaysia

A. Kumar et al. (eds.), *Computational Intelligence in Machine Learning*, Lecture Notes in Electrical Engineering 834, https://doi.org/10.1007/978-981-16-8484-5_34

military purposes such as surveillance, recognition, and communication, as well as educational purposes [1].

To meet the growing demands on quadrotor efficiency, a number of advanced control techniques have recently been adopted, such as sliding mode control [2, 3], feedback linearization [4, 5], and backstepping technique [6, 7]. Investigation of a standard quadrotor flight control system using linear controllers such as proportional-integral-derivative (PID) and linear quadratic regulators (LQR) has been reported in [7, 8].

Fuzzy logic controller (FLC) is one of the most popular strategies for addressing nonlinear dynamic uncertainties. This modern controller has been reported in [9, 10] as it can effectively deal with highly nonlinear systems. This paper focus on fuzzy logic controller [11] for altitude and attitude stabilization of micro-aerial vehicle (MAV). In addition, a slight perturbation was applied to observe the robustness of this controller during hovering.

2 Quadrotor Systems Modeling

Quadrotor MAVs are 6-DOF with four rotors which in either +-configuration or X-configuration. Using Newton–Euler formulation, the mathematical modeling of the quadrotor is divided into two parts. Equation (1) represent the dynamics model of the quadrotor

$$\begin{aligned}
m\ddot{x} &= U_1(S\psi S\phi + C\psi S\theta C\phi)\\
m\ddot{y} &= U_1(-C\psi S\phi + S\psi S\theta C\phi)\\
m\ddot{z} &= mg + U_1(C\theta C\phi)\\
I_{xx}\ddot{\phi} &= \left(I_{yy} - I_{zz}\right)\dot{\psi}\dot{\theta} - (J_r\Omega_d)\dot{\theta} + lU_2\\
I_{yy}\ddot{\theta} &= (I_{zz} - I_{xx})\dot{\psi}\dot{\phi} + (J_r\Omega_d)\dot{\phi} + lU_3\\
I_{zz}\ddot{\psi} &= \left(I_{xx} - I_{yy}\right)\dot{\theta}\dot{\phi} + U_4
\end{aligned} \tag{1}$$

where m is the mass, g is the gravitational coefficient, l is the lateral arm length, and U_1 is the total thrust force generated by four rotors. I is the moments of inertia for the quadrotor, the diagonal matrix 3-by-3 is defined as $I =$ diagonal $\left[I_{xx} I_{yy} I_{zz}\right]^{\mathrm{T}}$, J_r is the rotor inertia, and Ω_d is total rotor speeds generated from the two pairs of rotors. U_2, U_3, and U_4 are the total torque.

3 Fuzzy Logic Control

Fuzzy logic is designed to control MAV during hovering and attitude stabilization. Here, four variables state z for altitude, roll (ϕ), pitch (θ), and yaw (ψ) for attitude are used as feedback response, and four dedicated FLC have been designed to control each state. The errors and its derivatives are used as fuzzy controller inputs to generate require outputs to stabilize MAV during hovering and stabilized the attitude. In addition, an integral feedforward is applied to improve steady state of the response signal. The controller architecture used in this study is illustrated as Fig. 1.

In this Mamdani fuzzy model, three memberships' function is created for error (e) and derivative error (de). The triangle shape, spline-based S-shaped, a spline-based Z-shaped, and Gaussian membership function are used accordingly. The error and derivate error are set to normalize at range [−2 2]. For the output, five memberships function is created to cover more areas during centroid methods' defuzzification. These fuzzy memberships function for the output are mixed with triangles type and sigmoidal membership functions. The range for the output is set normalize at [−10 10]. The linguistic variables used for error and derivative error are NEGATIVE/ZERO/POSITIVE, while for the output SLOWER_DOWN, FAST_DOWN, HOVER, SLOWER_UP, and FAST_UP. Figure 2 shows the membership function developed for the error, derivative error, and output, respectively.

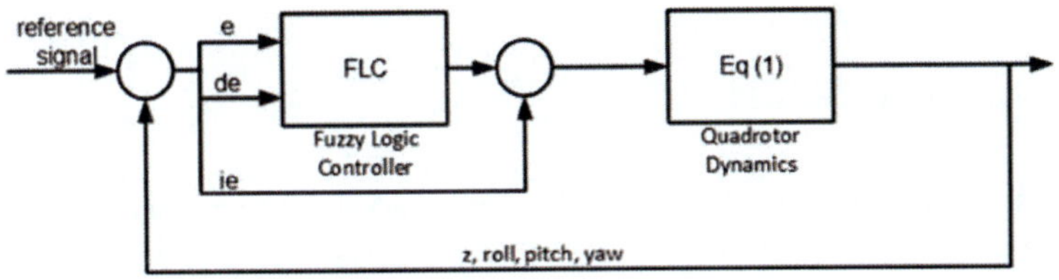

Fig. 1 Fuzzy logic control plus integral feed-forward

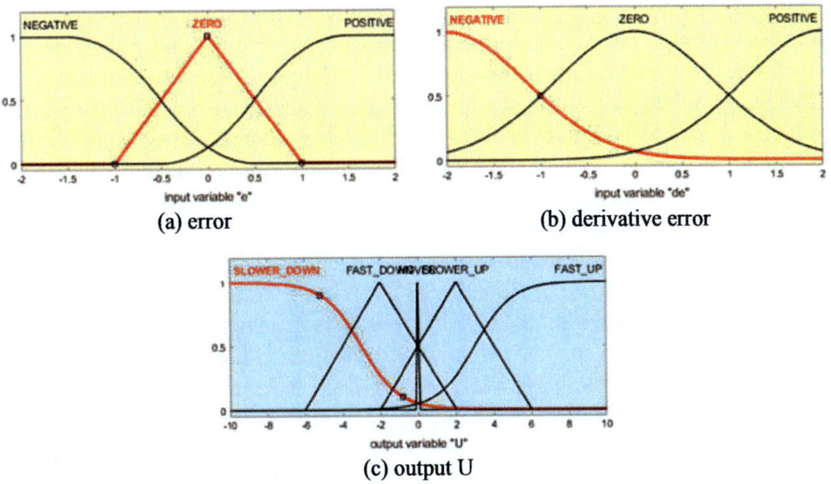

(a) error

(b) derivative error

(c) output U

Fig. 2 Membership function for **a** error, **b** derivative error, and **c** output U

Table 1 3× 3 fuzzy rules

de	e		
	Negative	Zero	Positive
NEGATIVE	FAST_DOWN (0.5) SLOWER_DOWN (0.5)	SLOWER_DOWN (1)	FAST_DOWN (0.5) HOVER (0.5)
ZERO	SLOWER_DOWN (0.5)	HOVER (1)	SLOWER_UP(0.5)
POSITIVE	HOVER (0.5) FAST_UP (0.5)	SLOWER_UP (1)	FAST_UP(1)

The 3 × 3 rules are created based on heuristics [12] using IF–THEN statements and combined with logical AND. For instance, IF error is "NEGATIVE," AND derivative error is "NEGATIVE," THEN output is FAST_DOWN (0.5). The number in the bracket is a weightage set for the fuzzy engine to determine probability of which rules need to be active in case ad there are two set of rules which are developed as given in Table 1.

4 Simulation Results

The quadrotor model is simulated in MATLAB Simulink. The parameters value used for quadrotor's dynamic is referred in [12]. A few tests were conducted in order to evaluate the performance of the designed fuzzy logic. Integral absolute error (IAE) is chosen as the index performance.

Initially, the quadrotor is set at $[0, 0.1, 0.1, 0.1]$ and settling at $[-2, 0, 0, 0]$ for z, ϕ, θ, and ψ, respectively. The gains were manually tuned to obtain satisfactory results and settling times. The simulation results presented in Figs. 3 and 4 demonstrate the satisfactory performance of the proposed FLC system. The proportional gain, derivative gain, integral gain, and output gain are manually tuned until the responses of proposed fuzzy logic control converges to the desired values. As normal PID gains, the proportional gains increase rise time, the derivative gains reduce the oscillation, and the integral gains will improve steady-state value of response signal. In the simulation, for the altitude control, the settling time, Ts, to reach 2 m is about 4 s, with no overshoot and undershoot occurs.

In the attitude control, the settling time, T_s, is about 1.00 ms, 2.00 ms, and 0.64 ms for roll, pitch, and yaw, respectively, with very least steady-state error. The integral absolute error (IAE) for the case without perturbation on the altitude z, is 4.6 and about 0.07 for the attitude, ϕ, θ, and ψ respectively. Meanwhile, for the case with perturbation as shown in Fig. 4, the IAE increase to 5.4 with a small steady-state error for the altitude but maintain the attitude. This shows the good performance of FLC during handling slightly force for small-scale MAV (less than 0.1 kg).

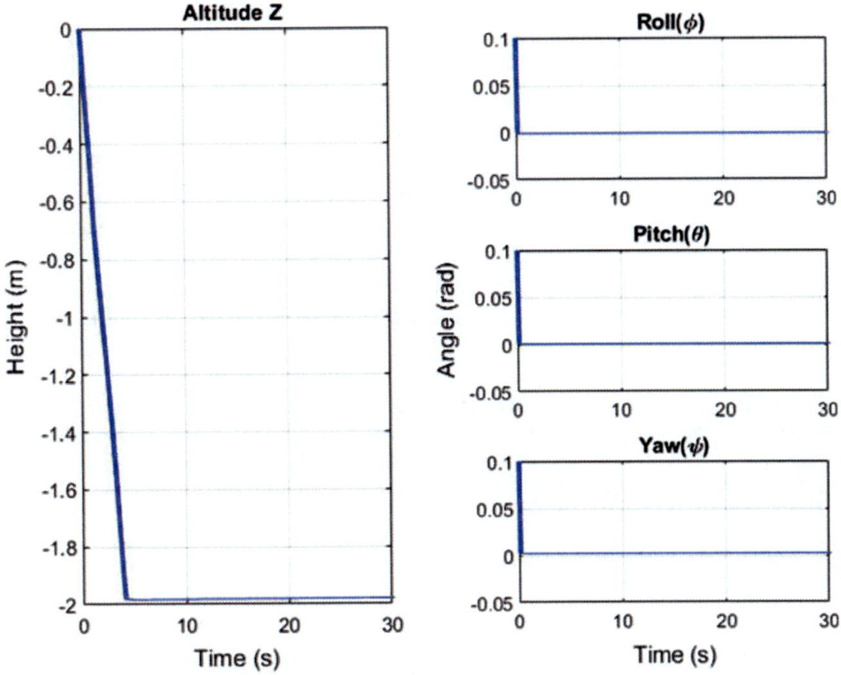

Fig. 3 Altitude and attitude control using proposed control

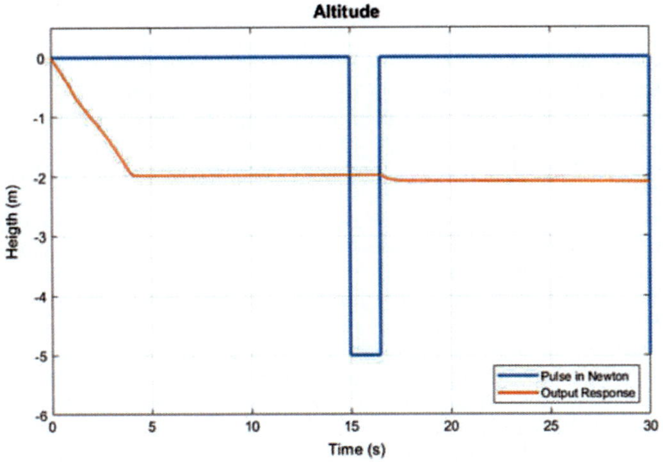

Fig. 4 Altitude with a slightly perturbation using proposed control

5 Conclusion

In this paper, a control law using fuzzy logic controller with 3×3 memberships function is proposed to control quadrotor MAV altitude (z) and the attitude (ϕ, θ, ψ) stabilization. The simulation of the proposed fuzzy logic controller system was conducted in a MATLAB Simulink environment, where the results demonstrated an acceptable performance without or with perturbation (slightly force) applied on altitude z. In the future, genetic algorithm or particle swarm algorithm can be applied to optimize the controller performance. Furthermore, investigation on position controller using proposed control scheme can be investigated to have completed control of 6-DOF quadrotor.

Acknowledgements The authors would like to thank Universiti Teknologi Malaysia (UTM) under the Research University Grant (R.J130000.2651.17J42), Universiti Teknikal Malaysia Melaka (UTeM), and Ministry of Education Malaysia for supporting this research.

References

1. J. Števek, M. Fikar, Teaching aids for laboratory experiments with AR.Drone2 Quadrotor. IFAC-PapersOnLine **49**, 236–241 (2016). https://doi.org/10.1016/j.ifacol.2016.07.183
2. A. Noordin, M.A.M. Basri, Z. Mohamed, Sliding mode control for altitude and attitude stabilization of quadrotor UAV with external disturbance. Indones J. Electr. Eng. Inf. **7**, 203–210 (2019). https://doi.org/10.11591/ijeei.v7i2.1149
3. W. Alqaisi, B. Brahmi, J. Ghommam, et al., Sliding mode controller and hierarchical perturbation compensator in a UAV quadrotor, in *CIVEMSA 2018—2018 IEEE International Conference on Computational Intelligence and Virtual Environments for Measurement Systems and Applications, Proceedings*, pp. 4–9 (2018). https://doi.org/10.1109/CIVEMSA.2018.8440003
4. I.M. Lazim, A. Rashid Husain, N. Adilla Mohd Subha, M. Ariffanan Mohd Basri, Intelligent observer-based feedback linearization for autonomous quadrotor control. Int. J. Eng. Technol. **7**, 904 (2018). https://doi.org/10.14419/ijet.v7i4.35.26280
5. A. Mokhtari, N.K. M'Sirdi, K. Meghriche, A. Belaidi, Feedback linearization and linear observer for a quadrotor unmanned aerial vehicle. Adv. Robot **20**, 71–91 (2006). https://doi.org/10.1163/156855306775275495
6. M.A. Mohd Basri, A.R. Husain, K. Danapalasingam, Intelligent adaptive backstepping control for MIMO uncertain non-linear quadrotor helicopter systems. Trans. Inst. Meas. Control **37**, 345–361 (2015). https://doi.org/10.1177/0142331214538900
7. Y. Kartal, P. Kolaric, V. Lopez et al., Backstepping approach for design of PID controller with guaranteed performance for micro-air UAV. Control Theory Technol. (2019). https://doi.org/10.1007/s11768-020-9145-y
8. S. Khatoon, D. Gupta, L.K. Das, PID & LQR control for a quadrotor: modeling and simulation, in International Conference on Advances in Computing, Communications and Informatics, *ICACCI 2014*, pp. 796–802 (2014). https://doi.org/10.1109/ICACCI.2014.6968232
9. K.C. Vijaykumar Sureshkumar, Intelligent fuzzy flight control of an autonomous. AIAA SciTech 1–10 (2014). https://doi.org/10.2514/6.2014-0992
10. M. Santos, V. López, Intelligent fuzzy controller of a quadrotor, in *2010 IEEE International Conference on Intelligent Systems and Knowledge Engineering* (2010), pp. 0–5. https://doi.org/10.1109/ISKE.2010.5680812

11. T. Brehm, K.S. Rattan, Hybrid fuzzy logic PID controller, in *Proceedings of 1994 IEEE 3rd International Fuzzy Systems Conference* (1993), pp. 1682–1687. https://doi.org/10.1109/FUZZY.1994.343602
12. A. Noordin, M.A.M. Basri, Z. Mohamed. Simulation and experimental study on PID control of a quadrotor MAV with perturbation. Bull. Electr. Eng. Inf. **9**, 1811–1818 (2020). https://doi.org/10.11591/eei.v9i5.2158

Deep Learning Based Model for COVID-19 Pneumonia Prediction with Pulmonary CT Images

Tanzida Begum Tani, Tamima Afroz, and Md. Khaliluzzaman

Abstract COVID-19 pneumonia prediction from computed tomography (CT) images has recently become a crucial part of the computer vision field. Computed tomography images can be used to predict COVID-19 pneumonia as an alternative to traditional testing such as RT-PCR mechanisms that helps to consume time and save more lives as well. The previous researcher faces the overfitting problem while working on a small dataset and deeper convolutional neural networks (CNNs). To overcome this problem, in this work, we consider a large dataset named Corona Hack-Chest X-Ray. And perform a comparative analysis based on fine-tuned CNN models, i.e., VGG16 and miniVGGNet. These models are used based on the transfer learning. For that, the models are initially trained with the ImageNet weights and re-trained with the dataset. The comparative analysis of the models for predicting COVID-19 from CT image is evaluated by the Corona Hack-Chest X-Ray dataset. We trained and test the models with 6368 CT images and evaluate the networks performance in terms of various machine learning metrics. This dataset shows 93% accuracy for the miniVGGNet model while showing 89% accuracy for the VGG16 model.

Keywords COVID-19 · Convolutional neural network · Computed tomography (CT) images · Corona hack-chest X-ray dataset · VGG16 · miniVGGNet

1 Introduction

The COVID-19 pandemic, additionally known as the coronavirus pandemic, is a continuous widespread of coronavirus malady 2019 (COVID-19) caused by extreme intense respiratory disorder coronavirus 2 (SARS-CoV-2). The infection was to begin with recognized in December 2019 in Wuhan, China. At the end of January 2020, the World Health Organization (WHO) issued an emergency international public health concern. The situation got much worse after the announcement, hence; WHO

T. B. Tani · T. Afroz · Md. Khaliluzzaman (✉)
Department of Computer Science and Engineering, International Islamic University Chittagong (IIUC), Chittagong 4318, Bangladesh

A. Kumar et al. (eds.), *Computational Intelligence in Machine Learning*, Lecture Notes in Electrical Engineering 834, https://doi.org/10.1007/978-981-16-8484-5_35

declared a pandemic on March 11 2020. According to the latest statistics, around 108 million cases have been reported in 219 countries which caused more than 2,300,000 deaths. On the other hand, more than 80 million people have already been recovered [1].

Today the whole world goes into a world pandemic situation because of the novel coronavirus. Therefore, the prediction of COVID-19 pneumonia is a significant and efficient research area in the field of computer vision at this current decline. It helps computers to extract, analyze, and understand useful details from a single as well as a series of images. There are many techniques in computer vision to perform many applications; among those one of the best techniques is image classification. Currently, the best application for image classification is COVID-19 pneumonia prediction using pulmonary CT images.

Recently, many convolutional models are created for predicting COVID-19 pneumonia with pulmonary CT images. The use of pulmonary CT images to predict COVID-19 pneumonia has a variety of applications. It analyzes the lung infection stage to monitor patients with COVID-19 and has a crucial role to play in the development of an effective coronavirus vaccine. It creates a telemedicine network for a particular area. It reduces the stress of our frontline clinical doctors and radiologists. All these applications help to contribute to the control of today's world pandemic situation because of the coronavirus.

The main contribution of this paper is making a comparative analysis between two identical models, i.e., VGG16 [2] and miniVGGNet [3] in predicting COVID-19 pneumonia. The convolutional neural network of these models helps to recognize and classify the features by focusing on the infected areas from CT images. We have used the binary classification process to categorise the CT images as COVID-19 and non-COVID (normal).

The paper continuation is arranged as follows. In Sect. 2, the review of the state-of-the-art articles are described. In the next section, we describe our work named as methodology. In Sect. 4, we talk about the results of our experiments. Finally, Sect. 5 discusses the conclusion.

2 Related Work

In different regions of interest, each computed tomography (CT) image has distinct characteristics. In this respect, feature extraction methods should be capable of extracting features from the CT images. Some pieces of research are designed to predict COVID-19 pneumonia in the current year using CT images.

We investigate concisely about a couple of recently published various deep learning-based works for the prediction of COVID-19. Such as, in [4] authors established a 3-D model based on the deep learning for screening coronavirus disease with 2019 pneumonia. This paper focuses to distinguish between the Influenza and COVID-19 pneumonia. A total number of 618 limited number of pulmonary CT images are used in this paper. Here, authors concern about two models that are

ResNet23 and another model is update of ResNet23 with location attention mechanism. Finally, the type of infection and the overall score of confidence are evaluated by noisy or Bayesian function.

In [5], authors proposed a model with a deep learning strategy, where COVID-19 prediction and detection were included. Here, three predictive techniques: auto-regressive integrated moving average model (ARIMA), prophetic algorithm (PA), and long short-term neural memory network (LSTM) use to forecast COVID-19 validation numbers using real-world dataset of 1000 X-ray images. In [6], a respiratory simulation model (RSM) is used in small scale application for prognosis, treatment, and screening of COVID-19 based on breathing pattern. Here, six clinically important respiratory structures, i.e., Tachypnea, Cheyne-Stokes, Eupnea, Bradypnea, Central-Apnea, and Biots, are classified.

A deep learning (DL)-based segmentation model is formed in [7]. Here authors utilize the 3-D convolutional neural network named as VB-Net which includes V-Net with the bottle-neck structure. For determining the efficiency, the Dice coefficient of similarity, the variations in volume, and the percentage of infection (POI) evaluate on the validation set between automatic and manual segmentation tests. In [8], the author proposed an assessment for coronavirus pandemic through AI development cycle. An agile two and 3-D deep learning methods build for the identification, quantification, and monitoring of coronavirus. The authors perform a 3-D volume analysis for producing a corona score which helps to track disease development over time. The ground glass opacities, explicitly peripheral ground glass and bilateral opacities, and consolidate pulmonary-opacities consider as a feature of COVID-19.

In [9], authors proposed a model based on deep learning for predicting COVID-19 from high-resolution computed tomography which is a prospective study. This paper focuses on the viral pneumonia on the input image. A novel and efficient UNet++ architecture extracts efficient features from input images. An XGBoost algorithm is used to predict the prognostic in [10]. A supervised XGBoost classifier works as the predictor because of the superb pattern characterization and feature selection capability. This model distinguishes three main clinical characteristics, i.e., high sensitivity C-reactive protein (hs-CRP), lymphocyte, and lactic dehydrogenases (LDH), from a collection of more than 300 features. The key features rank by Multi-tree XGBoost.

In [11], authors proposes a model to predict the positive COVID-19 cases through using recurrent neural network-based long-short term memory (LSTM), convolutional LSTM, and bi-directional LSTM. Here, the bi-directional LSTM shows higher accuracy then others. An investigation based on the chest CT images and RT-PCR for COVID-19 prediction is presented in [12]. Here, authors use Wilson's score method to provide a confidence interval. In [13], authors proposed a CNN model approach to predict and detect COVID-19 using deep learning. Here, authors use three pre-trained models, i.e., VGG16, VGG19, and Resnet50, for comparative analysis.

In [14], a parallel-dilated CNN is developed by using 2905 chest X-ray images for COVID-19 prediction. Here, authors choose most prominent dilated convolution through a parallel sequence of convolutional structure to catch and transmit important features across the network in parallel.

3 Methodology

In our work, a comparative analysis is performed based on the two deep learning-based models that are VGG16 and miniVGGNet for COVID-19 pneumonia prediction. These architectures automatically extract features of COVID-19 to predict COVID-19 pneumonia form an input pulmonary computed tomography (CT) image. Here, in our work, the COVID-19 pneumonia prediction methods work on a large number of a dataset on a small number of output classes. So, we demonstrate our work with a small number fully connected layers, where models can show promising outcomes. The work-flow architecture is shown as a flowchart in Fig. 1.

In Fig. 1, the overall work-flow architecture for COVID-19 pneumonia prediction is shown where we make a balanced dataset through data preprocessing and augmentation. After that, dataset is split into training (80%) and testing (20%). Then, we train the both models. Finally, we measure models performance through different machine learning metrics.

3.1 Data Preprocessing

To train and evaluating the networks, in this work, the Corona Hack-Chest X-Ray dataset [15] is used. In deep learning both the training and testing phases we need

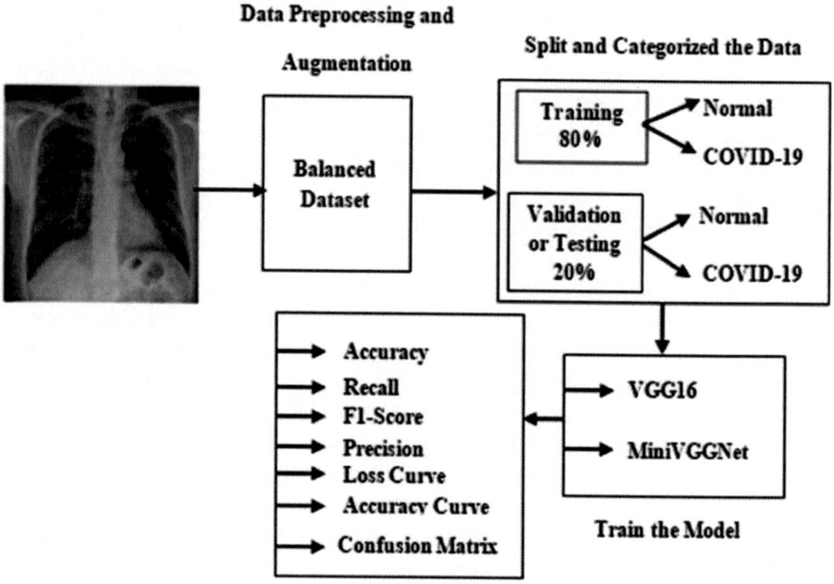

Fig. 1 Work-flow architecture for COVID-19 pneumonia prediction

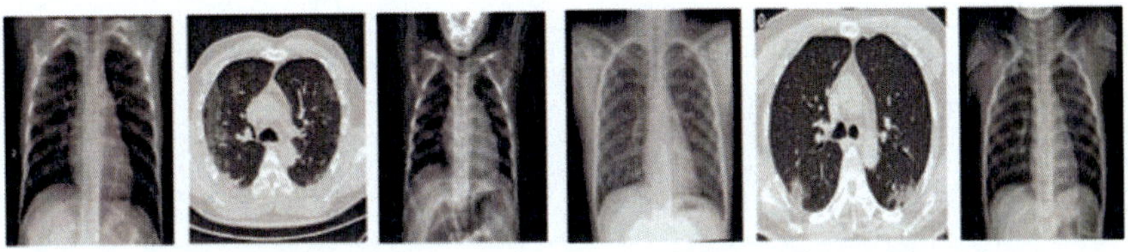

(a) COVID-19 (b) Non-COVID (Normal)

Fig. 2 Sample images of Corona Hack-Chest X-Ray dataset: **a** COVID-19, and **b** non-COVID-19 (normal)

the same size of images. For this regard, we resize all the images in 128 × 128 sizes. After that, we divide the images by 255 for normalizing image intensity values between 0 and 1 and reduce the computational time.

From this dataset, we use a total of 6368 CT images. We make those images into a balanced dataset. Out of these 6368 images, 3184 are labeled as COVID-19 samples and 3184 are labeled as non-COVID, i.e., normal samples. Some reference images of COVID-19 as well as non-COVID from this dataset are presented in Fig. 2.

3.2 Model Explanation

We have chosen two models in this work, where one model is VGG16 which is pretty large in structure. Another one is miniVGGNet [3] which is small in structure compared to the VGG16 [2]. These different models are chosen to find how well the dataset respond in the large and small models.

The VGG16 which is first used in 2014 is an excellent version of the CNN architectural model. The 16 refers to the layers numbers that contain the weights. The VGG16 architecture has low hyper-parameters with convolutional layer. These layers have same filter size 3 × 3. The stride of the filter is one. The VGG16 model has same padding and max-pooling layers of 2 × 2 filters. The stride size of the max-pooling layer is two. VGG16 architecture follows the aforementioned structure of convolution and max-pooling layers all over the model. In the dense layer, the VGG16 model contains three fully connected (FC) layers followed by the Softmax activation function. The block diagram of VGG16 model is presented in Fig. 3.

The miniVGGNet structure is small in model size compared to the VGG16. Here, miniVGGNet architecture has two block of convolutional layer. In the first block, it contains to convolutional layers with the 32 filters size of 3 × 3. In the second block, miniVGGNet architecture also consists of two convolutional layers with size of 3 × 3 64 filters. In the each blocks, each convolutional layer followed by an activation function and Batch Normalization to speed up the model. After each block, max-pooling is performed with the filter size of 2 × 2 and stride 2. A dropout is applied after the each block to regularized the model and recover from the overfitting problem. In the dense layer, the miniVGGNet model contains one fully connected layer with

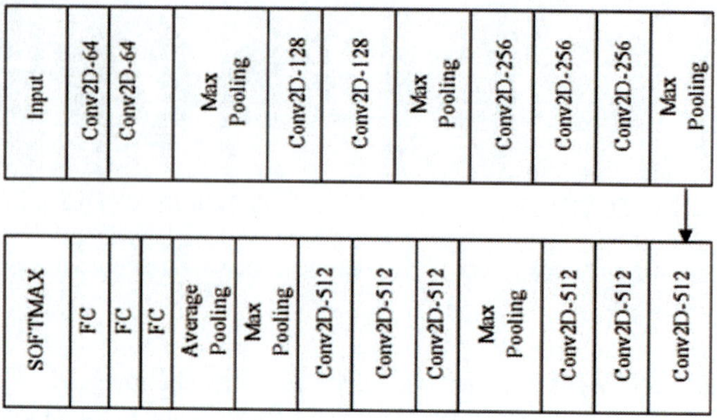

Fig. 3 Block diagram of VGG16

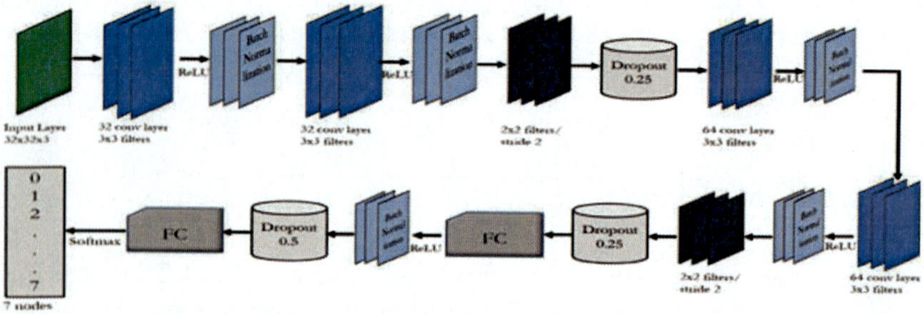

Fig. 4 The model architecture of miniVGGNet

1024 neuron is used which followed by the activation function that is Softmax. The block diagram of miniVGGNet model is presented in Fig. 4.

3.3 *Classification Loss Function*

To determine the efficiency of a classification model whose output is a probability value, a loss function is used. A cross-entropy loss is the loss element that we are going to use. It evaluates the efficiency of a model with a probability value of 0 to 1. Further the predicted value differs from the real value, the more the loss of cross-entropy becomes. For binary classification, the mathematical formula for cross-entropy loss is provided in Eq. 1.

$$\text{Loss} = -(y(lo(p)) + (1 - y)\log(1 - p) \tag{1}$$

where log = natural log, p = the number of predicted probability.

4 Experimental Results

4.1 Evaluation Platform

In this experimental section, we perform our analysis on the dataset described in Sect. 3. Here, a comparative analysis is made based on the VGG16 and miniVGGNet model. We tried to keep the architecture and hyper-parameters identical during training. On ImageNet, these models have been pre-trained instead of replicating the entire training procedure. We simply load the model parameters, i.e., weights during initialize the model. This consumes our computation time. In this work, 25, 50, and 100 epochs and batch size 32 are used for training. Using Keras framework the specification was written in Python in Google CoLab. For backend, we use TensorFlow.

4.2 Performance Evaluation

The performance is evaluated based on the different machine learning matrices such as accuracy, recall, precision, $F1$-score. The proportion of total samples correctly recognized by the classifier is known as accuracy present in (2). The recall is the total positive samples that the classifier correctly predicted as positives present in (3). The total predicted positive samples by the classifier that are true positives are known as precision present in (4). $F1$-score measures a balanced average outcome by combining precision and recall present in (5).

$$\text{Accuracy} = \frac{\text{TP} + \text{TN}}{\text{TP} + \text{FP} + \text{TN} + \text{FN}} \tag{2}$$

$$\text{Recall} = \frac{\text{TP}}{\text{TP} + \text{FN}} \tag{3}$$

$$\text{Precision} = \frac{\text{TP}}{\text{TP} + \text{FP}} \tag{4}$$

$$F1\text{ - score} = \frac{2 * \text{precision} * \text{recall}}{\text{precision} + \text{recall}} \tag{5}$$

The above equations show how to calculate different machine learning matrices with true positive (TP), false positive (FP), true negative (TN), and false negative (FN), respectively.

The models are evaluated based on the number of epochs using Corona Hack-Chest X-Ray dataset. Here, we consider standard decay and 0.1 as our initial learning rate with batch size 32. The result is represented in Table 1.

Table 1 VGG16 and miniVGGNet performance evaluation on basis of epochs

Epoch	VGGG16 (%)	miniVGGNet (%)
25	49	86
50	51	88
100	49	88

Table 2 The classification report of VGG16

	Precision	Recall	$F1$-score	Support
COVID	0.98	0.81	0.89	651
Normal	0.83	0.98	0.90	623
Accuracy			0.89	1274
Macro average	0.91	0.90	0.89	1274
Weighted average	0.91	0.89	0.89	1274

From Table 2, it is revealed that VGG16 shows best accuracy of 51% on the epoch of 50 and miniVGGNet shows best accuracy of 88% at the same epoch. The miniVGGNet accuracy remains constant after 50 epochs, where VGG16 accuracy is going to decrease.

For VGG16, the training and validation loss curve is shown in Fig. 5a. The training and validation accuracy curve for VGG16 is shown in Fig. 5b. For optimization of the models, an empirical learning rate is used, i.e., 0.001. As the Adam optimizer is an adaptive optimizer, it takes less time to train the model, in this work; we consider the Adam optimizer to train the model. The test data is used to see how well the model respond. Training and validation loss, and also training and validation accuracy, are both measured and plotted on the curve.

From the loss curve during training and testing of VGG16, it is revealed that both of the losses are reducing till the 18 epoch. After that, the training loss is constantly reduced. However, the validation loss is increased. That means, the model

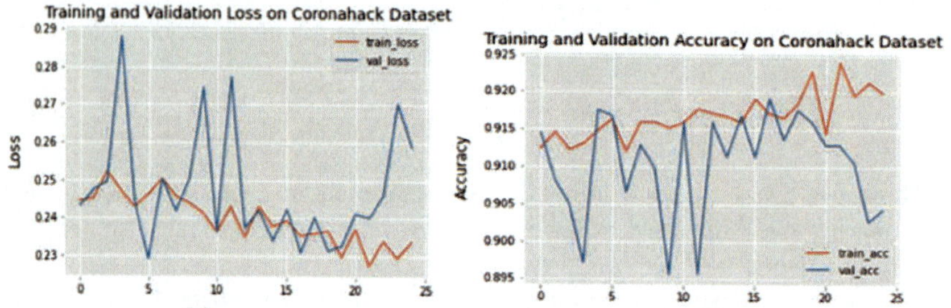

(a) (b)

Fig. 5 For Corona Hack-Chest X-Ray dataset loss and accuracy for VGG16 model: **a** training and validation loss curve and **b** training and validation accuracy curve is not evaluate the testing data perfectly. The same situations are also observed in the case of training and validation accuracy. After 20 epochs it is revealed that the loss difference between training and validation is increased. That means, the model is going to overfit after 20 epochs.

As the VGG16 model shows overfit for the Corona Hack-Chest X-Ray dataset. Now we are training and validating that dataset using the miniVGGNet model. The validation and training loss curve is shown in Fig. 6a. The validation and training accuracy curve is shown in Fig. 6b. The training is continuing to 50 epochs as the models training and validation losses are constant after 45 epochs. From the loss curve, it is revealed that the losses both of training and validation are continuously reduced. That means the model is learning the images accurately. After 30 epochs it also revealed that the model's losses both for training and for validation are almost constant. That means the model is slowly learned after the 30 epochs.

From Fig. 6b, it is also revealed that the miniVGGNet model is threaded as a regularized model and it is not overfitted like VGG16. In miniVGGNet, there are only four convolution layers where VGG16 consists of 13 convolutional layers. So, miniVGGNet takes less time to learn then VGG16 model. The miniVGGNet shows the accuracy of 93% where VGG16 shows 89% accuracy. That means miniVGGNet gives better accuracy than the VGG16 model when dealing with Corona Hack-Chest X-Ray dataset.

We also determine the confusion matrix of the expected outcomes which is often used on test data for which the true values are known to justify the outcome of a classification model. When it makes assumptions, the confusion matrix reveals how the classification model is uncertain. The graphical confusion matrix for VGG16 and miniVGGNet is presented in Fig. 7.

From Fig. 7, it is shown that the false positive and true negative values for VGG16 are 10 and 124, respectively. Whereas the false positive and true negative values for miniVGGNet are 12 and 86, respectively. From these data we can say that miniVGGNet reduces the true negative rate with respect to the VGG16 model. That means, miniVGGNet predicts non-COVID pneumonia little bit better than VGG16 model.

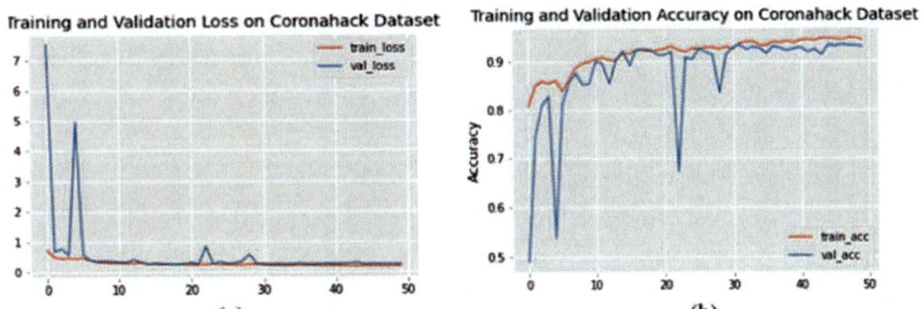

(a) (b)

Fig. 6 For Corona Hack-Chest X-Ray dataset loss and accuracy for miniVGGNet model: **a** training loss and validation loss curve and **b** training and validation accuracy curve

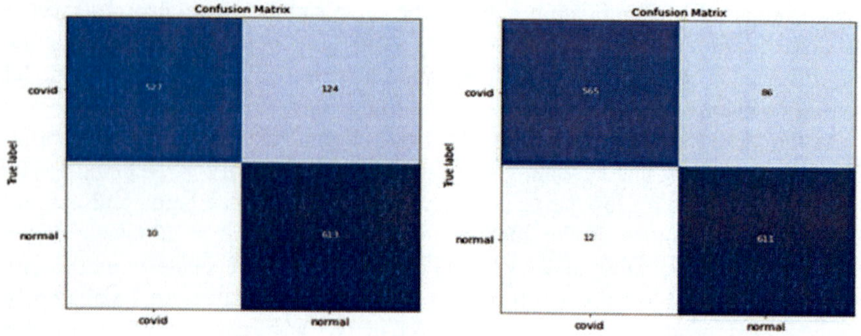

(a) (b)

Fig. 7 The graphical confusion matrix for Corona Hack-Chest X-Ray dataset: **a** VGG16 and **b** miniVGGNet

The classification report is generated from the confusion matrix. The classification report visualizes the recall, accuracy, $F1$-score, and assistance ratings of the model. Concerning the matrix below, we have decorated our experimental analysis. For VGG16, the classification report is presented in Table 2. The classification report for miniVGGNet is presented in Table 3.

From the classification report in Table 2, the recall, precision, $f1$-score, and support are 0.81, 0.98, 0.89, and 651, respectively, for COVID-19. And, for normal pneumonia the recall, precision, $F1$-score, and support are 0.98, 0.83, 0.90, and 623, respectively. The overall accuracy of VGG16 model is 89%.

From the classification report in Table 3, the recall, precision, F1-score, and support are 0.89, 0.97, 0.93, and 651, respectively, for COVID-19. And, for normal pneumonia the recall, precision, F1- score, and support are 0.97, 0.90, 0.93, and 623, respectively. The overall accuracy of miniVGGNet for the dataset is 93%.

We also evaluate the models by comparing with various initial learning rates which is shown in Table 4. Here, we consider 0.1, 0.01, and 0.001 as our initial learning rate. Standard decay with batch size 32 and 100 epochs are also considering for this evaluation.

From Table 4, it is revealed that miniVGGNet performs better with 93% accuracy at 0.001 initial learning rate compare to VGG16.

Table 3 The classification report of miniVGGNet

	Precision	Recall	$F1$-score	Support
COVID	0.97	0.89	0.93	651
Normal	0.90	0.97	0.93	623
Accuracy			0.93	1274
Macro average	0.93	0.93	0.93	1274
Weighted average	0.93	0.93	0.93	1274

Table 4 Performance evaluation on basis of initial learning rate

Initial learning rate	VGG16 (%)	miniVGGNet (%)
0.1	49	88
0.01	49	92
0.001	**89**	**93**

Table 5 Performance evaluation on basis of learning rate decay

Epoch	Learning rate decay	VGG16 (%)	miniVGGNet (%)
50	Step	51	**93**
	Linear	51	**93**
	Polynomial	**88**	92
	Standard	86%	91%

In Table 5, we evaluate our models using initial learning rate 0.01. Here, we consider step, linear, polynomial, and standard decay with batch size 32 and 50 epochs.

From Table 5, it is revealed that for miniVGGNet, we achieve 93% accuracy when we use step learning rate decay. However, in VGG16, we only get 51% accuracy for step learning rate decay.

Different learning rate decay analysis with and without data augmentation of VGG16 is shown in Table 6. Here, we set batch size 32, epoch 50 for training.

From Table 6, it is revealed that the VGG16 gets 89% accuracy when we do data augmentation with 0.001 initial learning rates.

For Table 7, we set our parameter batch size 32, epoch 50 for training with and without data augmentation. Training parameters are set to batch size 32, epoch 50.

Table 6 Data augmentation and learning rate decay evaluation for VGG16

VGG16			
Initial learning rate	Learning rate decay	Training with data augmentation (%)	Training without data augmentation (%)
0.1	Step	51	49
	Linear	51	49
	Polynomial	49	51
0.01	Step	51	88
	Linear	51	88
	Polynomial	87	88
0.001	Step	**89**	88
	Linear	**89**	88
	Polynomial	**89**	88

Table 7 Data augmentation and learning rate decay evaluation for miniVGGNet

miniVGGNet			
Initial learning rate	Learning rate decay	Training with data augmentation (%)	Training without data augmentation (%)
0.1	Step	82	84
	Linear	79	81
	Polynomial	82	86
0.01	Step	**93**	92
	Linear	**93**	92
	Polynomial	92	92
0.001	Step	**93**	92
	Linear	**93**	91
	Polynomial	92	91

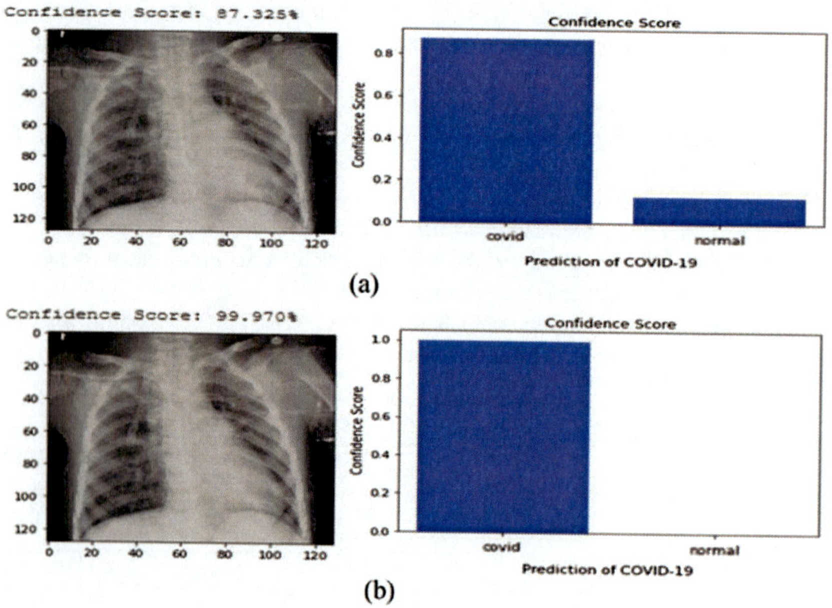

Fig. 8 For Corona Hack-Chest X-Ray dataset: **a** image test by VGG16 and **b** image test by miniVGGNet

The performance of miniVGGNet is shown in Table 7 when training our model with a data augmentation procedure. The model shows highest accuracy of 93%.

The trained model is evaluated by using the test images. The same images are tested by the VGG16 and miniVGGNet models. The test results for VGG16 and

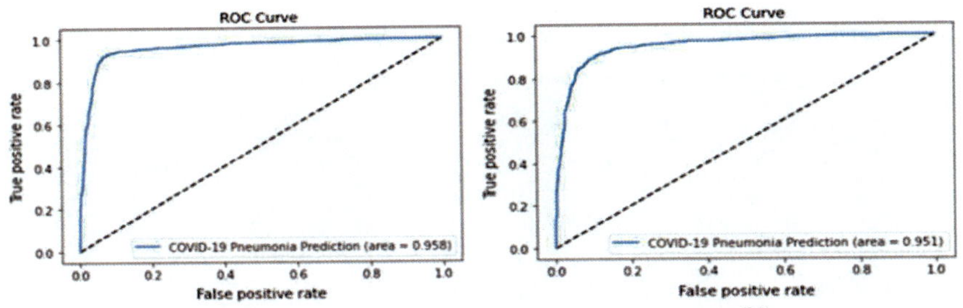

(a) (b)

Fig. 9 ROC Curve for Corona Hack-Chest X-Ray dataset: **a** VGG16 and **b** miniVGGNet

miniVGGNet are presented in Fig. 8a, b, respectively. We achieve better confidence score for the miniVGGNet.

We perform our analysis through receiver operating characteristic (ROC) curve. The ROC curve for our dataset is presented in Fig. 9.

For VGG16, the area value of the ROC curve is 0.951 which is shown in Fig. 9a. The area of the ROC curve of the same images through miniVGGNet is 0.958 that is shown in Fig. 9b.

We evaluate our prediction through some sample testing image for miniVGGNet model. The prediction of COVID-19 (covid) as well as non-COVID (normal) is presented in Fig. 10.

4.3 Discussions

From the experimental analysis it is observed that the miniVGGNet is performed better than the VGG16. The VGG16 has 16 weighted layers where as miniVGGNet has only 5 weighted layers. As the dataset used in this work is not too large, for the larger network the model going to be overfitted. If we are used the small network the model would be underfitted. To overcome this problem we should use such type of network that has moderate label of convolutional network and dense layers such as miniVGGNet. As miniVGGNet has only four convolutional layers and one fully connected layer, so, this model achieves regularized model on the Corona Hack-Chest X-Ray dataset. The miniVGGNet performs faster than the VGG16 because of the miniVGGNet use batch normalization after each block.

5 Conclusions

The analysis of COVID-19 pneumonia with pulmonary CT images has gained a lot of coverage and very helpful for COVID-19 identification. In this paper, a performance

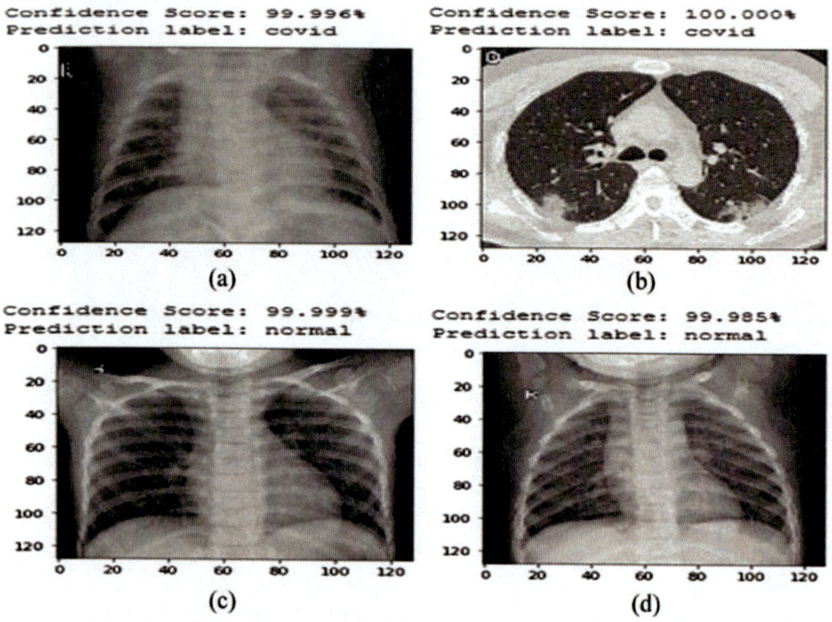

Fig. 10 Sample test images using miniVGGNet

comparison is explored between VGG16 and miniVGGNet model for COVID-19 prediction from CT images. For that, Corona Hack-Chest X-Ray dataset is used in this work. Both VGG16 and miniVGGNet models are trained and evaluated based on the Corona Hack-Chest X-Ray dataset. The miniVGGNet achieves 93% accuracy while, VGG16 achieves 89% accuracy. During training the VGG16 shows overfitting while the miniVGGNet shows regularized action. After comparison of both model, it is revealed that miniVGGNet model performs better than VGG16. In future, large different dataset will be used to create a regularized framework to predict COVID-19 from CT images.

References

1. Covid-19 corona virus pandemic. https://www.worldometers.info/coronavirus/. [Accessed 3 Oct. 2020]
2. K. Simonyan, A. Zisserman, Very deep convolutional networks for large-scale image recognition. arXiv preprint arXiv:1409.1556 (2014)
3. A. Ismail, S.A. Ahmad, A.C. Soh, K. Hassan, H.H. Harith, Improving convolutional neural network (CNN) architecture (miniVGGNet) with batch normalization and learning rate decay factor for image classification. Int. J. Integr. Eng. **11**(4) (2019)
4. X. Xu, X. Jiang, C. Ma, P. Du, X. Li, S. Lv, L. Yu, Y. Chen, J. Su, G. Lang, Y. Li, H. Zhao, K. Xu, L. Ruan, W. Wu, *Deep Learning System to Screen Coronavirus Disease 2019Pneumonia* (2020)

5. M. Alazab, A. Awajan, A. Mesleh, A. Abraham, V. Jatana, S. Alhyari, COVID-19 Prediction and detection using deep learning. Int J Comput Inf Syst Ind Manag Appl **12**, 168–181 (2020)
6. Y. Wang, M. Hu, Q. Li, X.-P. Zhang, G. Zhai, N. Yao, Abnormal Respiratory Patterns Classifier May Contribute to Large-Scale Screening of People Infected with COVID-19 in an Accurate and Unobtrusive Manner (Cornell University, 2020)
7. F. Shan, Y. Gao, J. Wang, W. Shi, N. Shi, M. Han, Z. Xue, D. Shen, Y. Shi, Lung Infection Quantification of COVID-19 in CT Images with Deep Learning (Cornell University, 2020)
8. O. Gozes, M. Frid-Adar, H. Greenspan, P.D. Browning, H. Zhang, W. Ji, A. Bernheim, E. Siegel, Rapid AI Development Cycle for the Coronavirus (COVID-19) Pandemic: Initial Results for Automated Detection & Patient Monitoring using Deep Learning CT Image Analysis (CornellUniversity, 2020)
9. J. Chen, L. Wu, J. Zhang, L. Zhang, D. Gong, Y. Zhao, S. Hu, Y. Wang, X. Hu, B. Zheng, K. Zhang, H. Wu, Z. Dong, Y. Xu, Y. Zhu, X. Chen, L. Yu, H. Yu, Deep learning-based model for detecting 2019 novel coronavirus pneumonia on high- resolution computed tomography: a prospective study (2) (2020)
10. L. Yan, H.-T. Zhang, Y. Xiao, M. Wang, C. Sun, J. Liang, S. Li, M. Zhang, Y. Guo, Y. Xiao, X. Tang, H. Cao, X. Tan, N. Huang, B. Jiao, A. Luo, Z. Cao, H. Xu, Y. Yuan, Prediction of criticality in patients with severe Covid-19 infection using three clinical features: a machine learning-based prognostic model with clinical data in Wuhan (2) (2020)
11. P. Arora, H. Kumar, B.K. Panigrahi, Prediction and analysis of COVID-19 positive cases using deep learning models: a descriptive case study of India. Chaos, Solitons Fractals **139**, 110017 (2020)
12. T. Ai, Z. Yang, H. Hou, C. Zhan, C. Chen, W. Lv, Q. Tao, Z. Sun, L. Xia, Correlation of chest CT and RT-PCR testing for coronavirus disease 2019 (COVID-19) in China: a report of 1014 cases. Radiology **296**(2), E32–E40 (2020)
13. K.F. Haque, A. Abdelgawad, A deep learning approach to detect COVID-19 patients from chest X-ray images. AI **1**(3), 418–435 (2020)
14. N.K. Chowdhury, Md. Muhtadir Rahman, M.A. Kabir, PDCOVIDNet: a parallel-dilated convolutional neural network architecture for detecting COVID-19 from chest X-ray images. Health Inf. Scie. Syst. (2020)
15. CoronaHack-Chest X-Ray-Dataset. [online] Kaggle. Available at: https://www.kaggle.com/praveengovi/coronahackchestxraydataset?select=Chest_xray_Coron_Metadata.csv. Accessed 3 Oct 2020

Object Detection with Occlusion Handling in Maritime Surveillance Using Histogram of Orientated Gradient

Rubeena Banu and M. H. Sidram

Abstract An automatic surveillance of maritime is becoming very important to improve the situation connecting to maritime security or safety. In maritime domain, surveillance operation is often more complex. The maritime threats such as illegal underwater activities, smuggling, illegal fishing, criminal, and terror operation are on the rise. Hence, it is very important to have an automatic surveillance system in order to secure the commercial as well naval ships from pirates and terrorist attack. This paper presented a system which detects the ship at the condition of occlusion using histogram of orientated gradient (HOG). This system requires a less memory and processing time. HOG is used as classifier and conducted experiments on standard dataset. Experimental outcomes show that the proposed method gives promising results.

Keywords Maritime surveillance · HOG · Horizon line detection · Occlusion · Object detection

1 Introduction

In the area of computer vision, detecting the moving object and tracking from a video is an indispensable and challenging job [1]. Video surveillance attempts to detect, identify, and track objects over a series of images, each of which is called as frame. Surveillance in maritime domain is very important for ensuring genuine riposte to any improving situation which is related to security or safety of maritime. Moreover, existence of problems related to environment protection and criminal activities such as illegal fishing and smuggling, oil spills and piracy, and immigration, motivates to develop an intelligent automatic system. However, the present surveillance of maritime systems still fail to provide effectual result due to the various reasons

R. Banu (✉) · M. H. Sidram
Department of Electrical and Electronics Engineering, Sri Jayachamarajendra College of Engineering, Mysuru, India
e-mail: drmhs@sjce.ac.in

A. Kumar et al. (eds.), *Computational Intelligence in Machine Learning*, Lecture Notes in Electrical Engineering 834, https://doi.org/10.1007/978-981-16-8484-5_36

like illusion, poor visibility, cast shadows, and very importantly object's occlusions. Occlusions take place when the faraway targets are hidden by objects which are nearby the viewer, the occlusion might be partial or full. Many algorithms of computer vision find it difficult due to the presence of occluded objects in a scene. It is the rudimentary element that restricts the information in an image. Hence, handling of occlusion plays an important role in object detection, but it is still very taxing because dynamic character of ocean and the targets required to be extricated in an environment where it is very difficult to predict [2–5]. The remaining part of the paper is assembled as follows. Related work described in Sect. 2. Proposed methodology is discussed in Sect. 3; results and discussion are explained in Sect. 4. Finally, the conclusion is reported in Sect. 5.

2 Related Work

Many researchers have looked into the detection of object in maritime video surveillance, but very less amount of work has been carried out in occlusion handling issues. Bao et al. [6] proposed a detection of moving ship method in video surveillance system. This method carried out an over-segmentation to classify and divide by support vector machine (SVM) segments into water or non-water. Contingent on water context system, non-water parts are amalgamated into regions with motion similarity. Then, identified moving boats by estimating the motion saliency of that regions. Context-based motion analysis is employed to separate ship even in the occlusion amid ships and vegetation. Bachoo et al. [7] implemented the system that is vigorous to platform vibration, changes in target aspects, and short period occlusion. This approach built an optical tracker employing a particle filter and an appearance model. If an occlusion takes place, the variance of the Gaussian is set to huge application-specific value. The greater variance scatters the particles and enhances the identification of target. Cane and Ferryman [8] presented an approach for finding the object based on visual saliency and utilized widely tuned, intensity decoupled red, blue, and green color channels which reduced the number of false detection from wave and reflections. For the purpose of traffic management, Wijnhoven et al. [9] developed an object detection system. In this system, HOG approach is used for object detection. Online and offline experiments have been performed. SVM classifier is utilized for offline training and stochastic gradient descent used as training algorithm which is given to an efficient object detection problem.

Bousetouane and Morris [10] presented a deep CNN- based FGC pipeline. This system is developed on top of the Fast-RCNN framework with main three steps. It uses a weak HOG-based identification object in first step. The second step uses the fine-tuned VGG-16 network utilizing a linear support vector machine (SVM). The last step gives region corroboration and allocates final class labels employing a simple confidence plan. In this world, the 80% of the trade is depended on safe maritime route, and providing effective maritime surveillance poses a significant challenge. So, it is very essential to be aware all the time that what is the status of the surroundings sea and coastal area to its security, safety, environment, and prosperity. Hence, to address this problem, this work used the online available maritime dataset in order to detect the boat in occlusion condition and achieved good true positive rate.

3 Proposed Methodology

The goal of the detection module is to detect the interested object in the present frame. Figure 1 shows the main component of detection module which consists of frames, horizon line detection, and histogram of orientation gradient.

3.1 Frames

Before starting the process, video must be fragmented into sequence of frames. This sequence of frames is given to the horizon line detection.

3.2 Horizon Line Detection

In order to avoid the false positives, it is important to separate sky and sea region. Hence, method has been used to find the horizon that will separate the region into sky and water region. There are chances of detecting the sky region object as object of interest during the detection process which increase the false positive. Therefore, in this work, horizon line is detected before the sequence of images given into the classifier in order to avoid false positive. The selection of candidate lines is done using Sobel operator [11, 12], and morphological operations [13] are performed. Sobel is considered as popular edge detection algorithm employed in this work. It executes a 2D spatial gradient quantification on frames. It utilizes pair of horizon and vertical gradient matrices whose dimensions are 3x3 for the operation of edge

Fig. 1 Object detection module

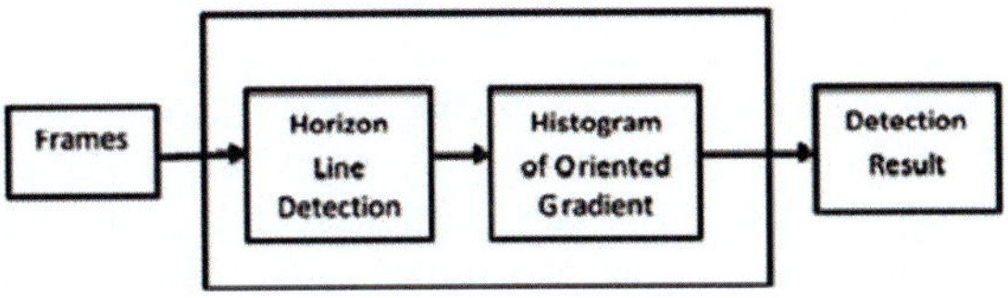

detection. The Sobel is the magnitude of the gradient which is calculated by using Eq. 1

$$M\sqrt{S_x^2 + S_y^2} \tag{1}$$

where the partial derivatives are computed by Eqs. 2 and 3

$$S_x = (a_2 + C_{a3} + a_4) - (a_0 + C_{a1} + a_6) \tag{2}$$

$$S_y = (a_0 + C_{a1} + a_2) - (a_6 + C_{a5} + a_4) \tag{3}$$

S_x and S_y are performed employing convolution mask.

$$\mathbf{S}_x = \begin{bmatrix} -1 & 0 & 1 \\ -2 & 0 & 2 \\ 1 & 0 & 1 \end{bmatrix} \qquad \mathbf{S}_y = \begin{bmatrix} 1 & 2 & 1 \\ 0 & 0 & 0 \\ -1 & -2 & -1 \end{bmatrix}$$

Sobel operator is put on an emphasizing pixels those are near to the mask at the center.

a_0	a_1	a_2
a_7	$a_{[i,j]}$	a_3
a_6	a_5	a_4

A convolution mask utilized is normally smaller than the original image. As an output, the mask is glide over the region of the input frame, changes that pixel's value, and later moves one pixel to the right, and same will continue until it reaches the end of a row. Then, it will start from the beginning of the next row. The advantage of using this Sobel edge detector is detecting edges, and orientation is simple due to the approximation of the gradient magnitude and also gives better performance under the noisy condition [14]. Once the edge is detected, the morphological operation is performed for the detected edge as post- processing. After the completion of post-processing, the horizon line will be drawn over the image where the sky and sea region get separated. Now, the sky region is neglected, and only the sea region is given for further processing.

3.3 *Histogram of Orientated Gradient (HOG)*

HOGs are feature descriptor, which is an effective way to extract feature out of pixel color that has been broadly and successfully utilized for the detection of object. It represents object as a single feature vector as opposed to group of feature vectors where each represents a segment of the image. It is computed by sliding window detector over an image, where a HOG descriptor is calculated for each position [15]. In order to compute a HOG descriptor, initially need to compute the horizontal and vertical gradients which is smoothly accomplished by filtering the image using following kernels.

$$\begin{bmatrix} -1 & 0 & 1 \end{bmatrix} \qquad \begin{bmatrix} -1 \\ 0 \\ 1 \end{bmatrix}$$

The gradient's magnitude and direction found by utilizing the below formula

$$g = \sqrt{g_x^2 + g_y^2}$$
$$\theta = \arctan \frac{g_y}{g_x}$$

At each pixel, the gradient has a magnitude and a direction. The gradients of the three channels are evaluated for color images. The magnitude of gradient at a pixel is the maximum of the magnitude of gradients of the three channels, and the angle is the angle corresponding to the maximum gradient. The window of image is divided in to small spatial region called 'cell.' Using an 8x8 pixel identification cell, calculate the gradient vector or edge orientation at every pixel. It will create $64(8 \times 8)$ gradient vectors and represente as a histogram. Each cell is then broken into angular bins (used 9 bins 0–180° (20° each bin), where every bin corresponds to a gradient direction. It will effectively reduce 64 vectors to just nine values. As it keeps gradient magnitude, it is relatively immune to deformations. Later, normalize the gradients to ensure invariance to illumination changes, i.e., brightness and contrast. Instead of individual window cell normalization, a block normalization method is employed and takes into account neighboring blocks. Figure 2 demonstrates the algorithm implementation scheme.

Fig. 2 Working of HOG

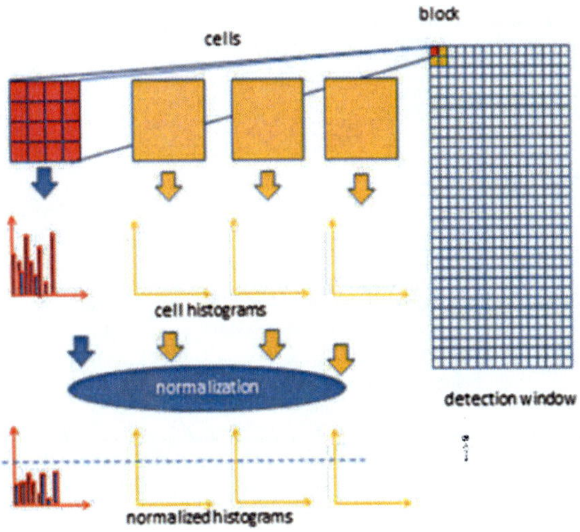

4 Result and Discussion

In order to assess the performance of the approach, MarDCT, maritime detection, classification, and tracking [16], dataset has been used which contains videos and pictures along with ground truth notations. The dataset used consists of 1532 positive samples and 4089 negative samples. The detection operation accuracy has been examined by using EO images. True positive rate (recall) and false alarm rate (false discovery rate) have been measured in order to evaluate detection accuracy, which is given in Eqs. 4 and 5

$$\text{TPR} = \frac{\text{TP}}{\text{TP} + \text{FN}} \tag{4}$$

$$\text{FAR} = \frac{\text{FP}}{\text{TP} + \text{FP}} \tag{5}$$

where TP is true positive, which represents ship which are detected correctly. FN is false negative, i.e., ships are not found. FP is false positive which means the number of background region incorrectly labeled as regions of interest. TPR computes amount of correct ships found over the total number of images. FAR measures the wrongly detected ships among the total detected ships which is generated by the systems. In this approach, three different occlusion videos are used such as occlusion 1, occlusion 2, and occlusion 3.

Figure 3 indicates the flowchart of the implementation. Initially, create the positive (which holds the object of interest) and negative images (which do not contain the interested object) database. The trian cascade object detector function trains a model for a cascade object detector and to train the custom classifier. The function computes the number of positive samples to be utilized at each stage using the following formula:

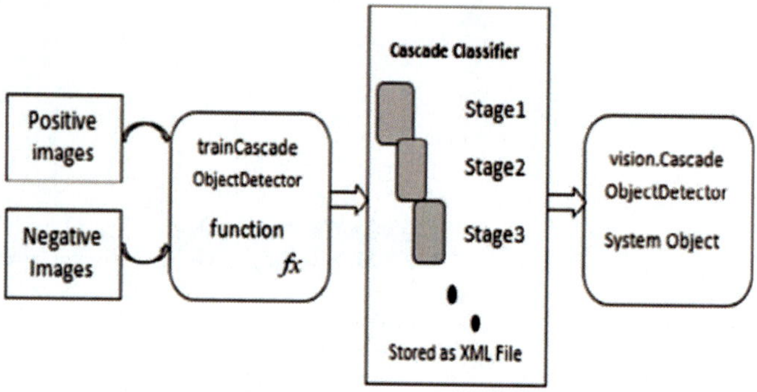

Fig. 3 Flowchart of implementation

$$\text{Number of positive samples} = \frac{\text{total Positive Samples}}{1 + (\text{Num Cascade Stage-1}) * \text{True Positive Rate}}$$

At stage 1, the number of positive samples which is less than the total number of user-given positive samples has been calculated. The negative samples will be produced from the user-given negative images. In the training stage 2, it utilizes the stage 1 and classifies each one of the positive samples. The samples which is misclassified as negative will be discarded. In this stage, make use of the same calculated number of positive samples from the remaining positive samples. Negative samples will be produced by processing the negative images with sliding window and utilizing false positive classified samples. The same procedure will be continued till the N stages. The vision cascade object detector system finds objects in images by sliding window over an image. The detector later utilizes cascade classifier to determine whether the window consists interested object. The window size varies to find object at various scales, but its aspect ratio remains same.

Table 1 presents the experimental results for three different occlusion videos and shows that HOG gives promising results. Figure 4a–c show the horizon line detected results, and Fig. 5a–c indicate the object detected results with handling the occlusion condition.

Table 1 Detection result

Video	Horizon detection	HAAR		HOG (proposed)	
		TPR	FAR	TPR	FAR
Occlusion 1	Yes	0.780	0.200	0.998	0.002
Occlusion 2	Yes	1.000	0.000	1.000	0.000
Occlusion 3	Yes	1.000	0.000	1.000	0.000

a b c

Fig. 4 .

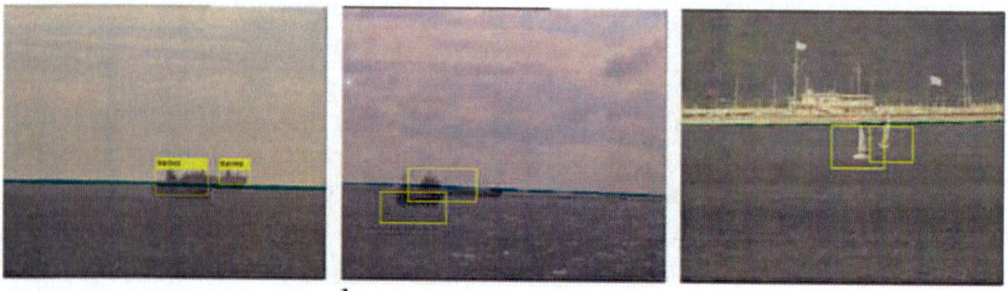

a b c

Fig. 5 .

5 Conclusion

In this experiment, a maritime object detection in occlusion condition has been presented. The Sobel operator is utilized to detect horizon line. It utilizes a set of 3 x 3 convolution masks, one approximate gradient in the x-direction, and other approximate gradient in the y-direction. Horizon line detected before the sequence of images is given to the classifier to avoid the number for false positive. Finally, the object has been detected using the HOG which captures edge or gradient structure which is very characteristics of local shape. Three different occlusion videos are used in the experiment and achieved promising results by evaluating performance using true positive rate and false alarm rate. This work reduced several false detection's which have been found above the horizon line and attained good computational speed.

The future work will focus on tracking the object in occlusion condition and comparing the results.

References

1. R.K. Rout, *A Survey on Object Detection and Tracking Algorithms*. Ph.D. dissertation, pp. 8290–8297 (2013)
2. D.D. Bloisi, F. Previtali, A. Pennisi, D. Nardi, M. Fiorini, Enhancing automatic maritime surveillance systems with visual information. IEEE Trans. Intel. Transp. Syst. **18**(4), 824–833 (2016)
3. G. Kumar, H. Saini, Novel non commutative cryptography scheme using extra special group. Secur. Commun. Netw. **5**, 1–21 (2017)
4. G. Kumar, H. Saini, Secure and Robust telemedicine using ECC on Radix-8 with formal verification. Int. J. Inform. Secur. Privacy (IJISP) **12**(1), 13–28 (2018)
5. D. Frost, J.-R. Tapamo, Detection and tracking of moving objects ina maritime environment using level set with shape priors. EURASIP J. Image Video Proces. **1**, 1–16 (2013)
6. X. Bao, S. Zinger, R. Wijnhoven et al.,Robust moving ship detection using context-based motion analysis and occlusion handling, in *Sixth International Conference on Machine Vision (ICMV), International Society for Optics and Photonics*, vol. 9067 (2013), p. 90670F
7. A.K. Bachoo, F. Le Roux, F. Nicolls, An optical tracker for the maritime environment. Signal Proces. Sens. Fus. Target Recogn. **8050**, 8050G (2011)

8. T. Cane, J. Ferryman, Saliency-based detection for maritime object tracking, in *Proceedings of the Conference on Computer Vision and Pattern Recognition Workshops* (IEEE, 2016), pp. 18–25
9. R. Wijnhoven, K. van Rens, E.G. Jaspers, P.H. de With, Online learning for ship detection in maritime surveillance, in *31th Proceedings of the Symposium on Information Theory* (Benelux, 2010), pp. 73–80
10. F. Bousetouane, B. Morris, Fast cnn surveillance pipeline for finegrained vessel classification and detection in maritime scenarios, in *13th IEEE International Conference on Advanced Video and Signal Based Surveillance (AVSS)* (IEEE, 2016), pp. 242–248
11. O.R. Vincent, O. Folorunso et al., A descriptive algorithm for sobel image edge detection, in *Proceedings of Informing Science and IT Education Conference (InSITE)* (California, 2009), pp. 97–107
12. D. Dusha, W.W. Boles, R. Walker, Fixed-wing attitude estimation using computer vision based horizon detection, in *2nd Proceedings of Australasian Unmanned Air Vehicles Conference (AIAC1)* (Australia, 2007), pp. 1–19
13. R.M. Haralick, S.R. Sternberg, X. Zhuang, Image analysis using mathematical morphology. IEEE Trans. Pattern Anal. Mach. Intel. **4**, 532–550 (1987)
14. G. Shrivakshan, C. Chandrasekar, A comparison of various edge detection techniques used in image processing. Int. J. Comput. Sci. Iss. (IJCSI) **9**(5), 269–274 (2012)
15. N. Dalal, B. Triggs, Histograms of oriented gradients for human detection. IEEE Computer Society Conference on Computer Vision and Pattern Recognition (CVRP) (2005), pp. 886–893
16. D.D. Bloisi, L. Iocchi, A. Pennisi, L. Tombolini, ARGOS-Venice boat classification. 12th International Conference on in Advanced Video and Signal Based Surveillance (AVSS). IEEE, Germany (2015), pp. 1–6

Machine Vison-Based System for Vehicle Classification and Counting Using YOLO

Elbaraa Refaie and Ahmad Athif Mohd Faudzi

Abstract This paper proposes a machine vision-based system that is used for vehicle classification and counting. Vehicle classifier and counter are a very crucial in road design to determine the road load capacity. It is also important to monitor, manage, and analyze the traffic flow. In this work, the input video is obtained from a static video camera and from drone-captured video to count the cars, determine its direction and to classify the vehicle types on the road. The data will be fed to the vision-based system that will pass the frames captured by the camera through a YOLO neural network to classify the vehicles. Then, the cars contours are used to determine the car path that determines in which direction the vehicle is moving before it is being counted. The system will display the total number of vehicles moved in each direction and while identifying each vehicle's type. This work is possible to be used for other application such as surveillance tracking and monitoring.

Keywords Machine vision · Yolo · Classification · Surveillance · Quadrotor · Machine learning · Deep learning · OpenCv

1 Introduction

Over the last decade, the industry witnessed a revolution in the technologies used under Industrial Revolution 4.0. These rapid and new technology is proposed to ease human daily lives. One of the fields that have geared a big portion of the renaissance is the area of machine vision and artificial intelligence. Currently, machine vision has a wide range of applications in various fields like agriculture [1], inspection [2], and medical equipment [3]. The demand on using machine vision and artificial intelligent (AI) has spread out dramatically to cover almost all the life aspects. Highway and road

E. Refaie · A. A. M. Faudzi (✉)
School of Electrical Engineering, Faculty of Engineering, UniversitiTeknologi Malaysia, Johor Bahru, Malaysia
e-mail: athif@utm.my

A. A. M. Faudzi
Center for Artificial Intelligence and Robotics, Universiti Teknologi Malaysia, Johor Bahru, Malaysia

A. Kumar et al. (eds.), *Computational Intelligence in Machine Learning*, Lecture Notes in Electrical Engineering 834, https://doi.org/10.1007/978-981-16-8484-5_37

management is one of the fields that applies machine vision and artificial intelligent to facilitate the work to get better and accurate results.

Machine vision system integrated with AI and machine learning is used not to only count the vehicles on the roads but also to classify them whether it is a car, truck, or bus. Such a system is very timesaving and accurate in compared to the manual method of vehicle counting. Normally, these vehicles are counted by human operators who will sit for a prolonged time to count the vehicles manually with low efficiency and prone to human error. The same system can be applied in different regions like traffic control and police surveillance [4, 5].

2 Machine Vision Traditional Methods

Machine vision has witnessed many improvements. Machine vision technology can be divided into three levels of are low, mid, and high. The low-level machine vision is focusing on the basic features of the image. The traditional methods of vehicles detection can be performed by using the low-level machine vision in three different methods which are difference method, thresholding method, and contour method [4]. One of the traditional detection methods is difference method that find the difference between two frames [6]. By finding the difference, a set of points will be detected as a difference between the two frames. Results of difference method can be seen from the taken drone video [7] as in Fig. 1.

Another method used in detection is thresholding method. In this method, image will be converted into a black and white image [8]. The pixels that have a value above the threshold will turned into white color. Otherwise, it turns to black color. Same approach is being used in color detection as in Fig. 2.

Finding contours is the third method of the low-level machine vision. The concern is about finding the boundaries of the region of interest (ROI). To find contours, we

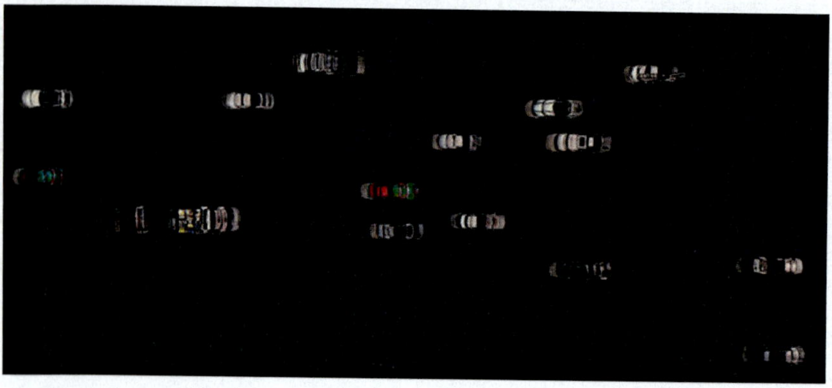

Fig. 1 Output of difference method detection

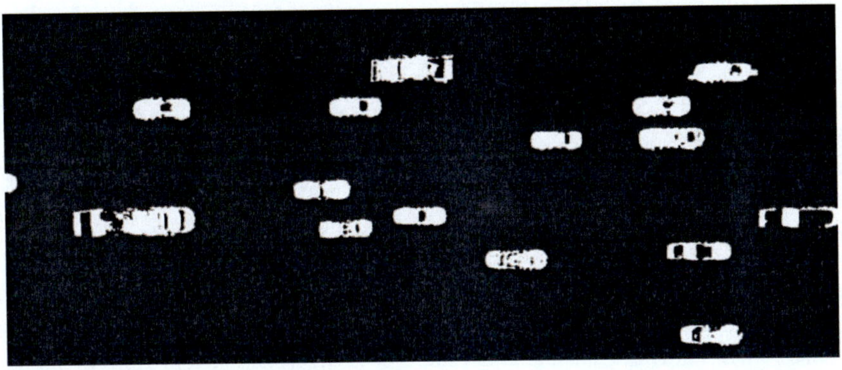

Fig. 2 Output of thresholding process

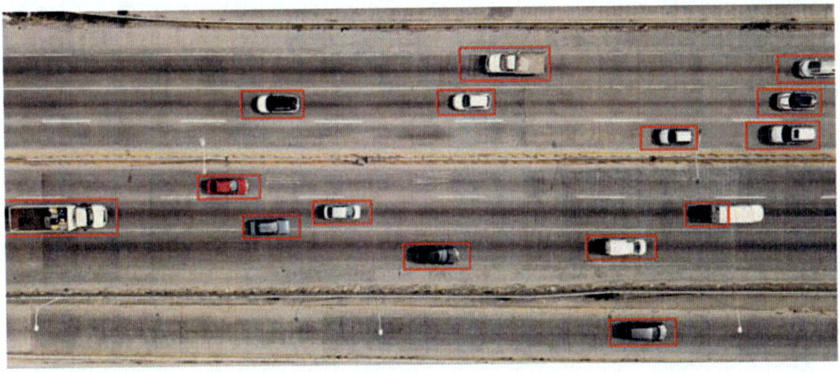

Fig. 3 Output for contour detection

are required to do thresholding first; then, we can apply our contour functions to find objects boundaries [8] as shown in Fig. 3. The main advantage of using the traditional methods is the high speed of detections that does not require a lot of algorithms or data training. However, it is inefficient in terms of classification, as it will count all the moving objects regardless whether it is a vehicle or human or animal. Moreover, the results are affected by the image exposure and brightness. Therefore, it is important to find an alternative method that can detect the objects and classify them.

3 Convolutional Neural Network (CNN)

CNN is a kind of deep neural networks that contains neurons as building blocks of the convolutional neural network [9]. The advantage of the convolutional neural network over the ordinary network is its ability of receiving the input in the form

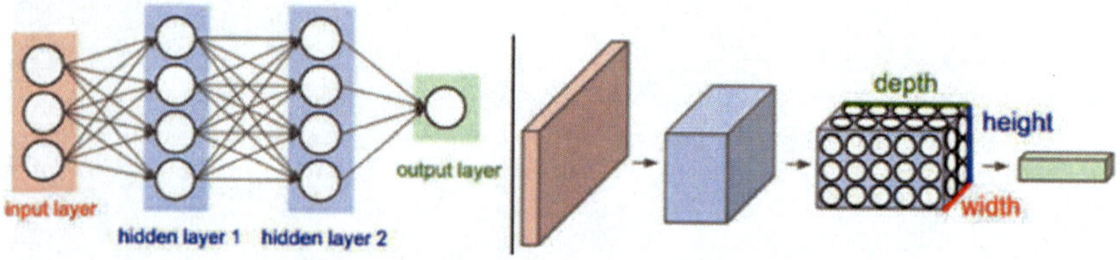

Fig. 4 A regular three-layer CNN structure [11]

of images. The CNN is designed in a way to imitate the work of the frontal lobe of human brain that is responsible about the processing visual information. It can be considered as a set of layers, which are input layer, hidden layers, and output layer. Each layer has a respective transformation from one activations volume to another. There are different types of layers like Pool and ReLU. Each layer has an input with three dimensions and do the respective transformation. The output of each layer will be fed to the next layer and so on as clarified in Fig. 4. In general, the input of the CNN is a low resolutions image (normally 32 * 32 * 32). Although using CNN is better than the traditional methods, it still has limitation as it is affected by scale changes [10].

4 YOLO

YOLO stand for (You Only Look Once). It is considered as one of the best and effective algorithms for object detection in real time. The main challenge in object detection is to determine what is the object type and where it is located with respect to your frame [12]. The most repeated words in machine vision are detections and classification. Although the difference is unnoticeable and confusing, it is important to be cleared to know how they are different. In terms of complexity, the detection problems are tougher. As classification has the ability to recognize the object only not to determine its location on the frame. So if we defined the detection by two words: what and where. We can see classification only achieves the first part only. Moreover, for multiple objects, classifications are not valid as it is working for the images that has only one object [13].

YOLO has fascinating advantages that make it one of the pioneering algorithms in object detection. In addition to, it has ability to do real-time detection; it has a high accuracy result. In YOLO, only one CNN can handle numerous bounding boxes that considered as the regions of interests and recognize the class. Moreover, it is detection speed is extremely fast compared to other algorithms [14]. Figure 5 shows the CNN structure for YOLO.

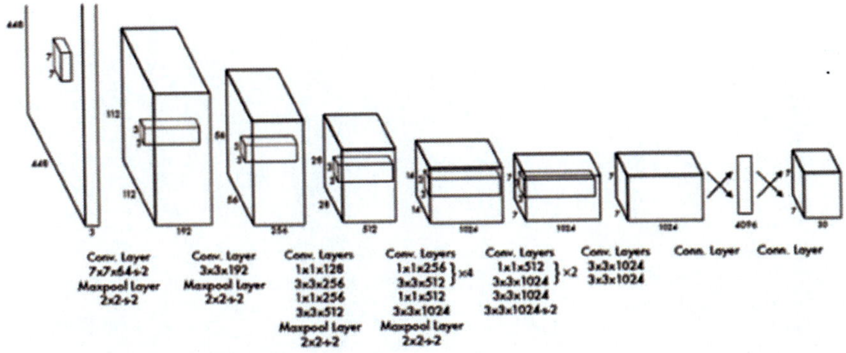

Fig. 5 CNN structure for YOLO [15]

5 Experimental Results: YOLO for Car Counting

YOLO has the ability to detect up to 80 different class with high speed of detection and accuracy. This gives YOLO a lot of advantages and wide range of applications. One of the applications is car-counting system. The vehicles detections is performed over a set of steps. Initially, the data come from the camera feed the neural network that will do a road extraction to give better results [16, 17]. Then, the road-extracted frame will be fed to the algorithms to detect the objects classes. Figure 6 shows the YOLO block diagram for car counting.

Although the output results of Yolo are accurate, there is around 10% in counting. One of the limitations of using Yolo is the difficulty in detecting the far way cars that are small camera frame or vehicles that moves with high speed [19] as traffic videos [20, 21] in Fig. 7.

The YOLO classification output is shown in Fig. 8 where the percentage of vehicle accuracy is highlighted for comparison between person, car, and truck. Figure 9 shows YOLO car-counting system with vehicles direction labeling

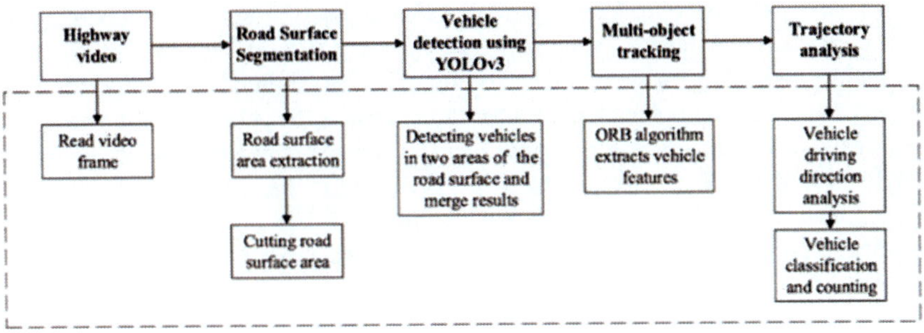

Fig. 6 YOLO for car counting [18]

Fig. 7 Yolo limitations in detecting small cars

Fig. 8 YOLO classification output

6 Conclusion

In this paper, we have introduced a machine vision-based system for car counting in both traditional and deep learning methods. For deep learning, the system is based on YOLO algorithm for detections and classification. Using YOLO algorithm, it gives the accuracy and speed required for the system as the vehicles are moving and required a fast detection. On the other hand, if the car is too fast or far, accuracy maybe drop down. For car counting, it is observed that the output results have an approximate error around 10%. Object tracking and detection still a challenging problem that requires more research to improve the accuracy.

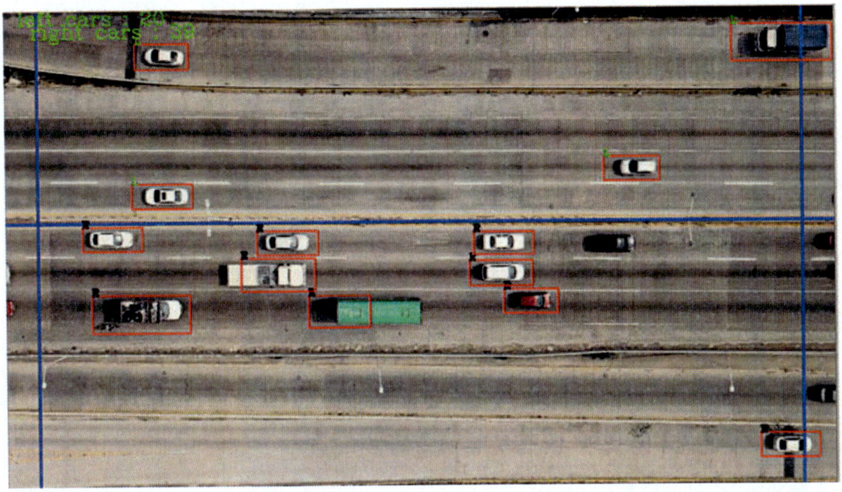

Fig. 9 YOLO car-counting system with vehicles direction labeling (Drone frame [7])

Acknowledgements The research has been carried out under program Research Excellence Consortium [JPT (BPKI) 1000/016/018/25 (57)] provided by Ministry of Higher Education Malaysia (MOHE). The author also would like to thank Universiti Teknologi Malaysia (UTM) for work and facilities support.

References

1. V. Dharmaraj, C. Vijayanand, Artificial intelligence (AI) in agriculture. Int. J. Curr. Microbiol. App. Sci. **7**(12), 2122–2128 (2018). https://doi.org/10.20546/ijcmas.2018.712.241
2. Z. Xue-Wu, D. Yan-Qiong, L. Yan-Yun, S. Ai-Ye, L. Rui-Yu, A vision inspection system for the surface defects of strongly reflected metal based on multi-class SVM. Expert Syst. Appl. **38**(5), 5930–5939 (2011)
3. B. Gholami, W.M. Haddad, J.M. Bailey, AI in the ICU: In the intensive care unit, artificial intelligence can keep watch. IEEE Spectrum **55**(10), 31–35 (2018). https://doi.org/10.1109/MSPEC.2018.8482421
4. K. Abdulrahim, R.A. Salam, Traffic surveillance: a review of vision-based vehicle detection, recognition and tracking. Int. J. Appl. Eng. Res. **11**(1), 713–726 (2016)
5. H. Asaidi, A. Aarab, M. Bellouki, Shadow elimination and vehicles classification approaches in traffic video surveillance context. J. Vis. Lang. Comput. **25**(4), 333–345 (2014)
6. L.I. Qiu-Lin, H.E. Jia-Feng, Vehicles detection based on three-frame-difference method and cross-entropy threshold method. Comput. Eng. **37**(4), 172–174 (2011)
7. Drone Traffic Monitoring by drone Footage Show Highway and Better Traffic Solution. https://www.youtube.com/watch?v=YxN6npu9EPw&ab_channel=Allin1Videos
8. J. Patreek, Build your own Vehicle Detection Model using OpenCV and Python (2020). https://www.analyticsvidhya.com/blog/2020/04/vehicle-detection-opencv-python
9. S. Albawi, T.A. Mohammed, S. Al-Zawi, Understanding of a convolutional neural network. Int. Conf. Eng. Technol. (ICET) **2017**, 1–6 (2017). https://doi.org/10.1109/ICEngTechnol.2017.8308186

10. X. Hu, X. Xu, Y. Xiao, C. Hao, S. He, Q. Jing, P.A. Heng, Sinet: a scale-insensitive convolutional neural network for fast vehicle detection. IEEE Trans. Intell. Transp. Syst. **99**, 1–10 (2018)
11. A. Karpathy, *CS231n Convolutional Neural Networks for Visual Recognition*. Stanford Education
12. Z.Q. Zhao, P. Zheng, S.T. Xu, X. Wu, Object detection with deep learning: a review (2018). arXiv:1807.05511
13. W. Luo, J. Xing, A. Milan, X. Zhang, W. Liu, X. Zhao, T.K. Kim, Multiple object tracking: a literature review (2014). arXiv:1409.7618
14. Z. Cai, Q. Fan, R.S. Feris, N. Vasconcelos, A unified multi-scale deep convolutional neural network for fast object detection, in *2016 European Conference on Computer Vision* (Springer International Publishing, 2016), pp. 354–370. https://doi.org/10.1007/978-3-319-46493-0_22
15. C. Kwan, B. Chou, J. Yang, A. Rangamani, T. Tran, J. Zhang, R. Etienne-Cummings, Deep learning-based target tracking and classification for low quality videos using coded aperture cameras. Sensors **19**, 3702 (2019). https://doi.org/10.3390/s19173702
16. M. Radhakrishnan, Video object extraction by using background subtraction techniques for sports applications. Digital Image Processing **5**(9), 91–97 (2013)
17. D. Rosenbaum, F. Kurz, U. Thomas, S. Suri, P. Reinartz, Towards automatic near real-time traffic monitoring with an airborne wide angle camera system. Eur. Transp. Res. Rev. **1**(1), 11–21 (2009)
18. H. Song, H. Liang, H. Li, Vision-based vehicle detection and counting system using deep learning in highway scenes. Eur. Transp. Res. Rev. **11**, 51 (2019). https://doi.org/10.1186/s12544-019-0390-4
19. G. Palubinskas, F. Kurz, P. Reinartz, Model based traffic congestion detection in optical remote sensing imagery. Eur. Transp. Res. Rev. **2**(2), 85–92 (2010)
20. WORST Cross Walk Accidents and CLOSE CALLS—BESTCompilation. https://www.youtube.com/watch?v=c1Hwm-vH9mo&has_verified=1&ab_channel=RedDFilm
21. A Road Traffic UK HD—Rush Hour—British Highway Traffic May 2017. https://www.youtube.com/watch?v=UM0hX7nomi8&ab_channel=bitgood

Automated Detection of Cataract Using a Deep Learning Technique

R. Angeline, R. Vani, A. Jeshron Sonali, and Dosapati Anhiti Rao

Abstract Eye sight is expressed in terms of visual acuity, more than which, it is the process of deriving meaning from what is seen. Cataract is one of the most pervasive reasons for visual deficiency in the industrialized world, representing more than half of visual impairment. It is a state when densely clouded form developed around the natural lens causing partial and blurred vision. Early recognition and treatment can lessen the enduring of cataract in patients and forestall visual weakness from transforming into visual impairment. In this study, we aim to classify the input eye image as cataract or normal. In this system, the input image is taken as a front viewed eye images captured with smart phone camera application instead of using the existing fundus images such that it makes the model user friendly and less expensive, leading to early detection of cataract. In this model, the VGG16, a pretrained model, is applied as on convolutional neural network (CNN) architecture which is deployed over the data set obtained. On modelling the system with collected samples, the neural network algorithm has resulted with an accuracy rate of 92.1% for cataract detection, and finally, the system is trained to classify an input eye image as cataract or normal eye.

Keywords VGG16 · Convolutional neural network architecture · Cataract detection

R. Angeline (✉) · D. A. Rao
Department of Computer Science and Engineering, SRM Institute of Science and Technology, Ramapuram, Chennai, India
e-mail: angelinr1@srmist.edu.in

R. Vani · A. Jeshron Sonali
Department of Electronics and Communication Engineering, SRM Institute of Science and Technology, Ramapuram, Chennai, India
e-mail: vanir@srmist.edu.in

A. Kumar et al. (eds.), *Computational Intelligence in Machine Learning*, Lecture Notes in Electrical Engineering 834, https://doi.org/10.1007/978-981-16-8484-5_38

1 Introduction

Cataract as defined is one of the major causes of eye vision loss apart from diseases like diabetic retinopathy, glaucoma, amblyopia, strabismus and so on. The natural lens of the eye are made out of a mix of water and proteins. The lens itself lack nerves, veins or connective tissue sometimes. The positioning of proteins will in general become increasingly insoluble as we age, and the lens dynamically lose its lucidity and focussing (accommodative) capacity. This restricts the amount of light passing through the lens to the retina, and the vision becomes hazy and blurred. Despite the fact that age is the main cause of cataract, there is a various lot of reasons behind this disorder. Congenital cataract might be caused in infants due to heredity or developed due to an infection (rubella) in the mother during pregnancy. It might slowly develop in adults and children due to exposure to radiations or drugs and over steroids intake, due to traumatic disturbance in the eye lens. The various cataract type includes nuclear, cortical, posterior capsular, congenital, secondary cataract, traumatic and radiation cataracts.

According to World Health Organization (WHO) reports and surveys, cataract is found to be the cause of vision impairment in over 62.5 million people [1]. Studies show that age-related reasons are more generic in most cases of cataract especially nuclear and cortical cataracts [2]. Early detection of cataract is of paramount importance to prevent it getting severe leading to blindness. Ophthalmology methods are used in general as cataract detection method for which medical expertise is essential, apart from this method being expensive, people without any prefatory medical knowledge would not know if their eyesight is affected because of cataract and would not take initiative to check the issue out. In such cases, it might lead to severity of cataract. To avoid such complexities, early analysis is more essential, which can be done using smartphone camera application by capturing the affected eye images. In order to detect, an effective neural network model must be trained with as many images as possible. So, in this system, using front view, eye image would be apt for early prediction purpose rather than using fundas images data.

2 Related Work

Cataract detection has always been a topic of research because of the fact that how complicated it can get when left undetected. Recently, it is detection and research has increased exponentially with the upcoming of new classification methodologies. In the existing models, the vast majority of prediction and classification are done using fundus images collected from reputed hospital database and the slit lamp based fundus images does not give a proper solution for detection of cataract at an earlier stage and processing. Some of the works also include android apps-based data collection. Majority of the models/systems employ image processing techniques, KNN, SVM which does not help much in attaining a state-of-the-art accuracy.

Zhou et al. [3] have modelled a paper which aims in automatic detection classification and grading of cataract which uses fundus image data set and a combination of improvised harr and visible structure as its features. By classifying as with and without prior medical knowledge, it uses multilayer perceptron with discrete state transition or exponential DST as classifiers with prior knowledge and residual neural networks with DST (DST-ResNet) or EDST (EDST-ResNet) for without prior medicine knowledge. With an state-of-art accuracy rate, they have graded the cataract into four stages successfully which are non-cataract, mild, moderate and severe.

Karamihan et al. [4] have modelled a work by developing cataract detection system using a Raspberry Pi single board-based detecting system. RPI camera module V2 is used for capturing the eye image. The model aims to detect cataract by not using the dilating drops and determines the grade, severity, colour and hardenss of cataract. Deep convolutional neural network is used along with GoogLeNet V4 transfer learning. They have also applied matrix laboratory image processing method based on LOCS III.

The experiment is done using RPI environment, and the transfer features were extracted and applied to the model. The model is implemented with an accuracy rate of 96.49% for cataract with a 5.5-h lasting experimentation, thereby giving the proper results and graph plot of accuracy using MATLAB.

Sahana et al. [5] have come up with a model to detect cataract using Inception V3 deep convolutional neural network architecture. The model was implemented on data set collected via various hospital databases which are originally obtained using slit lamp technique. Here, they have aimed to classify as mature and immature cataract with the aid of the collected data. On training this model, it has been shown to have obtained 87.5% accuracy using the Inception version 3 CNN architecture.

Agarwal et al. [6] have modelled a paper that proposes an android application-based cataract detection. The aim of this method is to make detection easier with the usage of smart phones. In this model, they have developed an Android App using which the data set is obtained which is subjected to image processing using OpenCV. The processed images are further classified using KNN and SVM for predicting whether the input eye image has cataract or not. On implementing, the results have shown KNN as optimum algorithm with 83.07% accuracy rates than SVM which gave an accuracy of 75.2%.

Zhang et al. [7] in their work have attempted to detect and grade cataract using DCNN method, where they have used the rear fundus image data. They have done the processing of image, and uneven illumination is overcome by using G-filter and then applying it to the deep convolutional network [8]. On the final edge, they have graded cataract as non-cataract, mild, moderate and severe with an accuracy of 93.52%, proving that the accuracy rate increases on usage of the G-filter as normal RGB set has given an accuracy rate of 89.92%. Finally, it has graded cataract with much efficient modelling.

3 Methodology

In this proposed system, we aim to detect cataract in an efficient way. In the sense, to make cataract detection with less intricacy and much lucid user interaction. We focus to develop a model that would detect cataract at early juncture such that future complications can be avoided. Rather than using a fundus image, which gives the rear view of eye, we have decided to use the front view of the eye, here, which involves several advantages such as low cost, easy to understand the usage, can easily reach common people who have no knowledge in the medical field. The stack of the model can be an Android App or any built device such as CATRA [9], and here, we concentrate on the method of implementation. Let us go through the methodology as follows.

3.1 Data

Gathering data were the first task of the process. The data, we use here, are an image data set. More the number of images more is the efficiency of the model. The original data were collected separately as cataract and normal eye image. Keeping that as goal, we collected data from various repositories online which include Kaggle, GitHub and various data sites. Apart from collecting it online, an offline data search was done to collect data for both cataract and normal images. Offline fetching was done through normal smart phone camera (which had good image resolution) and stored in memory card. On a total of 2120 images of cataract eye and 1600 images of normal eye, they were obtained separately. The data set collected had a lot of misinterpreted data which had to be preprocessed.

3.2 Preprocessing the Data

Mislabeling [occurrence of a non-cataract in the cataract set (Figs. 1 and 2)] of data was found to be one setback which was manually determined and moved to the respective set. Octoparse, an image scrapping API tool, which integrates with Python IDE is used to overcome the non-uniformity to restructure the data set. In the next step, the resizing of image is applied where the data set is re-sized into 224 * 224 pixels in this model using cv2.resize function.

In the next step, image augmentation is done. Augmentation is the process where the data set is expanded artificially to increase the training set data samples. Any deep learning model requires more amount of data for making the model perform better. This technique makes the image to be transformed in various forms like flipped horizontal and vertical fashion, shifting the image, random rotation, changing brightness, zooming the image and so on, thus creating increased number of samples

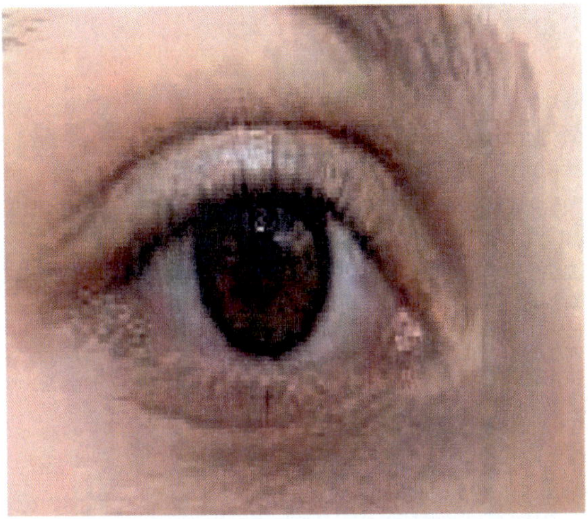

Fig. 1 Non-cataract front view

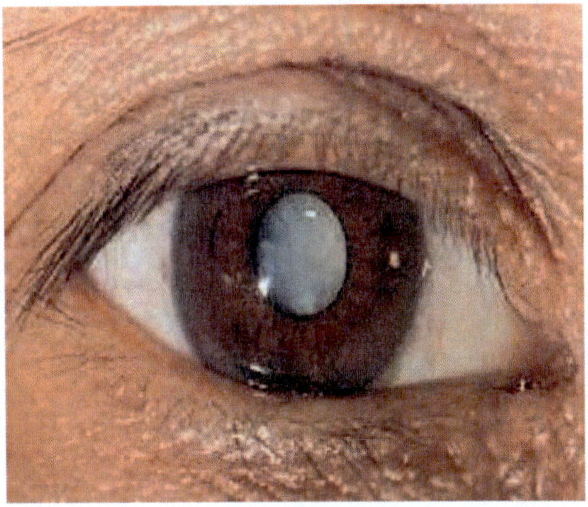

Fig. 2 Cataract front view

by nearly 3 to 4 times of the original number of images. Keras provides a way to implement this image augmentation using the class 'ImageDataGenerator' which uses the parameters such as re-scale, shear, zoom, horizontal flip in our model. On completing the data collection and preprocessing steps, the data set is totally fit to be applied to the neural network model.

3.3 Keras and TensorFlow

In the working of our model, keras was the major python libraries used. Keras is an API, an high level version written using the Python language which can run on top of platforms such as TensorFlow, Theano and so on. It is a powerful source yet user compatible and most robust and extensible. For instance, the previously mentioned ImageDataGenerator is implemented using keras. On the other hand, TensorFlow is the core platform to build, train and deploy machine learning and deep learning models in a facile manner.

3.4 Architecture

Deep convolutional neural network has been in practice for image classification. The primary motive of this model in Fig. 3 is to predict whether the loaded input image is an eye image with cataract or without cataract affected. There are several pretrained model which can be implemented and such a method is the transfer learning. It is expressed through pretrained models, that is, model is trained on existing huge volume of data set for resolving similar problems, for instance, classification of eye image in this model. Basic idea is to transfer the knowledge of pre-existing similar model to our model to attend to the new problem task, as data in the previous model is similar to the one which we are trying to get on with. On applying this concept, instead of training the models over and over again making the work tedious, existing trained patterns can be used for fine tuning such that time consumption is reduced. One such application is VGG16, which was trained over the existing ImageNet database that consists of million plus images for classification purpose. It is a 41 layered network on total containing 16 weighted, 13 convolutional and 3 fully connected layering and are separated as blocks of five. Instead of initializing all the layers randomly, the pretrained model N layers are added to the target model. Usually, either the top layer or the bottom layer is chopped off and is appended to new model, but the N layers exist as the same, which is termed as feature extraction or transfer parameters.

Apart from the pretrained parameters, rest of the new model layer is initialized. Here, the extracted feature of VGG16 (as in Fig. 4) is been applied over our model to classify the images as cataract and normal. In the course action of implementation, the VGG16 model was directly included due to the advantage of keras library. The features extracted using the pretrained VGG16 architecture is appended to our model. The process included saving and loading of the weights where the image input is applied to the model, and finally, prediction is depicted as 1 for a normal eye and 0 for a cataract affected person's eye image.

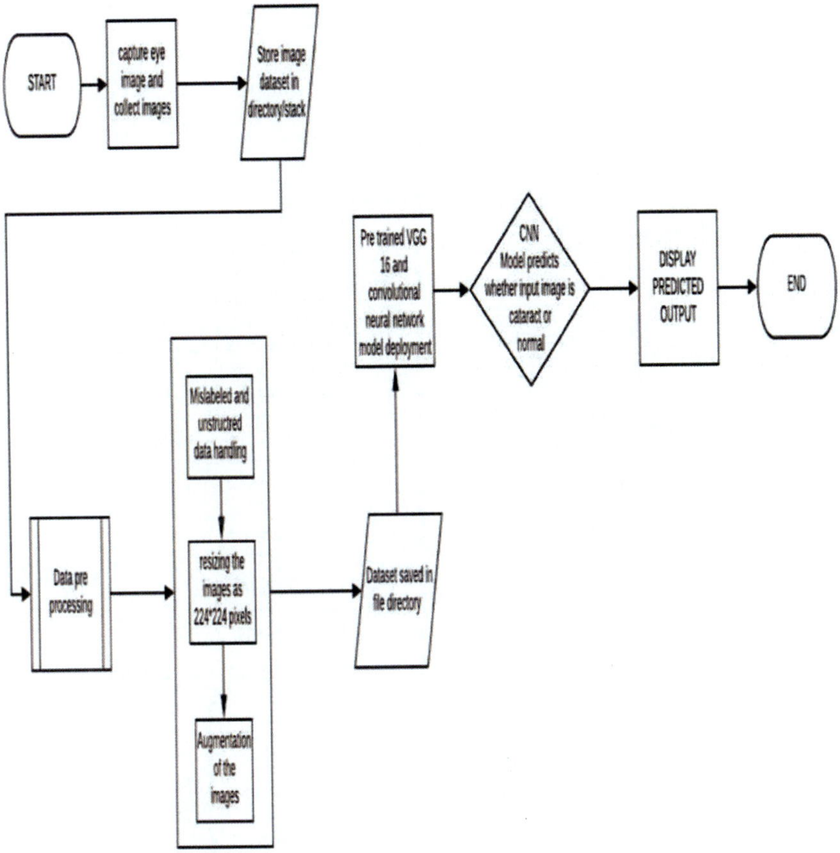

Fig. 3 Module flow diagram

4 Implementation and Results

On experimenting our idea of everything mentioned in the previous sections, we have come up with an efficient classification model for cataract disease detection at an early juncture, which gave an accuracy rate of 92.1%. On further note, the salient feature point of this experimentation is to classify an image as cataract eye or non-cataract eye when the use uploads the image to this model. The modelling of concept is as shown in the Fig. 5.

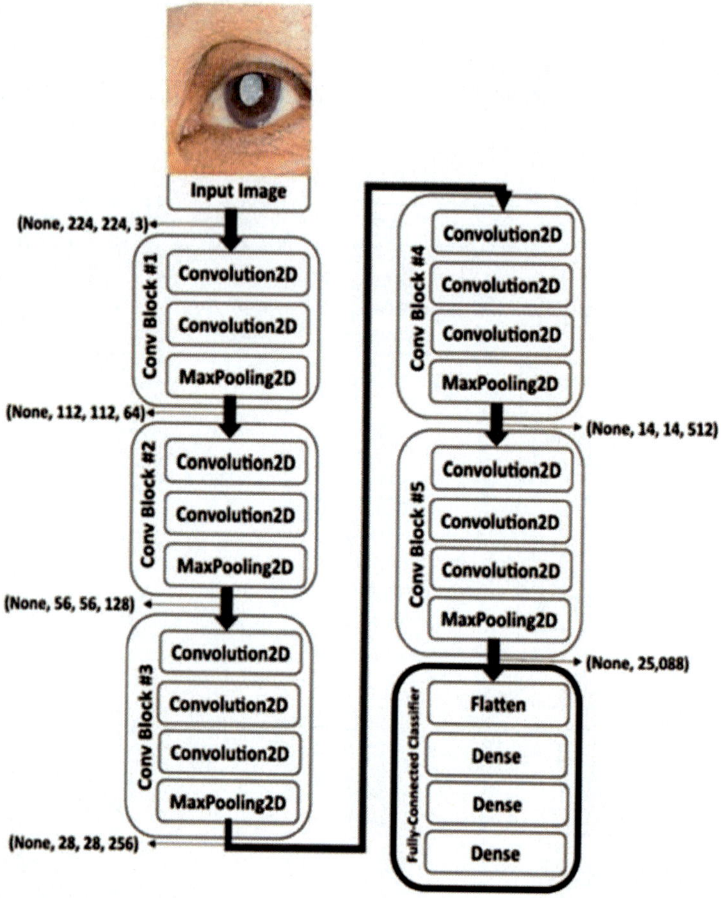

Fig. 4 VGG16 layer model

5 Conclusion

We can, thus, conclude that detection of cataract in human eye is of utter importance in accordance to its ramification. Our model has classified the image with an efficient model giving out an accuracy rate of 92.1% which validates the model as an efficient one. This can be used for future use by deploying it with a proper baseline and stack, such that it could reach market at a methodical way which would be economical.

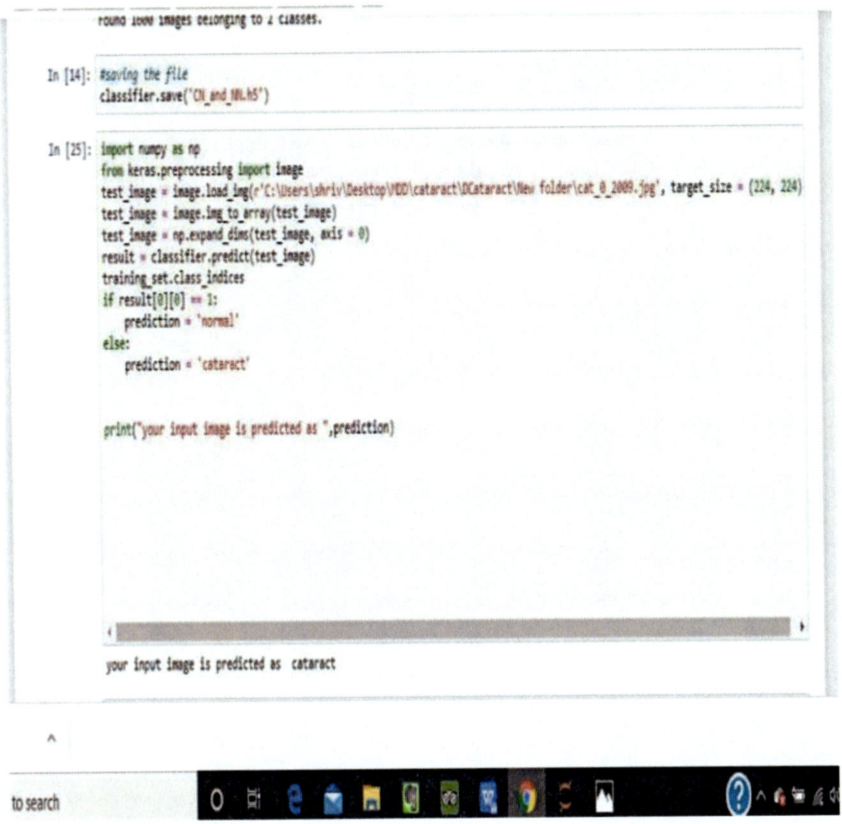

Fig. 5 Classification results

References

1. T.Y. Wong, S.C. Loon, S.M. Saw, The epidemiology of age related eye diseases in Asia. Br. J. Ophthalmol. **90**(4), 506–511 (2006)
2. World Health Organization. Global status report on alcohol and health 2018
3. Y. Zhou, G. Li, H. Li, Automatic cataract classification using deep neural network with discrete state transition. IEEE Trans. Med. Imaging **39**(2), 436–446 (2019)
4. K.C. Karamihan, I.D.F. Agustino, R.B.B. Bionesta, F.C. Tuason, S.V.E. Arellano, P.A.M. Esguerra, SBC-based cataract detection system using deep convolutional neural network with transfer learning algorithm. IJRTE **8**(2) (2019)
5. M. Sahana, S. Gowrishankar, Identification and classification of cataract stages in old age people using deep learning algorithm
6. V. Agarwal, G. Vaibhav, V.M. Vashisht, K. Sharma, N. Sharma, Mobile application based cataract detection system, in *2019 3rd International Conference on Trends in Electronics and Informatics (ICOEI)* (IEEE, 2019), pp. 780–787.

7. L. Zhang, J. Li, H. Han, B. Liu, J. Yang, Q. Wang, Automatic cataract detection and grading using deep convolutional neural network, in *2017 IEEE 14th International Conference on Networking, Sensing and Control (ICNSC)* (IEEE, 2017), pp. 60–65
8. J.C. Kavitha, R.Vani, D.Subitha, Novel approach for melanoma detection through iterative deep vector network. J. Ambient Intell. Humanized Comput. 1–10 (2021)
9. V.F. Pamplona, E.B. Passos, J. Zizka, M.M. Oliveira, E. Lawson, E. Clua, R. Raskar, CATRA: interactive measuring and modeling of cataracts. ACM Trans. Graph. (TOG) **30**(4), 1–8 (2011)

Program Synthesis—A Survey

S. Jacindha, G. Abishek, and P. Vasuki

Abstract The construction of programs from the user intends is called program synthesis. Program synthesis is being used in our daily life in various ways. For instance, Microsoft Excel has a feature called FlashFill that is capable of string manipulation on the data as per the users' interest. Depending on the specification of users intend, program synthesizers are classified into a deductive, inductive, and natural language-based synthesizer. Deductive synthesizers use specific descriptions or algorithms in generating programs. Inductive program synthesis uses inductive reasoning on generating programs out of this specification. Programming by examples (PBE) method comes under this type. The specification can be a set of exemplary data used ad input and output, use case diagrams, and so on. In natural language-based, normal English sentences are given as specifications. The synthesizers use artificial intelligence to synthesize programs. This survey provides an overview of program synthesis from its evolution to real-world applications. This survey also includes a comparison of theory and development of different approaches to program synthesis. ...

Keywords Program synthesis · Inductive program synthesis · Neural program synthesis · Programming by examples (PBE) · Domain-specific language (DSL)

1 Introduction

Modern life has introduced 'Automation' in every aspect. It has also influenced the daily and work-life greatly. For instance, auto-complete features have helped the programmers in fastening programming efficiently. Hence it would be useful to have a program generated automatically from the programmer's intend without him/her writing a code.

S. Jacindha (✉) · G. Abishek · P. Vasuki
Department of IT, Sri Sivasubramaniya Nadar College of Engineering, Chennai, India
e-mail: jacindhas16033@it.ssn.edu.in

G. Abishek
e-mail: AbishekG16001@it.ssn.edu.in

P. Vasuki
e-mail: VasukiP@ssn.edu.in

A. Kumar et al. (eds.), *Computational Intelligence in Machine Learning*, Lecture Notes in Electrical Engineering 834, https://doi.org/10.1007/978-981-16-8484-5_39

$$\text{Program Synthesis: } \forall x, \exists y \rightarrow P(x, f(x)) \text{ } where\ f(x) = y$$
$$\text{Program Induction:} \forall x, \exists y \rightarrow f(x)$$

Fig. 1 Inductive program synthesis

Induction programming is programming using reasoning. The equation in Fig. 1 describes the concept of inductive program and program synthesis. Program induction is identifying a function, which maps the output to the given input. Program synthesis is generating program with that maps input and output using appropriate function. The programmer could make use of automation to generate simple segments and thus enabling him/her to concentrate on code which would need complex logic. Program synthesis is generating programs automatically by the system in an underlying programming language that meets the user specification or purpose provided in some constraints' form. The constraint or specifications are used to guide the process of generating the right program. The constraints can be about the requirements, examples, or parameters like time or space complexity, and specifications can be input–output examples, natural language descriptions, algorithms, UML diagrams, and so on.

Program synthesis helps the programmer to generate simple modules in their project, which speeds up the development, not only in software development but program synthesis also used in data wrangling [16, 20] where it is tedious to extract data from text, string transformations in Microsoft Excel [10], data transformations, intelligent tutoring systems [5], competitive coding [2, 33], etc. Synthesis is also applied in database systems to write complex SQL queries [32]. Experts in data science can use synthesizers to clean the data which normally would consume most of their time. Repetitive tasks in text and string manipulations for writers and day-to-day office workers could be made automatically through program synthesizers. Computer-generated graphics and images are also newly welcomed targets for program synthesis.

The art of automatic program writing has been studied since the 1960s [30, 31]. The study started as a way of proving theorems by computers and then making use of theorem solving to program writing. The term program synthesis was introduced in "Towards automatic program synthesis" [22]. At some point, theorem solving and program synthesis are started to take different paths. And now we have program synthesis as a major field that is fast emerging in computer science.

Program induction is a kind of program synthesis, where this task is automatically finding the output when an input is given, by learning the inherent logic from the given user specifications. It learns to mimic the program rather than explicitly returning them [19]. For example, when input–output examples such as (15,225), (32,1024), (24,576), and (47,2209) are given as specifications, a program inducer learns the logic to be squaring the first number to produce the second number and answers the input from the logic learned. However, the program synthesizer gives a program in an underlying language as the output. However, interpretability and correctness of a program induction problem are difficult.

In the life-span of program synthesis, it evolved from deductive to inductive program synthesis. Later on, machine learning and deep learning techniques were used in synthesizing programs. This survey paper gives an elementary outline in the field of program synthesis. method.

2 Challenges of Program Synthesis

Synthesizing a program is a challenging task, which comprises of subtasks like understanding the requirement, creating a specification, interpreting the user's intent, and implementing synthesis based on intent. As every subtask is subjective, a generic synthesis is not achievable in all domains. This section describes some of the challenges faced in users intend, dataset, and search and ranking used in the existing system of program synthesis

2.1 Users Intend

Most of the synthesizers are based on the user intend, i.e., the specification of the program. Nevertheless, understanding the user intends through the specification is hard. Specifications can be a natural language sentence [16, 20], input–output examples [3, 7, 28], or demonstration [6, 23]. Incomplete and ambiguous specifications may lead to the generation of the wrong program. Whereas, correct interpretations of specification tends to increase the efficiency of the synthesizers. For instance, the context-free parser generates a parse tree for the given program. Program synthesis has been built to learn context-free parser from a set of programs (input) and corresponding parse tree (output) [31]. Wang et al. [32] use deep learning to convert natural language sentences to the corresponding DSL. Le et al. [20] proposed an interactive program synthesis model to reduce the cognitive difficulties that a user continuously trying to ensure that the system provides the aspired program.

2.2 Program Space

Program space, in the context of program synthesis, means the range of possible programs or solutions. Finding the right program from a huge space is one of the prime challenging aspects of program synthesis. It would take a lot of computational time to find a compatible program to satisfy the user intent. Also, the program that satisfies the given constraint may not be present in the given program space. Consequently, program space might require a large storage space to store the program and a good searching algorithm to retrieve the correct one efficiently.

```
Static void main(String[] args){
        int[] a = [2,7,-9,0,1];
        int max = 0;
        for(int var=0; var<=a.length;var++)
                if (max < a[var] *3)
                        max = a[var]*3;
        System.out.print(max);
}
```

```
Static void main(String[] args){
        int[] a = [2,7,-9,0,1];
        int var =0,max = 0;
        while(var <=a.length){
                if (max < a[var] *3)
                        max = a[var]*3;
                var++;
        }
        System.out.print(max);
}
```

Fig. 2 Program on finding largest number from the list when multiplied by 3. The Java code snippet on the left uses for loop, whereas, on the right, it uses while loop. The difference in function semantics on same logic

Fig. 3 This code logically differs from the problem of 2. Instead of multiplying every number by 3 and finding the greatest, it finds the largest number and multiplies by 3

```
Static void main(String[] args){
        int[] a = [2,7,-9,0,1];
        int max = 0;
        for(int var=0;var<=a.length;var++){
                if(max < a[var])
                        max = a[var];
        }
        System.out.println(max*3);
```

A program can be written in more than one way in a given target programming language. It can either be logically or semantically different. Let us consider the synthesis problem to be: Find the largest number in an array when multiplied by 3. Figure 1 shows a solution that can be written in two ways based on the kind of loop function used.

The problem can also be approached in more than one logic. Figure 3 shows an alternate logic for the problem in Fig. 2. Hence, the problem lies in the freedom of functions in any programming language. One of the solutions is to use domain-specific language (DSL), which is independent of programming language. It contains functions that are commonly used in the given domain or application. From Fig. 4, there can be different logic that produces the same result for the same set of inputs. In (A), we use sort function and then output the last element in the list, which is the largest. In (B), we use the MAXIMUM function, which greatly reduces the lines of code. MAP applies the given task, which in this case multiply by three on all the elements of the list. Therefore, the program has to be selected on some criteria as the final solution. Thus, the program has to be selected on some criteria as the ultimate solution.

2.3 *Program Search and Ranking*

The ranking has been used in synthesis from auto-completion of code to robotics auto-programming. These synthesizers may generate or identify more than one program

Fig. 4 DeepCoder DSL for the same problem of 2 and 3

```
a <- [int]
b <- MAP(*3) a
c <- SORT b
d <- TAIL c
```

a

```
a <- [int]
b <- MAP(*3) a
c <- MAXIMUM b
```

b

which could satisfy the given user specifications (i.e., input–output examples). The synthesis system executes top-down, bottom-up search or any stochastic search in the search space. Then, the synthesis system checks the program against the user constraints and lists the programs that satisfy the constraints. The resultant programs were ranked using any evaluation criteria like speed or coverage (Number of I/O exampled correctly decoded) [10, 28, 32] as well, the size of the program [25] using statistical methods or machine learning techniques. The correctness of the synthesized programs depends on the volume of the training set. If the training set is small in size, the integrated program may overfit to the training, and so, it may give correct output to trained cases but may fail for the unseen test case [28].

3 Type of Synthesizers

Based on the mode of synthesis from different kind of specifications, synthesizers can be classified into deductive, inductive type.

3.1 Deductive Program Synthesis

Deductive program synthesis is generating program from high-level description of programs.The deductive program synthesis involves mechanical theorem proving methods [8, 9] and formal methods for synthesizing programs. They require detailed specifications as opposed to the inductive program synthesis. The advantage of using theorem proving techniques with detailed specifications is that there is a minimal chance of logical errors. It has a varied application over education [29], robotics [7] and software engineering [14, 15]. STanford Research Institute Problem Solver (STRIPS) [7] is an automatic planner written in LISP that can solve robot problems. PROgram Writing (PROW) [31] is a program that produces code in LISP programming language when specifications are given in predicate calculus. It consists of two parts; in the first part, the theorem prover proves whether the specification(theorem) exists. In the second part, the output, which is the 'proof' gets fed into another program that produces the desired code. PROW is also capable of producing code in another language when the second part is modified. Manna and Waldinger [21] presented a theoretical deductive program synthesis approach that relies on a theorem proving method with a combination of the conversion rules, unification, and mathematical induction in the unique framework. The limitation of deductive program synthesis is its difficulty in specifying the user's interest. Many synthesizers have combined advantages of both deductive and inductive [17, 18, 26].

3.2 Inductive Program Synthesis

Inductive program synthesis is generating programs from the set of instances. Inductive programming observes the specific set of input and outputs and creates programs to generalize them. The terms inductive program synthesis and programming by examples (PBE) are alternatively used. This sub-field is user compatible approach and comparatively easier than deductive programming with respect to user intent. Thus, it had been explored the most. A user who is incapable of programming can easily explain to the computer through examples, and the PBE system comes with a program for it. The early works depend on search algorithms like enumerative, stochastic, constraint-based, and version-space algebra-based algorithms. But search-based techniques tend to be slow, and so, artificial intelligence techniques were used to guide and fasten the search for a suitable program. Counter-example-guided inductive synthesis (CEGIS) [4] is an approach under inductive synthesis, which makes use of verification oracles to produce counter-examples that are used to generate correct programs. The inductive program synthesis also tries to generalize from the incomplete examples. In recent years, researchers combine inductive and deductive programming approaches. Inductive programming creates candidate solutions from the specifications and verifies the functionality with the illustrative examples given by the user. For instance, FlashFill [10] and automate synthesis in spreadsheets [11, 13, 27] are the most popular real-world program synthesis application, synthesizing from very few input–output examples. PBE also finds its usage in relational databases for schema refactoring [32], which is a simplistic change in the schema which enhances its design, also maintaining both the behavior semantics and information semantics.

4 Program Synthesis Using Deep Learning

Deep learning solves the problems which are complicated by other methods. Thus, recently researcher started using deep learning for program synthesis. Though neural nets are weak in interoperability and consume more training time, they are capable of generalizing the problem, based on given training set and specification. As there are many developments happening in that field, an optimal system built on neural program synthesis is discussed below.

4.1 Case Study: DeepCoder and PC Coder

DeepCoder [2] is an important neural network model that synthesizes programs for competitive programming from input–output examples and utilizes neural networks to guide the search process. It follows an approach called learning inductive program synthesis (LIPS). The authors have defined a separate domain-specific language. The process is as follows:

- For the given maximum program size, datasets are generated from the defined DSL.
- DeepCoder [2] is one of the successful neural network models that used to synthesize programs. DeepCoder generates programs for competitive programming from input–output examples and utilizes neural networks to guide the search process. It follows an approach called learning inductive program synthesis (LIPS). The authors have defined a separate domain-specific language. The process is as follows:
- For the given maximum program size, datasets are generated from the defined DSL.
- Training of the neural network consists of two components, namely encoder and decoder.
    - First, each input–output example undergoes embeddings to convert into a single vector.
    - This vector is fed as input to the simple feed-forward architecture encoder. It maps them into a latent real-valued vector.
    - The decoder uses the output of the encoder to predict possible functions (or attributes) in the program.
- The predictions are used in helping the search the optimal program.

DeepCoder limits itself to small, simple competitive style programs due to their defined DSL. Zohar and Wolf created Predict and Collect Coder (PCCoder) [2] to generate long programs. They use an extended version of the DeepCoder DSL and train two neural networks. The first neural network predicts the statements in the program, both the function and the operators. The second network is the learned garbage.

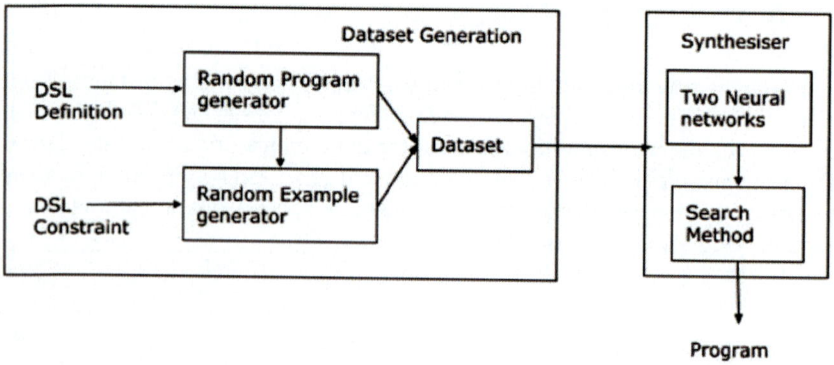

Fig. 5 Architecture of PC coder

DeepCoder limits itself to small, simple competitive style programs due to their defined DSL. Zohar and Wolf created PCCoder (Predict and Collect Coder) [2] to generate long programs. They use an extended version of the DeepCoder DSL and train two neural networks. The first neural network predicts the statements in the program, both the function and the operators. The second network is the learned garbage collector and redefines the program, which predicts the variables that are to be eliminated. PC Coder empirically proves as faster system than DeepCoder. The architecture of PC Coder is presented in Fig. 5

5 Domain-Specific Language

DSL is the programming languages defined for a specific application or scenario. The two most famous DSL are FlashFill DSL and DeepCoder DSL [2]. DeepCoder DSL focuses on integer and contains first-order functions like HEAD, LAST, TAKE, DROP, ACCESS, MINIMUM, MAXIMUM, REVERSE, SORT, SUM, and the higher-order functions like MAP, FILTER, COUNT, ZIPWITH, SCANL1, and higher-order lambda functions. FlashFill DSL [6] is for string manipulation and slightly similar to excel string functions. It has functions like concatenate, substring, loop, match, and switch. Some of the advantages of DSL are listed below.

- DSL is independent of the programming language. By adding a new module, DSL may be converted into the required language code.
- The universal programming language has many functions through which logic can be expressed differently. Thus, DSL minimizes the replication of the same module.
- Smaller representation makes learning easier.
- Restrictions on the language make it easier to search for the program space.

6 Real-World Applications

Some of the program synthesizers that have been made into products are discussed in this section.

6.1 PROSE SDK

Microsoft PROSE SDK is a framework of technologies developed for inductive programming environment, programming by examples: automatic generation of programs from input–output examples. It is used to generate the utilities like FlashFill, FlashExtract, and FlashRelate which are discussed in the subsequent sections. It includes a predefined suite of technologies for automatic programming and data wrangling. If the syntax for the domain-specific language is given, PROSE will create a synthesizer to generate DSL programs. It is based on 'FlashMeta': a framework for inductive program synthesis [24].

6.2 FlashFill

Microsoft for Excel, 'Quckcode' in the year 2010 and released the data tools 'FlashFill' in 2013 and release the revisions in the years 2016 and 2019. These tools in Microsoft Excel allow us to combine, extract, or transform data based on a few examples. FlashFill is a special tool that analyzes the information entered and automatically fills data when it identifies a pattern. It seizes a tedious task that would take hours to be performed manually and executes it automatically in a flash, based on the example given by the user. These program synthesis tools can perform, string, number, and data transformations.

Table 1 shows an example of a FlashFill, where the name has to be extracted from the email address. The first two rows were filled as the input–output example to FlashFill. For the rest of the rows has only input, then FlashFill generates the code

Table 1 Sample for FlashExtract

Email	**Name**
johndoe@gmail.com	John Doe
bill@gmail.com	Bill
anita@yahoo.com	Anita
ram@redmail.com	Ram
janedoe@gmail.com	Jane Doe
alice@gmail.com	Alice

to fill the corresponding output automatically. The FlashFill extracts the name which consists of one or more words using the list of functions in DSL like Switch, SubStr, Concatenate, and TokenSeq which might be used in the generated code.

6.3 FlashExtract

This technology helps in extracting structured data from various types of files like textfiles, webpage, spreadsheet, and logfiles semi-structured data through the programming by examples (PBE) paradigm [20]. For instance, FlashExtract may find the number of students who have got marks from 75 to 100 in a particular subject from university result file without user writes any query or code written by the use. It is featured in Microsoft Powershell (Convert From-String cmdlet) and Azure OMS (Operations Management Suite).

6.4 SyGuS Comp

Syntax-guided program synthesis problem (SyGuS) is to find an implementation that is syntactically and semantically correct and satisfies the user specification with the DSL program [1]. SyGUS gets the template of semantic constraints and logical specification given by the logical formula, set of candidate programs. SyGUS synthesis of the program generates a program based on the inputs and machine learning technology. The specification is logical SyGuS Comp [26] is an annually conducted competition to explore the field of program synthesis. They compare the solvers with a broad set of benchmark samples.

7 Other Program Synthesizers

There are many other program synthesizers exists with applications in different domains. One such utility based on program synthesis is heterogeneous data extraction framework (HDEF) [16] uses machine learning models along with noisy disjunctive program synthesis which is based on the FlashMeta framework, to extract structured information from heterogeneous data. They use conditional random fields (CRFs) and bidirectional long short-term memory augmented with CRFs (LSTM-CRFs) as the ML models. The model tries to label potential attributes that may be noisy. These labels are given as input to program synthesis to generate programs that could cover the input space. The outputs from the model and synthesizer are compared, and this is used to retrain the ML model. It follows an iterative process of retraining the ML models and re-synthesizing the programs so as to increase the quality of extraction and minimize user interaction.

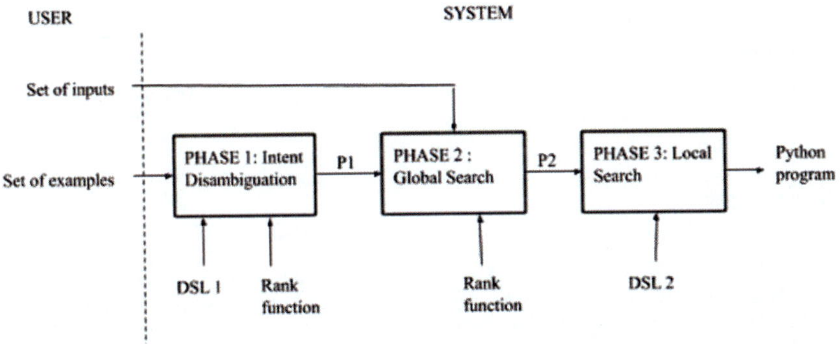

Fig. 6 Block diagram on how quantitative programming by examples works

Most of the approaches till far synthesize programs in DSL. Gulwni et al. [12] introduced the problem of quantitative programming by examples (qPBE) which is to produce the cost-optimized program in a real-world programming language. The approach involves three phases as shown in Figure fig:Phases. In the first phase, a standard PBE system such as FlashFill finds a program P1 over DSL 1 while minimizing the given input–output examples using a ranking function. The second phase applies a global search to find a program P2 equivalent to the one generated in the first phase. The program generated has to satisfy the set of given inputs optimized by another ranking function. The final phase rewrites P2 into DSL 2 whose grammar is similar to the target language.

8 Conclusion

Program synthesis is the generation of programs autonomously by a system that meets a given high-level formal specification. Even though program synthesis has progressed a lot in the following decade, still, the available algorithms are only able to synthesize small programs, and this field has lots of goals and challenges to overcome. From simple programming to robotics auto-programming, this field has shown its influence and significance. Through this literary survey, only an overview of this subject could be touched. The literature review encloses the two most common techniques in program synthesis that exists in state-of-art: inductive program synthesis, deductive program synthesis, which have been implemented using statistical and machine learning approaches. Deductive program synthesis generates the program based on formal specifications like constraints and algorithms. At the same time, inductive program synthesis generates the program based on the examples. Both the synthesis uses domain-specific languages, which may contain a set of functions suitable for that application. Nowadays, researchers use a hybrid synthesis approach using machine learning techniques. We reviewed some real-world example like FlashFill and FlashExtract, which expedite the process compared a human programmer to the greater extend. The applications of synthesis are ranging from a small spreadsheet application to data wrangling and robotic programming.

References

1. R. Alur, R. Bodik, G. Juniwal, M.M.K. Martin, M. Raghothaman, S.A. Seshia, R. Singh, A. Solar-Lezama, E. Torlak, A. Udupa, Syntax-guided synthesis, in *2013 Formal Methods in Computer-Aided Design* (2013), pp. 1–8
2. M. Balog, A.L. Gaunt, M. Brockschmidt, S. Nowozin, D. Tarlow, *Deepcoder: Learning to Write Programs*. arXiv preprint arXiv:1611.01989 (2016)
3. X. Chen, C. Liu, D. Song, *Towards Synthesizing Complex Programs from Input-Output Examples* (2017)
4. C. David, D. Kroening, Program synthesis: challenges and opportunities. Philos. Trans. R. Soc. A Math. Phys. Eng. Sci. **375**(2104), 20150403 (2017)
5. A. Desai, S. Gulwani, V. Hingorani, N. Jain, A. Karkare, M. Marron, R.S. Roy, *Program Synthesis Using Natural Language* (2015)
6. J. Devlin, J. Uesato, S. Bhupatiraju, R. Singh, A.R. Mohamed, P. Kohli, Robustfill: neural program learning under noisy i/o., in *Proceedings of the 34th International Conference on Machine Learning*, vol. 70. JMLR.org (2017), pp. 990–998
7. R.E. Fikes, N.J. Nilsson, Strips: a new approach to the application of theorem proving to problem solving. Artif. Intel. **2**(3–4), 189–208 (1971)
8. C. Green, Application of theorem proving to problem solving, in *Readings in Artificial Intelligence* (Elsevier, 1981), pp. 202–222
9. C. Green, Theorem proving by resolution as a basis for question-answering systems. Mach. Intel. **4**, 183–205 (1969)
10. S. Gulwani, Automating string processing in spreadsheets using input-output examples. ACM Sigplan Notices **46**(1), 317–330 (2011)
11. S. Gulwani, W.R. Harris, R. Singh, Spreadsheet data manipulation using examples. Commun. ACM **55**(8), 97–105 (2012)
12. S. Gulwani, K. Pathak, A. Radhakrishna, A. Tiwari, A. Udupa, Quantitative programming by examples (2019)
13. W.R. Harris, S. Gulwani, Spreadsheet table transformations from examples. ACM SIGPLAN Notices **46**(6), 317–328 (2011)
14. R.M. Hierons, K. Bogdanov, J.P. Bowen, R. Cleaveland, J. Derrick, J. Dick, M. Gheorghe, M. Harman, K. Kapoor, P. Krause et al., Using formal specifications to support testing. ACM Comput. Surv. (CSUR) **41**(2), 1–76 (2009)
15. H.M. Hoercher, J. Peleska, Using formal specifications to support software testing. Softw. Qual. J. **4**(4), 309–327 (1995)
16. A. Iyer, M. Jonnalagedda, S. Parthasarathy, A. Radhakrishna, S.K. Rajamani, Synthesis and machine learning for heterogeneous extraction, in *Proceedings of the 40th ACM SIGPLAN Conference on Programming Language Design and Implementation* (2019), pp. 301–315
17. S. Jha, S.A. Seshia, A theory of formal synthesis via inductive learning. Acta Inform. **54**(7), 693–726 (2017)
18. S.K. Jha, *Towards Automated System Synthesis Using Sciduction*. Ph.D. thesis, UC Berkeley (2011)
19. N. Kant, Recent advances in neural program synthesis. arXiv preprint arXiv:1802.02353 (2018)
20. V. Le, S. Gulwani, Flashextract: a framework for data extraction by examples, in *Proceedings of the 35th ACM SIGPLAN Conference on Programming Language Design and Implementation* (2014), pp. 542–553

21. Z. Manna, R. Waldinger, A deductive approach to program synthesis. ACM Trans. Program. Lang. Syst. (TOPLAS) **2**(1), 90–121 (1980)
22. Z. Manna, R.J. Waldinger, Toward automatic program synthesis. Communun. ACM **14**(3), 151–165 (1971)
23. E. Parisotto, A.R. Mohamed, R. Singh, L. Li, D. Zhou, P. Kohli, Neuro-symbolic program synthesis. arXiv preprint arXiv:1611.01855 (2016)
24. O. Polozov, S. Gulwani, Flashmeta: a framework for inductive program synthesis, in *Proceedings of the 2015 ACM SIGPLAN International Conference on Object-Oriented Programming, Systems, Languages, and Applications* (2015), pp. 107–126
25. M. Raza, S. Gulwani, N. Milic-Frayling, Compositional program synthesis from natural language and examples, in *Twenty-Fourth International Joint Conference on Artificial Intelligence* (2015)
26. S.A. Seshia, Sciduction: combining induction, deduction, and structure for verification and synthesis, in *DAC Design Automation Conference 2012* (IEEE, 2012), pp. 356–365
27. R. Singh, S. Gulwani, Synthesizing number transformations from input-output examples, in *International Conference on Computer Aided Verification* (Springer, 2012), pp. 634–651
28. R. Singh, S. Gulwani, Predicting a correct program in programming by example, in *International Conference on Computer Aided Verification* (Springer, 2015), pp. 398–414
29. J.R. Slagle, A heuristic program that solves symbolic integration problems in freshman calculus. J. ACM (JACM) **10**(4), 507–520 (1963)
30. J.R. Slagle, Experiments with a deductive question-answering program. Commun. ACM **8**(12), 792–798 (1965). https://doi.org/10.1145/365691.365960
31. R.J. Waldinger, R.C. Lee, Prow: a step toward automatic program writing, in *Proceedings of the 1st International Joint Conference on Artificial Intelligence* (1969), pp. 241–252
32. Y. Wang, J. Dong, R. Shah, I. Dillig, Synthesizing database programs for schema refactoring, in *Proceedings of the 40th ACM SIGPLAN Conference on Programming Language Design and Implementation* (2019), pp. 286–300
33. A. Zohar, L. Wolf, Automatic program synthesis of long programs with a learned garbage collector, in *Advances in Neural Information Processing Systems* (2018), pp. 2094–2103

Impedimetric Sensor for Cooking Oil Discernment Based on Planar Electromagnetic Sensor Array

Mohd Amri Md Yunus, Ruzairi Abdul Rahim, Sallehuddin Ibrahim, and Shafishuhaza Sahlan

Abstract This paper presents an investigation of the use of an impedimetric sensor which can be used to distinguish several types of cooking oils for three types of conditions such as characterization, adulteration, and the amount of usage for deep-frying conditions, utilizing the parallel configured planar electromagnetic sensor array with electrochemical impedance spectroscopy (EIS) technique. The planar electromagnetic sensor was designed by combining two types of sensors which are an interdigital sensor and a meander sensor as the interface to interact with the test samples. The meander type of coil was connected in series with the interdigital sensor coil. Several types of commercial cooking oil samples were prepared. However, since commercial palm oil is commonly used by Malaysian consumers, focus is on its samples preparation for adulteration and usage conditions. The results obtained showed that the sensor was able to differentiate the samples based on the impedance spectra in the range of frequencies from 100 Hz to 5 MHz. The equivalent circuit for the planar electromagnetic sensor was determined, and the equivalent parameter units were estimated with an estimation error of less than 8%. The conductivity and relative permittivity for the cooking oil samples were calculated, and a corrective factor was used to reduce the sensor measurement error. The reliability and repeatability of the planar electromagnetic sensor was also presented in this paper.

Keywords Impedimetric sensor · Planar electromagnetic sensors array · Impedance spectroscopy · Cooking oil discerning

1 Introduction

Cooking oils have been one of the major products produced by Malaysian manufacturing industries. However, to gain extra profit, these high-quality cooking oils are adulterated by manufacturers with low quality cooking oils [1]. Therefore, in

M. A. M. Yunus (✉) · R. A. Rahim · S. Ibrahim · S. Sahlan
Faculty of Engineering, Universiti Teknologi Malaysia, Skudai, Johor, Malaysia
e-mail: amri@utm.my

A. Kumar et al. (eds.), *Computational Intelligence in Machine Learning*, Lecture Notes in Electrical Engineering 834, https://doi.org/10.1007/978-981-16-8484-5_40

this research, an extremely sensitive and in-situ tool to distinguish cooking oil products is developed for a preventive action on this negative activity. From literature, various types of discerning techniques have been developed, for example, gas chromatography (GC) [2] and high-performance liquid chromatography (HPLC) [3] are common chemical techniques used by researchers to determine the quality of the cooking oils. These techniques measure the organic compounds in the solution via the retention time of the concentration of the ions using a chromatogram. However, these techniques are complex [4] and require a professional machine handler to operate the instrument that separates the ions in the solution using pressurization [5]. There are also other techniques that have been developed to distinguish the cooking and vegetable oil such as infrared spectroscopy [6, 7] and microwave spectroscopy [8] are based on chemometric analysis and microwave spectrum analysis. However, the analysis for the measurement data is difficult and requires a long duration of time for the analytical process.

Thus, this paper proposed a non-destructive, less complex system to distinguish four different products of commercial cooking oil samples for characterization, the adulteration, and the amount of usage conditions for a commercial palm oil samples using a planar electromagnetic sensor array (PESA) as the sensory part with the electrochemical impedance spectroscopy (EIS) technique for data analysis.

2 Impedance Spectroscopy Techniques

Electrochemical impedance spectroscopy (EIS) is a popular technique that utilizes a fringing electric and magnetic field sensor to investigate the properties of a material by applying an external electromagnetic field and assessing the output response in different frequency bands as a function of impedance [9]. The impedance spectroscopy technique has been used in much wider range of applications that include phthalates detection [10], protein binding monitoring [11], resin monitoring [12], and measuring salt levels in food [13]. Moreover, EIS is not limited to a single-side fringing field sensor as a detection tool but also includes sensors featuring two, three, or more electrodes which depend on the sensor geometry and materials. Furthermore, EIS also provides on-site measurements with less analytical time [14].

Theoretically, the EIS technique offers information about the changes in parameters in the electrochemical properties of the material by applying a low amplitude sinusoidal signal through the system and measures the impedance of the system in terms of amplitude and phase responses [12]. Then, the analysis of the complex value impedance of the system is interpreted in terms of the real and the imaginary parts of an impedance at different frequency bands that contain information about the system resistance, capacitance, and inductance of the bulk medium.

The analytical results of the system can be represented graphically by Nyquist plot where the real part of impedance is plotted on the X-axis and the imaginary part of impedance is plotted on the Y-axis for all applied frequencies. The completion of the EIS analysis is achieved by translating the electrochemical parameters of the system obtained from the impedance spectrum into an electrical equivalent circuit model which commonly consists of resistances, capacitances, and inductances. Typically, the electrical equivalent circuits are unique which differ from the geometry of the designed sensor. Moreover, the conductivity and permittivity of the material samples can be calculated by relating the equation from the system equivalent circuit defined in [15].

3 Materials and Experimental Setup

In this research, several samples are prepared based on the three types of conditions such as characterization, adulteration, and deep-frying usage conditions. For characterization, four types of cooking oil samples were used in the impedance measurement experiment namely sunflower oil, corn oil, palm oil, and canola oil. One liter of the oil was prepared for each sample. Next, a commercial palm cooking oil sample was a major focus for adulteration and usage conditions. The palm cooking oil was adulterated with a corn cooking oil sample using ultrasonic bath and the range of corn cooking oil concentration are 20, 30, and 40% from the total volume of the palm cooking oil (200 ml).

Then, one liter of palm cooking oil with a packet of chicken was used as the ingredient for deep-frying sample preparations. The ingredient was deep fried for one time, and the sample was collected at the end of the deep-frying process. The sample collection was continued until the palm cooking oil has been used for three times in the deep-frying process. Next, a custom-made oil container with three separate oil columns was filled with the cooking oil samples. Each of the column was occupied with one particular type of cooking oil sample as configured in Fig. 1.

The cooking oil samples were kept at a room temperature of 24 °C to prevent the sample from being hydrogenated.

Figure 2a, b show the top and bottom surfaces, respectively, of the fabricated FR-4 planar electromagnetic sensing array that consists of three identical parallel planar electromagnetic sensors connected to each other, namely S1, S2, and S3. An annealed copper cable with a diameter of 0.6 mm was wired to each of the planar sensor arrays to achieve better performance as the supplied current flows to the planar sensor array and three cables were cut with the same wire length of 15 mm. Moreover, the planar electromagnetic sensing array was coated with acrylic liquor to prevent the sensor circuit from shorting during measurements. The HIOKI IM3570 impedance analyzer was used to measure the impedance of the planar electromagnetic sensing array based on the cooking oil samples. The impedance analyzer user interface (UI) was utilized for data collection.

Before conducting the experiments, the impedance analyzer was calibrated to take into account the accuracy of the measurements by performing short and open test calibration. The procedures of the impedance measurements began by immersing the coated planar electromagnetic sensor array into the filled separated oil columns

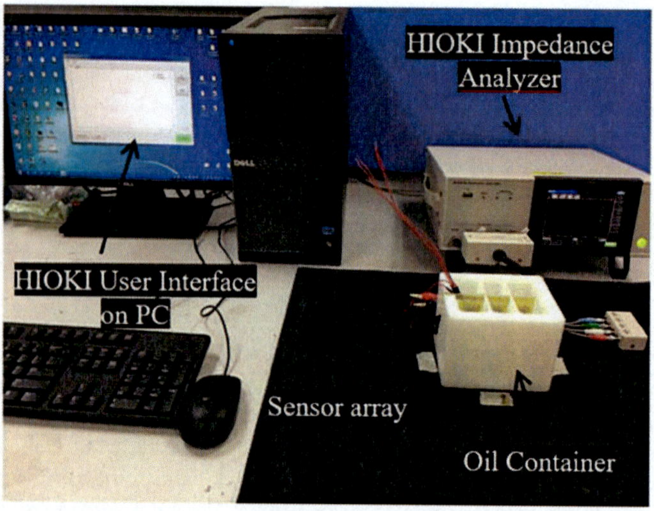

Fig. 1 EIS experimental setup for cooking oil impedance measurement

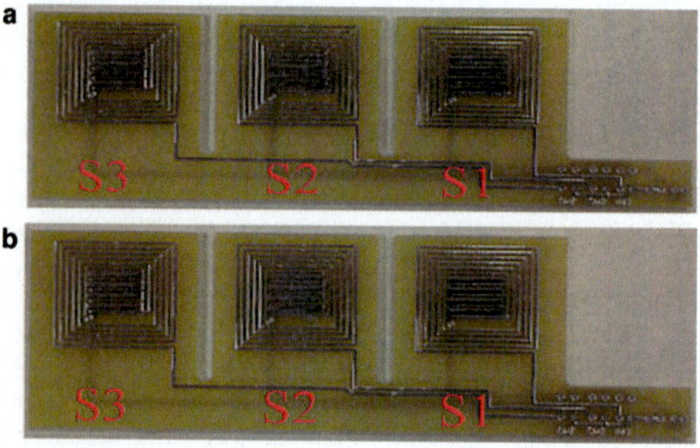

Fig. 2 **a** Top surface of planar sensor array. **b** Bottom surface of planar sensor array

as shown in Fig. 1. Then, the impedance analyzer was set at a low constant peak-to-peak voltage of 1V sinusoidal signal with a frequency sweep of 100 Hz–5 MHz with 801 data points on a log scale. The impedance was taken for 30 sets of data for each oil column to ensure the repeatability and reliability of the measurement results. Then, the oil samples were drained after all three separated oil columns were measured. Next, the planar sensor array and the oil container were cleaned with high concentration soap and rinsed with warm water to prevent the cooking oil sample from being contaminated for subsequent measurements. Lastly, an average of the datasets for each sample was presented graphically.

4 Results and Discussion

The impedance measurements for the four types of commercial cooking oils which are sunflower oil, corn oil, plan oil, and canola oil, are performed based on the impedance response, phase response, real impedance, and imaginary impedance. A total of 30 sets of impedance and phase responses were used for each sample, as shown in Figs. 3 and 4.

From Fig. 3, it can be observed that the results were repeatable which indicates that each time the impedance was calculated, it yields the same value. Similar results are observed in Fig. 4, which shows good repeatability.

The planar electromagnetic sensor equivalent circuit can be described by Equation (1):

$$Z = j\omega L + \frac{R_E}{1 + j\omega R_E C_E} \tag{1}$$

where Z is the total impedance, R_E denotes the material sample resistance, C_E is the capacitance of the material, and L is the total mutual inductance. To provide further information in distinguishing the cooking oil samples, conductivity and permittivity profile of the cooking oil samples are needed. The parameters of the electrical components in the proposed equivalent circuit for each sensor were estimated using the

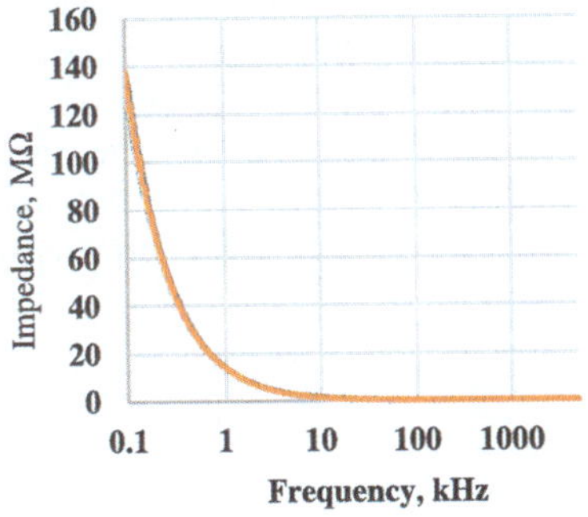

Fig. 3 30 sets of impedance measurements of a commercial palm cooking oil sample

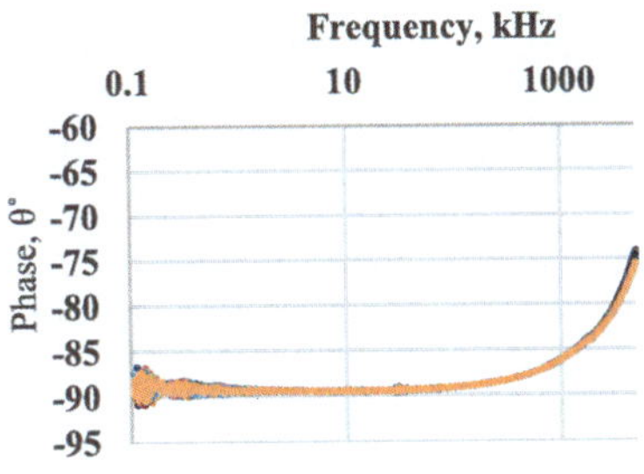

Fig. 4 30 sets of phase measurements of a commercial palm cooking oil sample

electrochemical spectrum analyzer with an error rate of less than 8% in 300 iterations. There is a variation in the values of parameters among the sensing arrays S1, S2, and S3 since the values in impedance and phase responses are differed between sensors. The equation from the system equivalent circuit with Eq. (2) is defined as:

$$\sigma = \frac{1}{\kappa R_E} \tag{2}$$

where R_E denotes the material sample resistance, σ is the conductivity of the material sample, and κ is the geometrically correlated constant of the interdigital sensor.

The permittivity of the material sample encompasses the relative permittivity, ε_r', and the complex permittivity, $\varepsilon_r'/$, which can be calculated using Eq. (3):

$$\varepsilon_r' = \frac{C_E}{\kappa \varepsilon_o} - \varepsilon_s \tag{3}$$

where C_E is the capacitance of the material, the free space permittivity, $\varepsilon_{o,}$ is 8.85×10^{-12} F/m, and the relative permittivity of the substrate, $\varepsilon_{s,}$ is 4.5.

In Figs. 5 and 6, the calculated conductivity and relative permittivity respective of the four cooking samples for sensors S1, S2, S3, and S4 are presented. In Fig. 5, the different values of the conductivities show that the cooking oil samples can be distinguished from each other. Similarly, in Fig. 6, the different values of relative permittivity show that the samples can be differentiated from each other. However, both conductivity and relative permittivity values tend to increase when looking at the values from each sample measured by sensor S1 until sensor S3. This behavior occurred due to the effect in estimating the equivalent circuit parameters where the conductivity and relative permittivity depend on the oil fats capacitance and resistance (C_E and R_E). Therefore, a corrective factor is needed to reduce the measurement error

Fig. 5 Estimated conductivity values of four types of commercial cooking oils

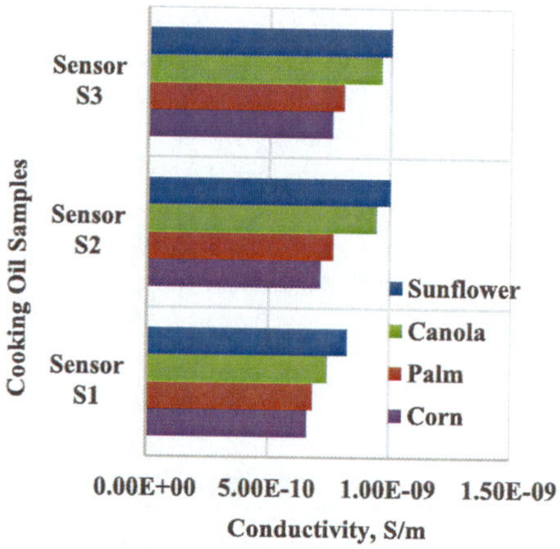

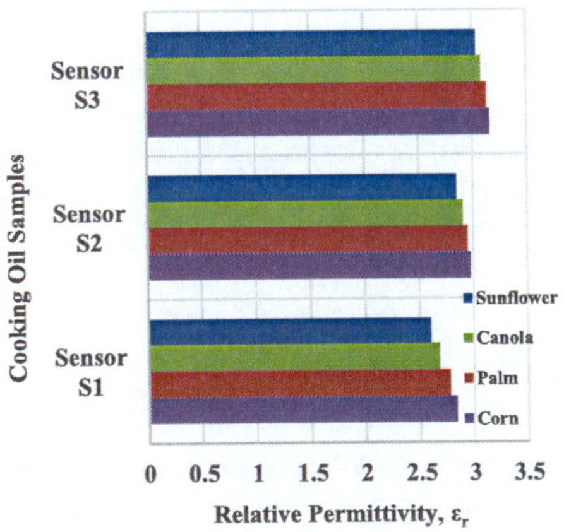

Fig. 6 Estimated permittivity values of four types of commercial cooking oils

and obtain similar results for all sensor arrays since the design of the sensor arrays is identical. The sensor array S2 was selected as the reference sensor since the generated electromagnetic field in the interfacial region of the sensor S2 is more dominant than sensors S1 and S3 as sensor S2 was in the center of the planar electromagnetic sensor array. The correction factor is determined by finding the relationship between the rate of change of the conductivity measured by the sensor arrays and the distance from the sensor surface to the point of measurement.

From the results obtained, the relative permittivity and conductivity values for the 30 measurements of each cooking oil sample were estimated. The highest relative permittivity value is corn cooking oil with a relative permittivity value of 2.9718 $\pm$ 0.0023, whereas the sunflower cooking oil is the lowest value of relative permittivity of 2.8431 $\pm$ 0.0015. Meanwhile, the electrical properties for adulterated palm cooking oil with corn cooking oil is shown in Fig. 7. From Fig. 7, the differences of the relative permittivity and conductivity values can be significantly seen where the value decreases when the concentration of corn oil increases. On the other hand, for

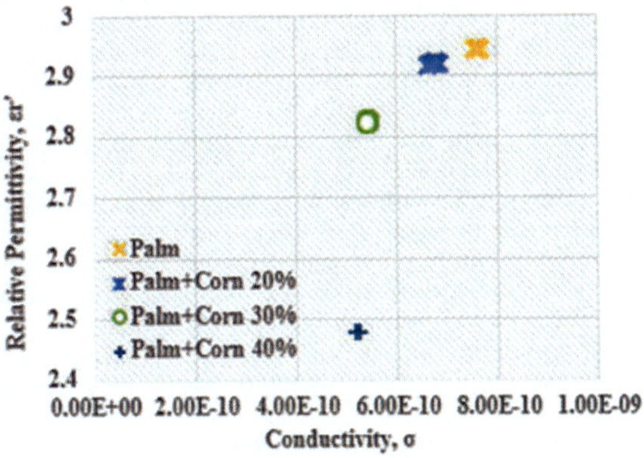

Fig. 7 Adulterations of palm cooking oil

deep-frying conditions, the electrical properties are shown in Fig. 7 where the relative permittivity value reduces while the conductivity value increases as the amount of use for palm cooking oil in deep-frying activity increases. Therefore, it can be concluded that the proposed method has shown the feasibility of the planar electromagnetic sensor in distinguishing several types of commercial cooking oil through the electrical profile obtained.

5 Conclusion

This paper shows that an impedimetric sensor combined with the electrochemical impedance spectroscopy technique has the ability to distinguish several types of cooking oils within three types of conditions such as characterization, adulteration, and deep-frying usages which based on impedance and phase information between the frequencies range of 100 Hz and 5 MHz. Furthermore, this paper also presents the conductivity and relative permittivity of the cooking oil samples using the estimated equivalent electrical circuit parameters with an estimation error of less than 8%.

Acknowledgements The authors would like to acknowledge the financial assistance under the Fundamental Research Scheme Grant (FRGS) from the Ministry of Higher Education Research (MOHE) Malaysia (Vote No. 4F594).

References

1. R.F. Stier, Tests to monitor quality of deep-frying fats and oils. Eur. J. Lipid Sci. Technol. **106**(11), 766–771 (2004)
2. H.-Y. Seo, J. Ha, D.-B. Shin, S.-L. Shim, K.-M. No, K.-S. Kim, K.-B. Lee, S.-B. Han, Detection of corn oil in adulterated sesame oil by chromatography and carbon isotope analysis. J. Am. Oil Chem. Soc. **87**(6), 621–626 (2010)
3. S. Cunha, M. Oliveira, Discrimination of vegetable oils by triacylglycerols evaluation of profile using HPLC/ELSD. Food Chem. **95**(3), 518–524 (2006)
4. H. Lizhi, K. Toyoda, I. Ihara, Discrimination of olive oil adulterated with vegetable oils using dielectric spectroscopy. J. Food Eng. **96**(2), 167–171 (2010)
5. A.I. Zia, A.R.M. Syaifudin, S.C. Mukhopadhyay, P.L. Yu, I.H. Al-Bahadly, C.P. Gooneratne, J. Kosel, T.-S. Liao, Electrochemical impedance spectroscopy based MEMS sensors for phthalates detection in water and juices. J. Phys. Conf. Ser. **439**(1), 12026 (2013)
6. S. Armenta, S. Garrigues, M. de la Guardia, Determination of edible oil parameters by near infrared spectrometry. Anal. Chim. Acta **596**(2), 330–337 (2007)
7. A.F.C. Pereira, M.J.C. Pontes, F.F.G. Neto, S.R.B. Santos, R.K.H. Galvão, M.C.U. Araújo, NIR spectrometric determination of quality parameters in vegetable oils using iPLS and variable selection. Food Res. Int. **41**(4), 341–348 (2008)
8. O. Korostynska, R. Blakey, A. Mason, A. Al-Shamma'a, Novel method for vegetable oil type verification based on real-time microwave sensing. Sensors Actuators A Phys. **202**, 211–216 (2013)
9. A. Azmi, A.A. Azman, K.K. Kaman, S. Ibrahim, S.C. Mukhopadhyay, S.W. Nawawi, M.A. Md Yunus, Performance of coating materials on planar electromagnetic sensing array to detect

water contamination. IEEE Sensors J. **17**(16), 5244–5251 (2017). https://doi.org/10.1109/JSEN.2017.2720701
10. A.I. Zia, A.R. Mohd Syaifudin, S.C. Mukhopadhyay, I.H. Al-Bahadly, P.L. Yu, C.P. Gooneratne, J. Kosel, T.-S. Liao, MEMS based impedimetric sensing of phthalates, in *2013 IEEE International Instrumentation and Measurement Technology Conference (I2MTC)*, 2013, pp. 855–860.
11. Z. Zou, J. Kai, M.J. Rust, J. Han, C.H. Ahn, Functionalized nano interdigitated electrodes arrays on polymer with integrated microfluidics for direct bio-affinity sensing using impedimetric measurement. Sensors Actuators A Phys. **136**(2), 518–526 (2007)
12. Y. Han, J. Wang, H. Zhang, S. Zhao, Q. Ma, Z. Wang, Electrochemical impedance spectroscopy (EIS): An efficiency method to monitor resin curing processes. Sensors Actuators A Phys. **250**, 78–86 (2016)
13. R. Masot, M. Alcañiz, A. Fuentes, F.C. Schmidt, J.M. Barat, L. Gil, D. Baigts, R. Martínez-Máñez, J. Soto, Design of a low-cost non-destructive system for punctual measurements of salt levels in food products using impedance spectroscopy. Sensors Actuators A Phys. **158**(2), 217–223 (2010)
14. K. Sundara-Rajan, A.V. Mamishev, M. Zahn, Fringing electric and magnetic field sensors. Encycl. Sensors **10**, 1–12 (2005)
15. K.G. Ong, C.A. Grimes, C.L. Robbins, R.S. Singh, Design and application of a wireless, passive, resonant-circuit environmental monitoring sensor. Sensors Actuators A Phys. **93**(1), 33–43 (2001)

Container ISO Code Recognition System Using Multiple View Based on Google LSTM Tesseract

Shaekh Mohammad Shithil, Ahmad Ridhwan Mohd Kamil, Sadat Tasnim, and Ahmad Athif Mohd Faudzi

Abstract Optical Character Recognition (OCR) system is vastly used to identify license plates, street signs, and other applications. However, it faces difficulties to recognize ISO code from natural images of shipping containers due to rough weather condition, varying color, illumination, etc. In this paper, these challenges were overcome by integrating deep learning-based OCR recognition from multiple view which increases both accuracy and reliability. Images are taken from three different views and based on the proposed algorithm it analyses all the images to detect ISO code format using sequence matching algorithm. Next, confidence level is calculated for each recognized code using Google LSTM neural-net based Tesseract engine model and identifies the one which has highest confidence level for the proposed OCR system to be robust in delivering high level accuracy in actual application.

Keywords OCR · Multiple view · Deep learning · LSTM Tesseract

1 Introduction

OCR has been extensively used for license plates, scanned documents, street signs, and other natural scene images. Localization of the regions of interest, preprocessing of the region of interest and character recognition are part of the standard pipeline for these processes. Various studies have been conducted on localizing the text region of interest (ROI) based on traditional statistical techniques by comparing the edges of text [1] and mathematical morphology [2]. Some classification models have been

S. M. Shithil · S. Tasnim · A. A. M. Faudzi (✉)
School of Mechanical Engineering, Faculty of Engineering, Universiti Teknologi Malaysia, Johor Bahru, Malaysia
e-mail: athif@utm.my

A. A. M. Faudzi
Centre for Artificial Intelligence and Robotics, Universiti Teknologi Malaysia, Johor Bahru, Malaysia

A. R. M. Kamil
Malaysia–Japan International Institute of Technology, Universiti Teknologi Malaysia, Kuala Lumpur, Malaysia

A. Kumar et al. (eds.), *Computational Intelligence in Machine Learning*, Lecture Notes in Electrical Engineering 834, https://doi.org/10.1007/978-981-16-8484-5_41

proposed to detect text regions of interest [1]. OCR recognition for natural images is challenging for damaged characters and unexpected noise. Therefore, preprocessing methods can increase the chance of correct OCR recognition [3]. Optical character recognition has been studied vastly for various applications from document analysis to license plate recognition. Different kind of approaches has been addressed such as statistical, machine learning, and deep neural-net based models [4].

Due to rise of import-export and ecommerce business worldwide, the usage of shipping containers increased a lot since it is the most convenience way to ship goods from one place to another. These containers were identified by a unique standardize ISO 6346 code [5] written at the exposed sides which separates a container from others. Till date, most of the ports are not fully automated and rely on manual recording system. This makes tracking and identification process consume a lot of time and could delay the discharging of shipping containers from the port. To speed up the tracking and identifying process, there are two solutions available to automate the process: Radio frequency identification (RFID) and optical character recognition (OCR). While the RFID has almost perfect accuracy, it has a huge installation cost in the system implementation. OCR on the other hand has less accuracy (95%) compared to RFID system due to unrecognized codes from rough weather condition, varying color, illumination, etc. Limited studies have been conducted on container code recognition. S. Kumano et al. worked on recognizing different character string layouts in container marks or numbers where a character recognition framework based on a dynamic design approach was used [6]. Another study has been conducted on step-based OCR localization, segmentation and recognition. Support vector machine (SVM) was proposed for character recognition [3]. Multiple views-based OCR recognition system has also been studying in container code recognition [7–9]. Therefore, in this paper, a deep learning-based multiple views container code recognition system is proposed using Google LSTM neural-net based Tesseract engine model. The deep learning-based method combines images from multiple views of a container to improve the accuracy of the OCR.

2 ISO Code ROI Localization, Extraction, and Preprocessing

One of the key challenges in natural images processing is to localize the region of interested text from the whole image. Since it is hard to localize text regions from natural images, deep learning method works effectively to localize text regions compare to the current machine learning and statistical models. Textual information extraction from the natural container images have been focused using an Efficient and Accurate Scene Text (EAST) detector which is based on a fully convolution network (FCN) as shown in Fig. 1. It is trained to directly predict the existence of text instances and their geometry from the full images. This FCN model has adapted for text detection that outputs dense per-pixel predictions of words and text lines

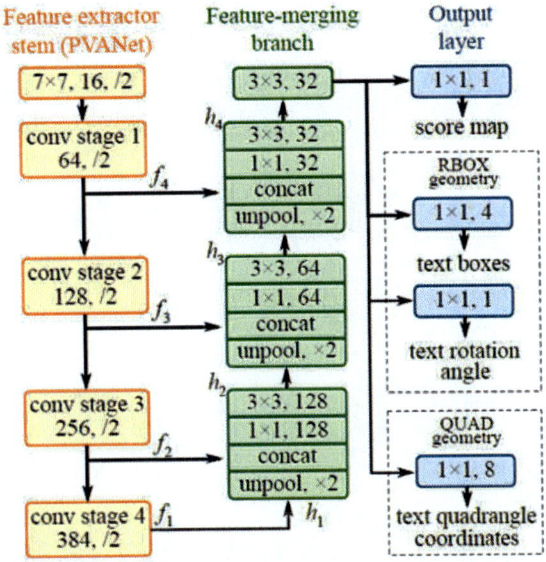

Fig. 1 EAST FCN model [10]

[10]. Hereby, it can predict text-line or word-level without slow and intermediate steps. The output of the model is a rectangle or quadrangles bounding box.

After the text section is localized, the accuracy improvement of the text recognition preprocessing methods is required. The recognition will be ineffective or can give incorrect results during text recognition if it is not properly preprocessed. The following techniques grayscale conversion, gaussian blurring, and binarization are used as noise filtering method [11].

3 ISO Code Recognition

After ROI extraction and preprocessing, the Google Tesseract engine will be used for text recognition. The image is converted using a computer vision system and then the characters are filtered out using Google Tesseract-OCR which was originally designed to scan hard copy documents and to filter out the characters from them [12]. Here, the deep learning-based LSTM, an artificial recurrent neural network (RNN) architecture Tesseract-OCR engine was used. The Tesseract method can be considered a pipeline in theory, but not in operation since there is a lot of revisiting of previous decisions. In Figs. 2 and 3, the Tesseract system architecture and the word recognizer pipeline are shown consecutively [13]. Figure 4 shows the results of the OCR recognition technique.

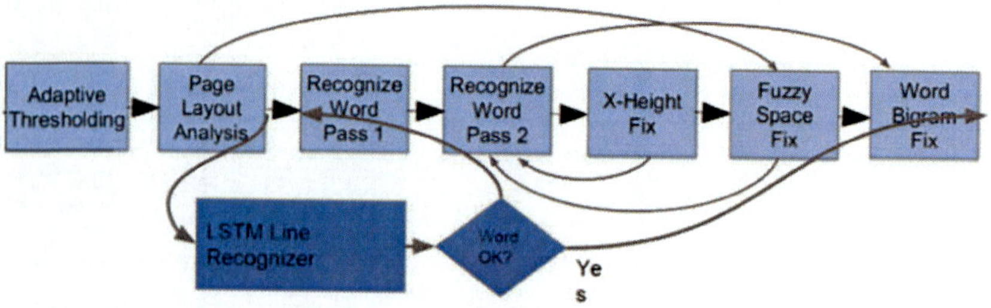

Fig. 2 System architecture of the Tesseract [13]

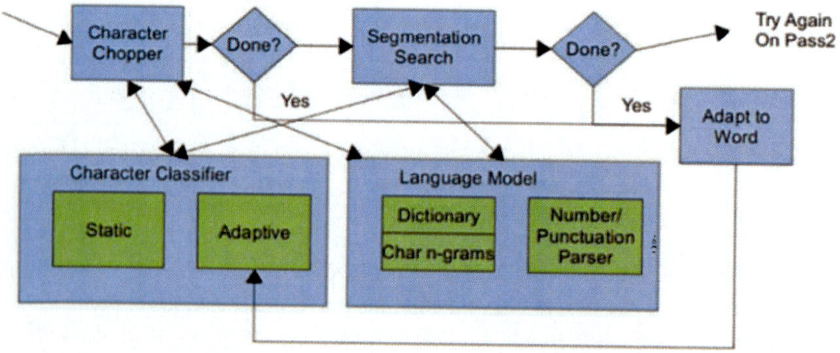

Fig. 3 Tesseract word recognizer [13]

Fig. 4 **i** ISO code region of interest (ROI) localization, extraction, and preprocessing, **ii** ISO code recognition

4 Combining Multiple Views to Improve OCR Accuracy

OCR engine is not 100% accurate especially in natural images where the accuracy of the OCR engine drops drastically. Therefore, combined multiple views based OCR system was proposed to improve the accuracy. Figure 5 shows the red dots indicating the camera positions for picture taken and the flow of the proposed system. Once the OCR code is recognized, the most important step is to check whether the recognized code is container ISO code or not. The container identification code follows a standard

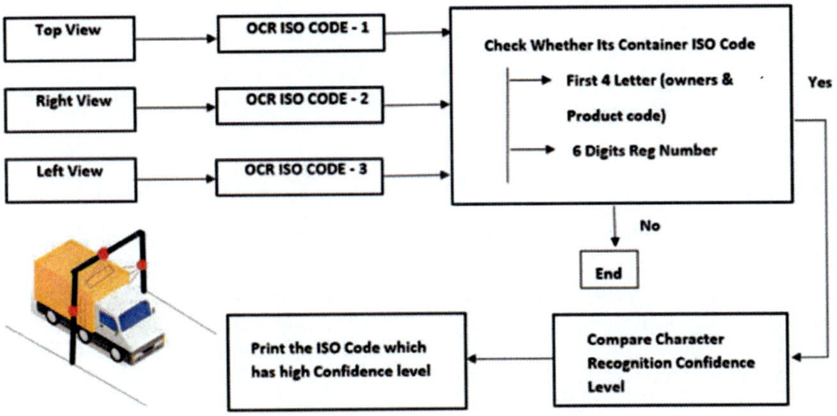

Fig. 5 Camera positions and the proposed multiview based OCR system flow

pattern. The first 4 letters represent the owner and product group code, and the next 6 digits represent the registration number of the container. A slice sequence matching method was proposed for the detected OCR code to validate if the first four digits are letters or not, then if the next six digits are numbers. After the confirmation of the valid ISO container code, the confidence level of the detected OCR code is compared. The code that has the highest confidence level is the final result of the system.

5 Experimental Results

This section represents the proposed multiple view OCR system performance accuracy. Table 1 shows the individual results of OCR recognition from the top, right, and left camera view. The best accuracy is shown in top view image. The ISO container code is also detected properly. On the other hand, right and left view image accuracy level is quite low due to the uneven pattern of the container—which makes it difficult for the OCR engine. In Table 1, the OCR recognition accuracy can be seen for multiple views.

As we can see from the accuracy percentage from different views of the container, OCR system mostly performed bad for the right and left views. It has been proposed to combine the results from all views to improve the accuracy level in order to make it a robust system. The proposed algorithm for combining multiple views is shown in Fig. 6.

The algorithm increases the accuracy of overall OCR recognition of the container ISO code. In this method, if one view fails to detect the right code, it will not affect in recognizing the correct code. It will make the system more redundant for OCR recognition. The results of combining multiple view algorithm are shown in Fig. 7.

Table 1 Proposed Multiview based OCR system flow

View	Detected code	Confidence level (%)	Detected Container ISO code
Top view	APHU 648709	91	APHU 648709
	0	80	
		92	
		Average: 87.67	
Right view	Null	0	Null
		Average: 0	
Left view	a	10	Null
	ae	15	
	fa	41	
	eas	48	
	a	18	
	ssr	4	
		Average: 22.67	

```
Input: ImageBundle [3 images of top, right, left view of a container]
for each X in ImageBundle
      Output OCR Code 1,2,3  [extracted OCR from 3different views]
end
letters = [A, …., Z]
numbers = [0, …., 9]

for each Code in Output OCR Code 1,2,3
      if Output OCR Code [0:3] in letters and Output OCR Code [4:9] in numbers
                  Print valid iso code
      else
                  end
if (valid iso code1> valid iso code2) and (valid iso code1> valid iso code3)

      correct_code = valid iso code1
elif (valid iso code2> valid iso code1) and (valid iso code2> valid iso code3)
      correct_code = valid iso code2
else
      correct_code = valid iso code3
end
```

Fig. 6 Proposed algorithm: combining multiple views to improve OCR accuracy

In the experiment the accuracy was tested on 1050 single view container images for the OCR recognition which gives an accuracy of 90%. However, this container recognition is a complex task to implement thus to make this system reliable for real usage, a multiple view based deep learning OCR recognition system has been proposed. A promising result was found when multiple views of containers images were tested

View	ROI	Detected Code
Top view		APHU 648709 0
Right view		Failed to localize
Left view	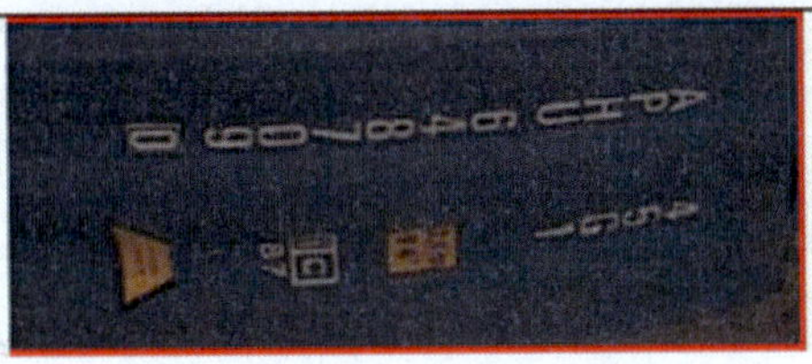	a ae fa eas a ssr
	Result from proposed algorithm	APHU 648709

Fig. 7 Results of combining multiple view algorithm

through our algorithm to find out if the proposed method can be implemented in real application.

6 Conclusion

In this paper, we proposed a novel deep learning-based multiple view OCR recognition system which will improve the current container recognition methods. An algorithm has been proposed which will detect the correct ISO code format and produce confidence level for each view. The deep learning-based algorithm will then analyze the confidence level and select the most convincing code from the raw results of the baseline algorithm. For proposed algorithm validation, experiments have been conducted on one shipping container images taken from multiple views as a sample. A promising results have shown that it is possible to implement the system in real life usage. To test the proposed system's reliability and accuracy, a set up this system will be implemented in the port to experiment it with a large data set. The result has shown a consistent and effective execution/performance/implementation of OCR system for real usage.

Acknowledgements he authors would like to acknowledge the support provided by Ministry of Higher Education Malaysia (MOHE) through support under Fundamental Research Grant Scheme (FRGS/1/2019/TK04/UTM/02/41). The author also would like to thank Universiti Teknologi Malaysia (UTM) for work and facilities support.

References

1. S. Bansal, M. Gupta, A.K. Tyagi, A necessary review on optical character recognition (OCR) system for vehicular applications, in *International Conference on Inventive Research in Computing Applications (ICIRCA)* (2020), pp. 918–922. https://doi.org/10.1109/ICIRCA48905.2020.9183330
2. K.M. Koo, E.Y. Cha, A novel container ISO-code recognition method using texture clustering with a spatial structure window. Int. J. Adv. Sci. Technol. **41**, 83–92 (2012)
3. W. Wu, Z. Liu, M. Chen, X. Yang, X. He, An automated vision system for container-code recognition. Expert Syst. Appl. **39**(3), 2842–2855 (2012)
4. R. Mittal, A. Garg, Text extraction using OCR: a systematic review, in *Second International Conference on Inventive Research in Computing Applications* (2020), pp. 357–362. https://doi.org/10.1109/ICIRCA48905.2020.9183326
5. ISO 6346 (1995): ISO (International Organization for Standardization). http://www.iso.org/iso/home/store/catalogue_tc/catalogue_detail.htm?csnumber=20453
6. S. Kumano, K. Miyamoto, M. Tamagawa, H. Ikeda, K. Kan, Development of a container identification mark recognition system, electron. Commun. Jpn. (Part II Electron.) **87**(12), 38–50 (2004)
7. A. Selinger, R.C. Nelson, Appearance-based object recognition using multiple views. Proc. IEEE Comput. Soc.Conf. Comput. Vis. Pattern Recogn. **1**, I-905–I-911 (2001)
8. G. Shakhnarovich, L. Lee, T. Darrell, Integrated face and gait recognition from multiple views. Proc. IEEE Comput.Soc. Conf. Comput. Vis. Pattern Recogn. **1**, I-439–I-446 (2001)
9. Y. Yoon, K. Ban, H. Yoon, J. Kim, Automatic container code recognition from multiple views. ETRI J. (2016)
10. X. Zhou, C. Yao, H. Wen, Y. Wang, An efficient and accurate scene text detector, in *IEEE Conference on Computer Vision and Pattern Recognition (CVPR)* (2017). arXiv:1704.03155
11. A.A.M. Faudzi, K.C. Loke, W.K. Tham, task learning utilizing curve fitting method for kinect based humanoid robot, in *IEEE International Conference on Robotics and Biomimetics*, 2015, pp. 1505–1511
12. R. Smith, An overview of the tesseract OCR engine, in *International Conference on Document Analysis and Recognition (ICDAR 2007)*, vol. 2 (2007)
13. I. Anuradha, Optical character recognition (OCR) for low resource languages with tesseract version, from https://medium.com/@isurianuradha96/training-sinhala-language-with-tesseract-4-1-version-3cd96ce84b6a (2019)

Path Planning Based on Inflated Medial Axis and Probabilistic Roadmap for Duct Environment

Ahmad Ridhwan Mohd Kamil, Shaekh Mohammad Shithil, Zool Hilmi Ismail, Mohd Saiful Azimi Mahmud, and Ahmad Athif Mohd Faudzi

Abstract This paper proposes an inflated medial axis based probabilistic roadmap (PRM) for the duct environment. The study intends to explore a sampling strategy for PRM toward clearance-driven path planning. Sampled nodes are constrained only inside the inflated medial axis area. The medial axis is generated from skeletonizing the map and inflate the skeletal diagram. To evaluate effectiveness of the proposed method, experiments were carried out on a reference map generated by simultaneous localization and mapping (SLAM) from an actual mockup duct environment. Experiments conducted showed the proposed method able to produce a path trajectory with clearance-driven behavior. The method is very useful in narrow spaces such as ducting, which required a safe distance when navigating autonomously in the environment.

Keywords Path planning · Medial axis · Roadmap · Skeletonized

1 Introduction

Most modern buildings nowadays are equipped with an air distribution system called as heating, ventilation, and air-conditioning (HVAC) system. HVAC ducts are the main structures for a modern building to provide ventilation inside the building. Poor

A. R. M. Kamil · Z. H. Ismail
Malaysia-Japan International Institute of Technology, Universiti Teknologi Malaysia, Kuala Lumpur, Malaysia

M. S. A. Mahmud · A. A. M. Faudzi (✉)
School of Electrical Engineering, Faculty of Engineering, Universiti Teknologi Malaysia, Johor Bahru, Malaysia
e-mail: athif@utm.my

S. M. Shithil
School of Mechanical Engineering, Faculty of Engineering, Universiti Teknologi Malaysia, Johor Bahru, Malaysia

Z. H. Ismail · A. A. M. Faudzi
Centre for Artificial Intelligence and Robotics, Universiti Teknologi Malaysia, Johor Bahru, Malaysia

A. Kumar et al. (eds.), *Computational Intelligence in Machine Learning*, Lecture Notes in Electrical Engineering 834, https://doi.org/10.1007/978-981-16-8484-5_42

HVAC systems can affect indoor air quality (IAQ) [1]. According to the guideline provided by the Department of Occupational Safety and Health Malaysia (DOSH) in the Industry Code of Practice on Indoor Air Quality 2010, inspection and cleaning of the HVAC system need to be conducted regularly to maintain the IAQ. As the demand for a larger and more complex building is increasing, ventilation duct structures also become more complicated.

Many mobile robots have been proposed to be used in substituting manual human labor [2], a small portable mobile robot proposed by [3, 4] to be used in ventilation duct cleaning operations. Many researchers have proposed autonomous robots to be used for the inspection and cleaning process [5]. Work by [6] proposed vision-based map-less navigation using a CCD camera and two laser sources to generate slit beams, and then by using the camera, it can acquire 3D information of the environment. To periodically maintaining the HVAC duct, prior knowledge of the environment can provide more efficient maintenance. PRM also offers an advantage over other methods, which is the multi-query planning [7, 8], where the sampling stage just needs to be done once, thus suitable to be used for repetitive works.

A classic variant of PRM tends to sample a set of configurations inside the space without considering the configuration's clearance distance, which will lead to a non-optimized path trajectory for maximum clearance, and some opted for optimized clearance during post-process [9]. According to [10], they propose a variant of PRM, which used a sampling strategy that randomly generates sampling toward the medial axis. It is successful when sampling inside narrow passages and generates more clearance trajectories from the obstacle. However, according to [11], the randomness of the technique used in [10] stated that sampling could be biased heavily toward a certain portion of the medial axis, sometimes ignoring critical parts to planning such as narrow passages. To overcome, [11] introduces uniform sampling on the medial axis variant of PRM. The uniformity proved to provide good planning for different environment possibly in duct environment which has narrow passages and static environment. Another work is done by [12] also uses the medial axis as a reference to do sampling. Instead of restricting only on the medial axis, it generates an inflated region around the medial axis. This provides less restrictions during sampling while still trying to generate sampling with maximum clearance possible. The method show that it can generate trajectory with high clearance, however the experiment only done in experiment and the environment does not represent an actual environment.

2 Skeletonized and Inflation of Medial Axis

The skeletonization of the configuration space is a process to estimate the medial axis of the environment [13]. The configuration space of 2D environment can be described as a binary image. The overall process of skeletonization and inflation is shown in Fig. 1. Skeletonized of the configuration space can be estimated using morphological process of the binary image using MATLAB 'bwmorph' with the operation set to 'skel' function from image processing toolbox. This skeletonized procedure able to

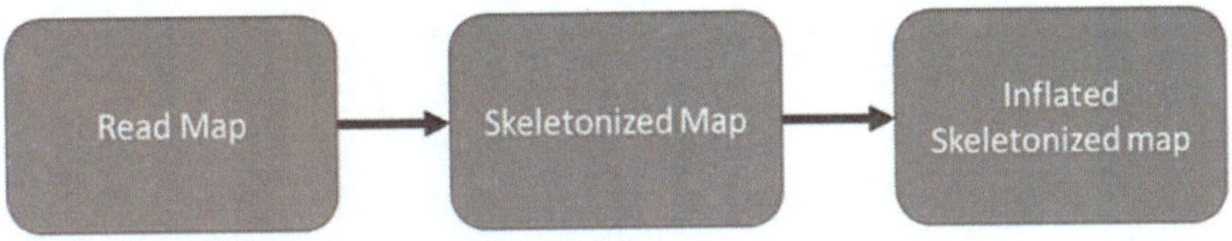

Fig. 1 Process of skeletonization and inflation

Fig. 2 Pixel *P* connectivity neighbors

X4	X3	X2
X5	P	X1
X6	X7	X8

find the skeleton of the configuration space by iteratively deleting the outer layer of the image pixel until only the skeleton remains [14]. The deletion process 'black' pixel (p) is depending on pixel neighbor connectivity configuration. MATLAB 'bwmorph' function used 8 connectivity. Figure 2 shows pixel P connectivity neighbors. The pixel $X1, X2, \ldots, X8$ is the 8 neighbors connectivity configurations.

Inflation of the medial axis is to a created buffer of configuration space that can be used to sample the PRM nodes inside it. This inflated area can be retrieved by using MATLAB function from Robotics System Toolbox, which is 'inflate' and the inflation radius with the unit of meter can be input as a function argument. This process uniformly inflated the skeletonized medial axis at a fixed radius from the input arguments. Figure 3 shows an example of the result of the proposed method.

Figure 3a shows the reference map, Fig. 3b shows the skeletonized free space with a size of 1 pixel. Figure 3c, d show the inflation result with 10 and 30 mm inflation radius. This skeletonized inflated map will later be used in the sampling process of the roadmap.

3 Inflated Medial Axis based Probabilistic Roadmap

The proposed method is similar to the classic PRM [15], where it can be divided into 2 steps process. First, the preprocessing is conducted where the sampling of node N, is sampled inside free space, $Cfree$ and build a connected roadmap as undirected

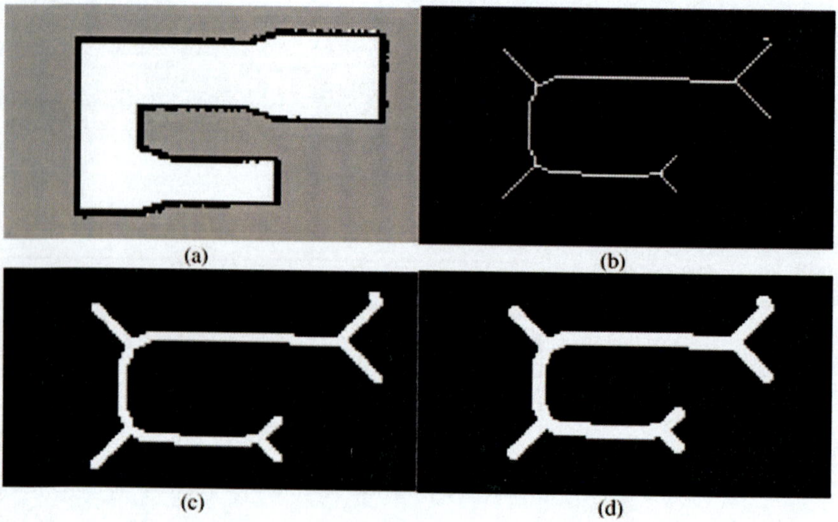

Fig. 3 **a** Reference map. **b** Skeletonized map. **c** 10 mm radius inflation. **d** 30 mm radius inflation

graph $R = (N, E)$ [15]. However, the proposed method only allows samples inside the inflated skeletonized space, $Cskel$ to be added and edges, E will be simplified by a straight feasible path connecting between nodes (a, b).

Second, a query stage is proposed to find the shortest path from the roadmap from start to goal point using A* algorithm. In the preprocessing stage, initially, the graph $R = (N, E)$ is empty, then the nodes are sample randomly in configuration space $Cspace$. For each sampling iteration, nodes will be verified inside $Cskel$. For each successful sampling, the node will be added to N. One of the advantages of PRM is that its multi-query planner generated roadmap can be used multiple times to query paths. Figure 4a, b show the difference between classic PRM and the proposed method.

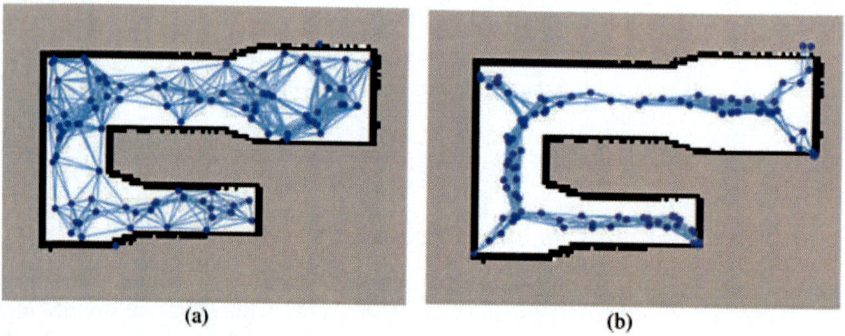

Fig. 4 **a** Classic PRM, **b** proposed method

4 Experimental Results

This section presents the experimental result of the proposed inflated medial axis based on PRM. Figure 5 shows the experiment environment where Fig. 5a show the actual mockup duct environment setup used in the experiment and Fig. 5b is the occupancy grid map of the mockup duct environment [16]. The occupancy map of the environment was acquired using the simultaneous localization and mapping (SLAM) toolbox in MATLAB. Algorithm 1 presents the inflated medial axis based PRM.

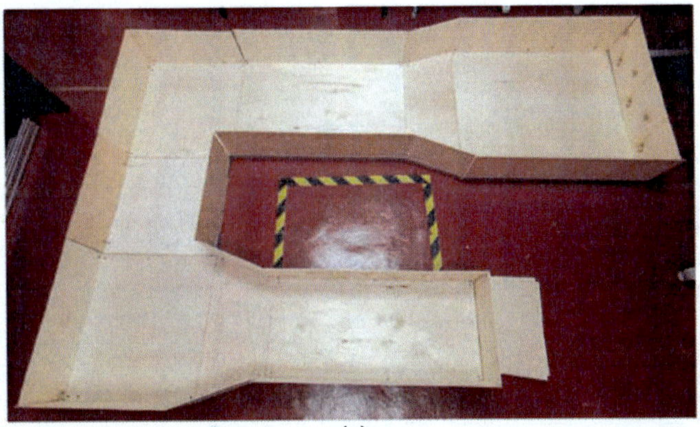

(a)

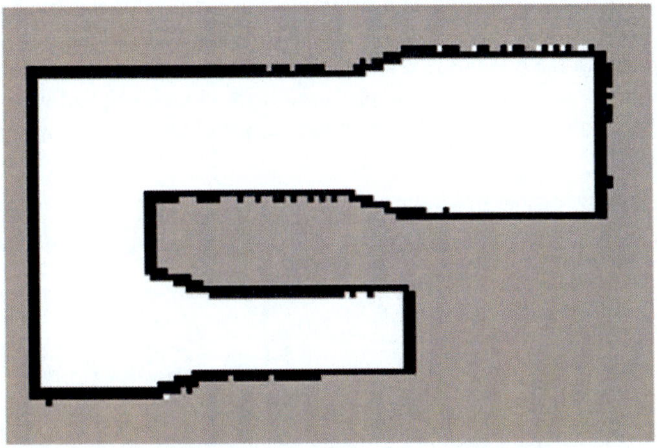

(b)

Fig. 5 Experiment environment **a** actual environment, **b** simulated environment

Algorithm 1: Inflated Medial Axis based PRM

Inputs: $\boldsymbol{n}$ (number of nodes in roadmap)
$\boldsymbol{d}$ (maximum connection distance)
$\boldsymbol{m}$ (occupancy grid map)
$\boldsymbol{r}$ (inflation radius)
Output: A roadmap $\boldsymbol{R} = (\boldsymbol{N}, \boldsymbol{E})$

$\boldsymbol{N} \leftarrow \emptyset$, $\boldsymbol{E} \leftarrow \emptyset$;
$\boldsymbol{Cskel}$ = Skeletonized ($\boldsymbol{m}$)
Inflated $\boldsymbol{Cskel}$ by radius $\boldsymbol{r}$
While $\boldsymbol{N} < \boldsymbol{n}$ do
    Repeat
        $\boldsymbol{x} \leftarrow$ a random configuration
    Until $\boldsymbol{x}$ is collision-free of $\boldsymbol{Cskel}$
    $\boldsymbol{N} \leftarrow \boldsymbol{N} \cup \{\boldsymbol{x}\}$
End While
For all $\boldsymbol{x} \in \boldsymbol{N}$, do
    $\boldsymbol{Nx} \leftarrow$ closest neighbour $< \boldsymbol{d}$
    For all $x' \in Nx$ do
        If $(\boldsymbol{x}, \boldsymbol{x}') \notin \boldsymbol{E}$ and $(\boldsymbol{xx}')$ is collision-free then
            $\boldsymbol{E} \leftarrow \boldsymbol{E} \cup \{(\boldsymbol{x}, \boldsymbol{x}')\}$
        End if
    End for
End for

This experiment was conducted to study the ability of algorithm 1 to generate clearance-driven path planning. The environment in Fig. 3 is generated from the actual mockup duct was selected as the experiment setup. A constant parameter that was used for each method is the maximum connection distance between nodes of 0.03 m. Both classical PRM and the proposed method were applied with different values of the number of nodes. The result of both path planning is shown in Fig. 6.

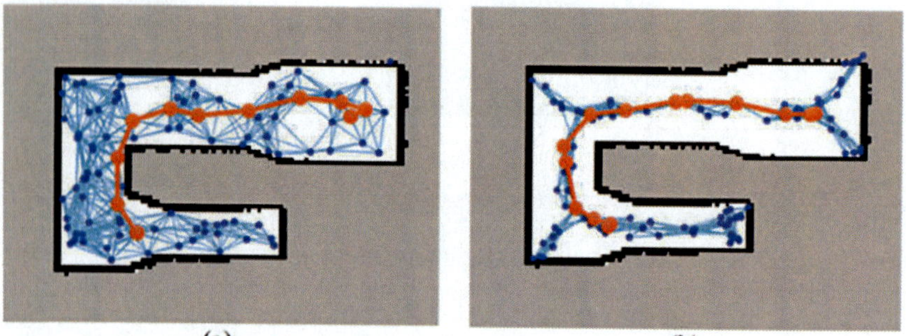

(a) (b)

Fig. 6 **a** Classical PRM. **b** Algorithm 1

Table 1 Result of path clearance and success rate

	PRM (100 nodes)	Alg. 1 (100 nodes)	PRM (200 nodes)	Alg. 1 (200 nodes)
Dist. to obs. (Min.)	0.047 m	0.136 m	0.046 m	0.136
Dist. to obs. (Max.)	0.237 m	0.251 m	0.237 m	0.253
Dist. to obs. (mean)	0.136 m	0.185 m	0.136 m	0.180
Success rate (%)	98.6	91.4	99.1	99.9

Table 2 Result of pre-processing and query computation time

	PRM (100 nodes) (s)	Alg. 1 (100 nodes) (s)	PRM (200 nodes) (s)	Alg.1 (200 nodes) (s)
Pre-processing	0.011	0.035	0.012	0.036
Query	0.011	0.032	0.013	0.099

The experiment was conducted for 1000 trials; 100 and 200 node sampling are executed for both methods. In Table 1, four different parameters were measured to prove the efficiency of the algorithm. The first three parameters are regarding the clearance of the path trajectory. In all cases, the proposed method able to generate a path with the largest distance toward the obstacle. In comparison, the classical PRM generated path that is nearer to the obstacle because it tends to find the shortest path as possible compared to the proposed method that is more constrained toward the medial axis of the map. The final parameter that was compared is the success rate of the planner. The proposed method shows a bit of disadvantages at 100 nodes sampled, but in 200 nodes sample, it gives 99.9% success rate.

Table 2 shows the time taken to execute the two stages of the planner, which are preprocessing and query stages. All the tests show a small difference between classical PRM and the proposed method, for preprocessing stages, even though the proposed method is a bit higher time taken to compute because it has an extra step to skeletonized and inflated the medial axis. Still, it only needs to be sampled once. Computation time for the query stage depends on the size of the roadmap. In 100 sampled nodes, classical PRM and the proposed method have the same performance, while in 200 sampled nodes, the proposed method takes a longer time to compute. It depends on the complexity of the roadmap graph, the proposed method focused on sampling the nodes toward the inflated area. This makes more edges can be constructed between nodes, thus increasing the roadmap complexity.

5 Conclusion

This paper presented on preprocessing method for PRM to sample the nodes with high clearance. The proposed method focused on path planning based on inflated medial axis and probabilistic roadmap for duct environment. The results show that the proposed method can compute a path planning trajectory with increased safety distance by focusing on sampling toward the inside of the inflated medial axis compared to the classical PRM. The future works will explore the postprocessing step to further optimized path planning and generate time-optimal trajectory.

Acknowledgements The research has been carried out under program Research Excellence Consortium [JPT (BPKI) 1000/016/018/25 (57)] provided by Ministry of Higher Education Malaysia (MOHE). The author also would like to thank Universiti Teknologi Malaysia (UTM) for work and facilities support.

References

1. S.A. Batterman, H. Burge, HVAC systems as emission sources affecting indoor air quality: a critical review. HVAC R Res. **1**, 61–78 (1995). https://doi.org/10.1080/10789669.1995.10391309
2. A. Athif, M. Faudzi, O. Tian-qiong, W. Liang, T. Weng, Development of ROV based water tank cleaning robot. J. Teknol. **2**, 103–110 (2015)
3. Y. Wang, Autonomous air duct cleaning robot system, in *2006 49th IEEE International Midwest Symposium on Circuits and Systems*, 2006, pp. 510–513
4. A. Aisha, B. Hisham, W. Liang, M. Hafis, I. Ishak, N. Hawani, A. Mohd, Mobile inspection robot for heating, ventilation and air conditioning. Jurnal Teknologi **6**, 89–93 (2015)
5. H. Sun, D.Y. Kim, J.H. Hwang, C.W. Park, Mobile robot navigation under duct environment. Proc. Appl. Imag. Pattern Recognit. Work 1–3 (2011).https://doi.org/10.1109/AIPR.2011.6176351
6. K. Koh, H.J. Choi, J.-S. Kim, K.W. Ko, H. Cho, Sensor based navigation of air-duct inspection mobile robots. Optomechatronic Syst. **4190**, 202–211 (2001). https://doi.org/10.1117/12.417227
7. L.E. Kavraki, P. Švestka, J.C. Latombe, M.H. Overmars, Probabilistic roadmaps for path planning in high-dimensional configuration spaces. IEEE Trans. Robot. Autom. **12**, 566–580 (1996). https://doi.org/10.1109/70.508439
8. K.E. Bekris, B.Y. Chen, A.M. Ladd, E. Plaku, L.E. Kavraki, Multiple query probabilistic roadmap planning using single query planning primitives. IEEE Int. Conf. Intell. Robot. Syst. **1**, 656–661 (2003). https://doi.org/10.1109/iros.2003.1250704
9. E. Plaku, E. Plaku, P. Simari, Clearance-driven motion planning for mobile robots with differential constraints. Robotica **36**, 971–993 (2018). https://doi.org/10.1017/S0263574718000164
10. S.A. Wilmarth, N.M. Amato, P.F. Stiller, MAPRM: a probabilistic roadmap planner with sampling on the medial axis of the space. Proc. IEEE Int. Conf. Robot. Autom. **2**, 1024–1031 (1999). https://doi.org/10.1109/robot.1999.772448
11. H.Y.C. Yeh, J. Denny, A. Lindsey, S. Thomas, N.M. Amato, UMAPRM: uniformly sampling the medial axis. Proc. IEEE Int. Conf. Robot. Autom. 5798–5803 (2014).https://doi.org/10.1109/ICRA.2014.6907711

12. M.R.H. Al-dahhan, K.W. Schmidt, Path planning based on Voronoi diagram and PRM for omnidirectinal mobile robots (2019)
13. L. Lam, S.W. Lee, Thinning methodologies—a comprehensive survey. IEEE Trans. Pattern Anal. Mach. Intell. **14**, 869–885 (1992)
14. M. Bao, S. Guo, Q. Tang, F. Zhang, Optimization of the bwmorph Function in the MATLAB image processing toolbox for binary skeleton computation. Proc. Int. Conf. Comput. Intell. Nat. Comput. CINC. 273–276 (2009). https://doi.org/10.1109/CINC.2009.96
15. W. Khaksar, T.S. Hong, M. Khaksar, O. Motlagh, A low dispersion probabilistic roadmaps (LD-PRM) algorithm for fast and efficient sampling-based motion planning. Int. J. Adv. Robot. Syst. **10**https://doi.org/10.5772/56973
16. L.X. Wong, A.A.M. Faudzi, M.H.I. Ishak, Mobile inspection robot for HVAC ducting system using embedded raspberry-PI controller, in *2nd Mini Symposium on Artificial Intelligence and Robotics*, 2014, pp. 1–5

Analytical Study on LSB-Based Image Steganography Approach

Oluwakemi Christiana Abikoye, Roseline Oluwaseun Ogundokun, Sanjay Misra, and Akasht Agrawal

Abstract Steganography has recently turned out to be an excellent field of science that is used to shield data from illegal entry. Steganography is described as the skill of communication which involves concealing knowledge in a different medium, for instance, textual, image, audio, video, or network in order not to provoke any trace, whereas steganalysis is the discipline of destroying the steganographic mechanism to expose the undisclosed communication. The efficiency of a steganography procedure is measured utilizing three key criteria, and they include payload efficiency, image excellence calculation, and degree of protection. This investigation is concentrated on image steganography which is extremely accepted in the steganography field. The least significant bit (LSB) approach is mainly the effectual method employed to entrench the undisclosed communication. The study, therefore, focused on the least significant bit steganography algorithm, thereby increasing its performance by proposing a modified least significant bit steganography called circular shift LSB. The study also elucidated the LSB entrenching procedure and presented the assessment outcomes for LSB text steganography for different cover object formats such as .png, .bmp, and .jpeg. This paper has more comprehensive information on different cover

O. Christiana Abikoye
Department of Computer Science, University of Ilorin, Ilorin, Nigeria
e-mail: abikoye.o@unilorin.edu.ng

R. Oluwaseun Ogundokun
Department of Computer Science, Landmark University Omu Aran, Omu-Aran, Nigeria
e-mail: ogundokun.roseline@lmu.edu.ng

S. Misra (✉)
Department of Computer science and Communication, Ostfold University College, Halden, Norway
e-mail: sanjay.misra@covenantuniversity.edu.ng

A. Agrawal
Amity University, Gurgaon, India

R. Oluwaseun Ogundokun
Center of ICT/ICE Research, Covenant University, Ota, Nigeria

A. Kumar et al. (eds.), *Computational Intelligence in Machine Learning*, Lecture Notes in Electrical Engineering 834, https://doi.org/10.1007/978-981-16-8484-5_43

image formats, focused on LSB. A comparative assessment was carried out to establish the usefulness of the suggested approaches. The success of the projected approach was appraised by the computation of mean square error (MSE) and peak signal-to-noise ratio (PSNR). The investigation showed that the LSB methods improved PSNR and MSE. Furthermore, PSNR error metrics were high and MSE was low indicating strong and robust cover image quality which makes the image highly imperceptible to human eyes or intruders.

Keywords Steganography · Least significant bit · PSNR · MSE · Comparative analysis

1 Introduction

Communication is a rudimentary inevitability of all rising zone in the world in recent trends. The proliferation of new networking systems, in particular, the case of data networks requires a specific means of protection mechanisms. Everyone wants their data to be communicated secretly and securely [1]. Data security is a significant topic of concern when sharing facts in an exposed link because cyberspace is not solitary a sole system but a large set of networks around the globe. Network protection is becoming increasingly relevant as the amount of communication shared via cyberspace intensified daily [2]. Data protection requires storing hidden documents, text, audio, or video files among any other archives that may be used to protect sensitive records. Recent literature works in steganography have the possibility of an anonymous individual remembering hidden knowledge. Owing to new knowledge, information performs a significant function in a wide variety of diverse areas [2, 3].

The steganography method involves masking concealed data into every single pixel's LSB in an image. Built on the LSB procedure, an 8- or 24-bit color image algorithm is established to boost the stego image accuracy of the color object proficient in generating a hidden concealed object that is fully imperceptible to the human eye [1]. The small, important parts of each pixel can be employed to entrench the hidden communication in the concealment medium. This approach increases adjustment sensitivity but degrades the stego image quality [4]. The LSB entrenching procedure implies that images could be concealed in the LSB of the concealment object so that the people's vision won't detect the concealed image in the concealment object [5]. This approach could as well be employed to conceal text in 24-bit or 8-bit or grayscale form. This research was, however, used to embed medical information into color cover file formats such as .bmp, .png, and .jpeg using a modified LSB steganography technique called circular shift LSB steganography algorithm. This paper also presents a comprehensive LSB-built image steganography knowledge in various image forms.

Several researchers have implemented the LSB steganography on different cover image formats. Some of their studies are discussed as follows:

Arya [1] has suggested an improved LSB substitution approach in this paper author to cover hidden image information files in a color image. For example, several different carrier file types should be used to show that this technique is ideal for such types, such as bitmap, jpeg, and PNG. The writers intended the performance assessment of hidden image steganography techniques using LSB data and image protection tools, their analysis on various sizes and file formats (.bmp; .jpg; .png) and the measurement of output parameters such as PSNR and MSE to evaluate payload ability. Hashim et al. [6] based on steganographic image, which is furthermost common in divisions of steganography. The LSB is usually the foremost successful method used to incorporate the hidden code. Therefore, this paper has more comprehensive information across different image formats dependent on the least significant bit LSB. All calculations are explained with arithmetical equations in this study, although few significant themes are also conferred at the finishing of the paper. Sharma et al. [7] provided an outline of different steganographic methods, their main forms, and steganography classification suggested in the literature over the past few years. They also objectively evaluated numerous theoretical approaches that demonstrate that the visual output of the image degrades as the hidden data rises to a threshold limit using methods focused on the LSB. Latika and Gulati [8] offered insight into the principles of cryptography and steganography. Besides, along with the different steganography methods, they have provided an outline of the work and developments in the arena of steganography. Their job also provides recommendations on potential steganography research. Champakamala et al. [9] suggested a technique for image steganography using hardware or electronics. They have shown that the integration of hidden knowledge into images involves detailed and in-depth calculations, and therefore, the implementation of steganography in hardware speeds up steganography. The study, therefore, focused on the least significant bit steganography algorithm, thereby increasing its performance by proposing a modified least significant bit steganography called circular shift LSB. The study also described the LSB entrenching procedure as well as offered the assessment outcomes for LSB text steganography for different cover object formats such as .png, .bmp, and .jpeg. This study further presented clear information built on LSB in numerous cover image forms. The key contribution in this research is to explain the specific of the various measurement criteria to achieve good image steganography efficiency. This paper as well as embraces extensive knowledge of LSB-based image steganography in different file formats.

The paper is structured as follows: Sect. 2 discusses the materials and methods used for the implementation and testing of the system. Section 3 discusses the outcome of the proposed studies, and the study was concluded in Sect. 4.

2 Material and Method

2.1 Modified LSB Steganography

Figure 1 shows the embedding phase of the proposed system while Fig. 2 shows the extraction phase of the system. In the embedding phase, the cover image and the undisclosed message are read into the system after which the undisclosed message is translated into ASCII. The LSB of cover image is replaced with the undisclosed message bits, and the result is stego image. In the extraction phase, the stego image

Fig. 1 Embedding phase

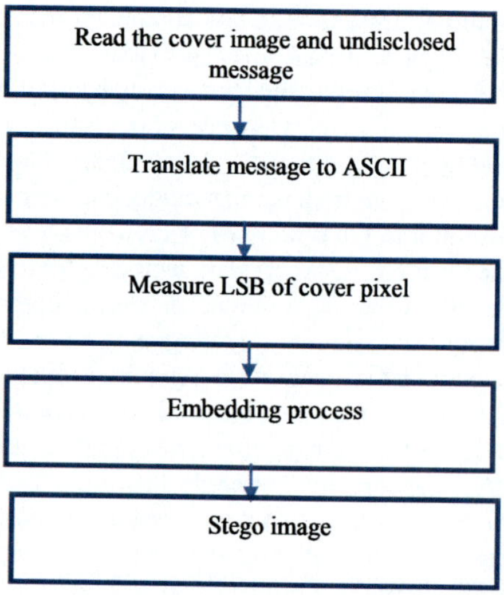

Fig. 2 Extraction phase

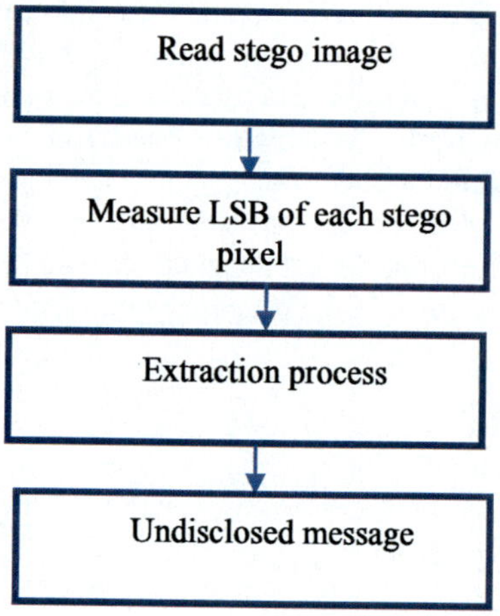

Table 1 Evaluation parameters

Capacity	Imperceptibility	Security
Maximum payload capacity Bit per pixel (bpp) or Bits per byte (BPB)	Root Mean Square Error (RMSE), Mean Squared Error (MSE), Signal Noise Ratio (SNR), Peak Signal-to-Noise Ratio (PSNR), Weighted Peak Signal-to-Noise Ratio (WPSNR), Average Difference (AD), R Measurement, Q index, Structural Information Change (SSIM), MSSIM, Image Fidelity (IF), Normalized Cross-Correlation (NCC)	Regular and Singular (RS), KLD, MMD, FLD, Histogram, Receiver Operating Characteristics (ROC), CM, others

is read in and the LSB of each stego pixel is measured after which the extraction process is conducted, and the result is the undisclosed message initially hidden.

3 Result and Discussion

3.1 *Image Steganography Evaluation*

When designing any steganography process, three goals must be considered, and they are quality of protection, capability, and image value. To analyze and quantify these targets, we will use the following metrics to determine the strengths and disadvantages of the program (Table 1).

3.2 *Different Evaluation of Image Formats*

Many papers have been published in image steganography with different algorithms, to select the best algorithm there must be an analysis of all factors within it. The different factors are selected to evaluate and analyze the file format for steganography. The alteration of the image should not be visual to the people's vision, where the payload capacity which is used to embed also plays a substantial role. Many testings within any algorithm are used to decide how much capacity of information could be used to embed. The steganalysis techniques are used to analyze the steganography methods and perceive the embedding material in the carrier object. The steganography methods should be more secure to guide against steganalysis algorithms. During sending the stego image, the intruder can check and manipulate the stego image to remove or alter the embed information.

Table 2 Performance parameters for steganography

Cover image	Format	PSNR	MSE
Walkway	.jpeg	83.219	0.0003099
Unilorin	.png	84.858	0.0002124
Barbara	.bmp	78.389	0.0009422

Table 3 Comparative analysis

Authors	Format	PSNR	MSE
Arya and Soni [1]	.jpeg .png .bmp	37.95 52.74 55.44	10.5833 0.34667 0.20000
Patel and Cheeran [10]	.jpeg .png .bmp	72.72 71.84 79.72	0.003645 0.003852 0.001013
Proposed system 2020	.jpeg .png .bmp	83.219 84.858 78.389	0.0003099 0.0002124 0.0009422

Observing the findings in Table 2 and testing the output based on modified LSB steganography, it was discovered that .png with 84.858 PSNR and 0.0002124 MSE is the best format for supporting steganography that is the best format for concealing undisclosed textual messages. It was also deduced that the minimum MSE and maximum PSNR requirements have been met in .png format together.

3.3 Discussion

Table 3 shows the comparative analysis between the proposed system and previous research works, and it was deduced from the table that the proposed system performed better than previous research works in terms of the formats, PSNR (84.86) and MSE (0.0002124) when compared with Arya and Soni [1] having 52.74 of PSNR and 0.34667 of MSE. Patel and Cheeran [10] also have 71.84 for the PSNR of .png and 0.003852 of MSE. The best cover image format that is suggested to be adopted from this study is the .png as it has the highest PSNR value and the lowest MSE values while Arya and Soni [1] suggested .bmp and Patel and Cheeran [10] work suggested .bmp as well. This means the stego image is highly imperceptible, and this leads to a high-security level in the process of hiding with the modified LSB steganography.

4 Conclusion

This study examined modified least significant bit (LSB) steganography for entrenching medical information in different concealment image formats such as .bmp, .png, and .jpeg. A comparison between the PSNR and MSE got from this study was compared, and the result was also evaluated with previous concealment image formats used by other researchers. The following outcomes were deduced from this study: Firstly, it was illustrated that this modified LSB method called circular shift algorithm performed better than previous researches when compared with them. Secondly, it was deduced that concealment image with .png format is more robust in masking textual information that is it hides textual information better when compared with other image formats because it had the highest PSNR and the lower MSE which are the two metrics used in evaluating the performance of the system.

Acknowledgements The authors appreciate the sponsorship from Covenant University through its Centre for Research, Innovation and Discovery, Covenant University, Ota Nigeria.

References

1. A. Arya, S. Soni, *Performance Evaluation of Secrete Image Steganography Techniques Using the Least Significant Bit (LSB) Method*, vol. 6 (2018), pp. 160–165
2. O. Rachael, S. Misra, R. Ahuja, A. Adewumi, F. Ayeni, R. Mmaskeliunas, Image steganography and steganalysis based on least significant bit (LSB), in *Lecture Notes in Electrical Engineering*, vol. 605 (2020), pp. 1100–1111
3. O.C. Abikoye, U.A. Ojo, J.B. Awotunde, R.O. Ogundokun, A safe and secured iris template using steganography and cryptography. Multimed. Tools Appl. **79**(31–32), 23483–23506 (2020)
4. M.S. Taha, et al., Combination of steganography and cryptography: a short survey, in *2nd International Conference on Sustainable Engineering Techniques (ICSET 2019), IOP Conference Series: Materials Science and Engineering*, vol. 518 (2019). https://doi.org/10.1088/1757-899X/518/5/05200
5. R.O. Ogundokun, O.C. Abikoye, S. Misra, J.B. Awotunde, Modified least significant bit technique for securing medical images. Lect. Notes Bus. Inform. Proces. **402**, 553–565 (2020)
6. M. Hashim, M. Rahim, M. Shafry, A.A. Alwan, A review and open issues of multifarious image steganography techniques in spatial Do-Main. J. Theor. Appl. Inform. Technol. **96**(4) (2018)
7. M.K. Sharma, A. Upadhyaya, S. Agarwal, Adaptive steganographic algorithm using cryptographic encryption RSA algorithms. J. Eng. Comput. Appl. Sci. (JEC&AS) vol. 2 (2013)
8. Latika, G. Yogita, A review of steganography research and development. Int. J. Adv. Res. Comput. Sci. Softw. Eng. (2015)
9. Champakamala, K. Padmini et al., Least significant bit algorithm for image steganography. Int. J. Adv. Comput. Technol. (IJACT)
10. F.R. Patel, A.N. Cheeran, Performance evaluation of steganography and AES encryption based on different formats of the image. Perform. Evaluat. **4**(5) (2015)

Classification of Lung Cancer Datasets Using Computational Intelligence Techniques

Roseline Oluwaseun Ogundokun, Sanjay Misra, Nerat Dazam, and Akasht Agrawal

Abstract Health technologies have led to an explosion of large data leading to the exponential increase in data in length, velocity, and diversity. Current data collection and management systems are unable to meet the demands of heterogeneous data. As a result, using computational intelligence (CI) methods in medical data will help with improved control, faster efficiency, and greater identification accuracy. Therefore, this study used two CI methods which are principal component analysis (PCA) and support vector machine (SVM) for the classification of the lung cancer dataset. These algorithms were used to detect whether a sick individual has lung cancer or not. The proposed system was established, trained, and validated using lung cancer datasets. The performance of the system was evaluated using a confusion matrix and the study used four performance metrics, for instance, accuracy, precision, F1 score, and recall. The system evaluation demonstrated that SVM with PCA can detect the presence or absence of lung cancer with 84% accuracy. This study assists in achieving sustainability goal six good health and wellbeing in developing countries.

Keywords Machine learning · Computational intelligence · Sustainability · SDG6

R. Oluwaseun Ogundokun · N. Dazam
Department of Computer Science, Landmark University Omu Aran, Omu-Aran, Kwara State, Nigeria
e-mail: ogundokun.roseline@lmu.edu.ng

N. Dazam
e-mail: dazam.nerat@lmu.edu.ng

R. Oluwaseun Ogundokun
Center of ICT/ICE Research, Covenant University, Ota, Nigeria

S. Misra (✉)
Department of Computer Science and Communication, Ostfold University College, Halden, Norway
e-mail: sanjay.misra@covenantuniversity.edu.ng

A. Agrawal
Amity University, Gurgaon, India

A. Kumar et al. (eds.), *Computational Intelligence in Machine Learning*, Lecture Notes in Electrical Engineering 834, https://doi.org/10.1007/978-981-16-8484-5_44

1 Introduction

Due to the huge-capacity, multifaceted and the increasing number of datasets created by numerous sources, for instance, the Internet of ihings (IoT) [1], management documents, medical archives, multimedia, phone logs, social network, and further digital technologies [2]. Furthermore, big data are being used to modernize public health and change patient practice. As a result, big data have unlimited capacity for storing, extracting, inquiring, and investigating health industry datasets effectively [3].

Machine learning classification techniques have been used for several analyzes [4–6]. Computational intelligence (CI) comprises numerous techniques, for instance, support vector machine (SVM), random forest (RF), deep neural network (DNN), K-nearest neighbor (KNN), Naïve Bayes (NB), gradient boosting machine (GBM), and so on. Although, these techniques function in a particular way. A couple of the most recently suggested CI techniques in the field of cancer detection are discussed as thus:

Lynch et al. [7] used the SEER database to rate lung cancer individuals according to their subsistences using supervised learning classification methods including linear regression (LR), DT, GBM, SVM, and a custom ensemble. The findings revealed that GBM, with a root, mean square error (RMSE) value of 15.32, was the utmost precision of the five individual models used. Hashi et al. [8] used DT and KNN procedures to identify diabetes ailment from the Pima Indians dataset, which included 768 data which point to eight features each and achieved a precision of 90.43 and 76.96%, respectively. Hayashi and Yukita [9], Sa'di et al. [10], Huang and Huang [11], who used the decision tree approach to achieve accuracies of 76.96, 83.83, 76.52, and 62.17%, respectively. On a heart ailment dataset with 303 cases, all of it with 14 traits, Khateeb and Usman [12] employed NB, KNN, J48, and catching classifiers/ML classification procedures. They split their data into six groups and found that the KNN classifier, which used all 14 attributes, has the best accuracy of 79.20%.

Therefore, this study used two CI methods which are principal component analysis (PCA) and support vector machine (SVM) for the classification of the lung cancer dataset. These algorithms were used to detect whether a sick individual has lung cancer or not. The proposed system was established, trained, and validated using lung cancer datasets. Additionally, a PCA is employed to discover the unsurpassed attributes.

The article remaining part is structured as thus: Sect. 2 discussed the materials and methods. The CI technique used for the execution of the system in this study was conferred in the section as well. Section 3 discussed the execution and testing of the system. Here, the findings discovered are also conferred in this unit. The study came to an end in Sect. 4.

2 Materials and Method

2.1 Dataset

Dataset was gotten from Kaggle. It has 16 columns and 309 rows. The column names are gender, age, smoking, yellow fingers, anxiety, peer pressure, chronic disease, fatigue, allergy, wheezing, alcohol consuming, coughing, shortness of breath, swallowing difficulty, chest pain, and lung cancer. Lung cancer is the target value.

2.2 The Proposed Approach

This study employed the use of support vector machine (SVM) learning algorithm as shown in Fig. 1. Python programming language was used for the implementation. The dataset is first loaded into the system after which PCA is used for dimensionality reduction. The reduced dataset is then passed into the SVM classifier, and the result gotten performance is evaluated using confusion matrix. This process is shown in Fig. 1.

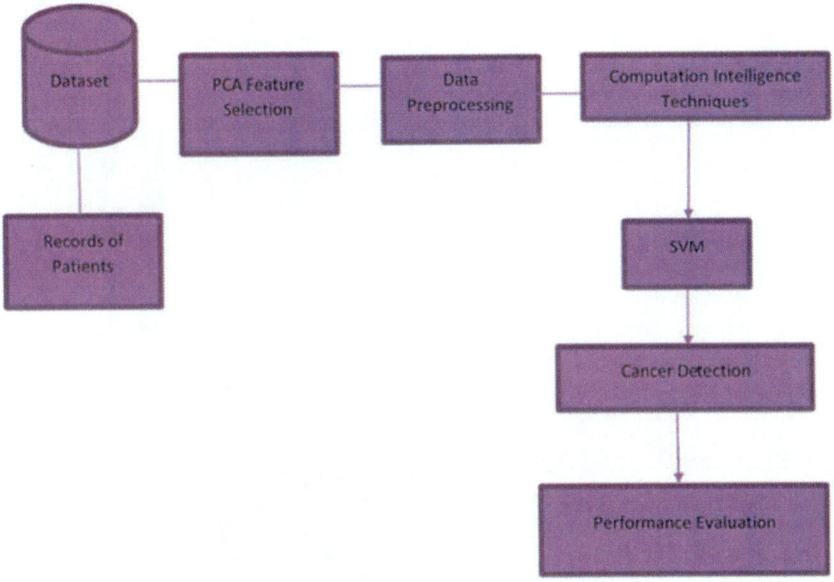

Fig. 1 Schematic illustration for the proposed technique

3 Results

The lung cancer dataset taken out was classified, and the outcomes display the features competency for categorizing the lung cancer status. The projected technique was employed in the openly accessible lung cancer database, the classification approach employed SVM employing Python programming language tools to implement the technique. PCA is based on the goal of reducing the dimensionality of a dataset composed of several variables that are associated with each other while preserving the dataset's maximum uncertainty [13]. The algorithm converts the variables in the dataset into a new set of orthogonal principal components that are sorted in such a way that the initial variables' variance preservation reduces as the order is traversed. As a consequence, the first principal component keeps the most variance from the initial components. In a covariation matrix, the principal components are orthogonal eigenvectors.

A simple supervised learning evaluation procedure is applied to each of the classifiers. To eradicate sampling bias, the preparation and research phases are tested using a confusion matrix method. The Python programming language is used to execute this protocol. The performance measures (accuracy, F1 score, precision, and recall) [14] are used to calculate the evaluation outcome. Figure 2 displays the confusion matrix table, which contains the TP, TN, FP, and FN, which are widely employed for measuring performance metrics.

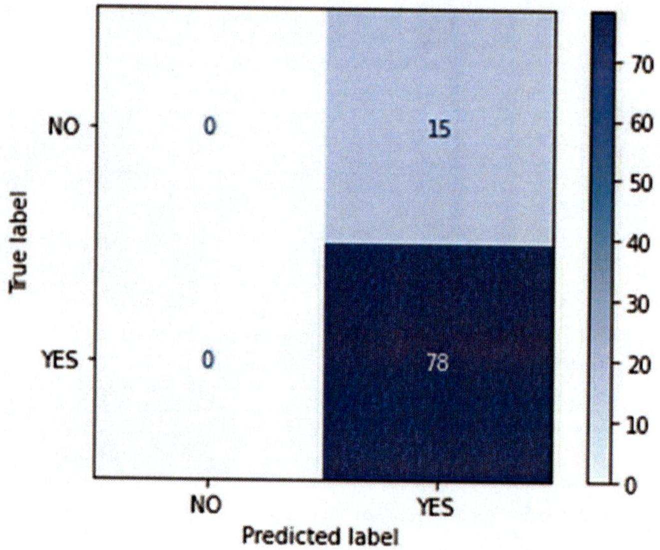

Fig. 2 PCA-SVM classification confusion matrix

Table 1 Performance analysis of the system

Performance metric		PCA-SVM
Precision	No cancer	0.00
	Yes cancer	0.84
	Weighted average	0.70
	Macro average	0.42
Recall	No cancer	0.00
	Yes cancer	1.00
	Weighted average	0.84
	Macro average	0.50
F1 score	No cancer	0.00
	Yes cancer	0.91
	Weighted average	0.77
	Macro average	0.46
Accuracy		0.84
Mean AUC score random state		0.95

Table 2 Comparative analysis with the previous researches

Authors	Disease	Accuracy
Han et al. [15]	Lung cancer	82.51%
Kusy et al. [16]	Cervical cancer	71.96
Proposed algorithm	Lung cancer	84%

Confusion matrix—The confusion matrix displays 0 true negatives (TN), 15 false positives (FP), 0 false negatives (FN), 78 true positives (TP). Table 1 shows the performance analysis of this proposed system.

Table 2 shows the comparative analysis with the previous researches. Han et al. [15] used computational intelligence algorithms on lung cancer and had an accuracy of 82.51%, Kusy et al. [16] used computational intelligence techniques on cervical cancer and had 71.96% accuracy while the proposed technique used computational intelligence techniques on lung cancer and outperformed the two previous pieces of research with an accuracy of 84%.

4 Conclusion

In recent years, substantial progress has been made in the development of medicinal datasets, as well as improvements in implementation metrics and efficiency, which are heavily driven by exploratory plans, activities, and data processing types. Cancer is a dangerous insect that comes in many forms. Numerous studies have been inspired by the importance of classifying lung cancer. Employing SVM on decreased dimensional data and the PCA algorithm, this research classifies lung cancer data. The experiment yielded a similar finding, indicating that PCA-SVM performs admirably, with an accuracy of 84%. More data and hybridized models can be used in future research to advance the accuracy of ML-based approaches.

Acknowledgements The authors appreciate the sponsorship from Covenant University through its Center for Research, Innovation and Discovery, Covenant University, Ota Nigeria

References

1. E.A. Adeniyi, R.O. Ogundokun, J.B. Awotunde, IoMT-based wearable body sensors network healthcare monitoring system. Stud. Comput. Intel. **933**, 103–121 (2021)
2. D. Boyd, K. Crawford, Critical questions for big data: provocations for a cultural, technological, and scholarly phenomenon. Inf. Commun. Soc. **15**(5), 662–679 (2012)
3. A. Gani, A. Siddiqa, S. Shamshirband, F. Hanum, A survey on indexing techniques for big data: taxonomy and performance evaluation. Knowl. Inf. Syst. **46**(2), 241–284 (2016)
4. R.O. Ogundokun, P.O. Sadiku, S. Misra, O.E. Ogundokun, J.B. Awotunde, V. Jaglan, Diagnosis of long sightedness using neural network and decision tree algorithms. J. Phys. Conf. Ser. **1767**(1), 012021 (2021, February)
5. R.K. Behera, S. Shukla, S.K. Rath, S. Misra, Software reliability assessment using machine learning technique, in *International Conference on Computational Science and Its Applications* (Springer, Cham, 2018), pp. 403–411
6. T.O. Oladele, R.O. Ogundokun, A.A. Kayode, A.A. Adegun, M.O. Adebiyi, Application of data mining algorithms for feature selection and prediction of diabetic retinopathy, in *Lecture Notes in Computer Science (including subseries Lecture Notes in Artificial Intelligence and Lecture Notes in Bioinformatics)*, vol. 11623 (LNCS, 2019), pp. 716–730
7. C.M. Lynch, V.H. van Berkel, H.B. Frieboes, Application of unsupervised analysis techniques to lung cancer patient data. PLoS One **12**(9), e0184370 (2017)
8. E.K. Hashi, M.S.U. Zaman, M.R. Hasan, An expert clinical decision support system to predict disease using classification techniques, in *2017 International Conference on Electrical, Computer and Communication Engineering (ECCE)* (IEEE, 2017), pp. 396–400
9. Y. Hayashi, S. Yukita, Rule extraction using Recursive-Rule extraction algorithm with J48graft combined with sampling selection techniques for the diagnosis of type 2 diabetes mellitus in the Pima Indian dataset. Inform. Med. Unlocked **2**, 92–104 (2016)
10. S. Sa'di, A. Maleki, R. Hashemi, Z. Panbechi, K. Chalabi, Comparison of data mining algorithms in the diagnosis of type II diabetes. Int. J. Comput. Sci. Appl. (IJCSA) **5**(5), 1–12 (2015)
11. N. Huang, S. Zhang, G. Cai, D. Xu, Power quality disturbances recognition based on a multiresolution generalized S-transform and a PSO-improved decision tree. Energies **8**(1), 549–572 (2015)
12. N. Khateeb, M. Usman, Efficient heart disease prediction system using K-nearest neighbor classification technique, in *Proceedings of the International Conference on Big Data and Internet of Thing* (2017, December), pp. 21–26
13. I.T. Jolliffe, J. Cadima, Principal component analysis: a review and recent developments. Philos. Trans. R. Soc. A: Math. Phys. Eng. Sci. **374**(2065), 20150202 (2016)
14. N.T. Johnson, A. Dhroso, K.J. Hughes, D. Korkin, Biological classification with RNA-Seq data: can alternative splicing enhance machine learning classifier? bioRxiv. 146340 (2017)

15. G. Han, X. Liu, G. Zheng, M. Wang, S. Huang, Automatic recognition of 3D GGO CT imaging signs through the fusion of hybrid resampling and layer-wise fine-tuning CNNs. Med. Biol. Eng. Comput. **56**(12), 2201–2212 (2018)
16. M. Kusy, B. Obrzut, J. Kluska, Application of gene expression programming and neural networks to predict adverse events of radical hysterectomy in cervical cancer patients. Med. Biol. Eng. Comput. **51**(12), 1357–1365 (2013)

Data Security in Cloud Computing Using a Hybrid Algorithm Approach

Kolawole Damilare Abel, Sanjay Misra, Oluranti Jonathan, Akshat Agrawal, Rytis Maskeliunas, and Robertas Damasevicius

Abstract Cloud computing is a technology that provides users with computing resources and storage. It removes the need for businesses and institutes to maintain expensive computing facilities and improves organizations by its services. This paper aims to use cryptography techniques to enhance data security in the cloud by implementing the provided algorithms. The work in this paper is majors on implementing a hybrid algorithm (symmetric and asymmetric algorithms). We proposed a hybrid cryptosystem that comprises symmetric and asymmetric algorithms using 3DES, RSA, and SHA-3 algorithms to enable tight security in the cloud. 3DES is for encryption RSA for authentication and SHA-3 for integrity.

Keywords Cryptography · Data · Security · Algorithms · Hybrid · Cloud computing · Encryption · Decryption · 3DES · RSA · SHA-3 · Computer security

1 Introduction

Despite the constant and rapid use of Internet and computer resources in this era, accessing such computer resources remains expensive. Cloud computing seeks to

K. D. Abel · O. Jonathan
Center of ICT/ICE, Covenant University Ota, Ogun, Nigeria
e-mail: jonathan.oluranti@covenantuniversity.edu.ng

S. Misra (✉)
Department of Computer Science and Communication, Ostfold University College, Halden, Norway
e-mail: sanjay.misra@hiof.no

A. Agrawal
Amity University Haryana, Gurgaon, India

R. Maskeliunas · R. Damasevicius
Silesian University of Technology, Gliwice, Poland
e-mail: rytis.maskeliunas@polsl.pl

R. Damasevicius
e-mail: robertas.damasevicius@polsl.pl

A. Kumar et al. (eds.), *Computational Intelligence in Machine Learning*, Lecture Notes in Electrical Engineering 834, https://doi.org/10.1007/978-981-16-8484-5_45

make computer resources cheap and available to all users based on their demands. It eliminates the burden associated with data storage and maintenance [1]. One of the challenges this technology face is data security; there have been little gaps faced by some of the existing algorithms like the DES, which can easily subject to brute force attack. In the Blowfish algorithm, key management serves as a deficiency as the number of users increases, while the AES algorithm makes use of an easy algebraic structure [2–4]. Data transmission security is an essential factor in network communication. Network communication is said to be safe so long the availability, privacy, and integrity of the data can be properly secured. In other, for us to maintain a high level of security, cryptography techniques were introduced. The symmetric-key algorithm uses a single key shared between two or more users. The common symmetric-key algorithms are data encryption standard which use a 64-bit plaintext and create a 64-bit cipher [5, 6]. Triple data encryption standard (3DES is a symmetric-key block cipher which triples the DES cipher to each data block) and advanced encryption standard (the key lengths in AES algorithm are 128, 192, and 256 which are strong enough to protect the transmitted data.) while Blowfish uses a variable-length key from 32 bits up to 44 bits and a 64-bit block size [7]. The Diffie–Hellman algorithm [8] has a key exchange method using a discrete logarithm. Both the sender and receiver will mutually set up a secret key to their symmetric-key system using an unsecured channel. The performance was based on processing time, operation of CPU, and encryption throughput. Nasarul Islam and Mohamed Riyas [9] discussed and highlighted various cryptography techniques that can be used in securing data over the cloud. Nanda et al. [10] studied the various algorithm approach of confidentiality in the cloud such as biometric encryption, secret sharing schema, and HPI-secure. Azougaghe et al. [11] Proposed a model of AES and ElGamal algorithm for data security. AES was used to encrypt the file, while ElGamal was used to encrypt AES for stronger security measures. Bhandari et al. [12], in this paper, a structured framework is suggested to maintain high security and privacy data in the cloud and improve security mechanisms like RSA using hybrid encryption RSA and advanced encryption standard (AES). A deep study was done by [13] for both symmetric and asymmetric algorithms and their respected families. The analysis report shows that different parameter determines the efficiency of each algorithm [14]. Focuses on AES and RSA algorithms for sharing files in the cloud securely [15, 16] This research paper delves into security challenges faced in the cloud; it also explains the current security approaches used to secure the cloud infrastructure, applications, and limitations. Then some essential research defiance of achieving a novel cloud-aware security solution caters for pre-emptive protection for complex and dynamic cloud infrastructure. Belguith and Lip [17], this research work proposed a flexible encryption algorithm that combines the symmetric algorithm for encryption and the asymmetric for the distribution of keys. The merging aids in benefiting from the reliable security of asymmetric algorithm encryption and the swift performance of symmetric encryption while limiting users to access data only the authorized way. Chandra et al. [18], in this paper, AES for data security in the Heroku cloud platform was implemented. The analysis result shows that AES cryptography can be used for data security.

2 Methodology

The proposed hybrid algorithm uses both asymmetric and symmetric algorithms for better security measures in the cloud. They all follow the process of both encryption and decryption. When the file is encrypted, the original data is turned into a ciphertext that humans cannot understand. To get back the original data, we decrypt the encrypted file by using the generated key. A prototype is given below using a server and a client.

Encryption

3DES algorithm (symmetric-key algorithm) turns the main data into a ciphertext. This method of encryption is related to DES but it is applied 3 times. One of the limitations of symmetric algorithm is transferring of key over the Internet [19], and in other to overcome this limitation, the RSA algorithm is applied.

SHA-3 algorithm is a digital signature that authenticate the message. It ensures that the data is only authorized by verified person (Fig. 1).

Proposed Algorithm

`Encryption`

1. `Get 3DES key (user defined or auto-generated)`
2. `Encrypt the 3DES key using RSA encryption (public key is used here)`

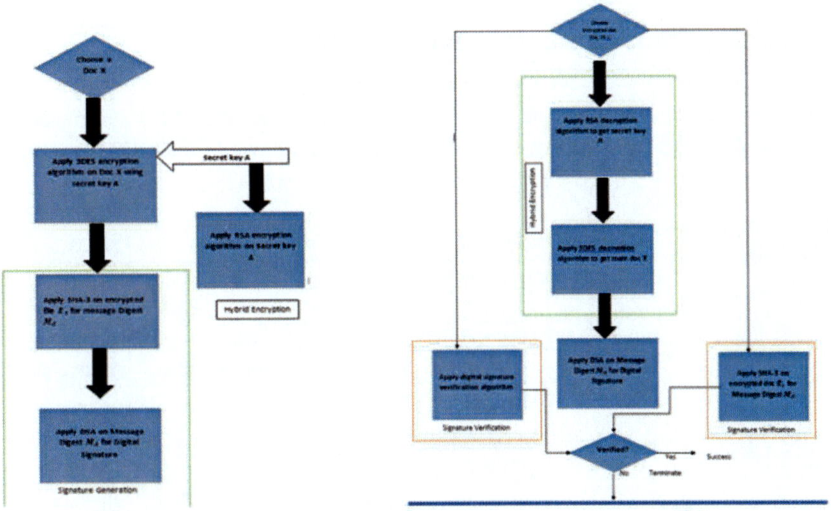

(a) (b)

Fig. 1 **a** Flowchart for encryption process, **b** Decryption process

3. Start the server by opening connection
4. If client has connected, send the encrypted 3DES key
5. Get the file to be securely transmitted
6. Encrypt the file using 3DES encryption
7. Compute the message digest of the file using SHA-3 algorithm
8. Send encrypted file data and message digest to client.

Decryption Algorithm

1. Get server IP and create a connection
2. If connected to server, receive the encrypted 3DES key
3. Decrypt the 3DES key using RSA decryption (Private Key is used here)
4. Wait for the server to send an encrypted file
5. If encrypted file is received, decrypt it using 3DES decryption, else repeat 4
6. Save decrypted file
7. Compute the message digest using SHA-3 algorithm and compare with the received one to check for data integrity
8. Repeat from 4.

3 Comparison, Discussion, and Results

A comparison of algorithms is given in Table 1. It shows that AES symmetric is the best algorithm in securing data. 3DES covers the flaws in the DES algorithm,

Table 1 Comparison of various algorithms

Factors	AES	DES	3DES
Key length	128, 192, 256 bits	112, 168 bits	56 bits
Cipher type	Symmetric block cipher	Symmetric block cipher	Symmetric block cipher
Block size	128, 192, 256 bits	64 bits	64 bits
Developed	2000	1978	1977
Attack	No known attack	Broken Brute force 1998	No known attack
Security	Considered secure	Only one weak which	Proven inadequate
Possible keys	2^{128} 2^{192}, 2^{256}	2^{112}, 2^{168}	2^{56}

which makes it to be more secure. In this paper, we combined 3DES with RSA and SHA-3 algorithm in other to build a strong hybrid system. If a hacker gets through with the encrypted file, he won't be able to get through the RSA algorithm of which the encrypted file is secured, and there will be no message digest performed on such data because the integrity has been broken.

Our model is therefore stronger than any single algorithm [18] used for data security. The limitation of this model is the encryption process which is 3DES of which can be improved on in the future works to be done. Our model has minimized data theft in transit and minimizes data intrusion and spying; this is done by the use of a secured IP address which will only be known to client connecting at the moment. A hybrid algorithm is a better for it strengthens the security system and makes it hard for hackers to breakthrough which is why this algorithm was proposed. Due to the limitation of single algorithms [13], we introduced the 3DES, RSA, and SHA-3 algorithm to provide an enhancement in the security department in the cloud. 3DES key length is 168-bit and follows an encrypt–decrypt–encrypt procedure as shown in Fig. 2.

Here, we programmed our application using Java and the prototype was made in form of a server and client interface as seen in Figs. 3 and 4. To send data, the server is launched and the file to be uploaded is chosen and then encrypted with 3DES

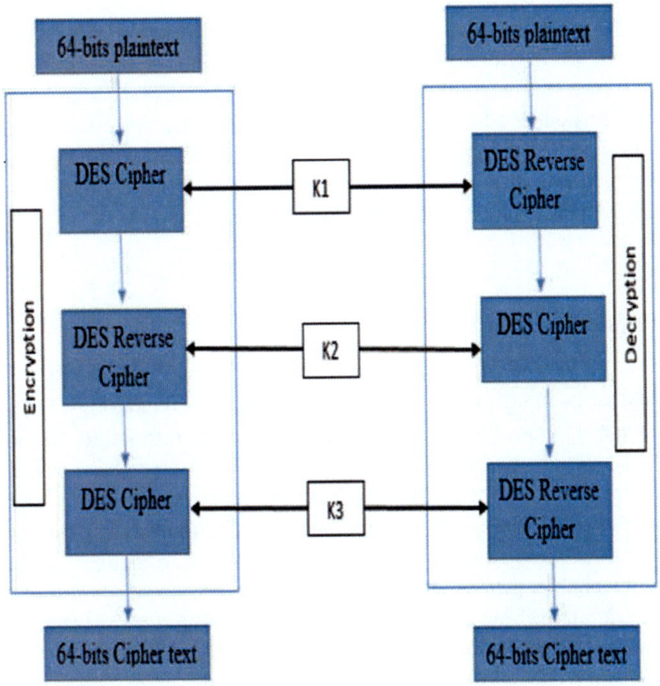

Fig. 2 Encrypt–decrypt procedure

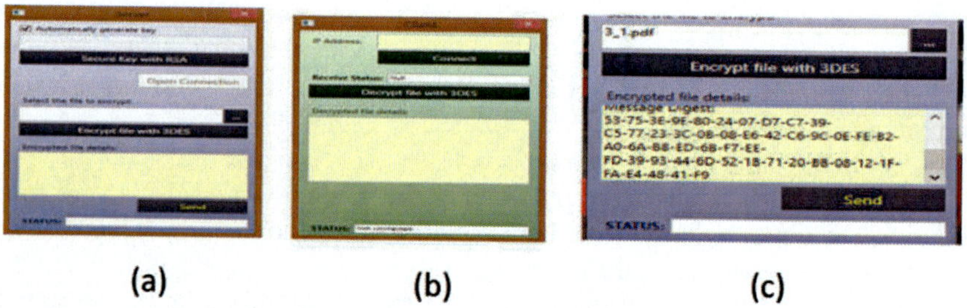

Fig. 3 **a** Server interface, **b** Client interface, **c** Message digest

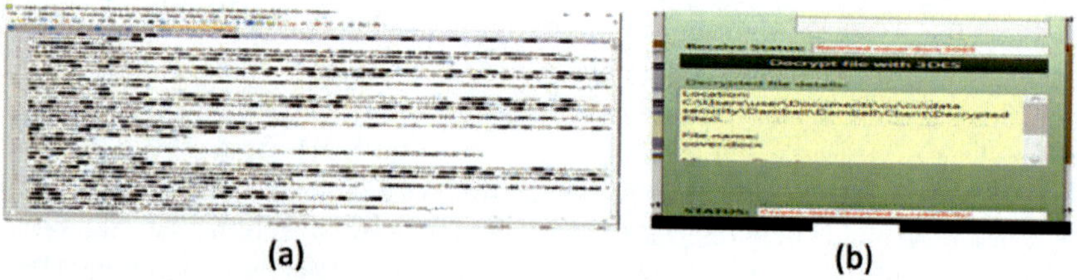

Fig. 4 **a** CipherText, **b** Data location after encryption

key after which we secure with RSA. Once this is done, the encrypted file details are shown as seen in Fig. 6, and then we open the connection for the client to join through a designated IP address in other to receive the transferred document.

Once the client joins the network, he receives the file and decrypts it with 3DES. Figure 7 is a pictorial view of the ciphertext, while Fig. 8 is the view of a client who has connected and received the document sent from the server.

4 Conclusion

Data security has been a pressing issue in cloud computing technology. Organizations do not embrace it because they feel they become more vulnerable, so they believe staying off the cloud would grant safety. Due to this, researchers have been able to do some deep learning or study in solving the problem of security in the cloud. Various techniques to make data secured have been proposed over time; among all these proposed techniques, cryptography has been found to be the most reliable. In this research paper, we proposed a hybrid encryption methodology using symmetric and asymmetric algorithms. In this proposed prototype, the data is encrypted using the 3DES symmetric algorithm; the RSA is used to authenticate or validate that data is being released to the right user, while the SHA-3 algorithm is for data integrity. This prototype is represented using a server and client-based portal, as shown in Figs. 3 and 4. Here the data is first secured with the 3DES key then secured with the RSA, after which the server is turned on for connection to the client. Once the client is connected to the server with the provided IP address given by the server, the encrypted 3DES key and a copy of the RSA private key are sent across from the server to the client, which will grant access for decryption. Since it is not a multi-client system, the server validates the client application immediately a connection is made. But the client only can decrypt the last file that was last received from the server. The RSA allows for key generation and can auto-generate its own key using random key generation algorithm; this generates a different key for every communication. This work can be improved by proposing a multi-client application so more users can connect and either a Blowfish, AES, or other algorithms due to their speed performance and easy interactions.

Acknowledgements The authors appreciate the sponsorship from Covenant University through its Centre for Research, Innovation and Discovery, Covenant University, Ota Nigeria.

References

1. I. Odun-Ayo, S. Misra, N. Omoregbe, E. Onibere, Y. Bulama, R. Damaševičius, Cloud-based security driven human resource management system. Front Artif Intel Appl Vol Adv Digit Technol **295**, 96–106 (2017)
2. K. Handa, U. Singh, Data security in cloud computing using encryption and steganography. Int. J. Comput. Sci. Mob. Comput. **4**(5), 786–791 (2015)
3. R. Chatterjee, Cryptography in cloud computing: a basic approach to ensure security in cloud cryptography in cloud computing: a basic approach to ensure security in cloud. Int. J. Eng. Sci. Comput. **7**(5), 11818–11821 (2017)
4. M. Olowu, C. Yinka-Banjo, S. Misra, H. Florez, A secured private-cloud computing system, in *Applied Informatics. ICAI 2019. Communications in Computer and Information Science*, vol. 1051, ed. by H. Florez, M. Leon, J. Diaz-Nafria (2019)
5. A. Bhardwaj, G.V.B. Subrahmanyam, V. Avasthi, H. Sastry, Security Algorithms for cloud computing. Procedia - Procedia Comput. Sci. **85**, 535–542 (2016)
6. S.D. Rihan, S.E.F. Osman, A performance comparison of encryption algorithms AES and DES. Int. J. Eng. Res. Technol. (IJERT) **4**(12), 151–154 (2015)
7. S. Rajendirakumar, A. Marimuthu, Cryptographic algorithms used in cloud computing—an analysis and comparison. Int. J. Res. Appl. Sci. Eng. Technol. **6**(I), 2718–2728 (2018)
8. P. Kalpana, Data security in cloud computing using RSA algorithm. Int. J. Res. Comput. Commun. Technol. **1**(4), 143–146 (2012)
9. K.V. Nasarul Islam, K.V. Mohamed Riyas, Analysis of various encryption algorithms in cloud computing. Int. J. Comput. Sci. Mob. Comput. **6**(7), 90–97 (2017)
10. D. Nanda et al., An efficient algorithm for data security in cloud storage, in *Proceedings of 2016 6th International Conference—Cloud System and Big Data Engineering (Confluence)*, vol. 3 (Nciccnda, 2017), pp. 517–525
11. A. Azougaghe, Z. Kartit, M. Hedabou, M. Belkasmi, M. El Marraki, An efficient algorithm for data security in cloud storage, in *IEEE International Conference on Communication International Conference on Intelligent System Design and Engineering Application ISDA*, vol. 2016-June, no. 15 (2016), pp. 421–427
12. A. Bhandari, A. Gupta, D. Das, Secure algorithm for cloud computing and its applications, in *IEEE Proceedings of 2016 6th International Conference on Cloud System and Big Data Engineering (Confluence)*, vol. 8, no. 16 (2016), pp. 188–192

13. M.B. Yassein, S. Aljawarneh, E. Qawasmeh, W. Mardini, Comprehensive study of symmetric key and asymmetric key encryption algorithms, in *IEEE ICET 2017*, Antalya, Turkey, vol. 17 (2017), pp. 1–7
14. K. Chachapara, S. Bhadlawala, Secure sharing with cryptography in cloud computing, in *IEEE 2013 Nirma University International Conference on Engineering NUiCONE 2013*, vol. 13, no. 4 (2013), pp. 1–3
15. G. Kaur, M. Mahajan, Analyzing data security for cloud computing using cryptographic algorithms. Int. J. Eng. Res. Appl. **3**(5), 782–786 (2013)
16. A. Behl, Emerging security challenges in cloud computing: an insight to cloud security challenges and their mitigation, in *IEEE Proceedings of 2011 World Congress on Information and Communication Technologies, WICT 2011*, vol. 8, no. 11 (2011), pp. 217–222
17. S. Belguith, L. Lip, Enhancing data security in cloud computing using a lightweight cryptographic algorithm, *Elev. Int. Conf. Auton. Auton. Syst. ICAS*, vol. 3, no. c (2015), pp. 98–103
18. S. Chandra, S. Paira, S.S. Alam, G. Sanyal, A comparative survey of symmetric and asymmetric key cryptography, in *2014 international conference on communication, computing engineering, ICECCE 2014*, vol. 4, no. 14 (2014), pp. 83–93
19. D.P. Timothy, A.K. Santra, A hybrid cryptography algorithm for cloud computing security, in *2017 IEEE International Conference on Microelectronic Devices, Circuits and Systems. ICMDCS 2017*, vol. 2017, no. 8, (2017) pp. 1–5

Data Security Using Cryptography and Steganography Technique on the Cloud

Kolawole Damilare Abel, Sanjay Misra, Akshat Agrawal, Rytis Maskeliunas, and Robertas Damasevicius

Abstract Cloud computing technology is on the rise in our world today. It has become a sought-after technology due to its flexibility for users to access their data at any time of the day. From anywhere they are provided, there is an Internet connection. One of the challenges this technology faces is security, due to this reason, some organization has refused to embrace this technology for the safety of their data. Researchers have done quite a lot of work in addressing this issue of which in this paper a novel approach is also introduced. Our proposed model has to do with a hybrid cryptography method using the blowfish algorithm for encryption and the RSA algorithm for the encryption of the secret key, and then the Replace R in RGB steganography algorithm will be used for the cover image in other to strengthen the security on the cloud.

Keywords Cryptography · Steganography · RSA · Blowfish · Data · Security · Algorithm · LSB

Present Address:
K. D. Abel
Center of ICT/ICE, Covenant University Ota, Ogun, Nigeria

S. Misra (✉)
Department of Computer Science and Communication, Ostfold University College, Halden, Norway
e-mail: sanjay.misra@hiof.no

A. Agrawal
Amity University Haryana, Gurgaon, India

R. Maskeliunas · R. Damasevicius
Silesian University of Technology, Gliwice, Poland
e-mail: rytis.maskeliunas@polsl.pl

R. Damasevicius
e-mail: robertas.damasevicius@polsl.pl

A. Kumar et al. (eds.), *Computational Intelligence in Machine Learning*, Lecture Notes in Electrical Engineering 834, https://doi.org/10.1007/978-981-16-8484-5_46

1 Introduction

In the modern world of today, storing and accessing stored data without the need of carrying an equipment around have become one of the needs of the society. This has now become possible through the use of cloud computing technology. Security is a challenge in the cloud, and to solve the prevailing challenges, various techniques and approaches have been proposed by various researchers amongst such solutions which are cryptography and steganography. Cryptography has two types of algorithms which are the symmetric and asymmetric algorithms. While on the other hand, steganography simply put is covered writing [1, 2]. It does this by encoding the data in a media format in other to divert the attention of the unwanted users. Both cryptography and steganography have the same aim of securing a message or data from unauthorized personnel.

In this paper, we proposed the combination of these two techniques in other to enhance data security [3] on the cloud and to avoid unauthorized users from accessing stored files. The blowfish and RSA cryptography algorithms are used for encryption and the Replace R in RGB steganography algorithm for covered writing. This study is organized as follows: Sect. 1 states the literature review, Sect. 2.1 explains the proposed algorithm, and Sect. 3 gives the result and discussion. The paper concludes with conclusion and future work in Sect. 4.

Singla [4] reviewed cryptography and steganography algorithms in which the critical analysis done reveals that these two techniques (cryptography and steganography) would provide stronger security and prevent unauthorized access. A new security technique was proposed by Maitri [5] of which the symmetric algorithm and steganography were used. Block-wise security to data was also implemented using AES, blowfish, RC6 and BRA algorithms with its key size at 128 bit. The files are divided into eight (8) different parts and are encrypted with the aid of different algorithm [6]. Obfuscation was used to convert the data into new form and then enshroud the main data while the existence of information is hidden with steganography. Olowu [7] proposed a novel approach of the least significant bit (LSB) steganography which enhanced the 1 byte least significance technique to produce quality images that cannot be easily recognized [3]. In this paper, the author introduced a steganography and cryptography technics in protecting cloud data. With the use of multilayer steganography and AES cryptography algorithm, a novel step for storing and sharing information is introduced. With the aid of MATLAB, Ranjan [8] developed a hybrid model of cryptography and steganography using a sequential algorithm for steganography and the symmetric XOR algorithm for cryptography. Saritha [9] proposed data secrecy through text steganography alongside cryptography. This text steganography is format based which is protected by a private cryptographic key for a higher security measure. The stego-text looks like the original text so as to make the hacker unaware of the specific hidden data [10]. To reduce the harm caused by hackers, Muttoo [11] proposed a security technique using the AES cryptography algorithm along with the steganography [12]. After encryption has been done by the AES algorithm, the ciphertext is saved as an image which is also the secret image and then embedded into the AES and least significant bit to produce the stego-image. Yogeswari and Eswaran [13] developed a two-level security hybrid system that allows high-quality stego-image and embedding capacity. In this model, to decrypt the ciphertext, the secret key must be known by the steganalyst. The encryption is done using AES_MPK and hidden using PVD-MSLDIP-MPK algorithm. In this paper, Dhamija [14] did a review of cloud computing basics and discussed security measures that can be applied in the cloud to secure user's data from unauthorized access. Saleh [15] discussed likely attacks on the cloud and the various measures that can be put in place to enhance security on the cloud. Biometric, confidentiality and the secret sharing schema techniques are being examined in which encryption and obfuscation handle the key. Patel [16] developed a new model that uses genetic algorithm with the aid of powerful features of the crossover and mutation operations to create a cryptographic method for encryption and decryption. Mandai [17], in this paper, the collaboration of encryption technique is recommended to meet adequate security of big data in the cloud [18, 19]. The proposed technique uses blowfish and RSA for encryption. The RSA algorithm is used here due to the risk symmetric algorithm face which is the transfer of secret key with the encrypted data to the receiver. Using the RSA algorithm, the secret key is encrypted due to its flexibility of different key for both encryption and decryption. After the encryption has been done, we use the Replace R in RGB steganography algorithm for our covered writing to hide our data into multimedia files.

2 Methodology

We used two systems for the experimental setup. System A with 12 GB RAM, 500 GB HDD, Processor: Intel (R) Core (TM) i5-6300U CPU @ 2.40 GHz 2.50 GHz was used as the server while System B which served as the client, featured: 4 GB RAM, 500 GB HDD and a processor of: Intel (R) Celeron (R) 2957u @ 1.40GHZ.

The client connects to the server via a valid IP address, inputs the image to be used, then applies RSA, then encrypts the message to be sent and sends for decryption.

The server receives the document, extracts the stego-image and then decrypts the file.

2.1 Proposed Algorithm

See Fig. 1.

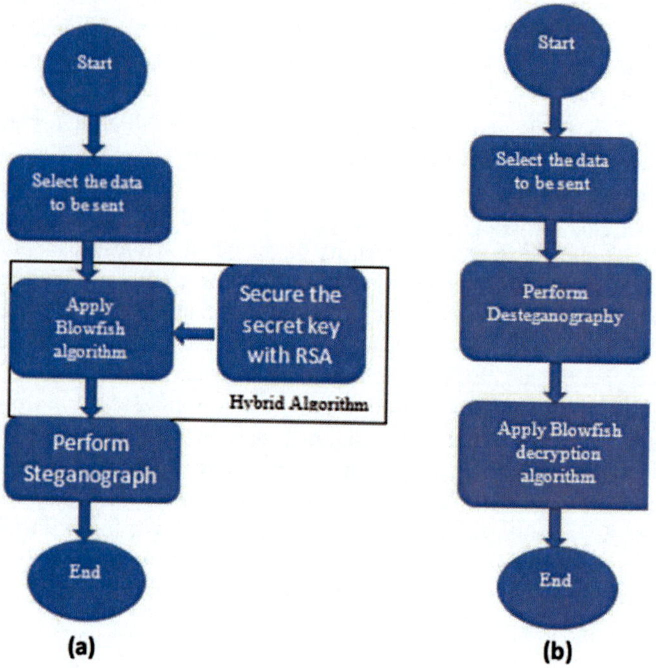

Fig. 1 Flowchart for **a** encryption process and **b** decryption process

2.2 *Steganography Algorithm*

```
Encryption
Start
Select the data to be transferred to the cloud
Encrypt the data using Blowfish algorithm
Encrypt the secret Key using RSA algorithm.

Perform steganography hide in a cover image using Replace R in RGB
algorithm
Save the stego-image to be sent out and send
End
Get the length (N) of message (M) to embed
Convert the image to bitmap
Store N in the first pixel of the bitmap
Set the COUNTER to value of N, and initialize x to 0 and y to 1
Replace R value of pixel at position Y with M(x) where M(x) is the
byte at position x in the message to embed.
Increment x and y and decrement COUNTER
If COUNTER = 0 then go to 8 else repeat from 4
Store the stego-image
Decryption
Start
Receive the sent data
```

```
Perform Desteganography to get the encrypted message
Decrypt using RSA algorithm to get the secret key
Decrypt using blowfish algorithm to get the original file
Save the original file decrypted
End
```

3 Discussion and Result

Our proposed model performed was able to perform encryption in milliseconds. It was tested using a client to server-based system. The server is first turned on, then the client secures his secret key using the RSA [3] algorithm, then connects to the server using the server's IP address, the status to show if connected or not is beneath preview image box, after which the preferred stego-image is selected and then we write the message to be sent in the message box encrypt, embed our data and then send to the server. Figure 1 shows a pictorial view of the client and server interface (Table 1).

The server is started, and then the client secures his secret key using the RSA algorithm and then connects to the server with the server's provided IP address. On the client interface, we select our preferred stego-image, and there is a check box to preview the selected image. After which, the message to be sent is then written in the message box and then encrypted using the blowfish algorithm and embedded using the Replace R in RGB algorithm. This algorithm processes faster and also makes the stego-image of the same quality like the original which will not make the hacker detect which exactly is the stego-image. Figure 2c shows the secret key being secured with RSA while Fig. 2d is a view of a message ready to be sent. The time taken for blowfish encryption was 213.46 ms, and the time taken for the embedding process is 3541.09 ms. The server then receives the message and can either decrypt manually or automatically. When decrpyting manually, the stego-image is first extracted and decrypted in other for the receiver to view what was sent (Fig. 3).

Table 1 Comparism table

Comparism	Our proposed model	Some existing model
Cryptography type	Blowfish (considered faster in symmetric algorithm)	AES, DES, 3DES, etc
Steganography algorithm	Replace R in RGB	Least significant bit
Speed	Uses lesser pixels and processes faster	Uses more pixel processes slower
Stego-image quality	Not blurry	A bit blurry

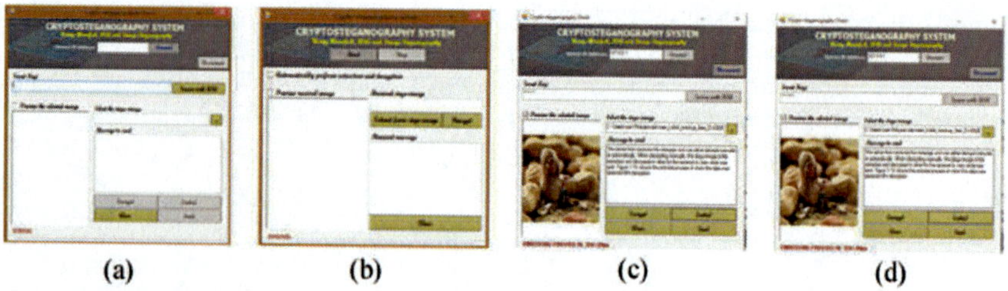

Fig. 2 Pictorial view of **a** client interface, **b** server interface, **c** secret key secured with RSA and **d** message ready to be sent

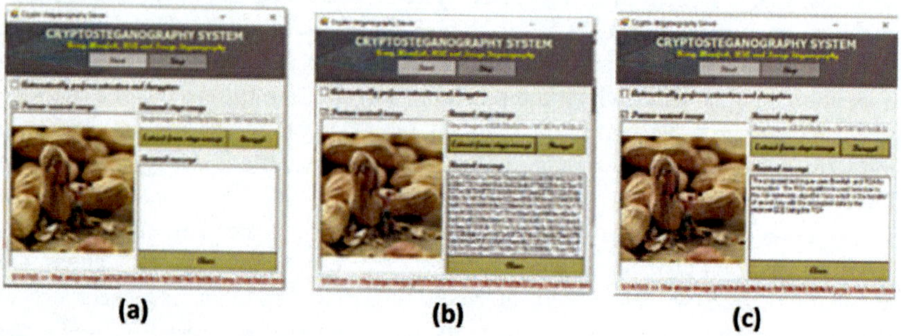

Fig. 3 **a** Stego-image received, **b** extraction process and **c** decrypted data

4 Conclusion

Data security in the cloud is essential for organizations and also individual to protect unauthorized access and hackers. Kessler et al. [20], in this paper, we have proposed a model that uses blowfish algorithm to encrypt the data, and RSA is used to secure the secret key while a stego-image is used in embedding the data in a pictorial format, and all this is achieved in milliseconds. The steganography algorithm used is the Replace R in RGB, the algorithm helps in achieving a quality stego-image which makes the stego-image and the original look the same and also make it hard for the hacker to identify which is the actual image for decryption, and it also uses lesser pixels which is also an added advantage to it process speed. The future works of the proposed model can be done by testing this algorithm on other media format such as audio and video and also improving on the speed on such media format. This approach will make a strong security system on the cloud.

Acknowledgements The authors appreciate the sponsorship from Covenant University through its Centre for Research, Innovation and Discovery, Covenant University, Ota Nigeria.

References

1. I. Odun-Ayo, S. Misra, N. Omoregbe, E. Onibere, Y. Bulama, R. Damaševičius, Cloud-based security driven human resource management system. Front. Artif. Intel. Appl. Vol. Adv. Digit. Technol. **295**, 96–106 (2017S)
2. Singla, A. Bala, A review: cryptography and steganography algorithm for cloud computing, in *2018 Second international conference on inventive communication and computational technologies*, no. ICICCT (2018), pp. 953–957
3. N.D. Jambhekar, S. Misra, C.A. Dhawale, Cloud computing security with collaborating encryption. Ind. J. Sci. Technol. **9**(21), 95293 (2016); S. Mittal, S. Arora, R. Jain, Data security using RSA encryption combined with image steganography, in *IEEE*, (2016), pp. 1–5.
4. P.V Maitri, Secure file storage in cloud computing using hybrid cryptography algorithm, in *IEEE WiSPNET 2016 Conference* (2016), pp. 1635–1638
5. P. Garg, M. Sharma, S. Agrawal, Y. Kumar, *Security on Cloud Computing Using Split Algorithm Along with Cryptography and Steganography* (Springer, Singapore, 2019)
6. A.Y. Alkhamese, I.M. Hanafy, Steganography: a review, in *2019 International Conference on Innovative Trends in Computer Engineering*, vol. 4, no. February (2019), pp. 549–558
7. M. Olowu, C. Yinka-Banjo, S. Misra, H. Florez, A secured private-cloud computing system, in *Applied Informatics. ICAI 2019. Communications in Computer and Information Science*, vol. 1051, ed. by H. Florez, M. Leon, J. Diaz-Nafria (2019)
8. A. Ranjan, Using multilayer steganography and cryptography, in *IEEE 2016 International Conference on Automatic Control and Dynamic Optimization Techniques International*, vol. 6, no. 5 (2016), pp. 35–41
9. M. Saritha, Using MATLAB, in *International Conference on Signal Processing, Communication, Power and Embedded System (SCOPES)* (2016), pp. 584–587
10. P. Uddin, M. Saha, Developing an efficient solution to information hiding through the text steganography along with a cryptography, in *IEEE 9th International Forum Strateg. Technol. (IFOST), Oct. 21–23, 2014, Cox's Bazar, Bangladesh Dev.* (2014), pp. 14–17
11. S.K. Muttoo, A multilayered secure, robust and high capacity image steganographic algorithm. World Comput. Sci. Inf. Technol. J. **1**(6), 239–246 (2011)
12. B. Pathankot, *Review Paper on Enhancing Data Security for Cloud Environment Cryptography and Steganography*, vol. 2, no. 1 (2016), pp. 44–48
13. G. Yogeswari, P. Eswaran, Enhancing data security for cloud environment based on AES algorithm and steganography technique. Int. J. Adv. Res. Trends Eng. Technol. **3**(20), 227–230 (2016)
14. A. Dhamija, A novel cryptographic and steganographic approach for secure cloud data migration, in *IEEE 2015 International Conference on Green Computing and Internet of Things*, vol. 6, no. 10 (2015), pp. 346–351
15. M.E. Saleh, Data security using cryptography and steganography techniques. Int. J. Adv. Comput. Sci. Appl. **7**(6), 390–397 (2016)
16. N.A. Patel, A survey on security techniques used for confidentiality in cloud computing, in *2018 IEEE International Conference on Circuits and Systems in Digital Enterprise Technology. ICCSDET 2018*, vol. 9, no. 18 (2018), pp. 1–6
17. S. Mandai, Secret data sharing in cloud environment using steganography and encryption using GA, in *IEEE 2015 International Conference on Green Computing and Internet of Things* (2015), pp. 1469–1474
18. P.G. Student, C. Science, D. Gurgaon, *Hybrid Protocol Employing Steganography & Cryptography for Cloud Storage Security*, vol. 4, no. 2 (2016)
19. W.A. Awadh, A.S. Hashim, *Using Steganography for Secure Data Storage in Cloud Computing Using Steganography for Secure Data Storage in Cloud Computing*, no. April, 2018
20. G.C. Kessler, G.K. Associates, A. Corporation, S. Carolina, *An Overview of Steganography*, 1st ed., vol. 83 (Elsevier Inc., 2011)

Design and Development of an Algorithm for Automatically Capturing and Recognizing the Number Plates of Vehicles—A Surveillance System to Support Indian Traffic Police

Syed Musthak Ahmed, Varsha Reddy, Rahul Chandrahas, Nanditha Reddy, Srinidhi, and Manohar

Abstract Automatic surveillance system plays an important role in controlling and reducing the road mishaps. As per census, it has been recorded that many mishaps have been taken place in our country due to harsh driving and/or violating traffic rules. Technological advancements have made the world of automation process ease, and one such area is an automatic number plate recognition system (ANPR). This system is a mass surveillance method used to read number plates of vehicles by the captured images. By this system, it is possible to extract the details of the vehicle number that has violated the rules or caused an accident and help the traffic officials to identify the details of the vehicle at a later stage. The present work describes an algorithm developed using MATLAB which helps to process a captured image and extract the desired information with high resolution. The algorithm developed helps to reduce human effort in identifying the violator by automatically analyzing the captured images. The work carried out is one such technique that saves time, money, and human effort.

Keywords Automatic number plate recognition (ANPR) · Surveillance system · Segmentation

1 Introduction

Automatic number plate recognition (ANPR) has become a necessary approach for monitoring vehicles these days. The application of this approach has eased various processes like authorizing vehicles, levying challans for traffic violations, toll tax collection, parking [1, 2]. The solution to the problem is to identify the particular vehicle, capture its image, and later gather the required information for identifying the vehicle. The number plate capturing is one such way to identify violation. It is a unique one to differentiate one vehicle from the other, which would be useful when both the vehicles are of same type of make and model. In ANPR approach,

S. M. Ahmed (✉) · V. Reddy · R. Chandrahas · N. Reddy · Srinidhi · Manohar
ECE, S. R. Engineering College, Warangal, Telangana, India

A. Kumar et al. (eds.), *Computational Intelligence in Machine Learning*, Lecture Notes in Electrical Engineering 834, https://doi.org/10.1007/978-981-16-8484-5_47

four steps are followed capturing vehicle image, capturing the vehicle number plates, segmenting the characters, and recognizing characters [3].

The first step sounds easy but, it is really difficult to capture a moving car in real time especially its number plate clearly. There are various approaches for locating vehicle number plates and extract their number. One approach of capturing is discussed in [4]. In this approach, they initially used different inbuilt functions and implemented some user-defined techniques related to image processing.

The cases in India have reported that Indian roads are the most notorious, and they claim over 1.5 lakh lives each year majorly due to the traffic violators [5, 6]. So without using technology mainly the Internet of Things, it is almost impossible to identify and rectify the mishaps thereby reduce violators efficiently [7]. ANPR can solve these problems up to a large extent.

2 Existing Method

The rapid increase in the traffic due to vehicles has caused the monitoring difficult. Vehicle tracking and monitoring system while drunk and drive and the ways to track the vehicle is discussed in [8]. Other errors include speed violation, inattentiveness, aggressiveness, age factors, lane maintenance, reaction time have been reported in research [9–12]. These errors and violation of rules have raised the accidental rate also. To overcome or minimize this, it is better to capture the license plate to find the defender and for further action. Manual monitoring is possible, but there is a need of dedicated persons to record and rectify the system. This is a tedious process and increased man power. When the traffic monitoring office notices illegal actions, it captures a picture of the vehicle number plate. The picture is uploaded to the database, and the fine is generated to the respective offender in the form of computer-generated challans against the offenses [13–17]. While the officer has to manually enter a lot of cases during a day, it is a time-consuming process and requires man power. The existing system is shown in Fig. 1.

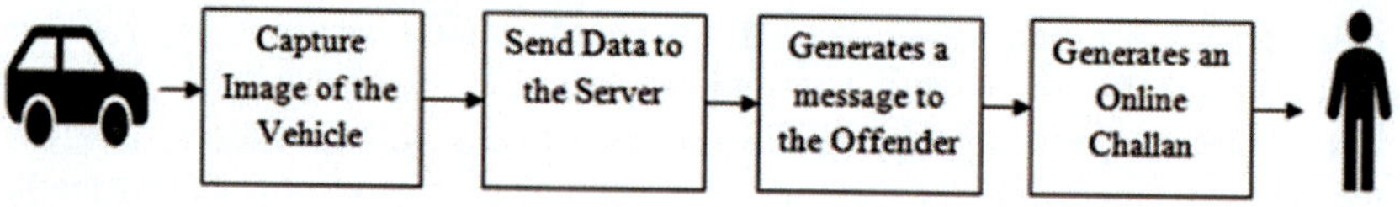

Fig. 1 Block diagram of existing method

3 Proposed Method

The present method reduces the manual work of the traffic monitoring officer. Both the monitoring and detection of vehicle are done by using the video surveillance system. The block diagram of the proposed implementation is shown in Fig. 2. By this proposed method, we are able to overcome the drawbacks of the few existing methods like man power and time-consuming.

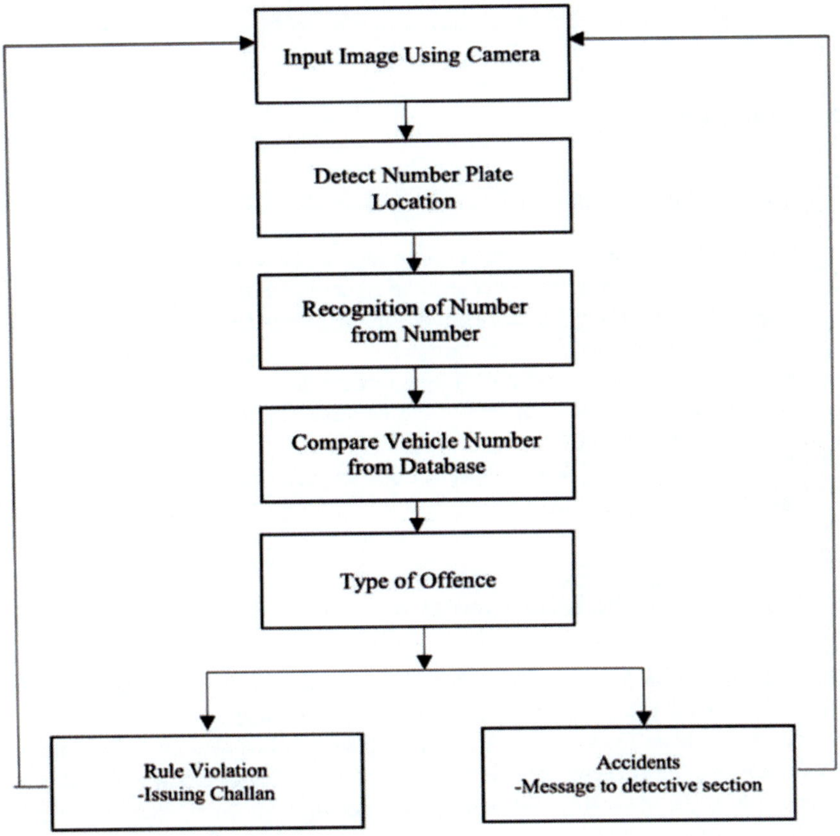

Fig. 2 Block diagram of proposed method

4 Objectives of the Proposed Work

The work is carried out to support the Indian traffic police and to see that road mishaps are reduced to a larger extent. The objectives of the project are framed with this motive, i.e.,

1. Develop an algorithm to identify authorized vehicle efficiently.
2. Identify the vehicle that violates the traffic rules.
3. Support government officials to trace the vehicle causing mishaps.
4. Develop an automated system which supports Make in India concept.

5 Methodology

In continuing with the said work, the following methodology is adopted.

1. Capturing the images of the vehicles through digital camera
2. Converting into image enhancement and removing noise using median filter
3. Edge detection finding the discontinues in image and detecting the rectangular shape of number plate from background
4. Portioning of number plate into objects or regions
5. Segmented characters compared to the template and find the best match and display characters in text
6. Depending upon the violation performed by the vehicles, certain action is initiated.

6 Flowchart of the Implementation

The implementation process carried out is described by a flowchart as shown in Fig. 3. The complete process is explained in previous section.

Here, the vehicle tracking and monitoring device is installed at various places where the probability of accidents commonly occurs. The device is now under surveillance checking the traffic flow. When an instance of violation takes place, the vehicle is shot by the camera. It is then processed for clear recognition of number plate which can later be used to locate, identify for further processing.

7 Results

The algorithm on ANPR is written in MATLAB. The complete system consists of a device to monitor the traffic violators which are being obtained using an image processing algorithm which captures and analyzes the number plates of vehicles.

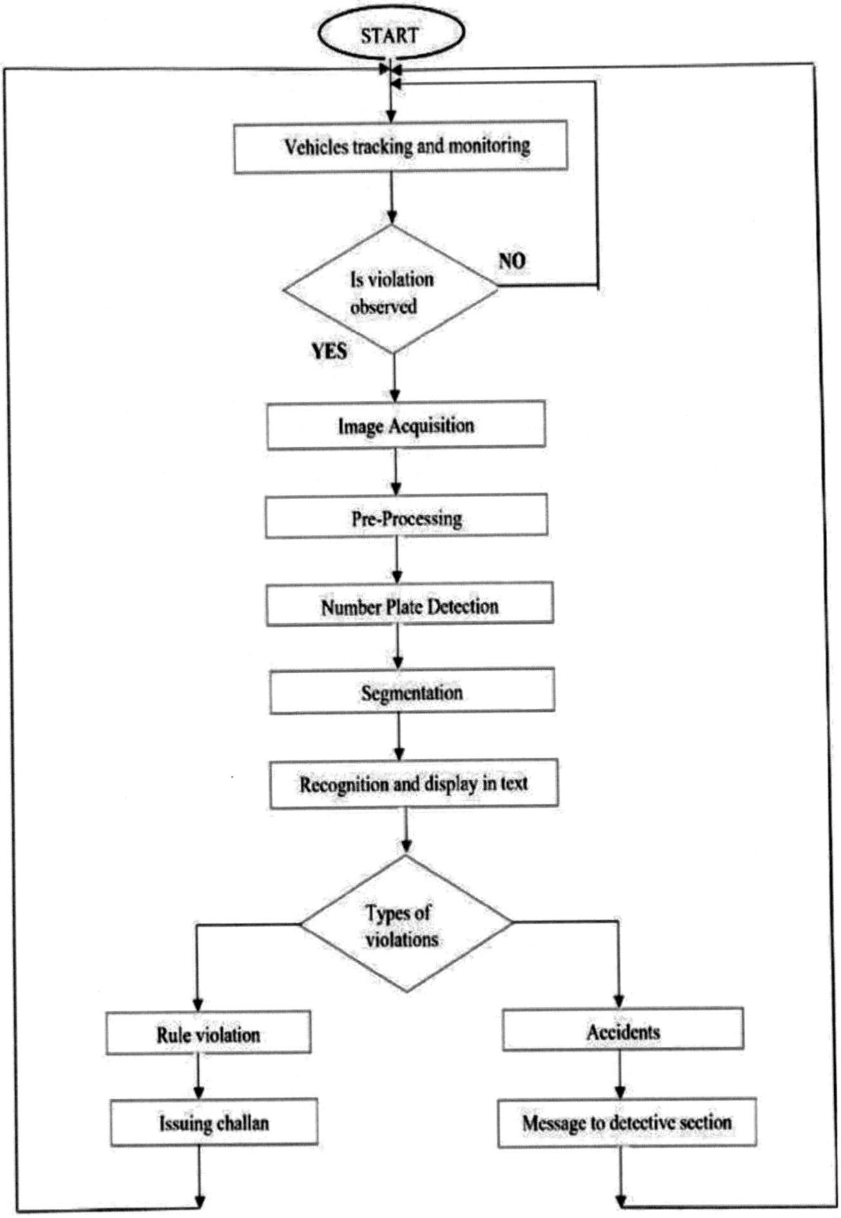

Fig. 3 Flowchart of proposed ANPR system

Fig. 4 Image taken using surveillance camera

After analyzing the number on the plate, correlation is done with the database, and when a perfect match is found, a message is sent to the supervisor. Action will be initiated depending upon the type of the violation. For example, if the rule is violated, an e-challan will be issued; during mishap, the captured number plate assists to take action as per law and norms. Here, the human intervention is completely eliminated, and efficiency of identifying the violators is substantially increased. Figure 4 shows the image taken by the surveillance camera of the vehicle which violates the rules. The image of the number plate that is extracted from the vehicle is shown in Fig. 5. The code written in executing the results was successfully achieved with high resolution.

8 Conclusion

The key purpose of this algorithm is to fix the data processing and exact complications of APNR by extracting the valid sequences of the problem statement, an algorithm that can easily surpasses the complexity of learning and recognition of characters, an approach in which the pre-defined functions are available for specific data processing in MATLAB. The developed algorithm is quite successful for most of the captured images, which were under different conditions. The present work is presented with a real-time system for detection and recognition of number plates. The more precise performance of the proposed system helps in compensation of the images with low resolution by considering acquired results of the successive frames. As the device is completely integrated on an embedded device, it operates without any manual intervention and only reports the final classification results to connected operators.

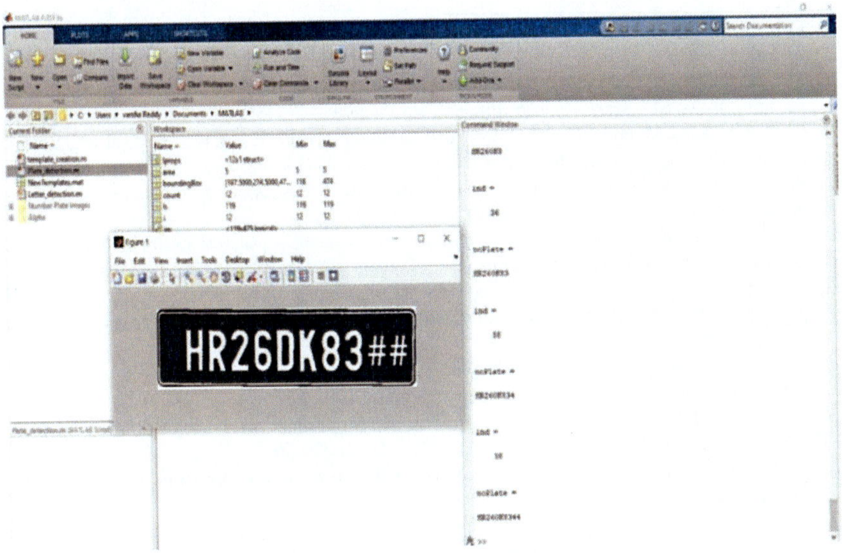

Fig. 5 Number plate extraction from the vehicle

Acknowledgements The authors acknowledge the Centre for Artificial Intelligence and Deep Learning, S. R. Engineering College, Warangal, for providing the necessary support in doing this project work. The authors also acknowledge the management and principal of the institution for providing necessary facilities to carry up the work.

References

1. M.T. Qadri, M. Asif, Automatic number plate recognition system for vehicle identification using optical character recognition, in *ICTE Conference on Education Technology and Computer*, vol. 4, Issue 6 (2017), pp. 2515–2520
2. M. El-Adawi, H.A. el MoneimKeshk, M. Mahmoud Haragi, Automatic license plate recognition. IEEE Trans. Intel. Transp. Syst. **5**, 42–53 (2004)
3. K. Vivek, S. Bhardwaj, Supervised machine learning and OCR implementation in system design of number plate recognition automatically. Int. J. Latest Trend Eng. Technol. **9**(3), 284–288 (2018)
4. K.K. Kim, K.I. Kim, J.B. Kim, H.J. Kim, Learning-based approach for license plate recognition. Proc. IEEE Signal Proces. Soc. Works. **2**, 614–623 (2000)
5. N.S. Pagad, B. Shetty, K.S. Nisarga, Sharada, S. Kulkarni, Automatic vehicle recognition system for enforcing security in restricted places, Int. J. Sci. Dev. Res. (USDR) **2**(6) (2017)
6. F. Yang, Z. Ma, Vehicle license plate location based on histogram ming and mathematical morphology, in *AUTOID: Proceedings of fourth IEEE Workshop on Automatic Identification Advanced Technologies*, vol. 4 (2005), pp. 89–94
7. C.H. Sandeep, K.S. Naresh, K.P. Pramod, Security challenges and issues of the IoT system. Ind. J. Publ. Health Res. Dev. (IJPHRD) **9**(11), 748–753 (2018)
8. Anusha, S.M. Ahmed, Vehicle tracking and monitoring system to enhance the safety and security driving using IoT, in *IEEE Conference on Recent Trends in Electrical, Electronics and*

Computing Technologies (ICRTEECT-2017) (2017). https://doi.org/10.1109/ICRTEECT.2017.35
9. N. Chakrabarty, K. Gupta, A. Bhatnagar, A survey on awareness of traffic safety among drivers in Delhi, India. SIT Trans. Industr. Financ. Bus. Manage. (IFBM) **1**(2), 106–110 (2013).
10. R. Dandona, A. Mishra, Deaths due to road traffic crashes in Hyderabad City in India: need for strengthening surveillance. Natl. Med. J. India **17**, 74–79 (2004)
11. K. Aghabayk, M. Sarvi, W. Young, Understanding the dynamics of heavy vehicle interactions in car-following. J. Transp. Eng. **138**(2), 1468–1475 (2012)
12. S.-S.J. Kass, K.E. Beede, S.J. Vodanovich, Report measures of distractibility as correlates of simulated driving performance. Accid. Anal. Prevent. (AAP) **42**, 874–880 (2010)
13. C. Anagnostopoulos, I.E. Anagnostopoulos, I.D. Psoroulas, V. Loumos, E. Kayafas, License plate recognition from still images and video sequences. IEEE Trans. Intel. Transp. Syst. **9**(3), 377–391 (2008)
14. B. Chidambararajan, M. Balakarthikeyan, Image processing for identifying vehicle license plate and traffic monitoring. CiiT Artif. Intel. Syst. Mach. Learn. **1**(2), (2009)
15. D.G. Bailey, D. Irecki, B.K. Lim, L. Yang, Test bed for number plate recognition applications. First IEEE Int. Works. Electron. Des. Test Appl. **1**, 501 (2002)
16. D. Zheng, Y. Zhao, J. Wang, An efficient method of license plate location. Pattern Recogn. Lett. **26**(15), 2431–2438 (2005)
17. S.L. Chang, L.S. Chen, Y.C. Chung, S.W. Chen, Automatic license plate recognition. IEEE Trans. Veh. Tech **5**(1), 42–53 (2004)

Applications of Machine Learning in Anomaly Detection

Gadilli Sai Nikhil, Arun Sekar Rajasekaran, Madupathi Parimala, and S. Velliangiri

Abstract Now-a-days, maximum number of application systems contains lot of middleware components like data base, search engines, storage, caches, and identity services. Each component is a separate package of physical or virtual machines which will produce lot of data in terms of logs and metrics. In order to process that data to get any inference for future analysis and anomaly detection, we need to have a separate system which will take more time to process data delivered by each component, and also, more computational cost is required. Here, we came up with two different methods, i.e., multivariate statistical analysis by finding Mob distance and by using auto-encoders which is one of the artificial neural networks method. Moreover, the results are showing that these applications are efficient that they are detecting anomaly in advance, i.e., ahead of time.

Keywords Anomaly · Multivariate statistical analysis · Principal component analysis · Condition monitoring

1 Introduction

The main goal of the trending concepts like digital transformation, digitization, industry 4.0 is the exploitation of data and technology to enhance productiveness and efficiency. In application systems, the relativity and connectivity of data among middleware. But, the important thing is how to process this huge amount of data and extract only fruitful information. As this preprocessing will result in reduction of cost and enhance capacity. To achieve these benefits, the recent trends data analytics and machine learning came into picture. One of the applications of the machine learning is anomaly detection. Anomaly is an abnormal or different event from normal flow

G. Sai Nikhil · A. S. Rajasekaran
GMR Institute of Technology, Razam, India

M. Parimala
CMR Institute of Technology, Bengaluru, India

S. Velliangiri (✉)
B. V. Raju Institute of Technology, Narsapur, Telangana, India

A. Kumar et al. (eds.), *Computational Intelligence in Machine Learning*, Lecture Notes in Electrical Engineering 834, https://doi.org/10.1007/978-981-16-8484-5_48

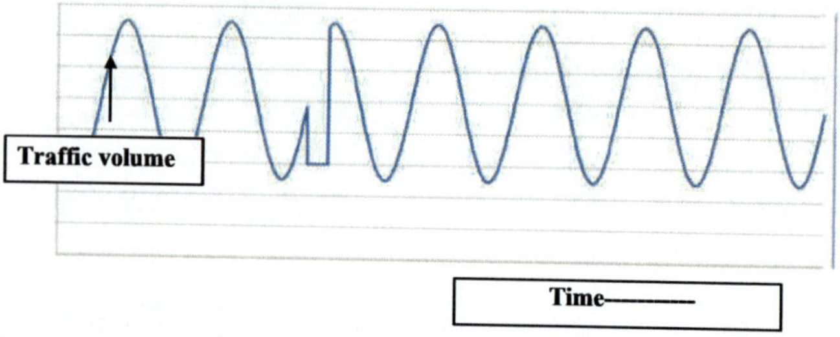

Fig. 1 Traffic volume (KPI) vs time series plot in a certain area

of a time series pattern [1]. Let us consider a situation to monitor traffic volume for every day, and we plotted a graph of time vs traffic volume, and it came as below.

Figure 1 shows the traffic volume vs time plot. Each cycle is a traffic volume per one day. Traffic volume is as usual on first two days, and on third day, it is too low, and this is different from normal pattern in other words called as anomaly. On normal days to control traffic, large number of policemen and other equipment's in that area are required. But on third day, if same number of policemen are there, then, it is waste of man power without much traffic. If they were in other area, then, it might be helpful to control traffic in other area. So, if we detect there will be not much third day traffic on second day or before few days itself. Then, the control room will take necessary precautions by sending the policemen to other place. Here, we can apply machine learning to detect the anomaly in this scenario. First, we will forecast the time series pattern for tomorrow traffic volume, and then, we will take the difference between the forecasted time series and actual time series pattern which help us in detecting the anomaly easily, and this will be reported to network operation center. Moreover, they will select the appropriate solution for this anomaly. So, in real-time scenarios, the detection of this anomaly may benefit us in saving lot of time, power, etc.

1.1 What is an Anomaly?

The term "anomaly" contains different meanings both mathematical and formal. In this time series and statistical concept, an anomaly is an outlier, i.e., it is a value which is outside of certain limits or thresholds. In-addition, the word anomaly can be applied to certain time points or longer time periods where we should observe a continuously abnormal behavior of time series. Below are the some examples of anomalies in different situations.

Figures 2 and 3 are the examples of anomalies in any series pattern in real-time scenarios. In each plot, the anomalies/abnormal cases are indicated by red-colored values, and normal behavior/pattern is indicated by green-colored values.

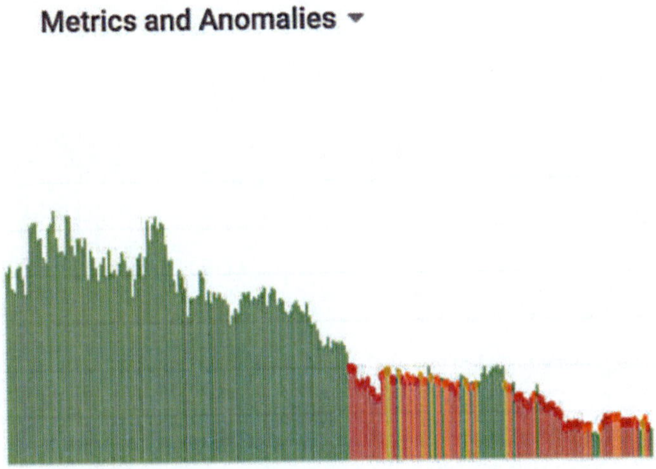

Fig. 2 My SQL DB throughput request per second

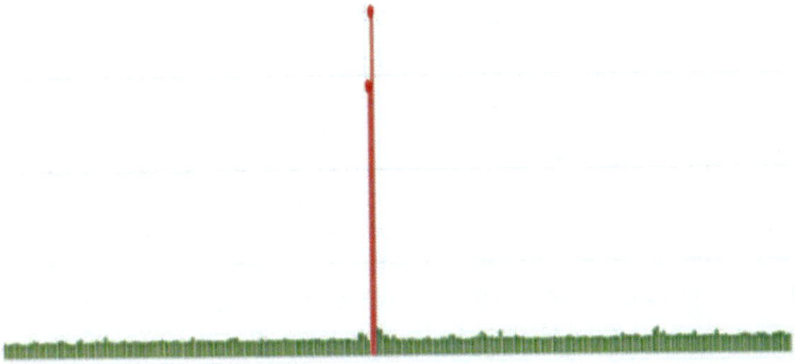

Fig. 3 MySQL cluster management node write bytes count

Anomaly Detection: It is also known as outlier detection [1, 2] which is the detection of occasional events, items, or data points which hike suspicious by altering consequently from the majority of the data. Generally, anomalous data can be related to some kind of abnormality or rare events like frauds in bank, medical problems, and breakdown of equipment. One of the best solution to detect an anomaly is data visualization, i.e., by drawing plot between any two variables and observing it and finding which data point is away from the distribution that is nothing but an anomaly (Fig. 5).

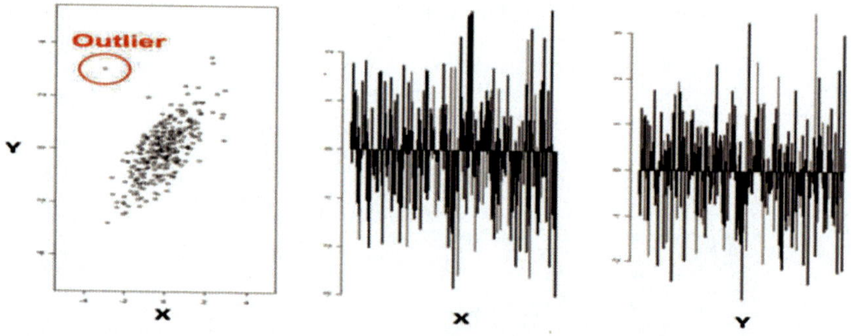

Fig. 5 Detecting of anomaly when two variables are there

2 Problem Statement

If we consider any equipment either it is of rotating (pump, gearing) or non-rotating (valve, heat exchanger) type, it will have lot of internal parts like gearings and sensors. Moreover, any machine will reach to point of failure which is not the actual failure, but the equipment is not in the condition to work normal or run as usual. If this problem suddenly occurs, then we can't replace it or repair it; in such case, it takes some more time to get back to original state. So, it should be detected in advance, i.e., passing signal in advance which tells that there may be need of some maintenance activity to be taken for equipment so that it will be debilitated. In other words observing the "health state" of our equipment is nothing but "condition monitoring".

The common solution is to conduct condition monitoring is to observe at every sensor measurement from the equipment and to fix maximum and minimum threshold for it. Then, at the time of testing, if the current value is not those limits for any middle ware sensor, then the equipment is declared as unhealthy and an indication is sent. But, this approach of putting limits to each and every sensor is having drawbacks that sometimes it will send false alarms in large number, i.e., when we consider any two sensors inside an equipment if the current value exceeds the maximum value for one sensor and current value is lower than minimum value for second sensor; then, also, the equipment is healthy only, but in this case alarm will be sent.

3 Anomaly Detection Techniques

3.1 Technique 1: Multivariate Statistical Analysis

This technique is combination of two approaches, i.e., dimensionality reduction using principal component analysis (PCA) and calculating of Mob distance.

(a) Dimensionality reduction using PCA

As discussed in data visualization section [3, 4], it is complicated while dealing with data distribution with more number of variables. So, in order to reduce complexity, we can use dimensionality reduction. For this dimension reduction, principal component analysis (PCA) is one of the techniques which conducts a linear mapping of the data to a lower-dimensional space. Moreover, here, the variance (expectation of the squared deviation of a variable from its mean, which tells us how far a set of data points is spread out from the average value) of the data is maximized. This variance should be as maximum as possible to retain the most important part of the data distribution for that Eigen vectors which correspond to largest Eigen values should be needed. In-addition, this Eigen vectors are calculated by constructing covariance matrix of the data. So that the Eigen vectors respective to the biggest Eigen values can be used to reframe a maximum portion of the variance of the original data.

3.2 PCA

PCA is the process of computing principal components and exploiting them to conduct a change of basis on the data. First, the data are distributed in a random manner, and the principal components are computed which are two vectors orthogonal to each other which tells us the direction and magnitude of distribution of data, and moreover, these two vectors are used to define the data distribution in a simple way. So, principal components of a collection of data points in a real-space are a sequence of direction vectors, where the nth vector is the direction of a line that represents the direction of data distribution and orthogonal to n-1 vectors. The flow of applying principal component analysis is demonstrated in Fig. 6.

3.3 Technique 2: Auto-Encoder Artificial Neural Network

Another method to detect anomaly is by using auto-encoder network [5]. It is similar approach as first technique with a slight difference. Auto-encoder is a type of artificial neural networks (ANNs) which are constructed by inspiring from biological neural networks. Basically, an ANN is a combination of connected units or nodes called artificial neurons. It has an input layer, one or two hidden layer, and then the output layer. And every layer contains a number of nodes, and each node is like a transfer function. ANNs are used to build a model which learn efficient data coding in an not supervised manner, and the main agenda of an auto-encoder is to learn (as it is a ML model, it has to learn in unsupervised manner) a depiction for a dataset, normally used for reduction of dimension.

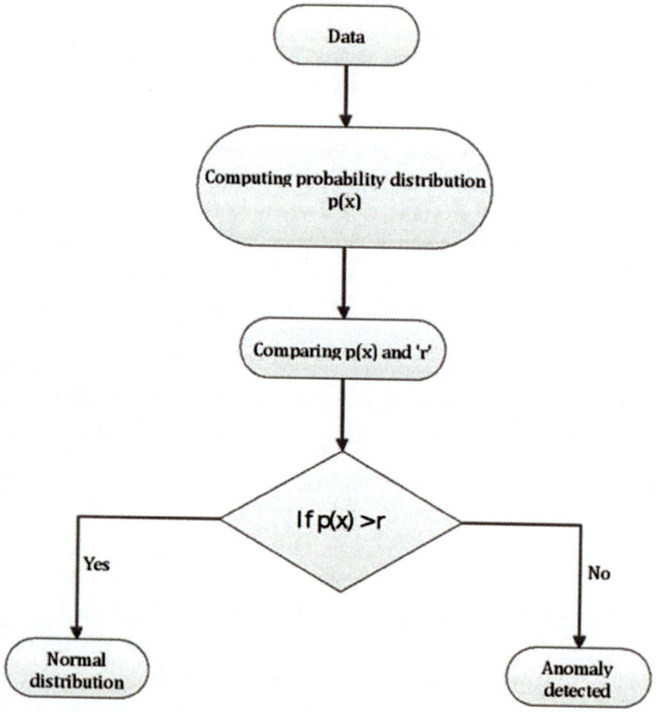

Fig. 6 Flow of anomaly detection by using PCA

First, it will reduce the dimensionality, and then, it reconstructs the original representation of data as similar as possible to its original input. Further, this auto-encoder is a feed forward neural network, i.e., from input layer, it directly goes to hidden layer, and then, it directly goes to output layer; if there will be no loops in the hidden layer (non-recursive), it has only one direction (forward) output layer and input layer both have same number of nodes for the purpose of reconstructing its own inputs. Figure 7 shows the auto-encoder artificial neural network.

4 Experimental Analysis

4.1 Applying ML Techniques as Condition Monitoring in Detecting Gear Bearing Failure

Here, in this part, we apply above discussed two techniques for an application. Here, we perform condition monitoring for equipment which contains "Gear bearing". It is a type of rolling-element bearing as shown in Fig. 8.

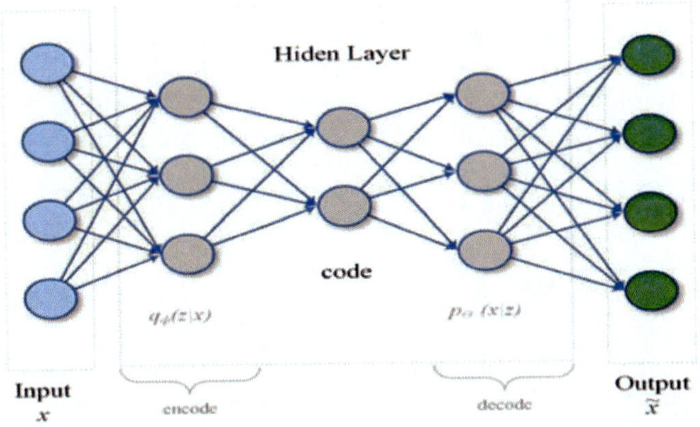

Fig. 7 Auto-encoder artificial neural network

Fig. 8 Gear bearing in an equipment

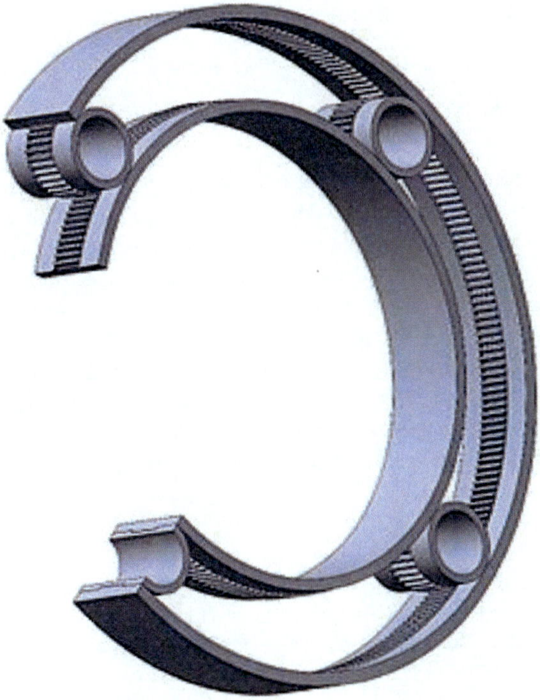

The main agenda is to detect failure of gear bearing in advance and send an indication in terms of warning that enables for predictive measures (either replacing/repairing). Here, in this scenario, an equipment containing four bearings was supposed to breakdown due to continuous load and continuous running conditions. We have three sets of data about four bearing; this data consist of vibration measurements over the lifetime of bearings until its breakdown and as per given dataset failure happened later 100 million cycles along with a crack in the external part.

5 Results Summary

So, by using these two techniques for anomaly detection, we can detect anomaly before the actual breakdown (in advance) and both the techniques achieved in detecting anomaly in advance (several days before only). So, in real-time scenario, we can use these techniques to send signals before any breakdown occurs to take any predictive measures, i.e., by repairing or replacing which results in both time and cost saving.

6 Conclusion

With the minimized cost of observing data through the sensors, and also increasing connectivity between devices, it is being important to extract valuable information from data and finding pattern in large amount of data is the main emerging feature of ML and statistics. Moreover, in this work, data preprocessing and machine learning are applied to detect anomaly in condition monitoring by using two techniques (i.e., PCA + Mob distance and auto encoders). These two approaches achieved anomaly detection in three days before the actual breakdown. In addition, in future, machine learning will develop even more advance than we imagined today. Further, ML can be used to detect anomaly even more faster in future, and also, we can use it for tracing where exactly breakdown will occur in future.

References

1. S.B. Wankhede,Anomaly detection using machine learning techniques, in *2019 IEEE 5th International Conference for Convergence in Technology (I2CT)*, Bombay, India (2019), pp. 1–3. https://doi.org/10.1109/I2CT45611.2019.9033532
2. W.S.J.D. Souza, B. Parvathavarthini, Machine learning based intrusion detection framework using recursive feature elimination method, in *2020 International Conference on System, Computation, Automation and Networking (ICSCAN)*, Pondicherry, India (2020), pp. 1–4. https://doi.org/10.1109/ICSCAN49426.2020.9262282
3. B.A.G. Oliveira, C.A.P.D.S. Martins, F. Magalhães, L.F.W. Góes, Difference based metrics for deep reinforcement learning algorithms. IEEE Access **7**, 159141–159149 (2019). https://doi.org/10.1109/ACCESS.2019.2945879

4. J. Klein, S. Bhulai, M. Hoogendoorn, R. Van Der Mei, R. Hinfelaar, Detecting network intrusion beyond 1999: applying machine learning techniques to a partially labeled cybersecurity dataset, in *2018 IEEE/WIC/ACM International Conference on Web Intelligence (WI)*, Santiago, Chile, pp. 784–787. https://doi.org/10.1109/WI.2018.00017
5. J. Zhang, M. Zulkernine, Anomaly based network intrusion detection with unsupervised outlier detection, in *2006 IEEE International Conference on Communications* (2006). https://doi.org/10.1109/icc.2006.255127

Using Data Mining to Preprocess Data for the Neural Network Model to Predict Water Level Applied for Northern Vietnam's Agriculture

Dang Trong Van, Le Hoang Lan, Nguyen Quang Dat, Do Duy Nhat, and Vijender Kumar Solanki

Abstract Water resource in agricultural canal is a necessary requirement in water resource system management. Nowadays, in Bac Hung Hai, a small area in Red River Delta (Northern Vietnam), it is essential to have a system for water level forecasting. In this research, we suggest several methods which are SARIMA, long short-term memory (LSTM) and hybrid seasonal-wavelet-LSTM to predict water level for water administration. We employed the data from Cau Cat station located on Red River (the water input of Bac Hung Hai canal system) to predict water level in 7 days. The results will be used to prepare for the accumulating or discharging of water for the canal. The study demonstrates that the seasonal-wavelet-LSTM model performs better than the single SARIMA model and the classic single LSTM model (in MSE and MAPE), respectively. The investigation provides a promising option for operating the water level of Bac Hung Hai canal, where the water flow is often unstable.

Keywords LSTM · Wavelet · Seasonal · Water level · Forecasting

D. T. Van (✉) · L. H. Lan · N. Q. Dat
High School for Gifted Students, Hanoi University of Science, Vietnam National University, Hanoi, Vietnam
e-mail: dangtrongvan_hsgs19@hus.edu.vn

L. H. Lan
e-mail: lehoanglan_hsgs19@hus.edu.vn

N. Q. Dat
e-mail: nguyenquangdat@hus.edu.vn

D. D. Nhat
International Cooperation Office, Le Quy Don University, Hanoi, Vietnam
e-mail: nhatdd@lqdtu.edu.vn

V. K. Solanki
Department of Computer Science and Engineering, CMR Institute of Technology, Hyderabad, India

A. Kumar et al. (eds.), *Computational Intelligence in Machine Learning*, Lecture Notes in Electrical Engineering 834, https://doi.org/10.1007/978-981-16-8484-5_49

1 Introduction

Accurate prediction of water flow is an essential demand in order to minimize flood damage, as well as to administer water resources system. It is a difficult problem since river flow is a nonlinear complex dynamic process which is influenced by several factors. In fact, Vietnam already has a long-term forecasting system for all rivers in the North side. However, as the system is made for a large-scale application, it might not provide accurate result when performing in small areas. According to [1–3, 7, 8], LSTM is a promising method of time series forecasting as it often performs better comparing to other method such as RNN or GRU. Due to the seasonal nature of the weather, the irrigation data in Vietnam are seasonally relevant. Therefore, using the seasonality of the data can greatly increase the accuracy of forecasts [3]. On the other hand, wavelet has the ability to solve the problems of denoising in preprocessing in data mining [4, 5]. Furthermore, wavelet can be employed to deal with bad noise which helps achieve more accurate results [6]. Therefore, we lead to a hypothesis that using hybrid model may implement more effective performance than single model structures.

2 Related Works

Lee et al. [2] divided dataset from the Namgang Dam basin into two groups: interval groups and time average group, applied the LSTM model on them to predict, and the results were that the prediction skill of group A is more stable than group B (The best predicted case of group A had a RMSE of 0.035, which was smaller than the best predicted case of group B with a RMSE of 0.0371).

In 2018, Zhang et al. [10] applied a LSTM-based model to water table depth prediction over the long term in agricultural areas. For the proposed four-year data set from five sub-areas of Hetao Irrigation District, China, the new model produced a range of RMSE from 0.070 to 0.184 (m), indicating its effectiveness.

Le et al. in 2019 suggested in one of his papers an LSTM neural network model for flood prediction. The enhanced model required only the data of daily discharge and precipitation as input. After taking into account two different scenarios, it was shown that both gave good results, with NSE value ranged from 86.2 to 94.5% [3].

In 2019, using RNN to determine the relationship between linear and nonlinear components, instead of simple addition, Xu et al. [9] proposed a new scheme based on ARIMA-RNN combined model for water level prediction. The results indicated that the combined model of ARIMA-RNN has a smaller RMSE of 0.021 in comparison with ARIMA or RNN alone, which has the RMSE of 0.047 and 0.033, respectively.

He and He [4] introduced an experiment with the Shanghai Composite Index closing prices dataset show that their proposed wavelet function was better than the soft and the non-negative garrote shrinkage functions, but a little worse than the hard shrinkage functions in terms of universal threshold rules, while it performed best among all the applied shrinkage functions for the minimax threshold rules, with the signal-to-noise ratio of 57.4802 and RMSE of 1.8495 in the final case.

The wavelet scaling approach in Joanna Bruzda's paper [5] outperformed traditional methods such as ARIMA though it did not occur without exceptions; the inaccuracy of this method did not exceed 2%; this approach's simple concept and low burden in computation convinced people that it was worth popularizing to apply for predicting.

3 Methodology

3.1 Seasonal Decompose

Seasonality is an important characteristic, but if the season components are separated, the accuracy of the forecasting model can be further increased [12, 13]. In the data on water level, data are taken over many years, and the season is shown very clearly.

STL, proposed by Cleverland in 1990 [11], is a very well known and highly applicable method for dividing data sequences into trend, seasonal and residual components. This method divides the time series into three components:$y_t = T_t + S_t + R_t$ with T_t is trend, S_t is seasonal, and R_t is residual.

STL is an iterative method which comprises two recursive procedures, the inner and outer loops. In each iteration through the inner loops, the seasonal component is updated by a seasonal-smoothing, and then a trend-smoothing updates the trend component, respectively. After each of the iteration passes the inner loops, robust weights calculated in the outer loops are then used to reduce the outliers' interference on updating the seasonal and the trend components in the subsequent inner loop.

After running process, the original time series y_t is decomposed into three components : seasonal component S_t, trending component T_t and remaining component R_t.

3.2 Wavelet Analysis

Wavelet analysis is one of the most powerful mathematical techniques in machine learning, which eliminates noise by transforming the original noisy signal to a silent signal. The wavelet transform invented by Mallat has played a key role in providing time-frequency representations of analyzed signals in the time series domains. Detailed coefficients are generated by high-pass filters, and approximate chains are generated by low-pass filters. The wavelet theory divides wavelet transform into three classes: continuous, discrete and multiresolution-based. In this study, we will take into account the discrete wavelet transform (DWT), which is a popular method to denoise a signal. Based on the enhancement of the wavelet theory, the DWT for discrete time series can be presented as shown in the following equation:

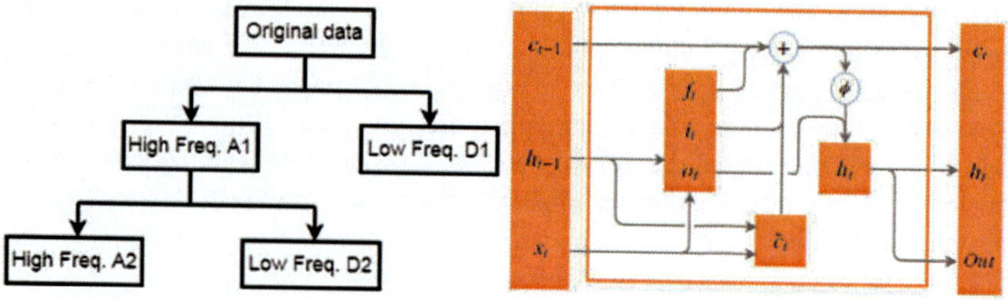

Fig. 1 Wavelet analysis (lelf) and LSTM unit (right)

$$W_{(m,n)} = 2^{-\frac{m}{n}} \sum_{t=0}^{N-1} \psi\left(\frac{t - 2^m \cdot n}{2^m}\right) \tag{1}$$

in which $2^m = s$ stands for the scaling parameter and $2^m \cdot n = \tau$ stands for the translation parameter of the DWT. The parameter s calculates the degree of compression. If $|s| > 1$, it can be told that the wavelet signal has been extended from the root signal. In contrast, if $0 < |s| < 1$, the wavelet signal has been compressed from the original.

In Fig. 1 $(left)$, from the input data, we use a first-order wavelet which splits into two components: low-freq D_1 and high-freq A1. With the high-freq part A_1, we can use a 2nd-order wavelet to split the low-freq D_2 and the high-freq A_2... After N^{th} orders, we obtain the time series (D_1, D_2, D_3 ,. .., D_N, A_N). The high-freq component A_N is noisy, and after some splits, we can delete this component. The wavelet number of N degrees can be calculated using the following formula [14] (with n is the size of the original time series):

$$N = \log_{10}(n) \tag{2}$$

3.3 *Long Short-Term Memory—LSTM*

Recurrent neural networks are capable of handling short-term dependencies in data series. The weakness of recurrent neural network is the ability to handle long-term dependencies [7, 8]. LSTM is a variant of recurrent neural network introduced by Hochreiter and Schmidhuber [8] to solve the vanishing problem of recurrent neural network, which occurs while training RNN networks with long-term dependencies. To solve this problem, LSTM has been nicely designed with three gates and memory cells to store information during the training period. Figure 1 $(right)$ shows an LSTM unit.

Input of LSTM unit at t-step is x_t and output at $t-1$-step is (h_{t-1}). The input will be filled by gates (with *sigmoid* function). The value of output is in range [0, 1]. If the output is 0, all inputs are removed. Otherwise, if output is 1, all information are passed. Then, the output at t-step (h_t) and the *cell state* at t-step will be solved in the following steps:
Step 1: Forget: Information from the *cell* of *state* is removed by forgotten gates:

$$f_t = \sigma(W_f.[h_{t-1}, x_t] + b_f) \tag{3}$$

Step 2: New information is stored in the *cell state*. In the first phase of this step, the input gate (with the function *sigmoid*) gets new values, and then, update this node. The second phase is that the *tanh* class gets a new vector $\tilde{c}_t$.

$$i_t = \sigma(W_i.[h_{t-1}, x_t] + b_i); \qquad \tilde{c}_t = \tanh(W_{\tilde{c}}.[h_{t-1}, x_t] + b_{\tilde{c}}) \tag{4}$$

Step 3: *cell* of *state* are updated from C_{t-1} to C_t:

$$c_t = f_t \otimes c_{t-1} \oplus i_t \otimes \tilde{c}_t \tag{5}$$

Step 4: Finds the output. Firstly, the output gate calculates the information which will be the outside of the LSTM unit. Then, *state* becomes a passed node which lies between -1 and 1. The final output will be calculated by this formula:

$$o_t = \sigma(W_o.[h_{t-1}, x_t] + b_o); \qquad h_t = o_t \otimes \tanh(h_t) \tag{6}$$

with W_f, W_i, $W_{\tilde{c}}$, W_o which are parameters of LSTM and b_f, b_i, $b_{\tilde{c}}$, b_o are biases of LSTM model.

3.4 Hybrid Model—LSTM Based on Seasonal and Wavelet-Denoising

By integrating LSTM models, seasonal and wavelet, it is hoped that the result of the hybrid model will be better than the results of the single methods.

The model hybrid LSTM based on wavelet-denoising can run in six steps as followed pseudocode:

Algorithm 1 The hybrid model LSTM based on seasonal and wavelet-denoising

Require: original_data = $\{y_i\}$
1: original_data = train_{set} + test_{set} ;
2: train_{set} = train_{trend} + $\text{train}_{seasonal}$ + $\text{train}_{residual}$;
3: Wavelet transform:
4: $\text{train}_{denoise}$ = DWT($\text{train}_{residual}$)
5: LSTM model:
6: $\text{train}_{residual\_predict}$ = LSTM($\text{train}_{denoise}$)
7: $\text{train}_{trend\_predict}$ = LSTM(train_{trend})
8: $\text{train}_{total\_predict}$ = Reconstruct($\text{train}_{trend\_predict}$,$\text{train}_{seasonal}$,$\text{train}_{residual\_predict}$)
9: $\text{test}_{total\_predict}$ = hybrid_model(test_{set})
10: MSE = MSE($\text{test}_{total\_predict}$); MAPE = MAPE($\text{test}_{total\_predict}$)

4 Experiment and Results

4.1 *Bac Hung Hai's Water Level Dataset*

This time series collects data every 6 hours from water level measurements of Cau Cat station in Bac Hung Hai agricultural area in Red River Delta located in North Vietnam, which dated from January 1, 2018, to December 31, 2020. The data are collected at 1 am, 7 am, 1 pm and 7 pm everyday.

4.2 *Criteria for Comparison*

2 criteria are used in order to compare the results: mean squared error (MSE) and mean absolute percentage error (MAPE):

$$\text{MSE} = \frac{1}{n} \cdot \sum_{i=1}^{n} (\hat{y}_i - y_i)^2; \quad \text{MAPE} = \sum_{i=1}^{n} \left| \frac{\hat{y}_i - \bar{\hat{y}}_i}{n} \right|$$

4.3 *Results on the Bac Hung Hai's Dataset*

Table 1 gives information on the result of the model after applying actual data of water level in Cau Cat station in Bac Hung Hai canal system (Fig. 2).

Table 1 Result of Bac Hung Hai data

Model	Compare		Note
	MSE	MAPE	
SARIMA	0.713307	0.699398	
LSTM	0.347144	0.449952	
ARIMA-RNN [9]	0.377597	0.444104	(rebuild)
Seasonal-Wavelet-LSTM	0.344907	0.443847	

The results from the SARIMA model, ARIMA-RNN model, LSTM model and seasonal-wavelet-LSTM model are shown in Table 1. We can point out that the predicted value can be applied for long-term forecasting. Therefore, the developed model can be employed to forecast the water level for the next 7 days (28-step-ahead). The MSE value of the SARIMA model, ARIMA-RNN model and LSTM is 0.713307, 0.377597 and 0.347144, respectively, whereas the MSE value of the proposed model is 0.344907. As for MAPE value, the seasonal-wavelet-LSTM model's MAPE is 0.443847, lower than all the other models in the table.

5 Conclusion

In this paper, we have introduced our method to develop hybrid models. Firstly, it is possible to use preprocessed data in order to eliminate noise that cannot be predicted from the input time series. One of the most popular methods to filter noise in preprocessing is called wavelet transform. The effectiveness and accuracy of the structure will be increased wildly if noises can be removed. Another direction is that we will apply robust online models. In this study, we employed wavelet transform to LSTM model instead of using pure traditional models. Therefore, the accuracy is expected to be higher. Nevertheless, to ensure its robustness, we should use all of the input data to test all used models. Furthermore, by using several different input time series, the model will receive more data to calculate the parameters. As a result, we will achieve better performances from the construction of models.

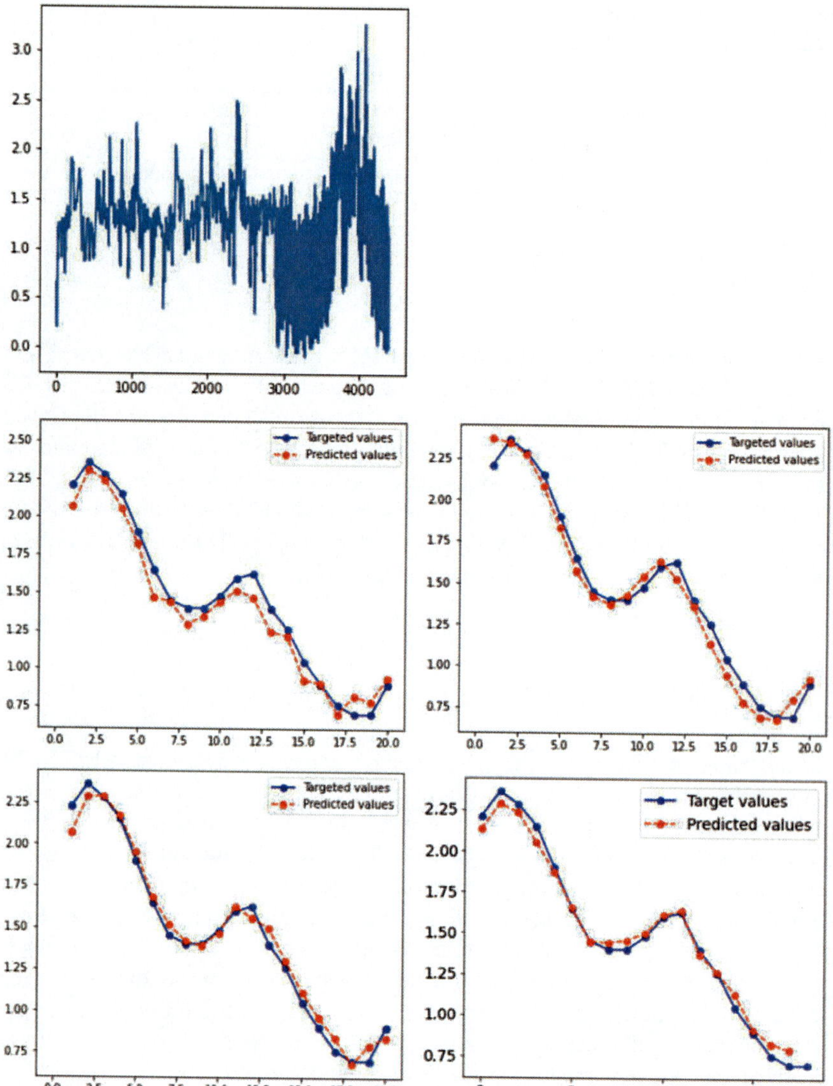

Fig. 2 Original data of Cau Cat station in Bac Hung Hai irrigation system (upper-left) and results of several models for water level of Cau Cat station: SARIMA (mid-left), LSTM (mid-right), Xu's model (ARIMA-RNN) (lower-left) and proposed model seasonal-wavelet-LSTM (lower-right)

Acknowledgements We are sincerely thankful to the members of WARM group for helping us in editing the earlier version of this article. This study was supported by WARM group, Thuy Loi University, Hanoi, Vietnam.

References

1. Y. Qi, Z. Zhou, L. Yang, Y. Quan, Q. Miao, A decomposition-ensemble learning model based on LSTM neural network for daily reservoir inflow forecasting. Water Resour. Manage. **33**, 4123–4139 (2019)
2. S. Lee, H. An, Y. Hur, Y. Kim, J. Byun, Flood prediction in the Namgang Dam basin using a long short-term memory (LSTM) algorithm. Korean J. Agric. Sci. **47**(3), 471–483 (2020)
3. X.-H. Le, H.V. Ho, G. Lee, S. Jung, Application of long short-term memory (LSTM) neural network for flood forecasting. Water **11**(7), 1387 (2019)
4. Fan He, Xuansen He, A continuous differentiable wavelet shrinkage function for economic data denoising. Comput. Econ. **54**, 729–761 (2019)
5. J. Bruzda, The wavelet scaling approach to forecasting: verification on a large set of noisy data. J. Forecast. **39**(3), 353–367 (2020)
6. H. Su, E. Zio, J. Zhang, M. Xu, X. Li, Z. Zhang, A hybrid hourly natural gas demand forecasting method based on the integration of wavelet transform and enhanced Deep-RNN model. Energy **178**, 585–597 (2019)
7. P. Razvan, M. Tomas, B. Yoshua, On the difficulty of training recurrent neural networks, in *International Conference on Machine Learning* (1982)
8. H. Sepp, S. Jürgen, Long short-term memory. Neural Comput. **9**(8), 1735–1780 (1997)
9. G. Xu, Y. Cheng, F. Liu, P. Ping, A water level prediction model based on ARIMA-RNN. BigDataService **38**, 221–226 (2019)
10. J. Zhang, Y. Zhu, X. Zhang, M. Ye, J. Yang, Developing a long short-term memory (LSTM) based model for predicting water table depth in agricultural areas. J. Hydrol. **205**, 918–929 (2018)
11. R.B. Cleveland, W.S. Cleveland, J.E. McRea, I. Tepenning, STL: a seasonal-trend decomposition produre base on Loess. J. Offic. Stat. **6**(1), 3–73 (1990)
12. Tao Xiong, Chongguang Li, Yukun Bao, Seasonal forecasting of agricultural commodity price using a hybrid STL and ELM method: evidence from the vegetable market in China. Neurocomputing **275**, 2831–2844 (2018)
13. L. Qin, W. Li, S. Li, Effective passenger flow forecasting using STL and ESN based on two improvement strategies. Neuralcomputing **356**, 244–256 (2019)
14. W. Wang, S. Ding, Wavelet network model and its application to the predication of hydrology. Nat. Sci. **1**(1), 67–71 (2003)

Hybrid Model Seasonal ARIMA-LSTM for Prediction Water Level in Black River on the Border Vietnam-China

Ngo Xuan An, Nguyen Nhat Minh, Nguyen Quang Dat, and Vijender Kumar Solanki

Abstract At Northern Viet Nam, on the Black River, lies three of the country's biggest hydroelectric dams. Along with producing electricity, the dams also regulate the water for agricultural activities on the Red River Delta. The lakes, rivers' inflows have a direct effect on the Vietnamese manufacturing section and economy. Another problem is the lack of inflow data of the river sections in China. As a result, water level forecasting in Northern Vietnam is very important. This research is conducted to propose a combined seasonal ARIMA and LSTM model to enhance the precision of the hydrological forecast activities on the Black River in Northern Vietnam and further develops the management of the water dams on this river. The proposed model was trained using the 4 years water level data at the Lai Chau hydroelectric dam on the Black River, generating the best result (MSE and MAPE) compared to the nearest models used on this area.

Keywords SARIMA · LSTM · Water level · Forecasting

1 Introduction

Hydroelectric dams in Northern Vietnam, apart from generating electricity, have another use-regulating water for agriculture activities on the Red River Delta. However, many rivers of the Delta have their upstream in China, who refuses to share

N. X. An (✉) · N. N. Minh · N. Q. Dat
High School for Gifted Students, Hanoi University of Science, Vietnam National University, Hanoi, Vietnam
e-mail: ngoxuanan_hsgs19@hus.edu.vn

N. N. Minh
e-mail: nguyennhatminh_hsgs19@hus.edu.vn

N. Q. Dat
e-mail: nguyenquangdat@hus.edu.vn

V. K. Solanki
Department of Computer Science and Engineering,
CMR Institute of Technology, Hyderabad, India

A. Kumar et al. (eds.), *Computational Intelligence in Machine Learning*, Lecture Notes in Electrical Engineering 834, https://doi.org/10.1007/978-981-16-8484-5_50

inflow data to Vietnam. As a result, researching algorithms for predicting inflow through the main rivers of the Delta is very important. The Black River, the second largest river of the Red River Delta, has three hydroelectric dams (at Lai Chau, Son La and Hoa Binh provinces). Forecasting inflows on the Black River has a big impact on the managing and operating these dams [1, 2]. With the use of deep learning on large data, we are trying to predict accurately inflow to the Lai Chau hydroelectric dam, the closest dam to China's border. There have been several research about forecasting inflow to the Black River [1] in order to optimize the dam's operating system. With the seasonal feature of the water data, we realize that the seasonal model is applicable in this case [3–5]. Hence, a forecasting model built on the seasonal ARIMA and LSTM model has been proposed. The seasonal ARIMA model can make use of the seasonal characteristics, while LSTM is a neural network model which is good for nonlinear components. The new model is expected to perform better than the single models used nowadays.

In this research, there are five main sections. Section 2 examines several researches related to predicting big data. Section 3 explores the proposed models and algorithms. Section 4 includes performance and results of the new model, as well as comparisons with several singular models and the hybrid model proposed by Xu et al. [2]. Finally, Sect. 5 contains evaluations of the model and future works that we intend to solve.

2 Related Works

Tadesse et al. [5] forecasted Waterval River flow using the SARIMA model. After evaluating several different SARIMA models, the SARIMA (3, 0, 2) × (3, 1, 3) was chosen, which produces the best parameters RMSE and MAE of 1.342 and 0.956, respectively.

In 2018, Zhang et al. [6] applied a LSTM model to water table depth prediction over the long term in agricultural areas. For the proposed four-year dataset from five sub-areas of Hetao Irrigation District, China, the new model produced a range of RMSE from 0.070 to 0.184 (m), indicating its effectiveness.

A LSTM model was developed by Le et al. in 2019 [1] for flood forecasting with the daily discharge and rainfall in the Black River basin in Vietnam. The developed model was proven to be an effective method, generating RMSE of 152.4, 360.7, 571.4 and RE of 6.6, 15.2, 28.2, respectively, for one day, two days and three days predictions.

Martínez-Acosta et al. in 2020 [7] proposed a SARIMA-based model for predicting of monsoon dataset at Doimukh, Arunachal Pradesh, India. The proposed model produced a MRE of 0.6 and ISE of 0.02 on the observed and forecasted data from 1988 to 2007.

Zhang in 2003 [8] proposed a hybrid approach to time series forecasting using an ARIMA-ANN model. He proved that the singular ARIMA, as well as the singular ANN model, is unable to handle all time series, and the developed hybrid model combines the advantages of each component model and should be used in both linear and nonlinear modeling.

Babu et al. in 2014 [9] came up with an ARIMA-ANN model with the help of the MA filter. Compared to the ANN model, the ARIMA model and some of the existing ARIMA-ANN models, the proposed model performed more accurately in both one-step-ahead and multi step-ahead forecasts. Considering the simulated sunspot data, the electricity price data and the stock market data, the new model produced a MAE of 0.1884 and MSE of 0.0507 for one-step-ahead prediction and a MAE of 0.2951 and MSE of 0.1445 for five-step-ahead prediction.

In a 2016 article [3], Hamid Moeeni and Hossein Bonakdari applied a SARIMA-ANN model to monthly inflow forecasting and applied it on the obtained data from the Jamishan River. The result indicated the efficiency of the hybrid model compared to each of the individual models with the best MARE, RMSE, obtained were 0.62, 1.02, respectively.

In a study by Shafaei et al. [10] in 2016, wavelet, SARIMA and hybrid ANN methods were tested for their accuracy in predicting monthly rainfall. The result indicated that the WA-SARIMA-ANN model outperformed both the wavelet-ANN and wavelet-SARIMA models.

Moeeni et al. [4] compared a SARIMA-GEP model with SARIMA, GEP and ANN alone and hybrid SARIMA-ANN models. As shown in the case study-forecasting monthly inflow to Jamishan Dam in west Iran, the developed model outperformed all the compared models, generating a RMSE of 0.89 and MAPE of 43.4.

In 2019, Xu et al. [2] proposed a new ARIMA-RNN combined model for water level forecasting. The results indicated that the proposed model of ARIMA-RNN has a smaller RMSE of 0.021 in comparison with ARIMA or RNN alone, which has the RMSE of 0.047 and 0.033, respectively.

Sahoo et al. [11] evaluated the suitability of the hybrid LSTM-RNN model for low-flow time series predicting in a paper from 2019. This forecasting model used daily data collected from the Basantapur station on the Mahanadi River basin, India. With the RMSE of 0.487, the LSTM-RNN model outperformed RNN model (RMSE = 0.516) and naïve method (RMSE = 0.793).

In 2020, Apaydin et al. [12] conducted a study involving a comparative analysis of the performance between RNN models (including Bi-LSTM, GRU and LSTM), singular RNN and two hidden layers ANN for reservoir inflow predicting. Results indicated that, with the RMSE of 17.92, the deep learning RNN methods generated better results than ANN (RMSE = 18.71). Among different RNN architectures, the LSTM method creates the best results, with the accuracy of 90 and 87% in the training stage and the testing stage, respectively.

3 Methodology

3.1 SARIMA

The auto-regressive integrated moving average (ARIMA) model was considered by Box and Jenkins [13] and was fully described in 1991 by Brockwell and Davis [14]. The ARIMA model was the most popular in the model creating process and in time series forecasting.

The auto-regressive method alone and the moving average method alone are suitable for forecasting. Their hybrid method (AR + MA) has been proven to perform better than component methods [8] and is suitable for a wider range of cases. In [15], based on the hybrid model idea, Salas et al. developed an estimation model for ARIMA with periodic coefficients.

The SARIMA model is an upgraded ARIMA model, containing an extra seasonal component. This model was popular applied in time series analysis and time series forecasting.

A time series $\{X_t\}$ with t = 1,2, . . , N is generated by a SARIMA (p, d, q)(P, D, Q) process by following equation:

$$\phi(L)(1-L)^d \Phi(L^s)(1-L^s)^D y_t = c + \theta(L)\Theta(L)\varepsilon_t \quad (1)$$

In this:

- $\phi(L) = 1 - \phi_1 L - \phi_2 L^2 - \cdots - \phi_p L^p$ is the auto-regressive operator (AR operator) of the order p.
- $\Phi(L) = 1 - \Phi_1 L^s - \Phi_2 L^{2s} - \cdots - \Phi_P L^P$ is the seasonal auto-regressive operator (S-AR operator) of the order P.
- $\theta(L) = 1 - \theta_1 L - \theta_2 L^2 - \cdots - \theta_q L^q$ is the moving average operator (MA operator) of the order q.
- $\Theta(L) = 1 - \Theta_1 L^s - \Theta_2 L^{2s} - \cdots - \Theta_Q L^Q$ is the seasonal moving average operator (S-MA operator) of the order Q.
- d—the number of regular differences. At 2006 [16], author Shumway et al. shown that $d \leq 2$;
- D—the number of seasonal differences. Usually, we set D = 1 in the case of seasonal data, and set D = 0 in the case of non-seasonal data.
- At moment t, ε_t is a white noise signal. This is the estimated residual that is identically and independently distributed as a normal random variable. Average value of ε_t equal 0 and variance of ε_t is σ_ε^2.

In the article [17], the researchers built and ran SARIMA model on the air passenger data, producing a MAPE of 1.5142% and a RMSE of 95.6635. In the research [18], the used SARIMA model generated a MAPE of 9.05% on data Beijing air passenger. In another study [19], the authors forecasted the solar PV power data from April 2017 to July 2017 and generated a RMSE of 16.1320 and a MAPE of 22.1810%.

In others articles, the researchers were combined the SARIMA with other model (or other models) to further increase the precision of the result. In 2014, Ruiz et al. produced better results than pure SARIMA [20] using a combined SARIMA-ANN model. Another hybrid model, the SARIMA-linear regression, proposed by t. Fang et al. [22], resulted with MAPE of 9.85% and RMSE of 22.85.

3.2 Long Short-Term Memory—LSTM

Recurrent neural networks are able to handle short-term dependencies in data series, but ineffective with long-term ones [21, 23]. Introduced by Hochreiter et al. [23], LSTM is a variant of recurrent neural network that is capable of solving the existing problems of recurrent neural network, which occurs during training RNN networks with long-term dependencies. To do so, LSTM has been designed with three gates and memory cells to store information during the training period.

Input of LSTM cell at t-step is x_t, and output at $t-1$-step is (h_{t-1}). The input will be filled with gates (with $sigmoid$ function). The output value is in range [0, 1]. If the output is 0, all inputs are removed. On the other hand, if output is 1, all information is passed. Consequently, the output at t-step (h_t) and the $cell\ state$ at t-step will be processed in the following steps:

- **Step 1**: Forget: Information from the $cell$ of $state$ is removed by forgotten gates:

$$f_t = \sigma(W_f.[h_{t-1}, x_t] + b_f) \tag{2}$$

- **Step 2**: New information is stored in the $cell\ state$. In the first phase of this step, the input gate (use $sigmoid$ function) gets new values, and then, update this node. The second phase is that the $tanh$ class gets a new vector $\tilde{c}_t$.

$$i_t = \sigma(W_i.[h_{t-1}, x_t] + b_i); \qquad \tilde{c}_t = \tanh(W_{\tilde{c}}.[h_{t-1}, x_t] + b_{\tilde{c}}) \tag{3}$$

- **Step 3**: $cell$ of $state$ are updated from C_{t-1} to C_t:

$$c_t = f_t \otimes c_{t-1} \oplus i_t \otimes \tilde{c}_t \tag{4}$$

- **Step 4**: Find the output. Firstly, the output gate calculates the information which will be the outside of the LSTM unit. After that, $state$ becomes a passed node which lies in range −1 to 1. The final output will be calculated by this formula:

$$o_t = \sigma(W_o.[h_{t-1}, x_t] + b_o); \qquad h_t = o_t \otimes \tanh(h_t) \tag{5}$$

with W_f, W_i, $W_{\tilde{c}}$ and W_o are parameters of LSTM and b_f, b_i, $b_{\tilde{c}}$ and b_o are biases of LSTM model.

3.3 Proposed Model: Hybrid SARIMA-LSTM Model

The SARIMA model has shown to be effective in forecasting linear time series, but not for nonlinear ones. Moreover, the model also consumes less computer resources. Meanwhile, the LSTM model can effectively handle nonlinear datasets but takes up more resources. Hence, by integrating SARIMA and LSTM models, it is hoped that the results of the combined model will be better than the singular methods'.

Pseudocode of the proposed model:

Algorithm 1 The hybrid model SARIMA-LSTM

Require: original_data = $\{y_i\}$;
1: $\{y_i\} = y_{train} + y_{test}$;
2: $y_{train} = \text{train}_{linear} + \text{train}_{non-linear}$;
3: SARIMA model:
4: $\text{train}_{linear\_predict} = \text{SARIMA}(\text{train}_{linear})$;
5: LSTM model:
6: $\text{train}_{nonlinear\_predict} = \text{LSTM}(\text{train}_{nonlinear})$;
7: $\text{train}_{total\_predict} = \text{Reconstruct}(\text{train}_{linear\_predict}, \text{train}_{non-linear\_predict})$;
8: $\text{test}_{total\_predict} = \text{hybrid\_model}(y_{test})$;
9: Evaluate:
10: $\text{MSE} = \text{MSE}(\text{test}_{total\_predict})$; $\text{MAPE} = \text{MAPE}(\text{test}_{total\_predict})$;

4 Experiment and Results

4.1 Data and Area

Daily inflow through Lai Chau hydroelectric dam The Lai Chau hydroelectric dam located on the main stream of the Black River, at Lai Chau province, Viet Nam. The reservoir is a part of a power plant which is important for electricity supplement and water distribution for the Red River Delta. The data series collects the plant's inflow data every day at 1am from June 14 to September 15 each year from 2017 to 2020 (Fig. 1).

Criteria for Comparison Two criteria are used in order to compare the results: mean squared error (MSE) and mean absolute percentage error (MAPE).

$$\text{MSE} = \sum_{i=1}^{n} \frac{\hat{y}_i - y_i)^2}{n} \quad \text{MAPE} = \sum_{i=1}^{n} \left| \frac{\hat{y}_i - \bar{\hat{y}}_i}{n} \right|$$

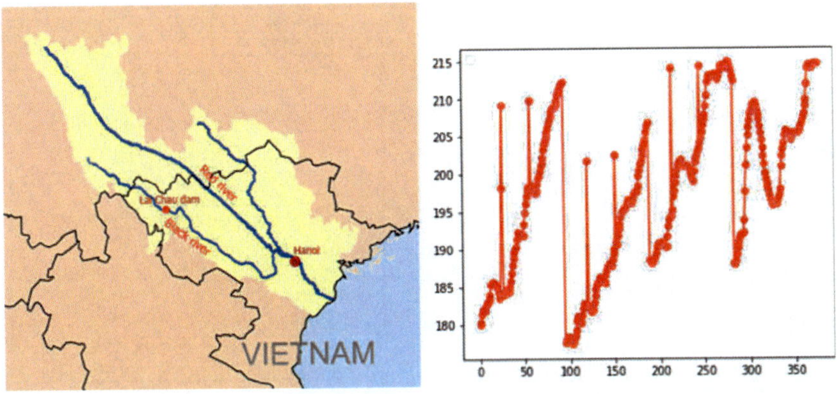

Fig. 1 Lai Chau hydroelectric dam on Black River (north Vietnam) and the daily inflow through Lai Chau dam

4.2 *Results on the Daily Inflow Through Lai Chau Hydroelectric Reservoir Dataset*

The dataset was trained with the proposed models, the rebuild ARIMA-RNN model from Xu et al. [2], the pure LSTM, SARIMA, ARIMA and RNN models (Fig. 2).

Table 1 gives information on the result of the model after applying actual data of inflow to Lai Chau hydroelectric dam.

In Table 1, it can be seen that the hybrid models produce better results than the singly built models. The simple RNN model has the largest MSE of 0.89719, followed by ARIMA, which has the MSE of 0.530260 and an exceptionally high MAPE of 1.855745, indicating its inaccuracy. Apart from ARIMA and RNN, the SARIMA model generates weaker result than others, in terms of MSE. However, the MAPE of this model was smaller than the SARIMA-LSTM combined models. This played a vital role in proving the high accuracy of the linear SARIMA model. There is a comparatively small difference of 0.001045 and 0.004543 in terms of both MSE and MAPE between LSTM and the hybrid model of ARIMA-RNN. The lowest MSE and MAPE values were obtained from the proposed SARIMA-LSTM model which has a 2.91% smaller MSE and a 9.13% better MAPE than its hybrid opponent.

5 Conclusion and Future Work

In this research, we have proposed a model that can meet the requirement of forecasting water level on the China–Vietnam border. With this model, we can take more advantages of the seasonal feature of the input data and hence increase the precision of the model. Moreover, the use of a strong neural network which is LSTM produces

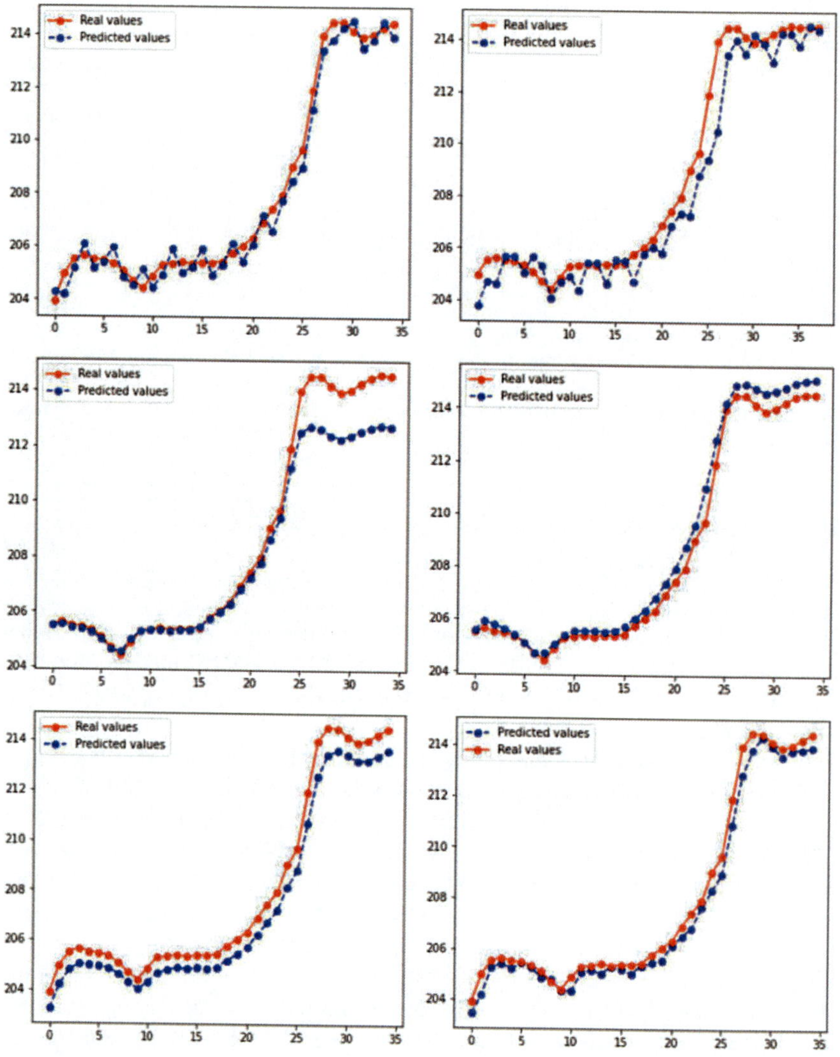

Fig. 2 Results of several models for water level of Lai Chau's dam: ARIMA (upper-left), SARIMA (upper-right), RNN (middle-left), LSTM (middle-right), Xu's model (ARIMA-RNN) (lower-left) and proposed model SARIMA-LSTM (lower-right)

Table 1 Result of Lai Chau data

Model	Compare		Note
	MSE	MAPE	
ARIMA	0.530260	1.855745	
SARIMA	0.247098	0.170136	
RNN	0.897196	0.273553	
LSTM	0.218315	0.201384	
ARIMA-RNN [2]	0.217270	0.196841	(rebuild)
SARIMA-LSTM	0.210942	0.17885	

better results than some researches before [2]. From the data evaluations, as well as the ability to use specific features of the data (which has not been used before), the hybrid model has produced the best compared to the pure models and the hybrid model from Xu et al. [2].

For the future work, we will try to solve some existing problems of the model: denoising (SAX, Wavelet, etc.), additional inputs (rainfall data, etc.) and online model. This could be a promising approach to this problem in Vietnam.

Acknowledgements We sincerely appreciate the help of the WARM group members during the editing phase of the earlier version of this article. This research was supported by WARM group of Thuy Loi University (Hanoi, Vietnam).

References

1. X.H. Le, H.V. Ho, G. Lee, S. Jung, Application of long short-term memory (LSTM) neural network for flood forecasting. Water **11**(7) (2019)
2. G. Xu, Y. Cheng, F. Liu, P. Ping, A water level prediction model based on ARIMA-RNN. BigDataService **38**, 221–226 (2019)
3. H. Moeeni, H. Bonakdari, Forecasting monthly inflow with extreme seasonal variation using the hybrid SARIMA-ANN model. Stochast. Environ. Res. Risk Assess. **11** (2016)
4. H. Moeeni, H. Bonakdari, I. Ebtehaj, Monthly reservoir inflow forecasting using a new hybrid SARIMA genetic programming approach. J. Earth Syst. Sci. **126**, Article 18 (2017)
5. K.B. Tadesse, M.O. Dinka, Application of SARIMA model to forecasting monthly flows in Waterval River. South Africa. J. Water Land Dev. **35**, 229–236 (2017)
6. J. Zhang, Y. Zhu, X. Zhang, M. Ye, J. Yang, Developing a long short-term memory (LSTM) based model for predicting water table depth in agricultural areas. J. Hydrol. **205**, 918–929 (2018)
7. L. Martínez-Acosta, J.P. Medrano-Barboza, Á. López-Ramos, J.F.R. López, Á.A. López-Lambraño1, SARIMA approach to generating synthetic monthly rainfall in the Sinú River Watershed in Colombia. Atmosphere **11**(6) (2020)
8. G.P. Zhang, Time series forecasting using a hybrid ARIMA and neural network model. Neurocomputing **50**, 159–175 (2003)
9. C.N. Babu, E. Reddy, A moving-average filter based hybrid ARIMA-ANN model for forecasting time series data. Appl. Soft Comput. **23**, 27–28 (2014)

10. M. Shafaei, J. Adamowski, M. Fakheri-Fard, Y. Dinpashoh, K. Adamowski, A wavelet-SARIMA-ANN hybrid model for precipitation forecasting. J. Water Land Dev. **28**, 27–36 (2016)
11. B.B. Sahoo, R. Jha, A. Singh, D. Kumar, Long short-term memory (LSTM) recurrent neural network for low-flow hydrological time series forecasting. Acta Geophysica **67**, 1471–1481 (2019)
12. H. Apaydin, H. Feizi., M.T. Sattari, M.S. Colak,S. Shamshirband, K-W Chau, Comparative analysis of recurrent neural network architectures for reservoir inflow forecasting. Water 12(5) (2020)
13. G.E.P. Box, G. Jenkins, *Time Series Analysis, Forecasting and Control* (Holden-Day, San Francisco, CA, 1970)
14. J.B. Peter, A.D. Richard, Time series: theory and methods, in *International Conference on Machine Learning* (1991)
15. J.D. Salas, D.C. Boes, R.A. Smith, Estimation of ARMA models with seasonal parameters. Water Resour. Res. **18**(4), 1006–1010 (1982)
16. R.H. Shumway, D.S. Stoffer, Very short-term solar PV generation forecast using SARIMA model: a case study. Time series analysis and its applications. With R Exemples (2006), pp. 47–82
17. L. Qin, W. Li, S. Li, A.S. Ahmad, Effective passenger flow forecasting using STL and ESN based on two improvement strategies. Neurocomputing **356**, 244–256 (2019)
18. S. Sun, H. Lu, K.-L. Tsui, S. Wang, Nonlinear vector auto-regression neural network for forecasting air passenger flow. J. Air Transp. Manage. **78**, 54–62 (2019)
19. V. Kushwaha, N.M. Pindoriya, Very short-term solar PV generation forecast using SARIMA model: a case study, in *2017 7th International Conference on Power Systems (ICPS)* (2017), pp. 430–435
20. J.J. Ruiz-Aguilar, I.J. Turias, M.J. Jiménez-Come, Hybrid approaches based on SARIMA and artificial neural networks for inspection time series forecasting. Transp. Res. Part E: Logist. Transp. Rev. **67**, 1–13 (2014)
21. P. Razvan, M. Tomas, B. Yoshua, On the difficulty of training recurrent neural networks, in *International Conference on Machine Learning* (2013), pp. 1310–1318
22. T. Fang, R. Lahdelma, Evaluation of a multiple linear regression model and SARIMA model in forecasting heat demand for district heating system. Appl. Energy **179**, 544–552 (2016)
23. S. Hochreiter, J. Schmidhuber, Long short-term memory. Neural Comput. **9**(8), 1735–1780 (1997)

Design of a High-Frequency Transformer Using Genetic Algorithm

S. Tamil Selvi, P. T. Nandh Kishore, Vishnu Ajith, M. K. Akshaya, and N. Karuppiah

Abstract In this paper, a 24 V/12 V, 1 kHz high-frequency transformer is designed optimally using genetic algorithm, with the aim of maximizing the power density and efficiency, while maintaining a reasonably small size and low cost. The advantage of designing a transformer using genetic algorithm is that it proposes to the designer a set of optimal designs instead of a single design solution. Such multiple possible solutions allow the user to choose the best fit for the application. But, the main challenges of high frequency transformer design are the number of design parameters, constraints, objectives, and the difficulty in confirming them to a single design methodology. In the proposed work, form factor, current density coefficient, specific volume coefficient, area constant, and Steinmetz constants are chosen as fixed parameters. The free parameters that are varied and optimized in the work are the window utilization factor K_u and flux density. The constraints that define the boundaries of the possible solutions are the maximum temperature rise and efficiency. Utilizing the optimized design data, a model has been simulated for 24 V/12 V, 1 kHz transformer in order to validate the optimized results. A downscaled prototype model of a 24 V/12 V, 1 kHz transformer has been constructed and the results are studied in this paper.

Keywords High-frequency transformer · Dual active bridge · Leakage inductance · Optimization · Genetic algorithm

1 Introduction

In recent times, the world of electronics has been moving toward higher power densities, particularly in applications such as offshore wind farms and traction. Increasing

S. Tamil Selvi (✉) · P. T. Nandh Kishore · V. Ajith · M. K. Akshaya
Department of Electrical and Electronics Engineering, Sri Sivasubramaniya Nadar College of Engineering, Chennai, India
e-mail: tamilselvis@ssn.edu.in

N. Karuppiah
Department of Electrical and Electronics Engineering, Vardhaman College of Engineering, Kacharam, Telangana, India

A. Kumar et al. (eds.), *Computational Intelligence in Machine Learning*, Lecture Notes in Electrical Engineering 834, https://doi.org/10.1007/978-981-16-8484-5_51

the operational frequency is the most common solution to achieve higher power densities. It also results in decreased weight and volume. Furthermore, high-frequency operation allows precise control over the operating parameters such as output voltage level, current level, voltage rise times, and response to variations in load demand. In addition, increasing the operation frequency reduces the volume of the magnetic device. High-frequency electronic transformers open up new possibilities for AC-AC and DC-DC conversion systems, characterized by high efficiency, small volume, low weight, and cost-effective features.

Much of the previous works is devoted to focus on the possible methods to overcome the core losses attributed to high frequencies [1]. Core loss at high frequency is estimated using Steinmetz method and other modified methods [2]. Classical approach is suggested to find out the high-frequency ferrite core losses in any magnetic component [3]. But the authors fail to suggest any design or optimization methodologies required to design a high-frequency transformer [2, 3]. Going a step further, researchers of [4] describe a transformer design considering the losses. Losses due to the skin effect and proximity effect increase at high frequencies. Along with the high-frequency transformer design, losses due to permeability and hysteresis were calculated [5]. At sinusoidal excitation, the core losses are calculated using the supplied core datasheets. However, considering the nonlinear magnetic characteristics of the core materials and non-sinusoidal waveforms, it is a challenge to estimate the losses of the high-frequency (HF) transformer used in a DC-DC converter [6]. Use of shell-type transformer is advocated for high-frequency non-sinusoidal wave-shaped applications [7].

The classical method of using the saturation constraint resulted in an overheated transformer due to excessive core loss at high frequency [8]. By optimizing the flux density, total transformer losses have been minimized [9]. An optimum design methodology for HF transformers is presented in the literature [10] that is applicable only to ferrite cored transformers. A step-by-step design for high-frequency transformers has been presented with high isolation requirements. The design considerations used are flux density limits, isolation, and thermal management. The performance of a core-type transformer is analyzed using the proposed design procedure [11]. Rather than obtaining different feasible solutions using different methodologies, a simple design with basic constraints is applied for HF transformer power design. The feasible solution is obtained by successive iterations [12]. Furthermore, the inverse methodology tries to determine the efficiency and the number of turns for an optimized transformer. Then its core material and utilization factor are selected based on the requirement [13]. An analytical expression intended to accurately evaluate the leakage inductance of transformers is provided in the high-frequency range. However, the conventional formulae for the calculation of leakage inductances fail, as the frequency increases. Also, it projects higher values since they have not addressed the change in behavior of the magnetic field at high frequencies. However, this work successfully addresses this issue of concern [14].

This work [15] aims to design the best HF transformer, for a given power converter topology, by optimizing transformer efficiency, weight, and also, transformer leakage and magnetizing inductances. Genetic algorithms are search and optimization algorithms that work on the principle of natural selection and genetic evolution [16, 17]. The authors of this research [18] highlight the transformer design optimization using genetic algorithm (GA) and simulated annealing (SA). The results obtained through bio-inspired algorithms are compared with the results obtained using geometric programming (GP) technique. The constraints considered are limits on efficiency, voltage regulation, temperature rise, no-load current, and winding fill factor. İt concentrates on the physical dimensions of the transformer rather than its leakage inductance, which is the most important parameter in designing a high-frequency transformer [18]. The authors proposed a novel design procedure for HF transformer optimization. İt is based on the core geometry which influences the regulation and copper loss. An isolated half-bridge DC–DC converter topology is chosen for implementing the proposed design [19].

This work [20] deals with the multi-objective optimization of HF transformers using genetic algorithm. The main aim of this work is to minimize the dimensions of the core and the windings. Also, the losses has to be reduced. A set of constraints for the design is also formulated. In this work, the area product and the power loss are used as the objective functions. The operating frequency and the maximum flux density are the optimization variables. The constraints are limits on efficiency, maximum surface temperature rise, and maximum ratio of no-load/full-load current. The design of a high-power density and medium-frequency transformer is studied. This can be used as a replacement for the rectifier transformer used in an electrostatic precipitator. Narrowing down to the scope of this work, the transformer is modeled by the use of McLyman's design approach. The variation of the RMS current and the phase shift necessary for bidirectional power flow in DAB with leakage inductance are also discussed. This work also testifies the supremacy of DAB over other DC-DC topologies based on the efficiency, switch ratings, and the ripples. It studies the effectiveness with which DAB can be an ideal choice for EV charging stations and energy storage applications. The optimized design data has been used to develop a prototype model of a 24 V/12 V, 1 kHz transformer.

2 HF Transformer Design Using Genetic Algorithm

The objective is to maximize the power density of the transformer using genetic algorithm. In its most general form, the design problem requires minimizing the mass or overall dimensions of the core and windings as well as the loss of the transformer while ensuring the satisfaction of a number of constraints. The design parameters of the transformers are classified as system requirements, free parameters, and fixed parameters. The system requirements are the user inputs to the program/design methodology. Those parameters that are constants and cannot be varied are classified as fixed parameters. The parameters that can be varied and selected are the free parameters. After obtaining the ranges and options over which the free parameters can be varied, the genetic algorithm is coded to genetically select the fittest free parameters. These parameters are used to complete the design calculations and obtain the design and specifications of the optimized transformer. The fixed parameters are form factor (k), saturation flux density of core (B_{sat}), current density coefficient (K_j), specific volume coefficient (K_v), area constant (K_s), and Steinmetz constants, α and β. The constraints for the optimization problem are maximum surface temperature rise (ΔT), efficiency (η_{max}), upper limit on area product, and upper limit on flux density. The system requirements are primary voltage (V_p), secondary voltage (V_s), output power (P_o), output frequency (f), allowed temperature rise ($\Delta T_{allowed}$), and minimum efficiency required ($\eta_{required}$). Finally, the free parameters are window utilization factor (K_u) and maximum flux density (B_{max}).

3 Design of a HF Transformer: 20 VA, 1 Phase, 24/12 V, 1 kHz

The optimization problem is to design a 20 VA, 24/12 V, 1 kHz high-frequency transformer with maximum power density while meeting thermal, electrical, and physical constraints. The maximum temperature rise and minimum efficiency were set at 50 °C and 98%, respectively. The various assumptions that are made in the design are listed in Table 1.

3.1 Design of Core

Current density co-eff	403.0000
Volume Coeff	13.1000
Surface area co-eff	32.5000
Steinmetz constant	2.480000e+00
Alpha	1.800000e+00
Beta	2.0800
Area product	2.495131e+00 cm^4
Core volume	26.0071 cm^3
Transformer current density	359.4747 A/sq.cm
Core area	0.6740 sq.cm
Length of mean turn	4.600000 cm
Inner diameter	2.240000 cm

Table 1 Assumptions considered in HF transformer design

Type of magnetic material used	4 (Nano crystal)
Type of core structure used	2 (Powder core)
Allowable temperature rise	50.0000 °C
Minimum efficiency	98.0000
Saturation maximum fux density	1.2000 Wb/m^2

Outer diameter	3.590000 cm
Height of core	1.050000 cm
Core flux density	0.5194 Tesla
Factor Ku	0.2120
Core loss	0.0165 W

3.2 *Design of Primary Winding (LV)*

Total number of turns	171
Current in primary winding	0.813333 A
Area of primary winding	0.226256 cm^2
Resistance of primary winding	0.523876 Ω
Total primary copper loss	0.346550 W

3.3 *Design of Secondary Winding (HV)*

Total number of turns	86
Current in secondary winding	1.626667 A
Area of secondary winding	0.452512 cm^2
Resistance of secondary winding	0.131458 Ω
Total secondary copper loss	0.347843 W

3.4 *Performance Figures*

Expected TEMP RISE	13.121801 °C
Leakage inductance	0.00002011042 Henry
Total core loss	0.0165 W
Total load loss	0.6944 W
Total loss	0.7109 W
Percentage efficiency	96.358071

3.5 *Output of Transformer*

Power density_fitness	0.7232 W/cm^3
Penalty	0

4 Hardware Modeling

The dual active bridge is a bidirectional, controllable, DC-DC converter that has high-power capabilities. It is comprised of eight semiconductor devices, a high-frequency transformer, an energy transfer inductor, and DC-link capacitors. It is essentially a full-bridge converter with a controllable rectifier. Due to the symmetry exhibited by this converter, due to its identical primary and secondary bridges, it is capable of bidirectional power flow and is hence very useful in isolated converter applications.

Similar to the methodology of obtaining the design parameters by optimization for values used in the reference papers, a separate calculation was carried out and optimized for the implementation of hardware design. The voltage levels adopted are 24 V on the primary side and 12 V on the secondary side. A prototype is developed on a PCB to ON a small load (LED).

The hardware apparatus consists of a DC power supply of 24 V given to a single-phase full-bridge MOSFET inverter. The gating is given by pulses of frequency 1 kHz obtained from Arduino kit. The output of the MOSFET inverter is a bidirectional wave that can be approximated nearly to that of a square wave. The wave obtained is influenced by the charging and discharging of the input and output capacitances of the MOSFET.

The bidirectional wave obtained from the inverter is given to the stepped down in the ratio of 2:1 by use of a ferrite-cored inductor. The hardware version of the inductor is modeled by making use of the bio-optimized values from the genetic algorithm including the core material, leakage inductance, utilization factor, and flux density. This inductor part provides the necessary isolation between the two DC links and reportedly promotes bidirectional power flow. The output of the AC connecting link is given to a single-phase full-wave bridge rectifier. The resultant wave is a pulsating square-shaped DC waveform that pulsates at a frequency of about 2 kHz.

The hardware circuit simulated is shown in Fig. 1 and its prototype in Fig. 2.

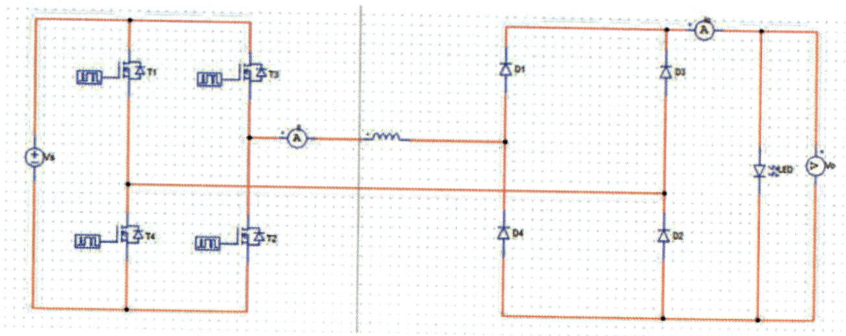

Fig. 1 Hardware circuit simulated model

Fig. 2 Inverter and the inductor hardware model

5 Conclusion

The design of high-frequency transformers is optimized by implementing genetic algorithm taking into consideration the relevant constraints. A working dual active bridge converter is designed and simulated using the previously determined high-frequency transformer specifications. The converter exhibits low power loss and good output regulation. A hardware model of the dual active bridge converter is implemented with the approximation of inductor instead of transformer. This design optimizes the power density of the transformer for practical applications. Further, optimization can be achieved by taking into account the real-time industry requirements of the transformer and implementing multi-objective optimization of parameters. The hardware could be made to do synchronized bidirectional power flow by implementing complex control using PIC or STM-32 controllers.

References

1. M.A. Bahmani, *Design Methodology and Optimization of HF Transformers for High Power DC-DC Applications*. Thesis for the degree of licentiate of engineering, Chalmers University of Technology, Sweden (2014)
2. S. Vaisambhayana, C. Dincan, C. Shuyu, A. Tripathi, T. Haonan, Karthikeya, State of art survey for design of medium frequency high power transformer, in *2016 Asian Conference on Energy, Power and Transportation Electrification (ACEPT)*, Singapore (2016)
3. W. Roshen, Ferrite core loss for power magnetic components design. IEEE Trans. Magn. **27**(6), (1991)

4. M.A. Bahmani, T. Thiringer, *Optimization and Experimental Validation of Medium-Frequency High Power Transformers in Solid-State Transformer Applications* (Chalmers University of Technology, Sweden, 2014)
5. A.F. Goldberg, J.G. Kassakian, M.F. Schlecht, Issues related to 1–10 MHz transformer design. IEEE Trans. Power Electron. **4**(1) (1989)
6. E. Agheb, M.A. Bahmani, H. Hoidalen, T. Thiringer, Core loss behaviour in high frequency high power transformers—II: Arbitrary excitation. J. Renew. Sustain. Energy (2012)
7. M. Kharezy, Constructional *Design, Manufacturing and Evaluation of High Power Density Medium Frequency Transformer Prototypes*. Master of Science Thesis, Department of Energy and Environment, Electric Power Engineering Dept. Chalmers University of Technology, Sweden (2014)
8. N.R. Coonrod, Transformer computer design aid for higher frequency switching power supplies. IEEE Trans. Power Electron. **Pe-1**(4) (1986)
9. K.D.T. Ngo, R.P. Alley, A.J. Yerman, R.J. Charles, M.H. Kuo, Design issues for the transformer in a low-voltage power supply with high efficiency and high power density. IEEE Trans. Power Electron. 7(3) (1992)
10. R. Petkov, Optimum design of a high-power, high-frequency transformer. IEEE Trans. Power Electron. **11**(1) (1996)
11. G. Ortiz, J. Biela, J.W. Kolar, *Optimized Design of Medium Frequency Transformers with High Isolation Requirements* (Power Electronic Systems Laboratory, ETH Zurich, Zurich, Switzerland, 2010)
12. I. Villar, A.B. Garcia, U. Viscarret, I.E. Otadui, *Proposal and Validation of Medium-Frequency Power Transformer Design Methodology* (Control Engineering and Power Electronics Area Technological Research Centre, Mondragon, Spain, 2011)
13. W.G. Hurley, W.H. Wolfle, J.G. Breslin, Optimized transformer design: inclusive of high-frequency effects. IEEE Trans. Power Electron. **13**(4), (1998)
14. M.A. Bahmani, T. Thiringer, *Accurate Evaluation of Leakage Inductance in High Frequency Transformers Using an Improved Frequency-Dependent Expression* (Chalmers University of Technology, Sweden, 2014)
15. A.G. Bediaga, I. Villar, A. Rujas, L. Mir, A. Rufer, *Multi objective optimization of medium-frequency transformers for isolated soft-switching converters using a genetic algorithm* (Control Engineering and Power Electronics Area Technological Research Centre, Mondragon, Spain, 2017)
16. D. Hermawanto, *Genetic Algorithm for Solving Simple Mathematical Equality Problem* (Indonesian Institute of Sciences (LIPI), Indonesia, 2013)
17. S.K. Suman, V.K. Giri, *Genetic Algorithms: Basic Concepts and Real World Applications* (2015)
18. A.K. Yadav, O.P. Rahi, H. Malik, A. Azeem, Design optimization of high-frequency power transformer by genetic algorithm and simulated annealing. Int. J. Electr. Comput. Eng. (IJECE) **1**(2), (2011)
19. J. Ravichandran, Banumathy, R. Veeraraghavalu, *High Frequency Transformer Design and Optimization using Bio-inspired Algorithms* (2016)
20. C. Versele, O. Deblecker, J. Lobry, Multiobjective optimal design of high frequency transformers using genetic algorithm, in *2009 13th European Conference on Power Electronics and Applications*, Barcelona, Spain (2009), pp. 1–10